The STUDY GUIDE

Designed by top Alberta Teachers to be compatible with all classroom resources
Wherever you are taking Math 30-1, this resource will help you succeed.

This Study Guide is best used with the **Math 30-1 EDGE Volume 2 – Practice Questions Workbook**

- On-point questions you need to build your skills and get your **best class mark**!
- Original **diploma-exam style questions** you won't find anywhere else!

Practice Questions Sample, from section 7.1 and 7.3 (Trig Identities)

*OR: **math30-1edge.com** for information on purchasing a digital copy of Volume 2 (which includes detailed solutions **plus** over 50 hours of Math 30-1 e-learning for just $49)*

Publisher: RTD Learning Publications
1914 13th st sw
Calgary, Alberta T2T 3P6
1-888-665-8803 **www.math30-1edge.com** • **edge@rtdlearning.com**

Author: Marc Lambert

Contributors: Daniel Shi, David McDiarmid, Lina Grager, and Dean Walls

ISBN: 978-1-7777496-0-6

Thank you for choosing the EDGE and all the best in your studies!

-Marc marc@rtdlearning.com

I welcome your feedback - let me know if you find any errors in this book!

TABLE of CONTENTS

When looking through this book - why are some pages "missing"?

Great question! We appreciate your eye for detail. The answer is: This Study Guide was originally 550 pages, which was beyond the page restriction for most online marketplaces.

So we removed the Practice Questions and put them in a separate book, which is also available as part of the set! But to keep the book aligned with our accompanying e-learning, we left the pages numbers as they were – leaving gaps!

TRANSFORMATIONS OF FUNCTIONS

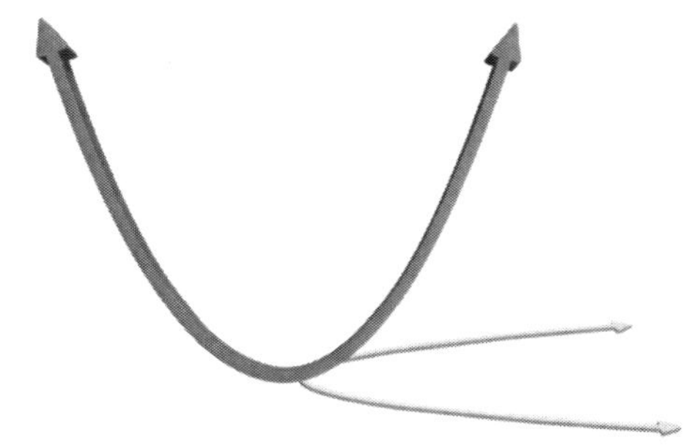

1.1 Prerequisite Skills + Translations of Functions

Part A – Prerequisite Skills

Review of Two Familiar Functions

In this unit we'll take some known and new functions and apply various transformations. And that means, if you're eager with anticipation, to alter the function's equation or graph.

However before we get into all of that – over the next few pages (and 6 warm-ups), we'll brush up on some key concepts we'll need in this first unit and throughout this course. Starting with – some functions from Math 20!

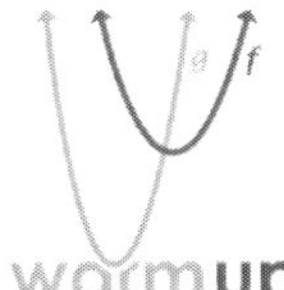

Warm-up
Exploration #1

The Quadratic Function – The Graph of $y = x^2$

1 **Complete** the **table of values** on the right, and **plot** the points to **sketch** the graph.

2 **State** the **domain** and **range** of the function.

______ Domain ______ Range

3 On the same grid, **sketch the graph** of $y = x^2 + 3$. *Add 3 to all y-coordinates, verify on your graphing calc.*

x	$y = x^2$
-3	$(-3)^2 = 9$
-2	
-1	
0	
1	
2	
3	

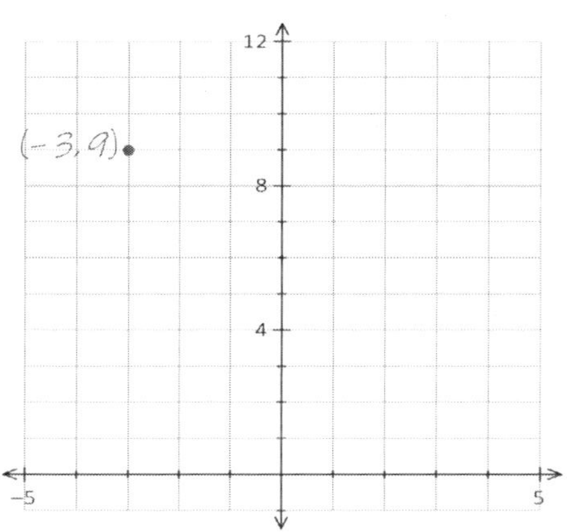

video solution

The Absolute Value Function – the Graph of $y = |x|$

4 **Complete** the **table of values** on the right, and **plot** the points to **sketch** the graph.

5 **State** the **domain** and **range** of the function.

______ Domain ______ Range

6 On the same grid, **sketch the graph** of $y = |x| - 2$.
Explain how the graph compares to $y = |x|$.

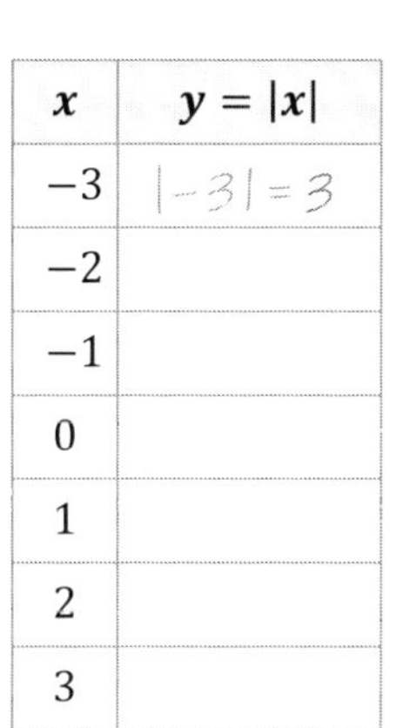

x	$y = \|x\|$
-3	$\|-3\| = 3$
-2	
-1	
0	
1	
2	
3	

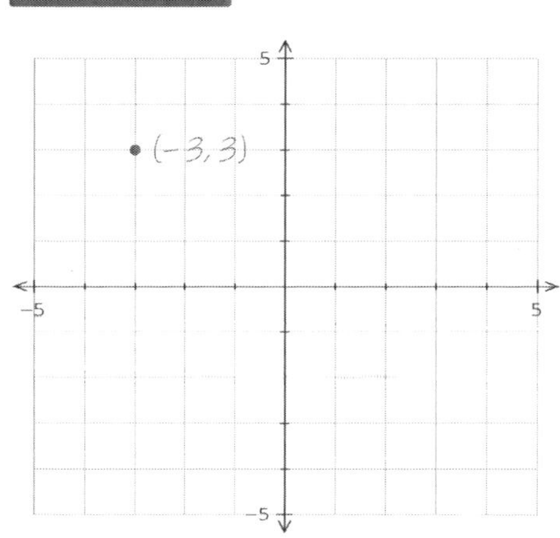

CALC TIP

Try yourself – Confirm the graphs on your calculator.

Plot1 Plot2 Plot3
$\backslash Y_1 = X^2$
$\backslash Y_2 = X^2+3$

Match the window to the given grid

WINDOW
Xmin=-5
Xmax=5
Xscl=1
Ymin=-1
Ymax=12
Yscl=5

Toggle between the two graphs using the arrow buttons

Confirm points using: trace

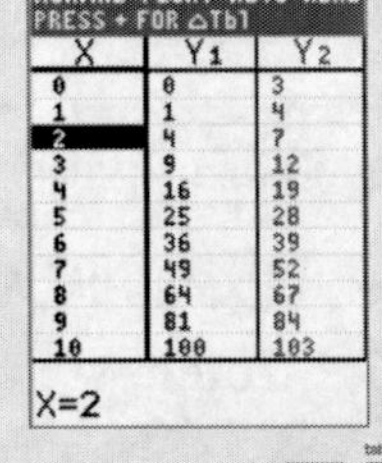
NORMAL FLOAT AUTO REAL
PRESS + FOR △Tbl

X	Y1	Y2
0	0	3
1	1	4
2	4	7
3	9	12
4	16	19
5	25	28
6	36	39
7	49	52
8	64	67
9	81	84
10	100	103

X=2

Also confirm points using ***table*** (2nd, graph)

To graph $y = |x|$, you'll need to find the ***abs(*** function.

Key in "MATH" then choose the first option from the "NUM" menu

NORMAL FLOAT AUT
MATH NUM CM
1:abs(
2:round(
3:iPart(
4:fPart(
5:int(
6:min(
7:max(
8:lcm(
9↓gcd(

Match the window again....

WINDOW
Xmin=-5
Xmax=5
Xscl=1
Ymin=-5
Ymax=5
Yscl=1
Xres=1

Plot1 Plot2
$\backslash Y_1 = |X|$
$\backslash Y_2 = |X|-2$

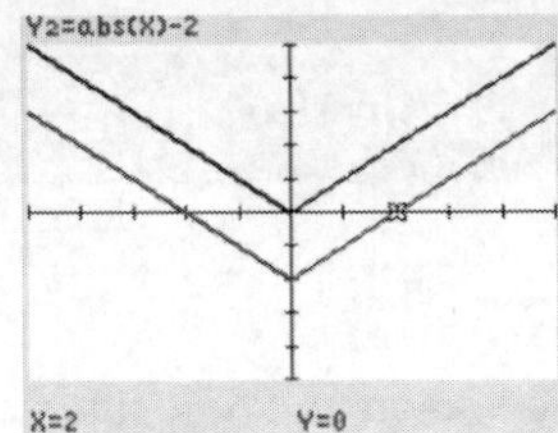

Try changing the constant term, how does that change things?

Preview of a Few New Functions

Over the next two warm-ups, we'll preview three functions we'll see much of later in the course.

Warm-up Exploration #2

The Cubic Function – The Graph of $y = x^3$

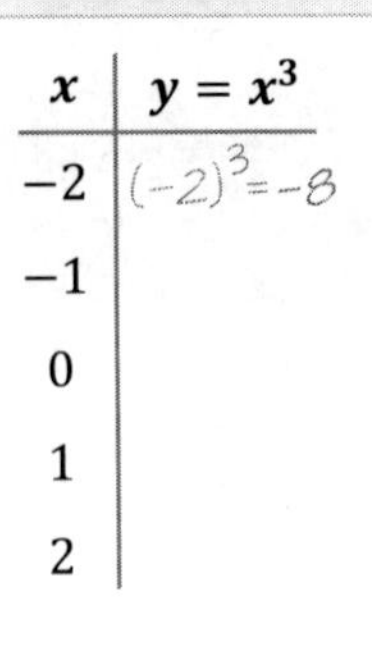

x	$y = x^3$
−2	$(-2)^3 = -8$
−1	
0	
1	
2	

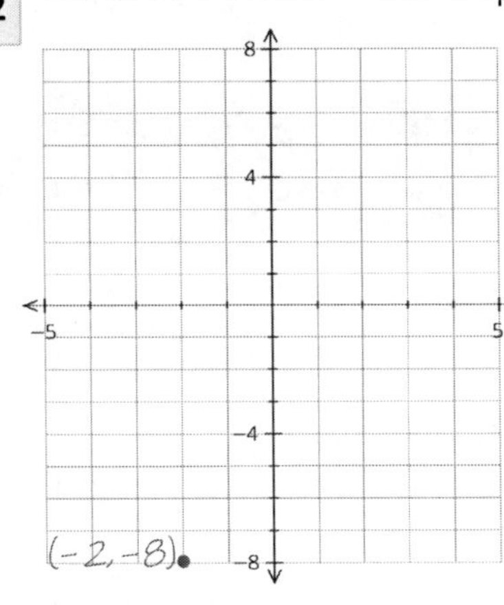

1 ➡ **Complete** the **table of values** on the left and **plot** the points to **sketch** the graph of the function. *Use your graphing calculator to confirm.*

2 ➡ **State** the function's **domain** and **range.**

__________ Domain __________ Range

The Radical Function – The Graph of $y = \sqrt{x}$

x	$y = \sqrt{x}$
0	$\sqrt{0} = 0$
1	
4	
9	
−1	

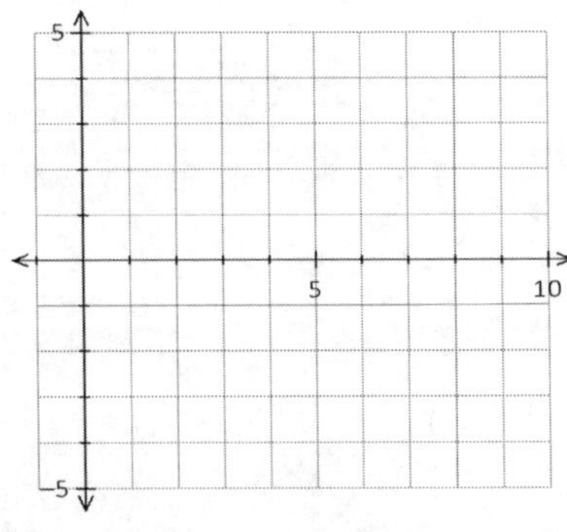

3 ➡ **Complete** the **table of values** on the left and **plot** the points to **sketch** the graph of the function. *Use your graphing calculator to confirm.*

4 ➡ **State** the function's **domain** and **range.**

__________ Domain __________ Range

5 ➡ **Compare** this table of values with that of $y = x^2$. What do you notice?

6 ➡ On the same grid, **sketch** the graph of $y = \sqrt{x} - 4$. **State** the **domain** and **range** of this new function.

__________ Domain __________ Range

Exploration #3 The Rational Function – The Graph of $y = \dfrac{1}{x}$

1 ➡ **Complete** the (partial) **table of values** on the left. *The points are already plotted on the graph.*

2 ➡ **Graph** $y = 1/x$ on your graphing calculator, using the window shown.

x: $[-5, 5, 1]$ y: $[-5, 5, 1]$

min max scl

3 ➡ With the help of you calculator, **sketch** the graph by connecting the plotted points in a smooth curve.

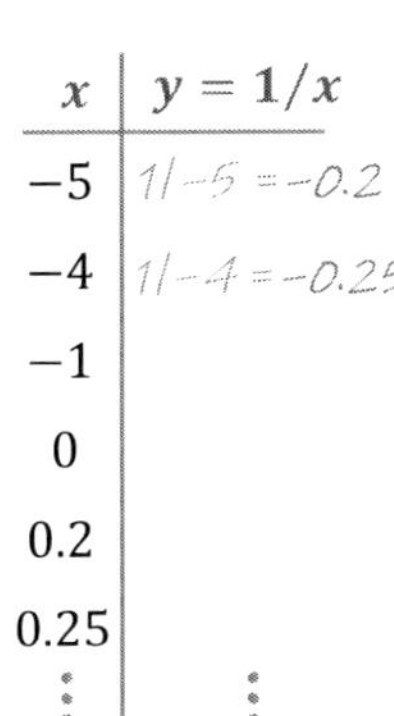

x	$y = 1/x$
-5	$1/-5 = -0.2$
-4	$1/-4 = -0.25$
-1	
0	
0.2	
0.25	
⋮	⋮

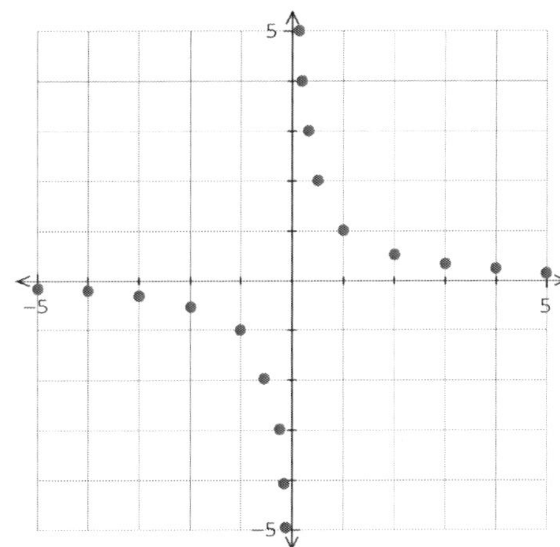

4 ➡ *Fill in the blanks:* The graph of $y = \dfrac{1}{x}$ has a vertical asymptote at _______ and a horizontal asymptote at _______. The domain of the function is ___________ and the range is ___________.

Can't divide by zero

Some of the **basic functions** graphs we should be familiar with are:

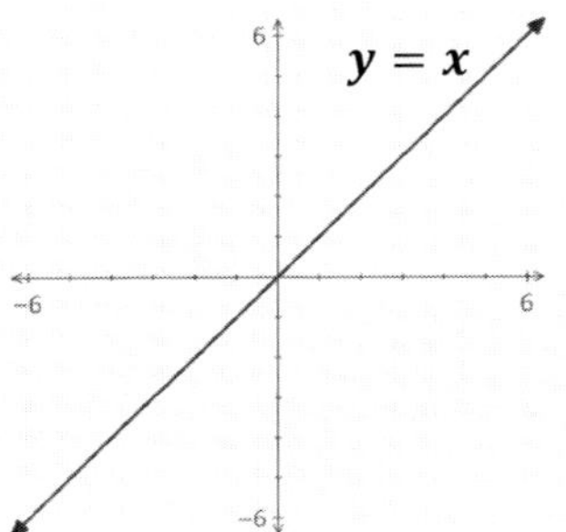

Domain: $\{x \in \mathbb{R}\}$

Range: $\{y \in \mathbb{R}\}$

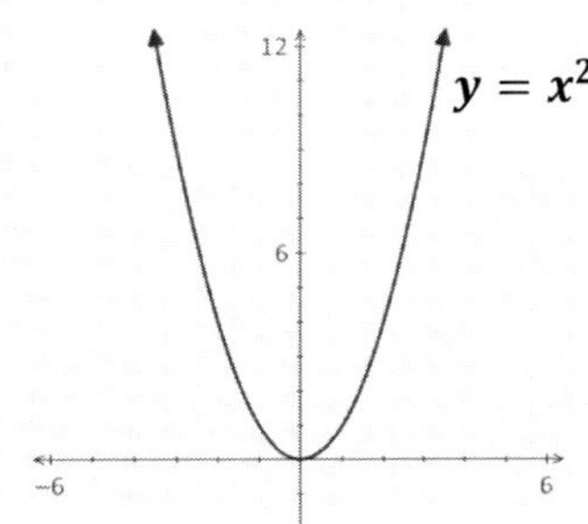

Domain: $\{x \in \mathbb{R}\}$

Range: $\{y | y \geq 0, y \in \mathbb{R}\}$

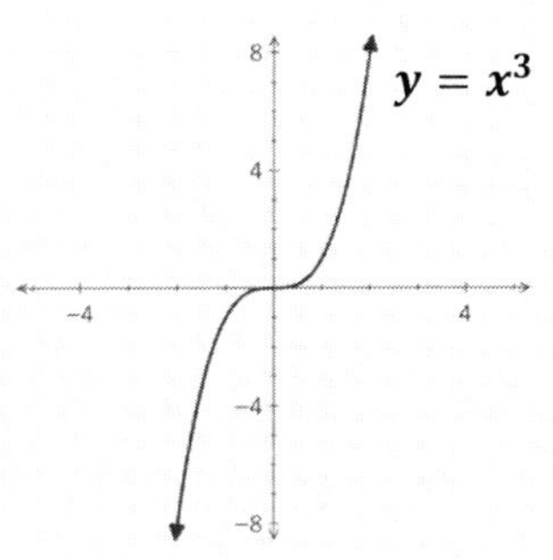

Domain: $\{x \in \mathbb{R}\}$

Range: $\{y \in \mathbb{R}\}$

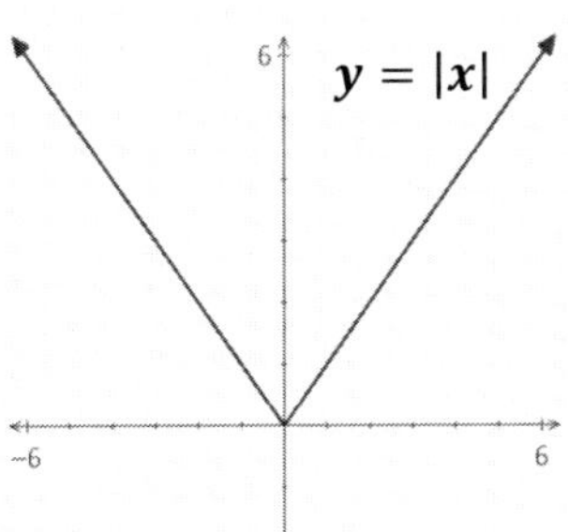

Domain: $\{x \in \mathbb{R}\}$

Range: $\{y | y \geq 0, y \in \mathbb{R}\}$

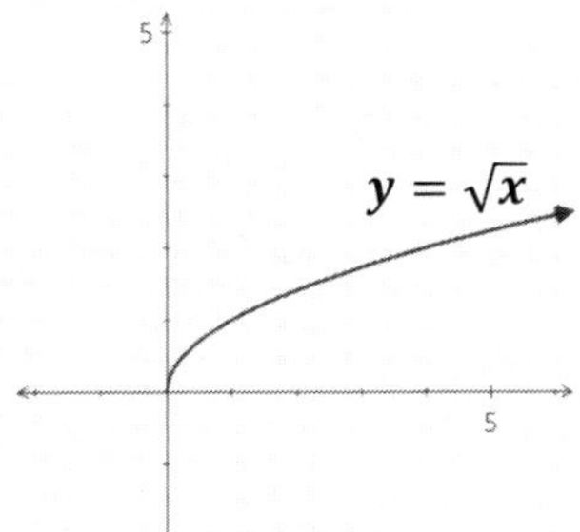

Domain: $\{x \geq 0, x \in \mathbb{R}\}$

Range: $\{y | y \geq 0, y \in \mathbb{R}\}$

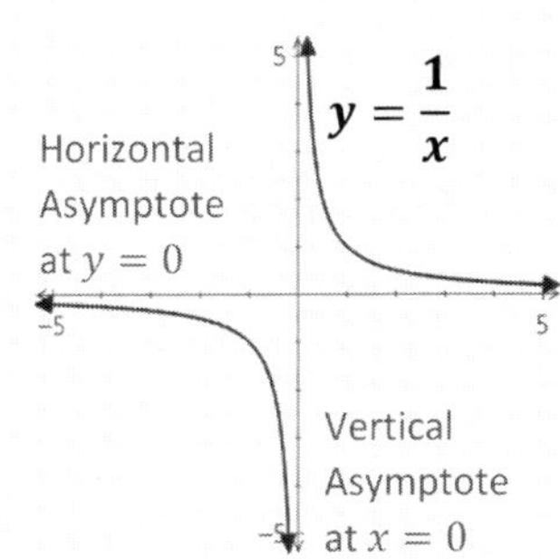

Domain: $\{x \neq 0, x \in \mathbb{R}\}$

Range: $\{y | y \neq 0, y \in \mathbb{R}\}$

Interval Notation of Domain and Range

You are likely familiar with the formats above, *set notation*. In this course we also use *interval notation*:

So, Domain: $\{x \in \mathbb{R}\}$ can be written in interval notation: $(-\infty, \infty)$ Read as: *"from $-\infty$ to ∞"*

Rounded brackets, do not include endpoints

And, Range: $\{y | y \geq 0, y \in \mathbb{R}\}$ can be written: $[0, \infty)$ Read as: *"from 0 to ∞, including 0"*

Square bracket, endpoint is included

Exploration #4 Domain and Range Fundamentals

The concept of **domain** and **range** is highly important in this course. In this next warm-up, we'll look at how to determine the domain and range from a **graph**, and how to determine the domain from the **equation** of a **function**.

Determining Domain and Range from the graph of a function

For **domain**, we consider all input, or x values.

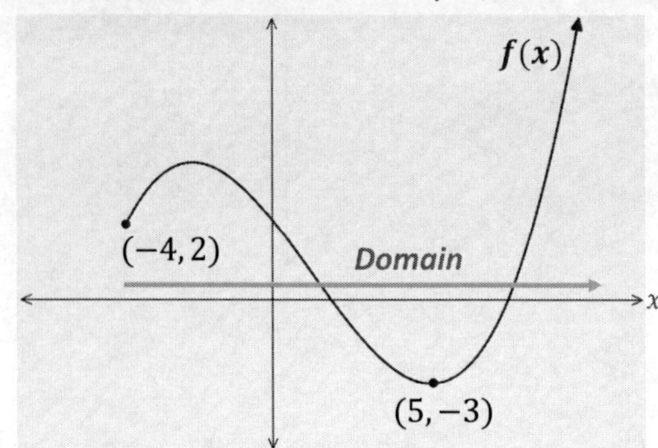

Domain is: $\{x|x \geq -4, x \in \mathbb{R}\}$ *Set notation*

or, alternatively: $[-4, \infty)$ *Interval notation*

For **range**, we consider all output, or y values.

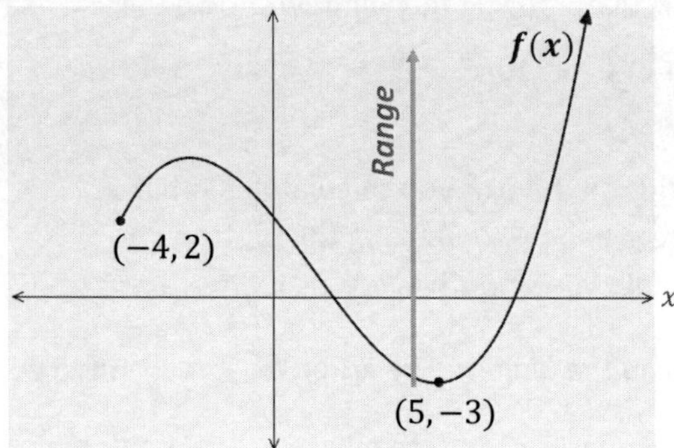

Range is: $\{y|y \geq -3, y \in \mathbb{R}\}$ *Set notation*

or, alternatively: $[-3, \infty)$ *Interval notation*

Determining Domain from the equation of a function

Given the equation of a function, we need to exclude any non-permissible values. (That is, we need to state any restrictions)

But what of these ***"restrictions"****?*

Which is actually easier than might sound! Because in this course, there are only three restrictions we need to consider. Ready? Just remember that we can't:

1 Divide by Zero *For example, what is the domain of....*

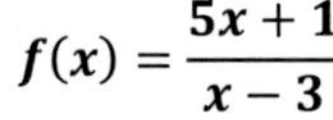

We do not need to graph on our calculators, or even consider what the graph might look like. Simply think .... For what value(s) x would the denominator ("bottom") be zero?

Domain is given by: $x - 3 \neq 0$ — Set the denominator *not equal to zero*, and isolate x

$\{x|x \neq 3, x \in \mathbb{R}\}$ — This domain is not suitable for interval notation, but if we chose to, it would be: $(-\infty, -3) \cup (-3, \infty)$

"Union" (think – *"combined with"*)

Note that this is not in the curriculum

2 Square Root Negatives *For example, what is the domain of....*

$$g(x) = \sqrt{4x + 3}$$

Again, we need not concern ourselves with the graph! (And like $f(x)$ above, we won't even get to the graphs until unit 7) Instead, think .... For what value(s) x would we be square-rooting negatives?

Domain is given by: $4x + 3 \geq 0$ — Set what's under the root sign greater than or equal to zero, and isolate x

$4x \geq -3$

$\{x|x \geq -3/4, x \in \mathbb{R}\}$ *In interval notation:* $[-3/4, \infty)$

3 Take the Logarithm* of 0 or Negatives

*Let's pin this for now – we'll come back in unit 3!

Before we continue our warm-ups and into transformations, let's do some practice with **domain** and **range**.

Class Example 1.11 *Obtaining Domain and Range from a Graph*

Given the graphs below, state the domain (D) and range (R) for each function, in both set and interval notation

(a)

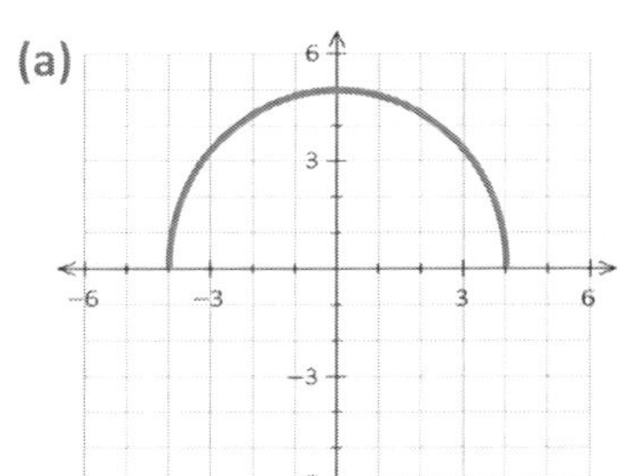

D: ______________ ______________
Set Notation *Interval Notation*

R: ______________ ______________
Set Notation *Interval Notation*

(b)

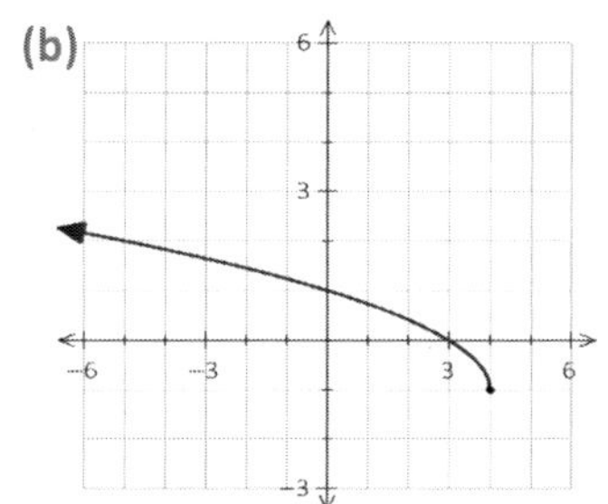

D: ______________ ______________
Set *Interval*

R: ______________ ______________
Set *Interval*

(c)

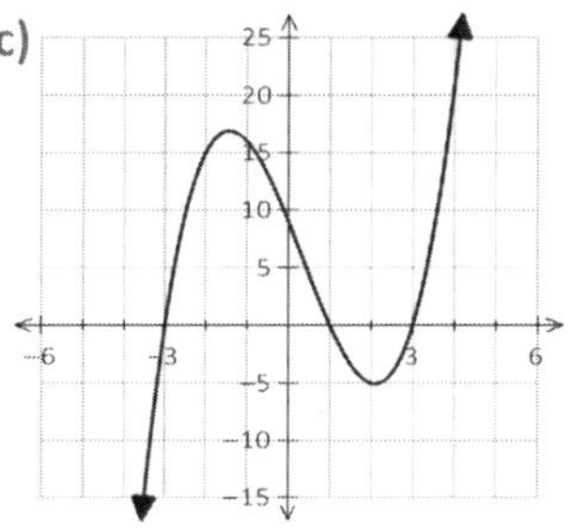

D: ______________ ______________
Set *Interval*

R: ______________ ______________
Set *Interval*

(d)

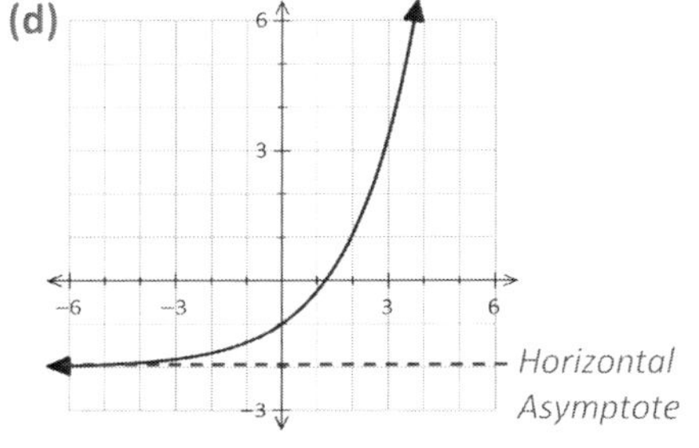

D: ______________ ______________
Set *Interval*

R: ______________ ______________
Set *Interval*

(e)

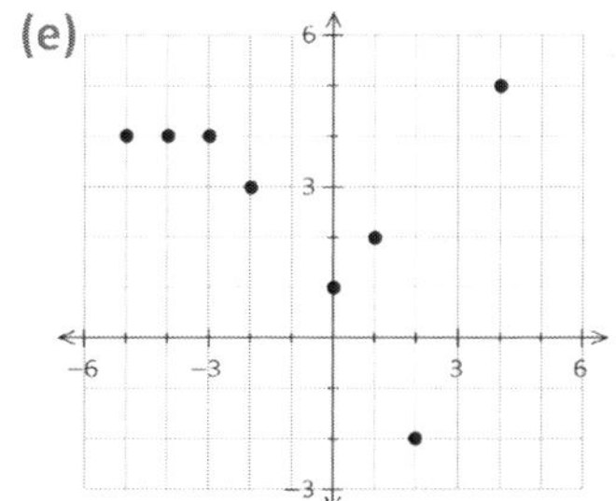

D: ______________
Set

R: ______________
Set

(f)

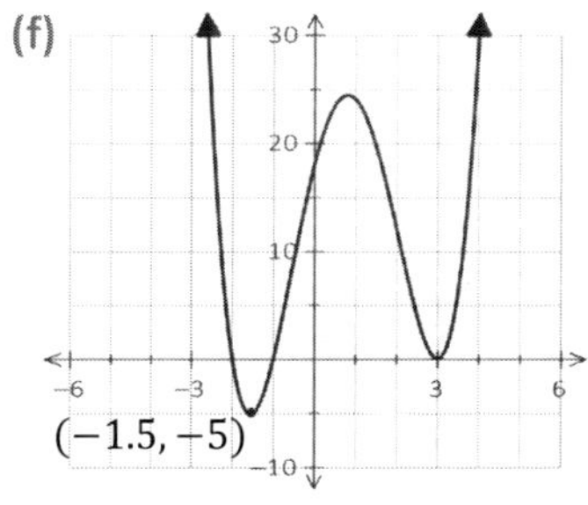

D: ______________ ______________
Set *Interval*

R: ______________ ______________
Set *Interval*

Class Example 1.12 *Obtaining Domain from a function equation*

Visit **math30-1edge.com** for solutions to all warm-ups and class examples

Without graphing, determine the domain of each function below.

(a) $f(x) = \sqrt{5 - 3x}$ Provide in Set *and* Interval notation

(b) $p(x) = 3x^4 - x^3 + 2x - 1$ Set *and* Interval notation

(c) $y = -|x + 4| + 1$

(d) $g(x) = \dfrac{x + 1}{5x + 7}$

(e) $f(x) = \dfrac{1}{x^2 - 9}$

(f) $f(x) = \dfrac{1}{x^2 + 1}$

In Math 20, we saw how a **quadratic function** could be expressed $y = a(x-p)^2 + q$, where the coordinates of the vertex are (p, q).

In Math 30, the vertex form equation is $\boldsymbol{y = a(x-h)^2 + k}$

Where the coordinates of the vertex are (h, k).

We also saw how while $\boldsymbol{h}$ and $\boldsymbol{k}$ affect the **position** of the graph, $\boldsymbol{a}$ affects the parabola **shape** (how wide / narrow) and the **orientation**. (opens up / down)

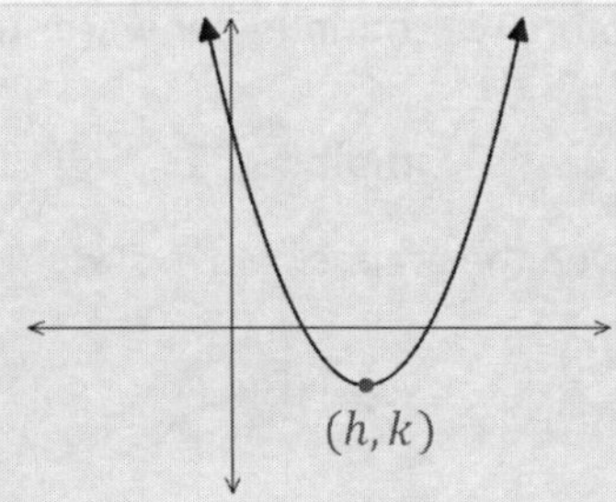

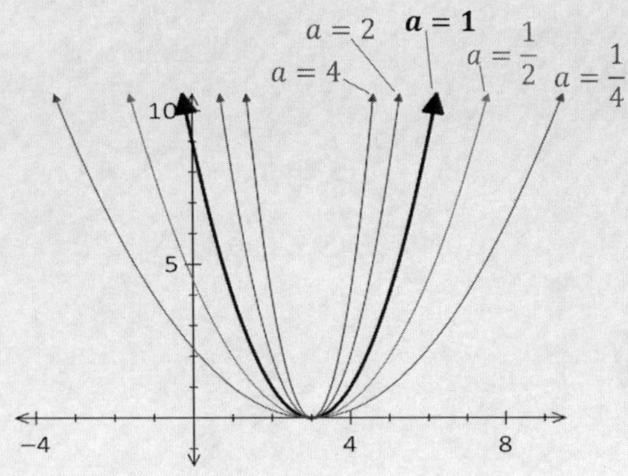

← For example, consider the graph of $\boldsymbol{y = a(x-3)^2}$, for different values of $\boldsymbol{a}$.

The center function is where $a = 1$, that is, $\boldsymbol{y = (x-3)^2}$

When $a > 1$, such as with $\boldsymbol{y = 2(x-3)^2}$, the graph is *vertically stretched by a factor of 2.* (The graph is narrower)

And when $0 < a < 1$, such as with $\boldsymbol{y = \frac{1}{4}(x-3)^2}$, the graph is *vertically stretched by a factor of 1/4.* (graph is wider)

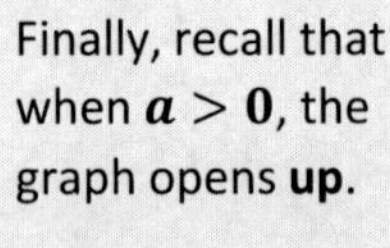

Finally, recall that when $\boldsymbol{a > 0}$, the graph opens **up**.

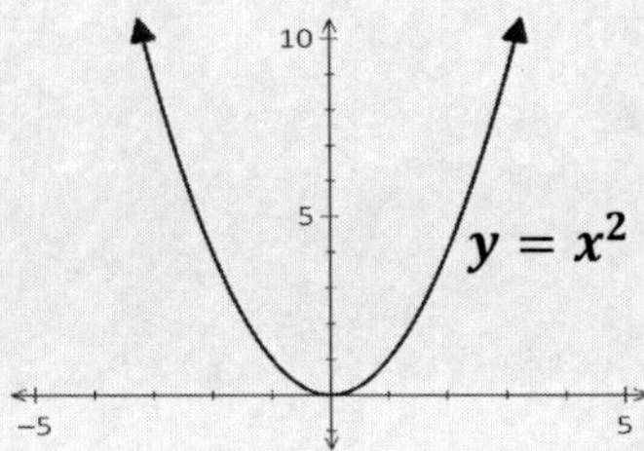

And when $\boldsymbol{a < 0}$, the graph opens **down**.

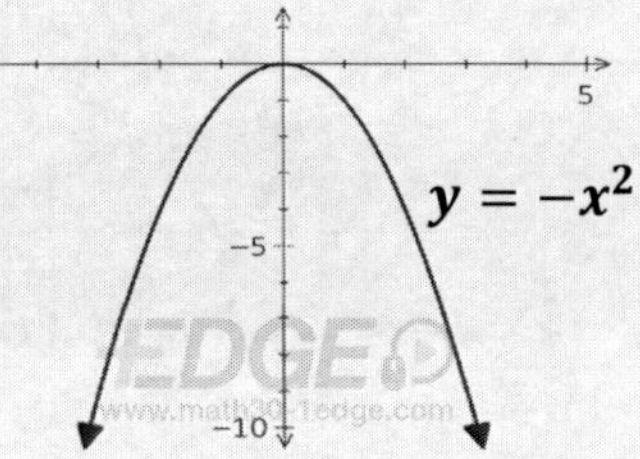

Exploration #5 Vertex Form of a Quadratic Function – The Graph of $\boldsymbol{y = a(x-h)^2 + k}$

1 ➡ Each of the graphs in group 1 represent a quadratic function in the form $\boldsymbol{y = a(x-h)^2 + k}$, where $a = 1$ and $h, k \in I$
Determine an equation for each graph.

❶ ______________ ❷ ______________ ❸ ______________

Group 1

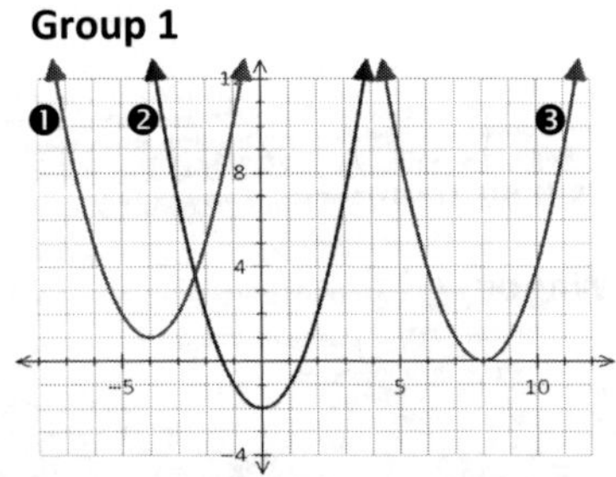

2 ➡ Each of the graphs in group 1 represent a quadratic function in the form $\boldsymbol{y = a(x-h)^2 + k}$, where $a = 1$ *or* $a = -1$, and $h, k \in I$
Determine an equation for each graph.

❶ ______________ ❷ ______________ ❸ ______________

Group 2

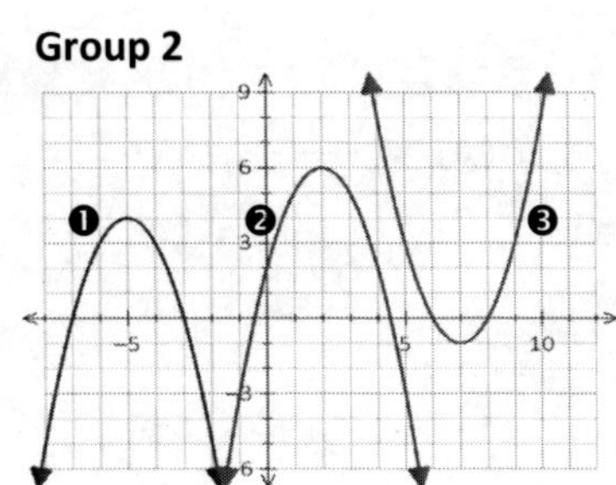

Part B – Horizontal and Vertical Translations

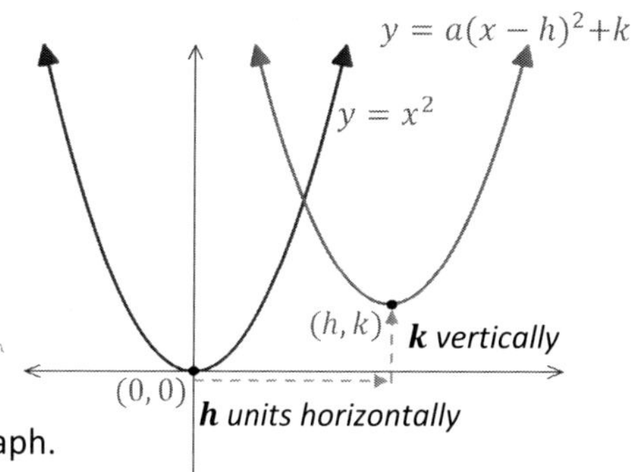

On the previous page, we saw how the parameters a, h, and k affected the graph of $\boldsymbol{y = a(x-h)^2+k}$.

We can think of the vertex as having shifted, or **translated**, from:

- $(0, 0)$ on the graph of $y = x^2$ to (h, k)
- (h, k) on the graph of $y = a(x-h)^2+k$

A function **transformation** alters the location, shape or orientation of graph.

A horizontal or vertical **translation** is a "shift", or change to the graph position. (Think of picking up and moving a graph left / right and up / down)

A function $\boldsymbol{y = f(x)}$, transformed to $\boldsymbol{y = f(x-h)+k}$, is

- **Horizontally** translated $\boldsymbol{h}$ units: RIGHT if $h > 0$ LEFT if $h < 0$
- **Vertically** translated $\boldsymbol{k}$ units: UP if $k > 0$ DOWN if $k < 0$

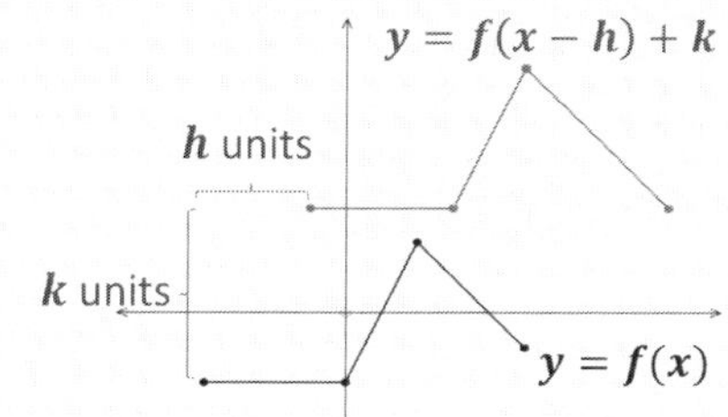

A **mapping rule** describes the effect on each point on the original function to the transformed function. Here, it's: $(x, y) \rightarrow (x+h, y+k)$

Note that the direction of the horizontal translation is the opposite of the sign

Class Example 1.13 *Determining the Horizontal / Vertical Translation from a graph*

For each pair of functions below, $y = g(x)$ is obtained by horizontally and / or vertically translating the graph of $y = f(x)$. Provide the indicated equations / mapping rule below.

(a)

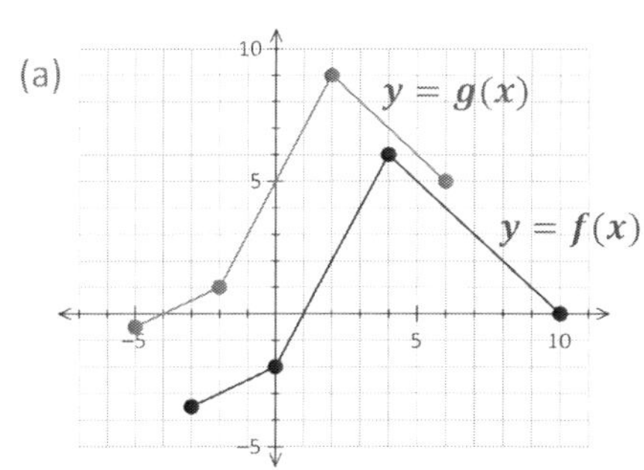

i ________
Equation of $g(x)$ in terms of $f(x)$

ii ________
Mapping rule of $y = f(x) \rightarrow y = g(x)$

(c)

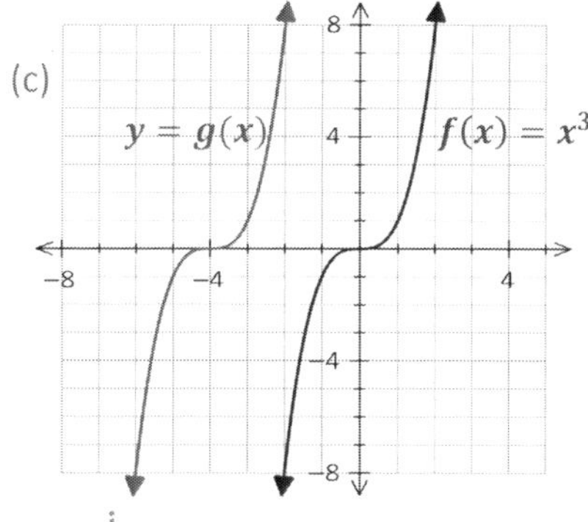

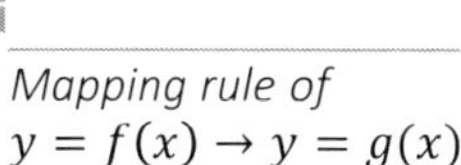

i ________
Equation of $g(x)$ in terms of $f(x)$

ii ________
Equation of $g(x)$ in terms of x

iii ________
Mapping rule of $y = f(x) \rightarrow y = g(x)$

(b)

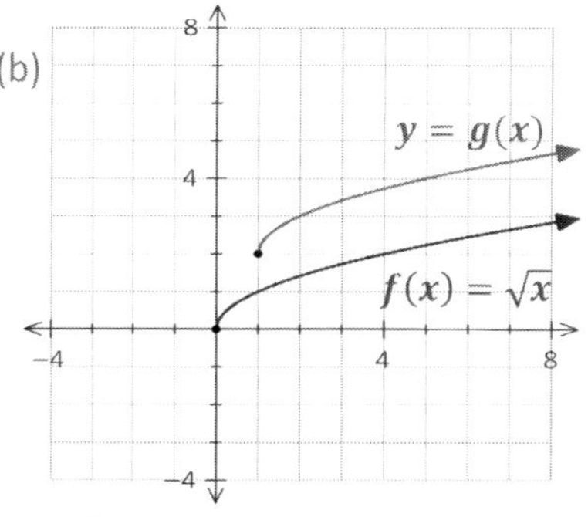

i ________
Equation of $g(x)$ in terms of $f(x)$

ii ________
Equation of $g(x)$ in terms of x

iii ________
Mapping rule of $y = f(x) \rightarrow y = g(x)$

Class Example 1.14 *Determining the Horizontal / Vertical Translation from the equation*

For each pair of functions below,

i - Describe how the graph of function ② can be obtained by transforming the graph of function ①.
ii – Provide a mapping rule for each.
iii – State the domain or range as prompted below

(a) ① $y = f(x)$
② $y = f(x+7) - 1$

i *Description of transformations from ① to ②:*

ii *Mapping rule:*

(b) ① $y = x^2$
② $y = (x-6)^2 + 4$

i *Description of transformations from ① to ②:*

ii *Mapping rule:*

iii *Domain of function ②:*

(c) ① $f(x) = \sqrt{x+2}$
② $y = f(x-4) + 1$

i *Description of transformations from ① to ②:*

ii *Mapping rule:*

iii *Domain of function ②:*

Class Example 1.15 *Sketching a graph using translations*

Given each basic graph below, use transformations to **sketch** the indicated function on the same grid, and provide a mapping rule. Be sure to carefully transform each point indicated (•).
Indicate the domain and range of each sketched function. (Use either set or interval notation)

(a) Sketch $y = \sqrt{x+1} + 3$

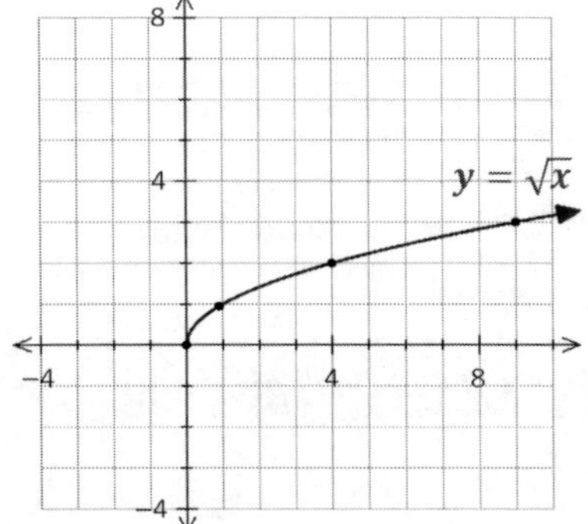

Mapping rule:

Domain:

Range:

(b) Sketch $y = |x-2| + 4$

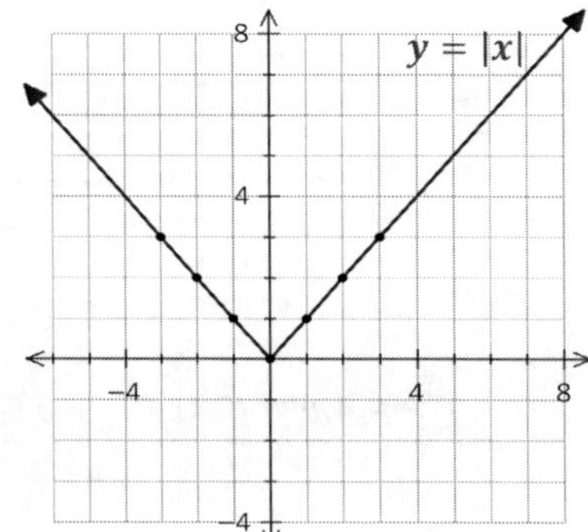

Mapping rule:

Domain:

Range:

(c) Sketch $y = f(x-2) + 3$

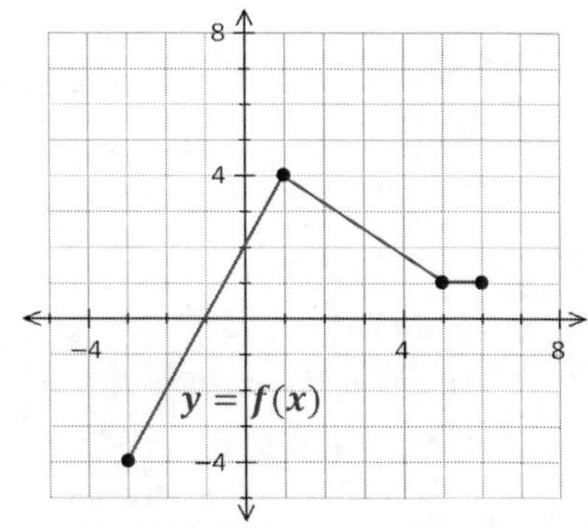

Mapping rule:

Domain:

Range:

1.2 Reflections

In the last section we looked at **translations** – in which the position of a graph is altered. *Think – picking up a graph and moving it left / right or up / down .... and then dropping it.*

We'll next consider **reflections**, which affects a graphs orientation. *Think of – lifting the top or bottom (or left side or right side) of a graph and flipping it over. (Vertically or horizontally)*

Warm-up Exploration #1 Vertical Reflections

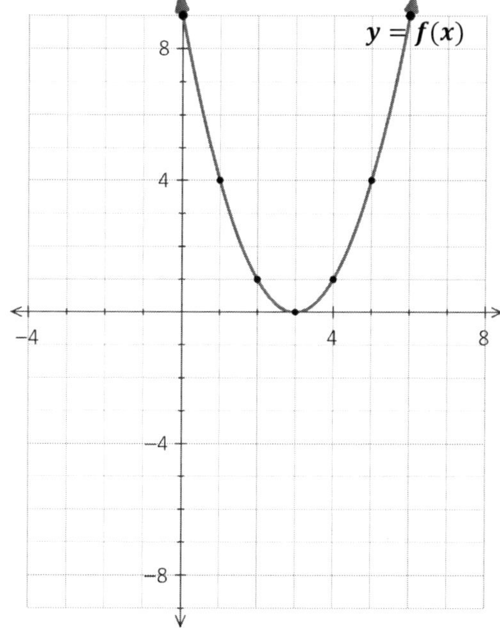

1 ➡ The graph on the function $f(x) = (x - 3)^2$ is on the right. **Sketch** the mirror image of the graph, reflected in the x-axis. *Be sure to indicate the placement of the 7 indicated points.*

Label the new graph $y = g(x)$.

2 ➡ **Describe** how the coordinates change from the graph of $y = f(x)$ to $y = g(x)$.

3 ➡ **State** the equation of $g(x)$, in terms of $f(x)$.

4 ➡ **Explain** how to change the equation of a function, such as $y = (x - 3)^2$, so that the graph is reflected about the x-axis. Determine an equation of the reflected graph, in terms of x.

Exploration #2

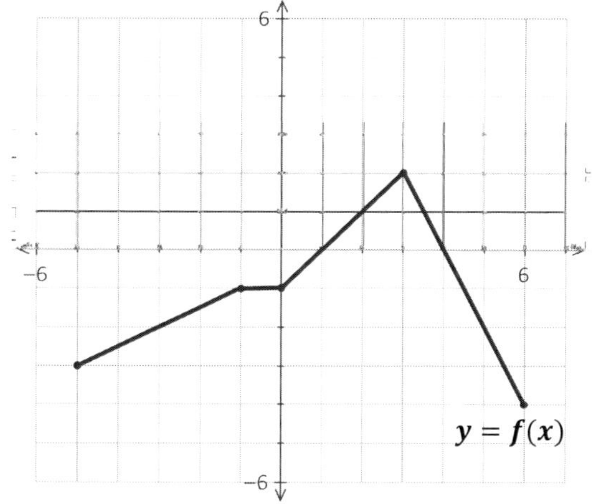

1 ➡ The graph of $y = f(x)$ is on the right. **Sketch** the mirror image graph, again reflected in the x-axis.

Label the new graph $y = g(x)$, and label the new coordinates of the five indicated points.

2 ➡ **State** the mapping rule that describes the change in all coordinates from $y = f(x)$ to $y = g(x)$.

3 ➡ **State** the coordinates of the two points that are on both the graph of $y = f(x)$ and $y = g(x)$. What do they have in common?

Exploration #3

1 ➡ Sketch function $y_1 = x^2 - 4x + 5$ in your graphing calculator. State the coordinates of the vertex.

2 ➡ Sketch function $y_2 = -(x^2 - 4x + 5)$ in your graphing calculator and state the coordinates of the vertex. How do the two graphs compare?

A function $\boldsymbol{y = f(x)}$, transformed to $\boldsymbol{y = -f(x)}$, is **vertically reflected in the x-axis**. *We can alternatively say "vertically reflected about the line $y = 0$"*

All points are transformed $(x, y) \rightarrow (x, -y)$. Points on the x-axis are **invariant**. (they don't change)

To obtain the new equation of a graph vertically reflected about the x-axis, replace $\boldsymbol{y}$ with $\boldsymbol{-y}$.

Let's dive in on that last note, to obtain the equation - replace "y" with "$-y$".

We saw this in action in **Warm-Up #3** where we graphed $y_1 = x^2 - 4x + 5$ and $y_2 = -(x^2 - 4x + 5)$. ➡

Start with: $y = x^2 - 4x + 5$

Replace $\boldsymbol{y}$ with $\boldsymbol{-y}$: $-y = x^2 - 4x + 5$

Isolate $\boldsymbol{y}$: $y = -(x^2 - 4x + 5)$

Now *simplify*: $y = -x^2 + 4x - 5$

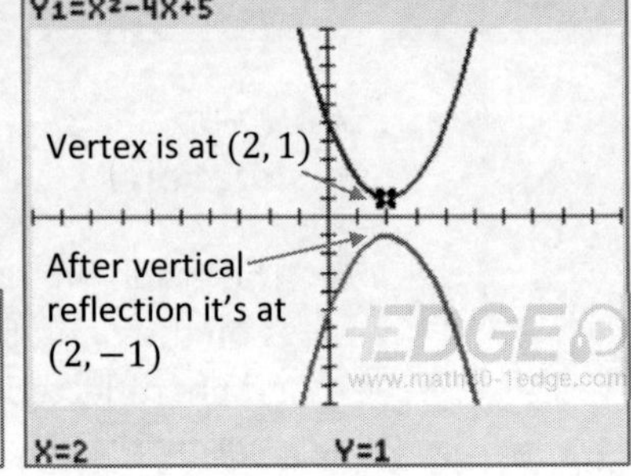

To obtain the equation of a graph that's been vertically reflected about the line $y = 0$, either replace $\boldsymbol{y}$ in the original equation with $\boldsymbol{-y}$... OR (equivalently) *make the entire original equation negative.*

Same result – both simplify to $y = -x^2 + 4x - 5$

$-y = x^2 - 4x + 5$ ⟷ $y = -(x^2 - 4x + 5)$

Exploration #4 Horizontal Reflections

1 ➡ The graph on the function $f(x) = \sqrt{x} - 2$ is on the right.
Sketch the mirror image of the graph, reflected in the y-axis.
Be sure to indicate the placement of the 4 indicated points.
Label the new graph $y = g(x)$.

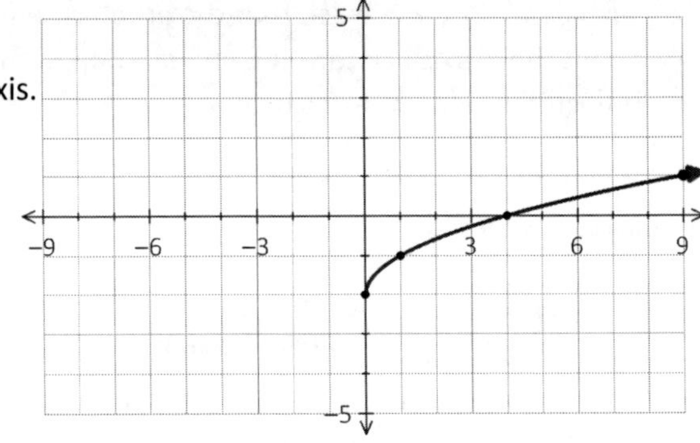

2 ➡ **Describe** how the coordinates change from the graph of $y = f(x)$ to $y = g(x)$.

3 ➡ **State** the equation, in terms of x, of $y = g(x)$.
Explore using your graphing calculator!
Explain how to change the equation of a function so that the graph is reflected about the y-axis.

Exploration #5

1 ➡ The graph of $y = f(x)$ is on the right. **Sketch** the mirror image graph, this time reflected in the y-axis.
Label the new graph $y = h(x)$, and label the new coordinates of the five indicated points.

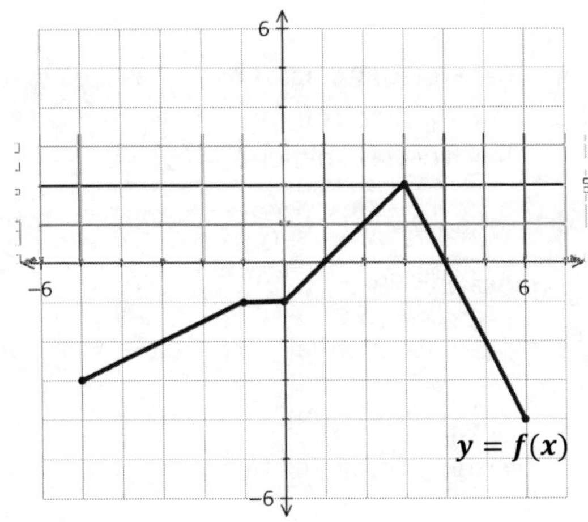

2 ➡ **State** the mapping rule that describes the change in all coordinates from $y = f(x)$ to $y = h(x)$.

3 ➡ **State** the coordinates of the point that is on both the graph of $y = f(x)$ and $y = h(x)$.

Exploration #6

1 ➡ Sketch function $y_1 = x^2 - 4x + 5$ in your graphing calculator. State the coordinates of the vertex.

2 ➡ Sketch function $y_2 = (-x)^2 - 4(-x) + 5$ in your graphing calculator and state the coordinates of the vertex. How do the two graphs compare?

A function $\boldsymbol{y = f(x)}$, transformed to $\boldsymbol{y = f(-x)}$, is **horizontally reflected in the $\boldsymbol{y}$-axis**. *We can alternatively say "horizontally reflected about the line $x = 0$"*

All points are transformed $(x, y) \rightarrow (-x, y)$. Points on the y-axis are **invariant**.

To obtain the new equation of a graph horizontally reflected about the y-axis, replace $\boldsymbol{x}$ with $\boldsymbol{-x}$.

Let's dive in on that last note, to obtain the equation - replace "x" with "$-x$".

In **Warm-Up #6** above we graphed $y_1 = x^2 - 4x + 5$ and $y_2 = (-x)^2 - 4(-x) + 5$

Which simplifies to: $y = x^2 + 4x + 5$

NORMAL FLOAT AUTO REAL DEGREE

Plot1 Plot2 Plot3

Y1=X²-4X+5

Y2=(-X)²-4(-X)+5

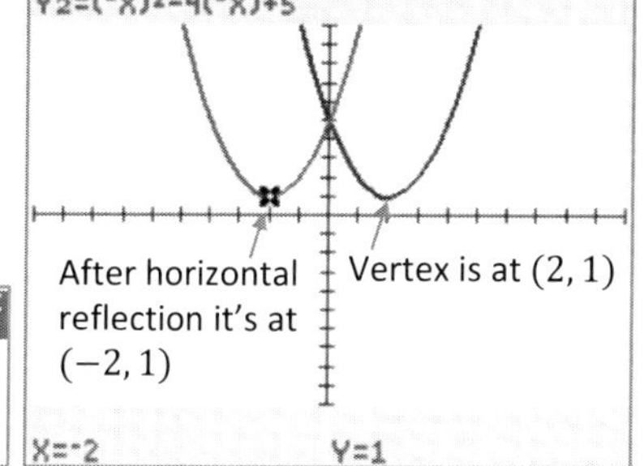

Exploration #7 Summary of Reflection Types

1 ➡ For each reflected graph below, where indicated (✣), state the function equation in terms of $f(x)$, then indicate the mapping rule and invariant points.

2 ➡ For each reflected graph where indicated (◆), state the function equation in terms of x.

Vertical Reflection

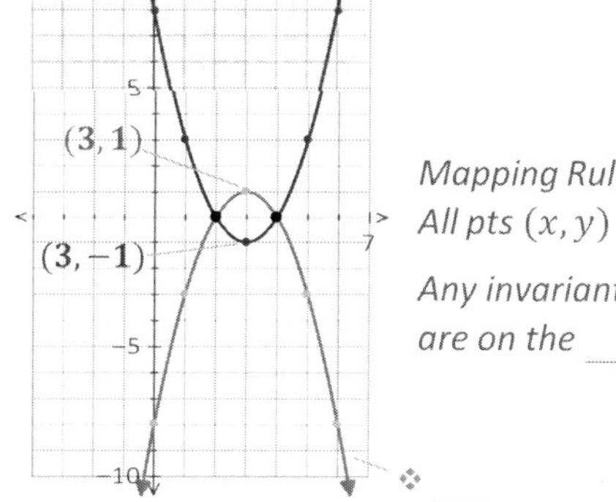

Mapping Rule:

All pts $(x, y) \rightarrow$ ________

Any invariant point(s) are on the ________

✣ ________ ◆ ________

Horizontal Reflection

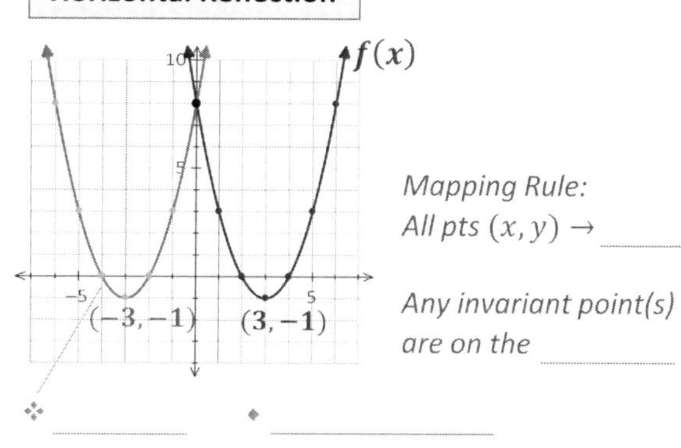

Mapping Rule:

All pts $(x, y) \rightarrow$ ________

Any invariant point(s) are on the ________

✣ ________ ◆ ________

Math foreshadowing time: There's actually a third type of reflection we consider in this course – a reflection about the line $y = x$.

Here, for all points on the graph and the equation, we ***switch x and y***.

We'll revisit this type of reflection at the end of the unit!

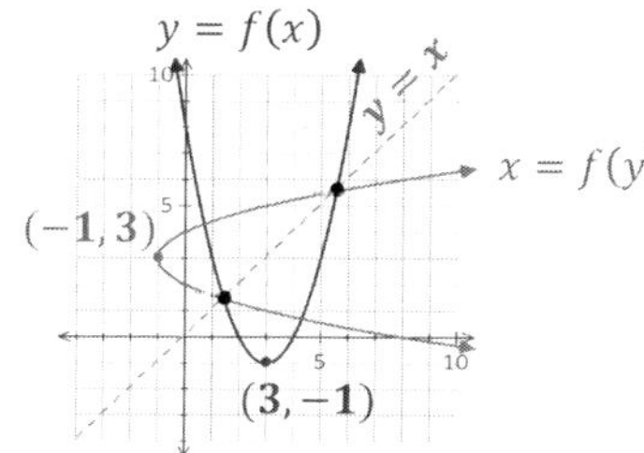

Class Example 1.21 *Sketching a Reflected Graph*

Visit math30-1edge.com for solutions to all warm-ups and class examples

Given the graph of $y = f(x)$ below, i – describe the type of reflection, ii - **sketch** each indicated graph, iii - provide a corresponding mapping rule, and iv - describe the location of any invariant point(s).

(a) Sketch the graph of $y = -f(x)$

i - *Type of Reflection:*

ii

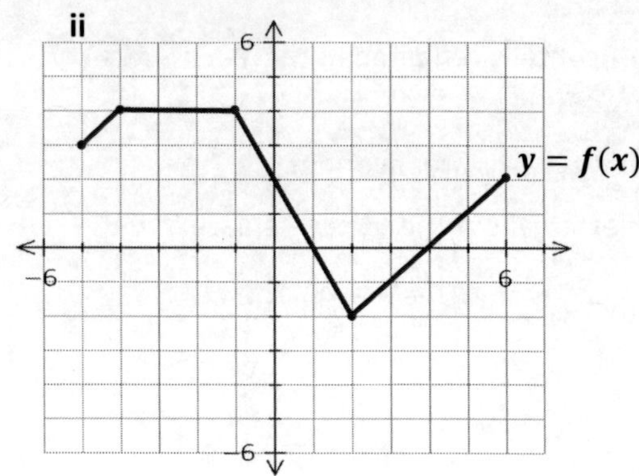

iii - *Mapping Rule:* ________

iv – *Invariant Point(s):* ________

(b) Sketch the graph of $y = f(-x)$

i - *Type of Reflection:*

ii

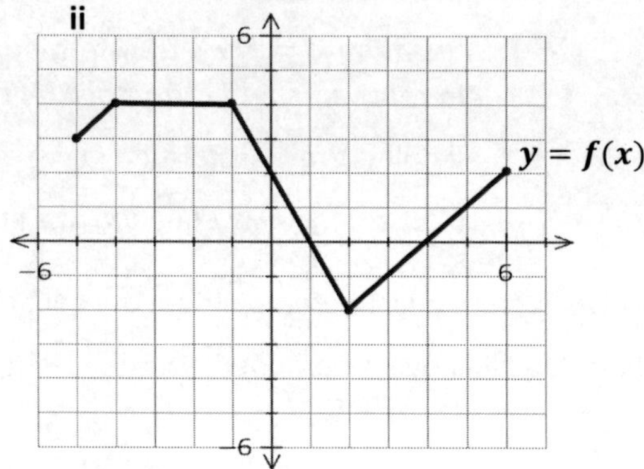

iii - *Mapping Rule:* ________

iv – *Invariant Point(s):* ________

Worked Example

The graph of $\boldsymbol{f(x) = \sqrt{x+4} - 1}$ is transformed by a vertical and horizontal reflection, as described below. For each scenario, sketch the resulting graph and:

(i) State an equation for the reflected function in terms of $f(x)$ and in terms of x

(ii) Determine the new coordinates for the point $P(5, 2)$ after the transformation

(iii) State the location of an coordinates of any invariant point(s)

(iv) State the domain and range of each transformed function

(a) The graph of $y = f(x)$ is vertically reflected about the line $y = 0$.

(b) The graph of y $= f(x)$ is horizontally reflected about the line $x = 0$.

Sol: (a) *Vertical reflection involves making "y", that is the entire expression,* ***negative****.*

i - *Equation in terms of* f: $\boldsymbol{y = -f(x)}$

Equation in terms of x: $y = -(\sqrt{x+4} - 1)$

Simplifies to.... $\boldsymbol{y = -\sqrt{x+4} + 1}$

ii – *New coordinates of* $P(5,2)$: $\boldsymbol{(5, -2)}$

iii – *Invariant pts are on the* x*-axis:* $\boldsymbol{(-3, 0)}$

iv – *Domain of* $f(x)$ *is* $\{x | x \geq -4, x \in \mathbb{R}\}$ — **No change** in the **domain** after vertical reflection

Range of $f(x)$ *is* $\{y | y \geq -1, y \in \mathbb{R}\}$ — **Range** after vertical reflection is $\boldsymbol{\{y | y \leq 1, y \in \mathbb{R}\}}$

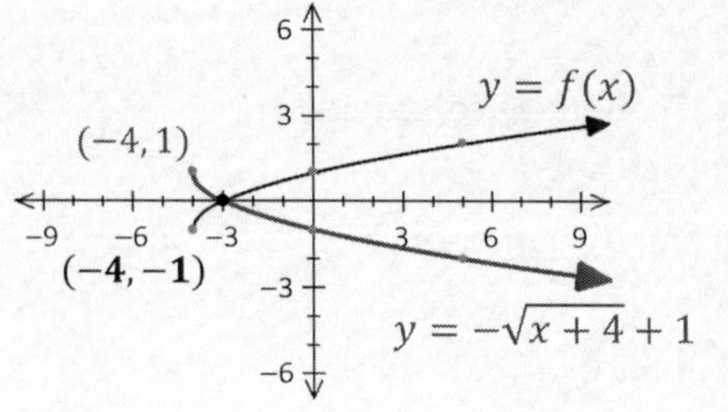

(b) *Horizontal reflection involves making only "x"* ***negative****.*

i - *Equation in terms of* f: $\boldsymbol{y = f(-x)}$

Equation in terms of x: $y = \sqrt{-x+4} - 1$

OPTION - Simplifies to.... $\boldsymbol{y = \sqrt{-(x-4)} - 1}$

ii – *New coordinates of* $P(5,2)$: $\boldsymbol{(-5, 2)}$

iii – *Invariant pts are on the* y*-axis:* $\boldsymbol{(0, 1)}$

iv – *Domain of* $y = f(-x)$ *is* $\boldsymbol{\{x | x \leq 4, x \in \mathbb{R}\}}$

No change to the Range of $y = f(-x)$*, it is still* $\{y | y \geq -1, y \in \mathbb{R}\}$

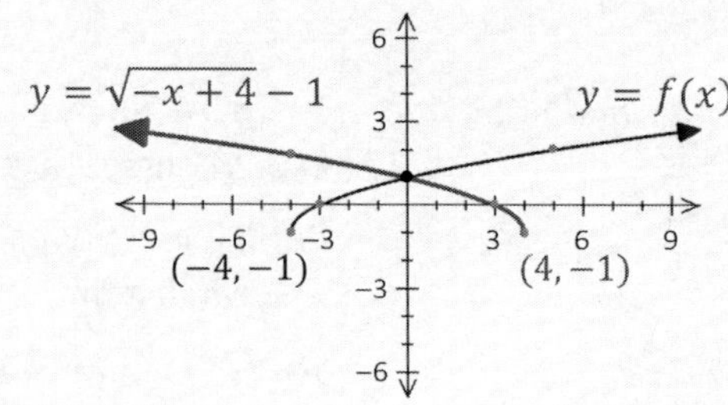

Class Example 1.22 *Applying Reflections to a Function and Graph*

The graph of $\boldsymbol{f(x) = x^2 + 8x + 12}$ is transformed by a vertical and horizontal reflection, as described below. Answer all questions as prompted.

(a) The graph of $y = f(x)$ is **vertically reflected** about the line $y = 0$.

i *Equation of transformed function, in terms of $f(x)$:*

Equation, in terms of x:

ii *Given the graph of $y = f(x)$, sketch the transformed function on the same grid: (Be sure to plot the new location of any indicated points)*

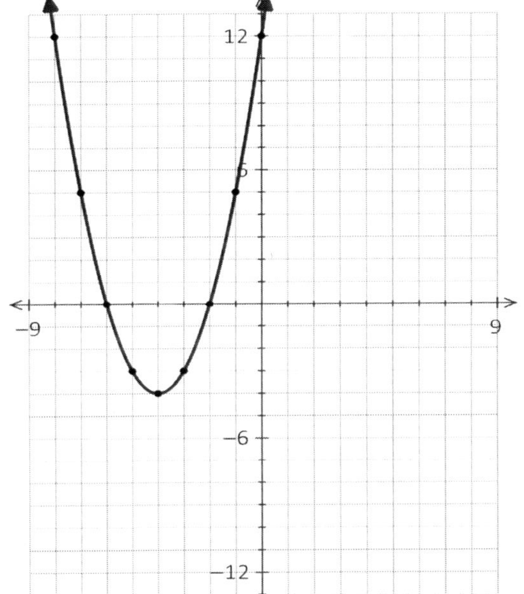

iii *Coordinates of invariant point(s):*

iv *Domain of transformed function:*

Range of transformed function:

(b) The graph of $y = f(x)$ is **horizontally reflected** about the line $x = 0$.

i *Equation of transformed function, in terms of $f(x)$:*

Equation, in terms of x:

ii *Given the graph of $y = f(x)$, sketch the transformed function on the same grid: (Be sure to plot the new location of any indicated points)*

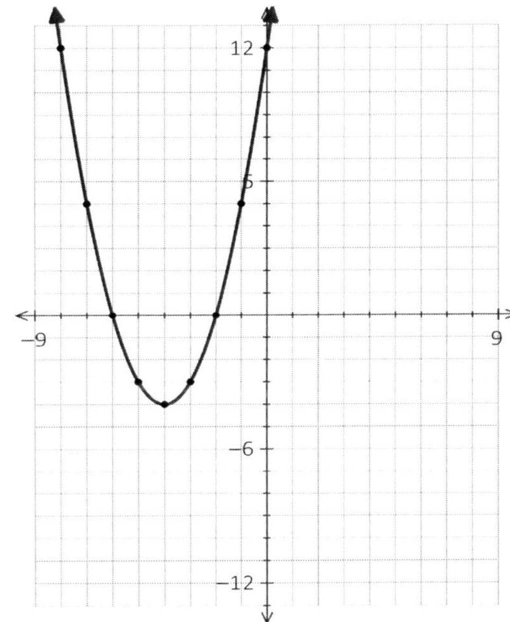

iii *Coordinates of invariant point(s):*

iv *Domain of transformed function:*

Range of transformed function:

note When considering reflections of different types of functions, we encounter some **interesting cases**!

- First, consider $\boldsymbol{f(x) = x^2}$.

We can reflect this graph vertically, with the resulting equation $\boldsymbol{y = -x^2}$.

What if we tried to reflect the graph of $\boldsymbol{f(x) = x^2}$ horizontally?
Would this change the graph? How about the equation?

$$y = (-x)^2$$

simplifies to... $\boldsymbol{y = x^2}$ *(same function!)*

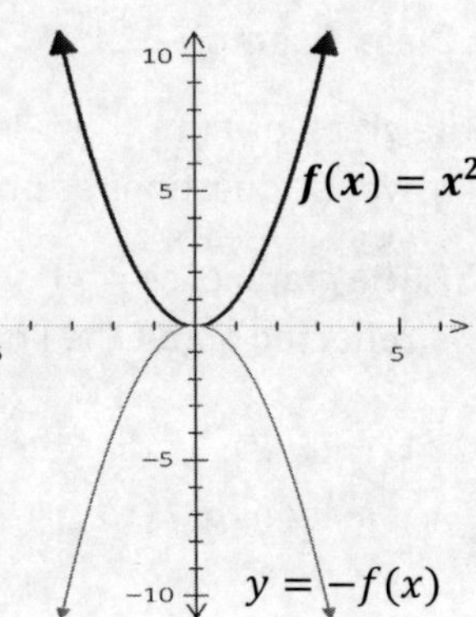

- Next, consider the graphs of $\boldsymbol{g(x) = \frac{1}{x}}$ and $\boldsymbol{h(x) = x^3}$.

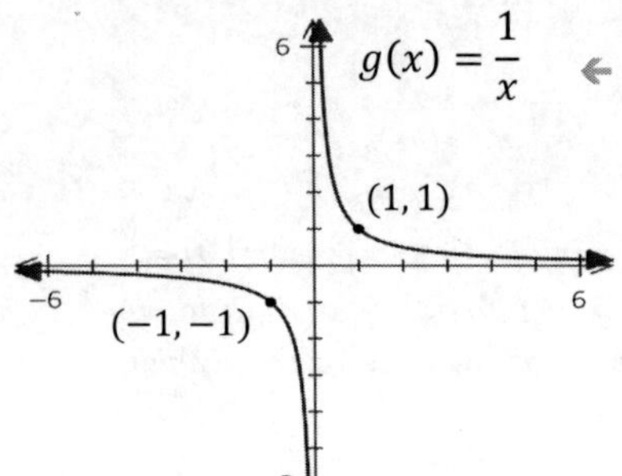

For both of these graphs - notice either a vertical or horizontal reflection gives the *same graph*!

Vertical reflection: ... *& Horizontal:*

$y = -\left(\frac{1}{x}\right)$ ⟺ *same!* ⟺ $y = \frac{1}{-x}$

$y = -(x^3)$ ⟺ *same!* ⟺ $y = (-x)^3$

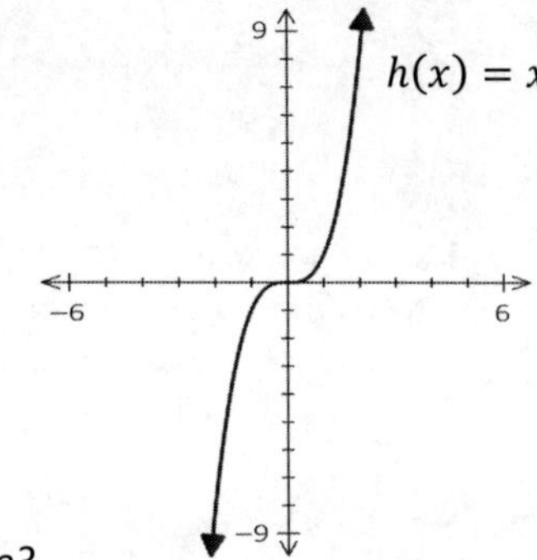

Can you see visualize how the graphs would also be the same?

***Also note** that applying both reflections returns the original function:* $y = -(-x)^3$ ➡ $\boldsymbol{y = x^3}$
(Vert. reflection, Horizontal)

Class Example 1.23 *Applying two types of Reflections*

The graph of $\boldsymbol{f(x) = \sqrt{x - 3} + 1}$, as shown on the right, is transformed to $\boldsymbol{y = g(x)}$ by applying both a vertical and horizontal reflection.

(a) Sketch the graph of $y = g(x)$ on the same grid.
Does the order in which the reflections are applied matter?

(b) State an equation for $y = g(x)$, both in terms of $f(x)$ and in terms of x.

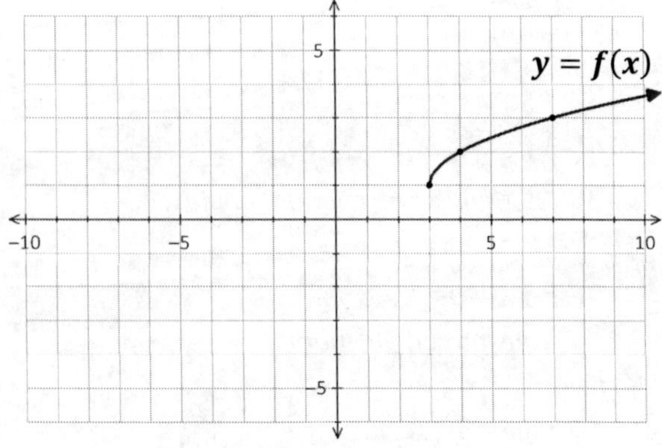

(c) State the domain and range for $y = g(x)$.

1.3 Stretches

Once again – let's first look back where we've been....

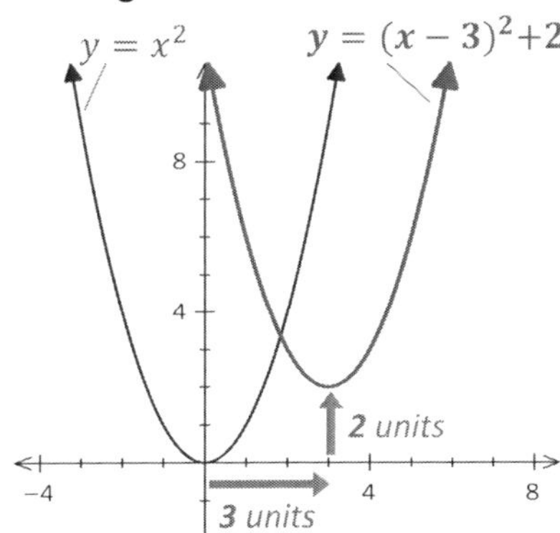

First, we saw how **translations** occur when we *add (or subtract) numbers.*

For example, a function $y = x^2$ can be transformed to $y = (x-3)^2+2$

Here we subtract 3 from $\boldsymbol{x}$

Graph horizontally translates 3 units right
(opposite direction of the sign)

And we subtract 2 from $\boldsymbol{y}$*

And vertically translates 2 units up
(same direction* of the sign)

Another way we can look at this is to apply *replacements:*

- To ***horizontally translate*** the graph 3 units right, replace "x" with "$x-3$"
- To ***vertically translate*** the graph 2 units up, replace "y" with "$y-2$"

So, $y = x^2$ *becomes* $y - 2 = (x-3)^2$ → *Simplifies to* $\boldsymbol{y = (x-3)^2 + 2}$

When we think of vertical translations this way, we can treat it "the same" as horizontal!
**That is, the opposite direction of the sign in the equation:* $y = f(x) \rightarrow \boldsymbol{y - k = f(x-h)}$

We can similarly think of **reflections** with replacements.

For example, given the graph of $y = (x-4)^2$ ...

- To ***horizontally reflect***, replace "x" with "$-x$"

$\boldsymbol{y = (-x-4)^2}$

We can optionally simplify to:
$y = [-1(x+4)]^2$
$y = (-1)^2(x+4)^2$
→ $y = (x+4)^2$

- To ***vertically reflect***, replace "y" with "$-y$"

$\boldsymbol{-y = (x-4)^2}$

Now divide both sides by -1 *to isolate "y"!*
→ $y = -(x-4)^2$

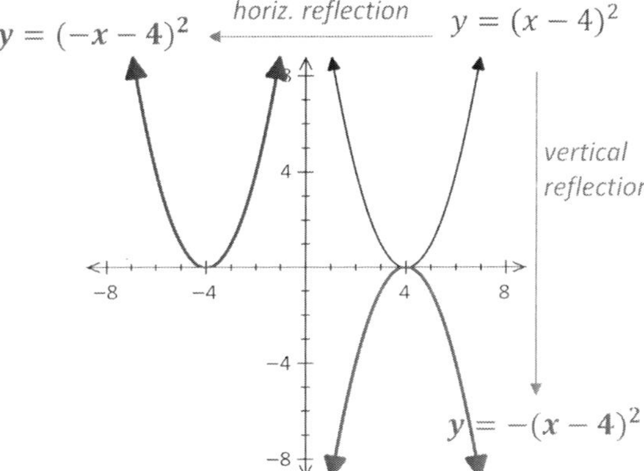

Replacing "y" with "−y" is identical to "making the entire right side negative"

$-y = f(x) \Leftrightarrow y = -f(x)$

→ Vertical reflection about the x-axis

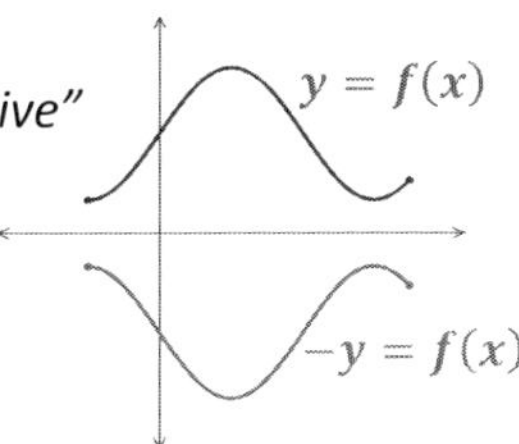

And replacing "y" with "y − k" is identical to "adding k to the function"

$y - k = f(x) \Leftrightarrow y = f(x) + k$

→ Vertical translation k units
- up, if $k > 0$ - down, if $k < 0$

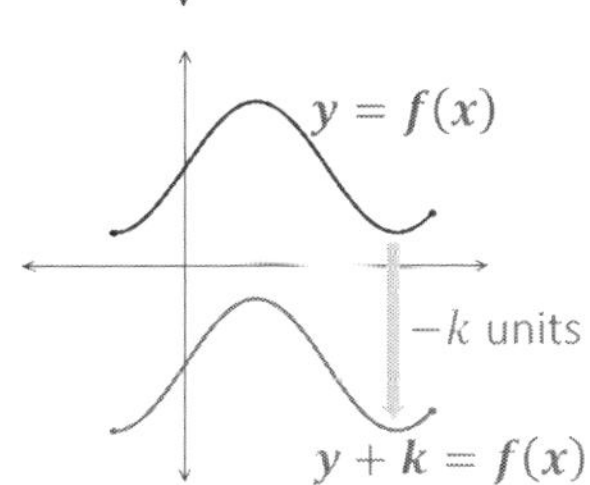

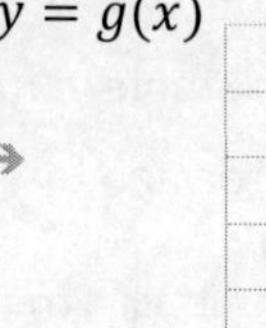

Exploration #1

The graph of $f(x) = \sqrt{x+4} - 1$ is on the right

warmup

1 ➡ On the same grid, construct a new graph of $y = g(x)$ by multiplying all of the y-coordinates by 2.
Be sure to transform all points indicated (•) →

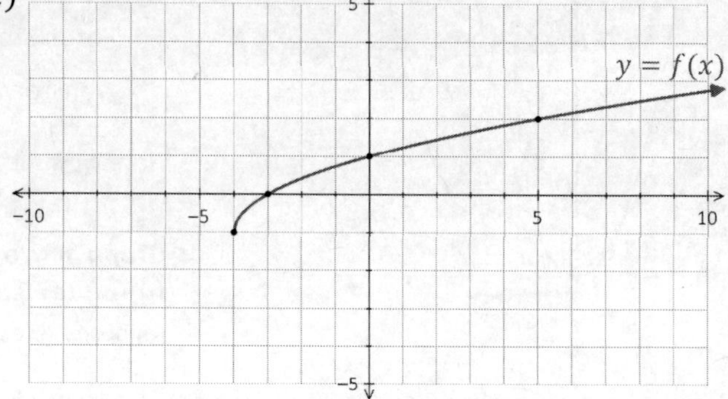

2 ➡ Is the resulting stretch horizontal or vertical?

3 ➡ How could this transformation be described with a mapping rule?

4 ➡ Describe the location and coordinates of any invariant point(s). *(Points that don't change)*

5 ➡ Use your graphing calculator to confirm the equation of $g(x)$ is: $g(x) = 2(\sqrt{x+4} - 1)$

Exploration #2

The graph of $f(x) = \sqrt{x+4} - 1$ is again on the right

1 ➡ On the same grid, construct a new graph of $y = k(x)$ by multiplying all x-coordinates by 2.

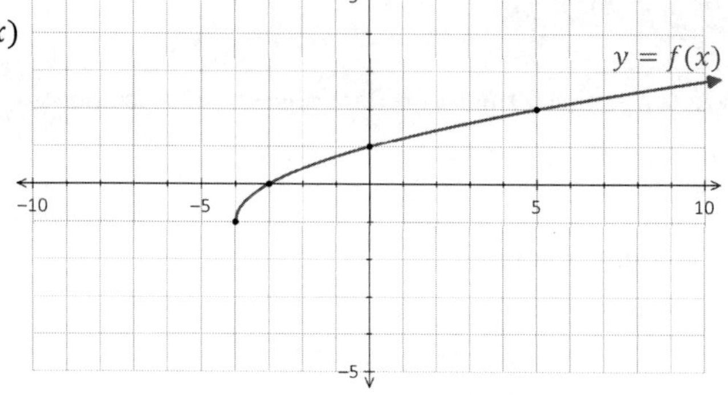

2 ➡ Is the resulting stretch horizontal or vertical?

3 ➡ How could this transformation be described with a mapping rule?

4 ➡ Describe the location and coordinates of any invariant point(s).

5 ➡ Use your graphing calculator to confirm the equation of $k(x)$ is: $k(x) = \sqrt{(1/2)x + 4} - 1$

Exploration #3

The graph of ① $g(x) = x^2 - 4$ is shown on the right.

1 ➡ Use your graphing calculator to help sketch the graphs of ② $y = 2(x^2 - 4)$ and ③ $y = 0.5(x^2 - 4)$ on the same grid.
Graph ① and ② together to compare, then graph ① and ③.
Be sure to indicate the points corresponding to those indicated (•) →

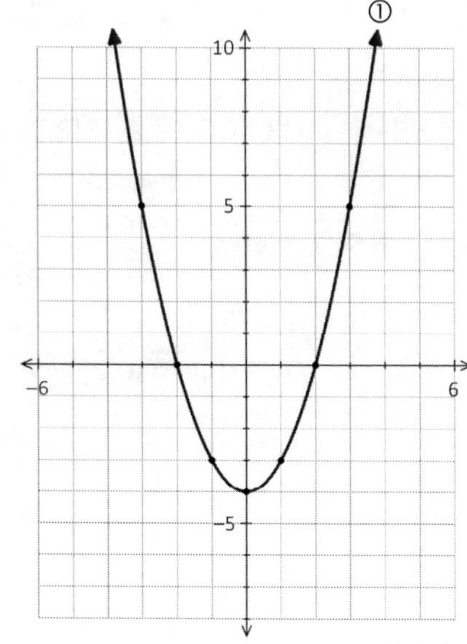

2 ➡ Explain the effect of the "2" and "0.5" in terms of stretching.

3 ➡ Describe the location and coordinates of any invariant point(s).

Exploration #4

The graph of ① $f(x) = x^2 - 4$ is shown on the right.

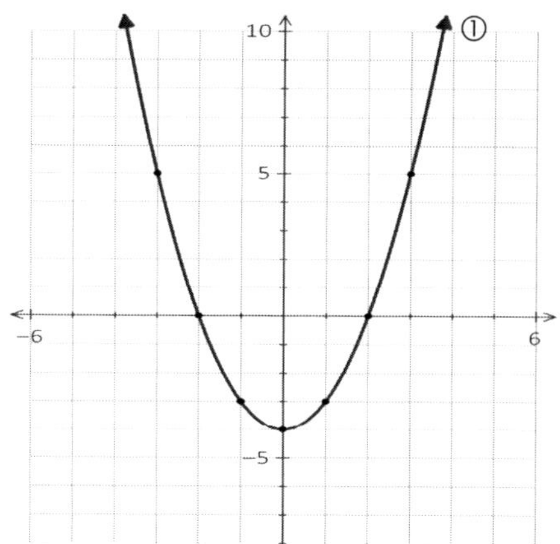

1 ➡ Use your graphing calculator to help sketch the graphs of ② $y = (2x)^2 - 4$ and ③ $y = (0.5x)^2 - 4$ on the same grid.

Be sure to indicate the points corresponding to those indicated (•) →

2 ➡ Explain the effect of the "2" and "0.5" in terms of stretching.

The graph of a function $\boldsymbol{y = f(x)}$, transformed to $\boldsymbol{y = af(x)}$, is **vertically stretched** about the x-axis by **a factor of $\boldsymbol{a}$**.

All points are transformed $(x, y) \to (x, ay)$.

All points on the graph of $y = af(x)$ are "$\boldsymbol{a}$" times further from the x-axis

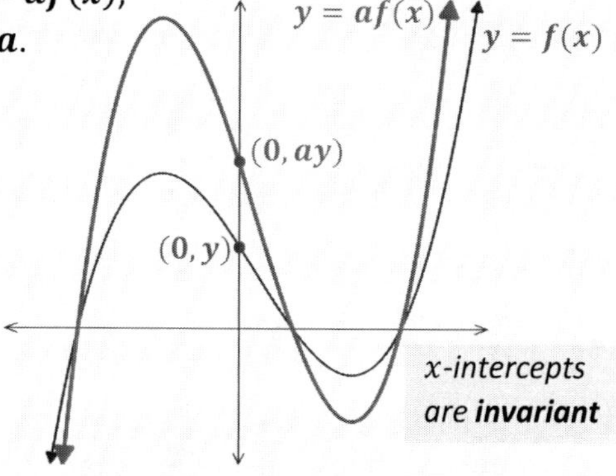

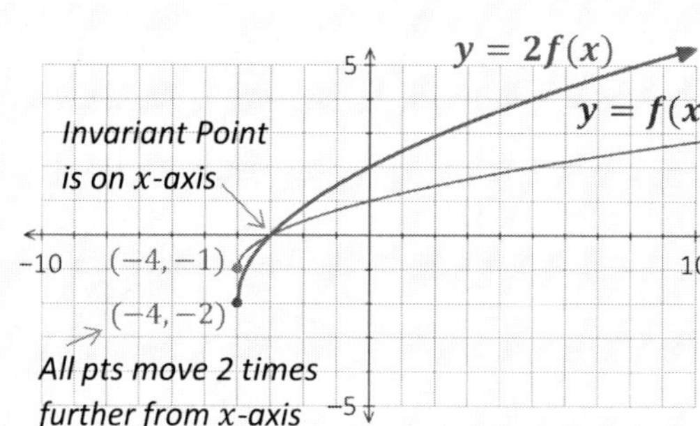

← *For example*, the graph of a function $f(x) = \sqrt{x+4} - 1$ can be vertically stretched by a factor of 2,

Giving an equation $y = \boldsymbol{2}(\sqrt{x+4} - 1)$

Which can be simplified to: $y = 2\sqrt{x+4} - 2$

The graph of a function $\boldsymbol{y = f(x)}$, transformed to $\boldsymbol{y = f(bx)}$, is **horizontally stretched** about the x-axis by **a factor of $\boldsymbol{1/b}$**.

All points are transformed $(x, y) \to (\frac{1}{b}x, y)$

Reciprocal

All points on the graph of $y = f(bx)$ are "$\boldsymbol{1/b}$" times further from the y-axis

*y-intercept is **invariant***

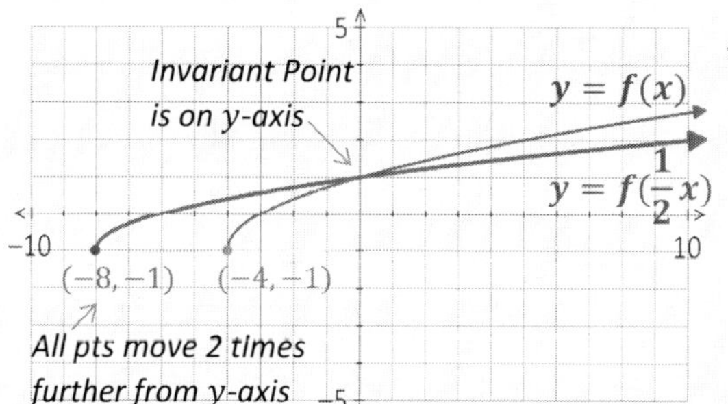

← *For example*, the graph of a function $f(x) = \sqrt{x+4} - 1$ can be horizontally stretched by a factor of 2,

Giving an equation $\boldsymbol{y = \sqrt{(1/2)x + 4} - 1}$

Which can be simplified to: $y = \sqrt{1/2(x+8)} - 2$

In the transformation $y = f(x) \rightarrow y = af(bx)$

*The **vertical** stretch is "a" (straightforward!)*

*The **horizontal** stretch is "$1/b$" (**reciprocal!**)*

Another way to look at this is to treat both the vertical and horizontal stretches the same.

So if, for example, a function $y = f(x)$ is horizontally stretched by a factor of 3 and vertically stretched as a factor of 4, we "could" ....

- Replace x with $\frac{1}{3}x$

and, similarly...
- Replace y with $\frac{1}{4}y$

$$\frac{1}{4}y = f(\frac{1}{3}x)$$

So here both the horizontal and vertical stretches are reciprocals in the equation

Note that this can be simplified to:

$$y = 4f(\frac{1}{3}x)$$

Which is how we would "normally" view this!

Class Example **1.31** *Stretching a Graph* — *Visit **math30-1edge.com** for solutions*

Given the graph of $y = f(x)$ on the right,

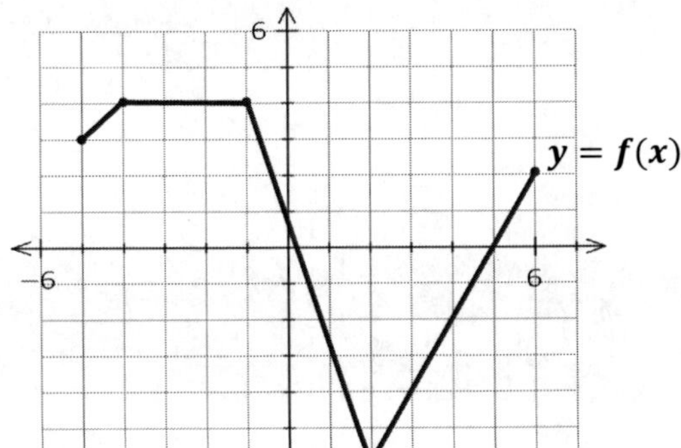

(a) Construct a mapping rule to sketch the graph of $g(x) = \frac{1}{2}f(x)$

(b) State the location of any invariant points.

(c) Compare the domain and range of $y = f(x)$ and $y = g(x)$. Describe which is affected, and how.

$y = f(x)$ *D:* __________ $y = g(x)$ *D:* __________

R: __________ *R:* __________

Class Example **1.32** *More Graph Stretching*

Given the graph of $y = f(x)$ on the right,

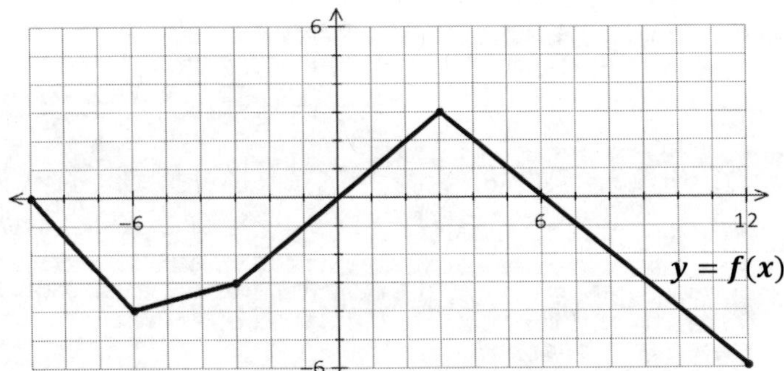

(a) Construct a mapping rule to sketch the graph of $g(x) = f(3x)$

(b) State the location of any invariant points.

(c) Compare the domain and range of $y = f(x)$ and $y = g(x)$. Describe which is affected, and how.

$y = f(x)$ *D:* __________ $y = g(x)$ *D:* __________

R: __________ *R:* __________

Worked Example

The graph of $\boldsymbol{f(x) = x^2 - 6x + 5}$ is shown on the right.

- $y = g(x)$ is obtained by vertically stretching the graph of $f(x)$ about the line $y = 0$, by a factor of 3, and reflecting the graph about the x-axis.
- $y = h(x)$ is obtained by horizontally stretching the graph of $f(x)$ about the line $x = 0$, by a factor of 2.

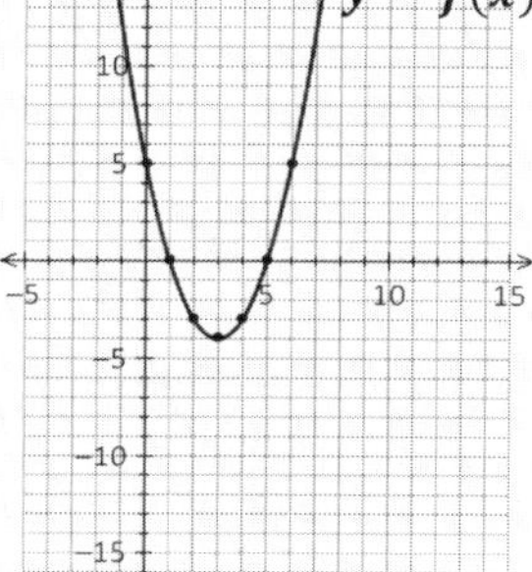

(a) State an equation for and sketch the graph of $y = g(x)$, on the same grid. *Be sure to indicate the new location of all indicated points.* (•)

(b) State an equation and sketch the graph of $y = h(x)$, on the same grid.

Sol: (a) *Equation in terms of* $f(x)$: $\boldsymbol{g(x) = -3f(x)}$

Vertical reflection Vertical stretch

Equation in terms of x: $y = -3(x^2 - 6x + 5)$

Simplifies to: $\boldsymbol{y = -3x^2 + 18x - 15}$

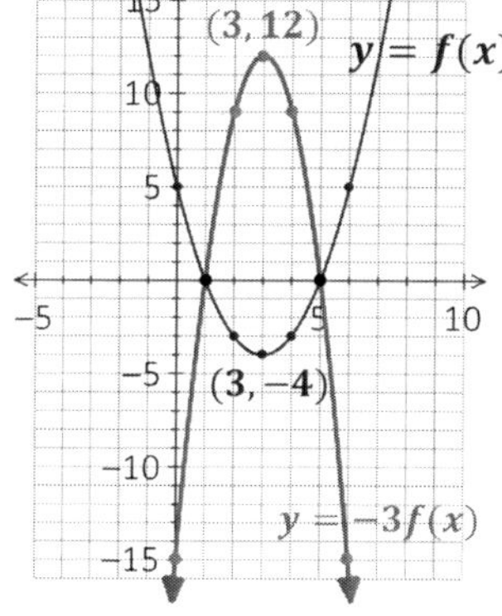

For the equation, multiply all y*-coordinates by* -3.
Mapping rule: All points $(x, y) \to (x, -3y)$

Invariant points *are on the* x*-axis, at* $(1, 0)$ *and* $(5, 0)$.

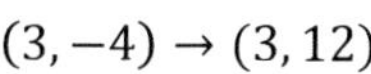

$(3, -4) \to (3, 12)$

$(2, -3) \to (2, 9)$ *and so on...*

(b) *Equation in terms of* $f(x)$: $\boldsymbol{h(x) = f(\frac{1}{2}x)}$

Horizontal stretch (reciprocal)

Equation in terms of x: $y = (\frac{1}{2}x)^2 - 6(\frac{1}{2}x) + 5$

Simplifies to: $\boldsymbol{y = \frac{1}{4}x^2 - 3x + 5}$

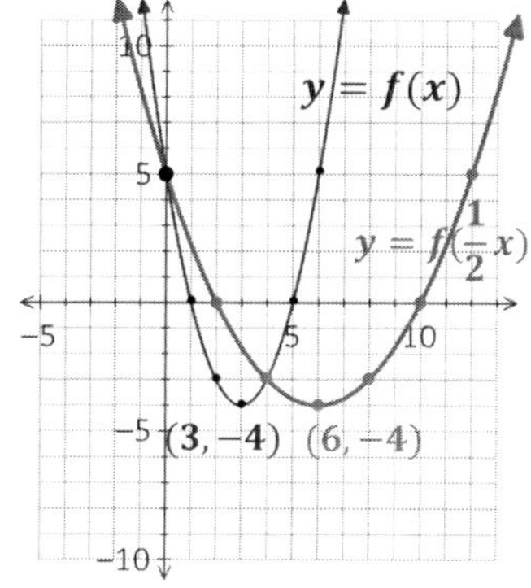

For the equation, multiply all x*-coordinates by* 2.
Mapping rule: All points $(x, y) \to (2x, y)$

Invariant point *is on the* y*-axis, at* $(0, 5)$.

$(3, -4) \to (6, -4)$

$(2, -3) \to (4, -3)$ *and so on...*

Class Example 1.33 *Applying a Stretch and Reflection to a Function Equation*

A function $f(x) = |x - 5| + 8$ has a range of $\{y \geq 8, y \in \mathbb{R}\}$. A function $y = g(x)$ is obtained by reflecting the graph of $y = f(x)$ about the line $y = 0$, and applying a vertical stretch by a factor of 1/4.

(a) Determine an equation for $y = g(x)$, both in terms of $f(x)$ and in terms of x.

(b) State the range of $g(x)$.

Class Example 1.34 *Applying a Stretch to a Function Graph and Equation*

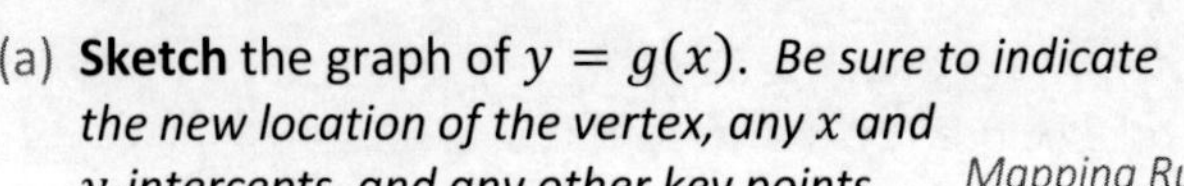

The graph of $f(x) = -(x-2)^2 + 9$ is shown. The graph of $y = g(x)$ is obtained by vertically stretching the graph of $f(x)$ about the line $y = 0$ by a factor of 3, and reflecting it about the line $y = 0$.

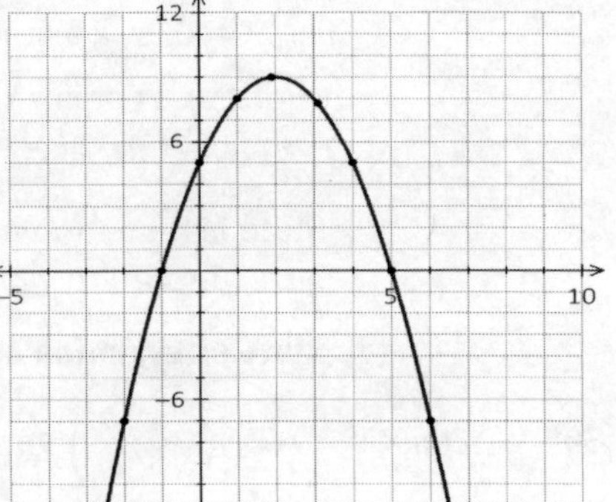

(a) **Sketch** the graph of $y = g(x)$. *Be sure to indicate the new location of the vertex, any x and y-intercepts, and any other key points.* *Mapping Rule:* ____________

(b) Determine an equation for $y = g(x)$,

In terms of $f(x)$:

In terms of x:

(c) State the location and coordinates of any invariant point(s).

(d) State the range of both $y = f(x)$ and $y = g(x)$.

$y = f(x)$ *R:* ____________ $y = g(x)$ *R:* ____________

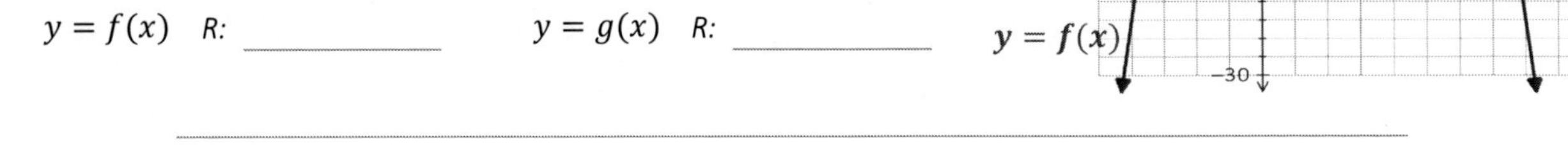

The graph of $f(x) = -(x-2)^2 + 9$ is again shown below. The graph of $y = h(x)$ is obtained by horizontally stretching the graph of $f(x)$ about the line $x = 0$ by a factor of 3.

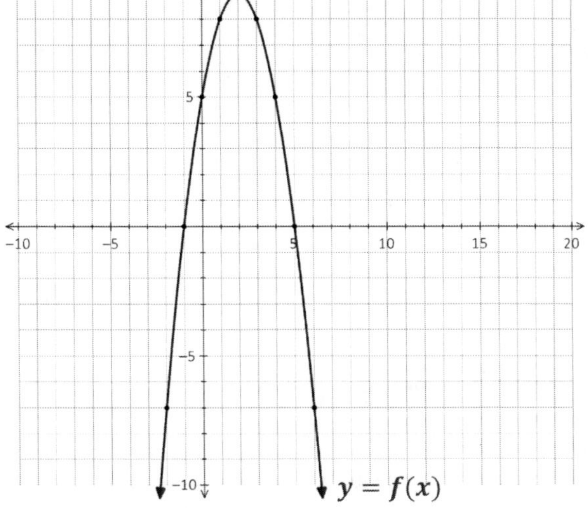

(e) **Sketch** the graph of $y = h(x)$. *Be sure to indicate the new location of the vertex, and any x and y-intercepts.*

Mapping Rule: ____________

(f) Determine an equation for $y = h(x)$,

In terms of $f(x)$:

In terms of x:

(g) State the location and coordinates of any invariant point(s).

Class Example 1.35 *Applying a Stretch to a Function Graph and Equation*

The graph of $f(x) = |2x + 6| - 12$ is shown on the right. A new function $g(x)$ is defined $g(x) = f(3x)$.

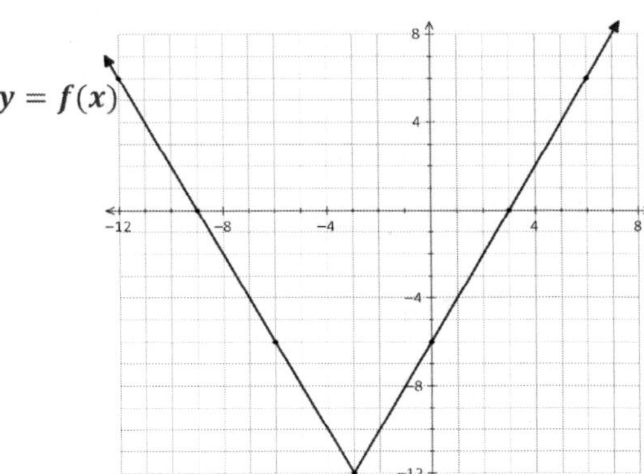

(a) **Sketch** the graph of $y = g(x)$, on the same grid.

Mapping Rule: ____________

(b) Determine an equation for $y = g(x)$, in terms of x.

Worked Example For each pair of graphs below, the graph of $y = g(x)$ is obtained by stretching the graph of $f(x)$. For each, determine an equation for $y = g(x)$.

(a)

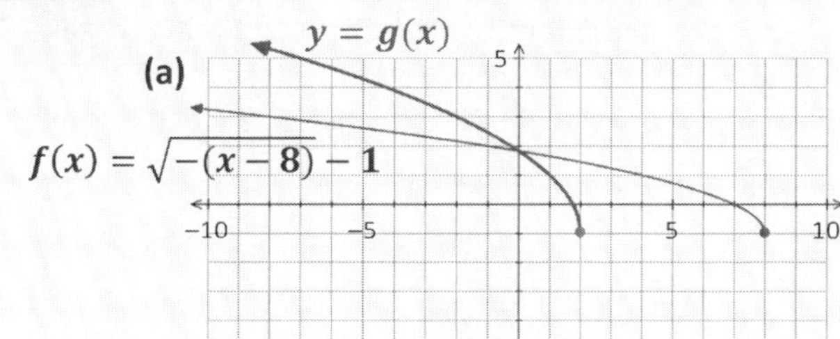

(b)

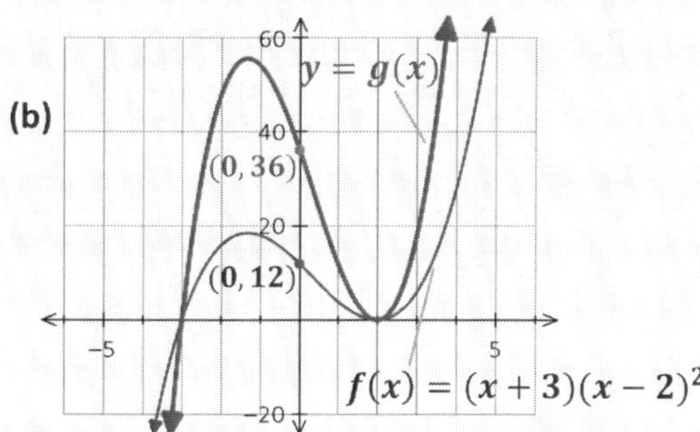

Sol: (a)

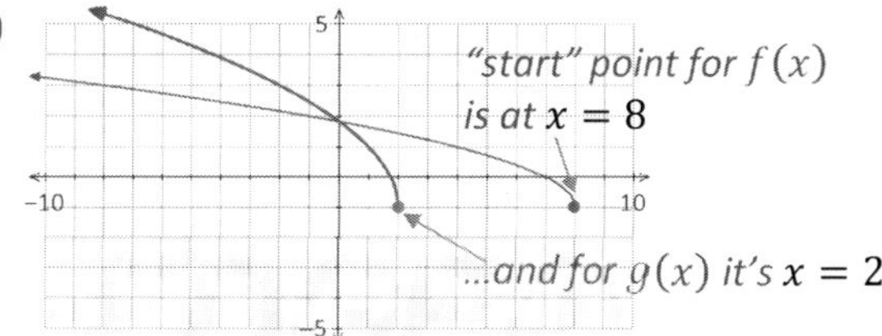

The start point (as with all other points) is stretched **horizontally**, as it moves closer to the y-axis.
Also note that the invariant point is on the y-axis

➔ $8 \times$ [horiz. stretch] $= 2$ ➔ [horiz. stretch] $= 2 \div 8$

horizontal stretch factor $= \dfrac{1}{4}$

So, equation is: $\boldsymbol{g(x) = f(4x)}$

Reciprocal of stretch factor goes in front of x

Equation in terms of x: $g(x) = \sqrt{-(4x-8)} - 1$

Optionally simplify to: $\boldsymbol{g(x) = \sqrt{-4(x-2)} - 1}$

(b)

Then on $g(x)$ it's at $y = 36$

(0, 36)

(0, 12)

y-intercept for $f(x)$ is $y = 12$

The y-intercept (as with all other points) is stretched **vertically**, as it moves further from the y-axis.
Also note that the invariant points are on the x-axis

➔ $12 \times$ [vert. stretch] $= 36$

➔ [vert. stretch] $= 36 \div 12$

vertical stretch factor $= 3$

So, equation is: $\boldsymbol{g(x) = 3f(x)}$

Stretch factor goes in front of all of $f(x)$

In terms of x: $\boldsymbol{g(x) = 3(x+3)(x-2)^2}$

Class Example 1.36 *Determining the Stretch from a Graph*

For each pair of graphs below, the graph of $y = g(x)$ is obtained by stretching the graph of $f(x)$.
Determine an equation for $y = g(x)$, in terms of $f(x)$.

(a)

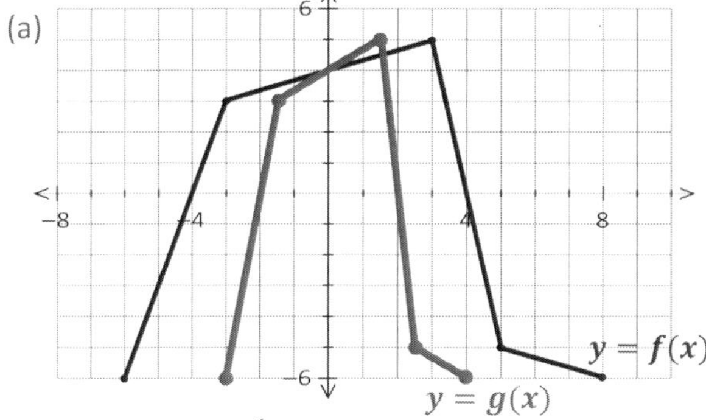

(b)

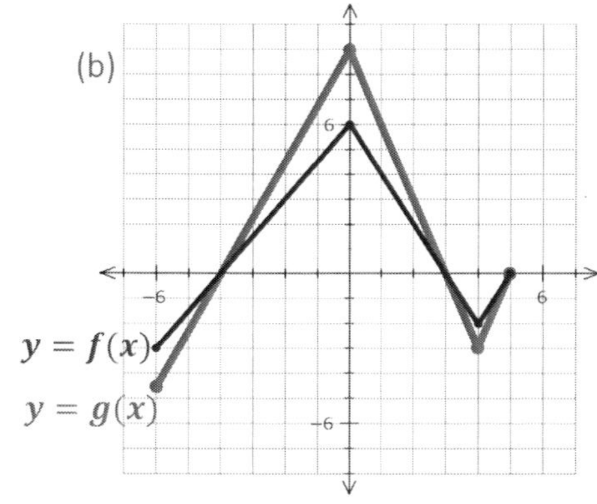

Mapping Rule: ____________

Equation: ____________

Mapping Rule: ____________

Equation: ____________

Class Example 1.37 *Determining the Stretch from a Graph*

For each graph below, the graph of $y = g(x)$ is obtained by stretching the graph of $f(x)$. Points indicated (•) have integer coordinates. Determine an equation for $g(x)$ for each, and identify all indicated characteristics.

(a)

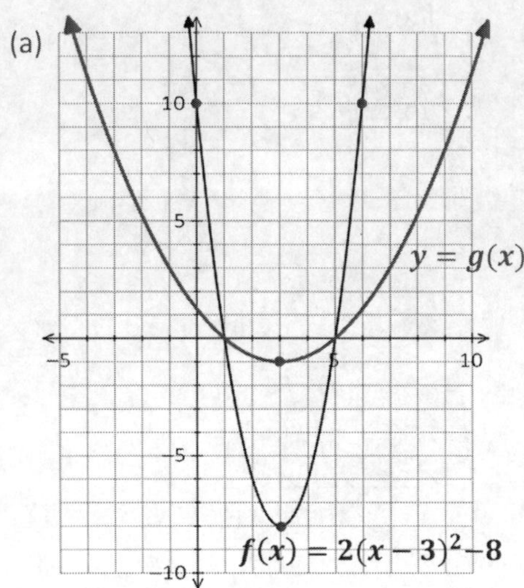

Equation in terms of $f(x)$: ____________

Equation in terms of x: ____________

y*-intercept of* $g(x)$: ____________

(b)

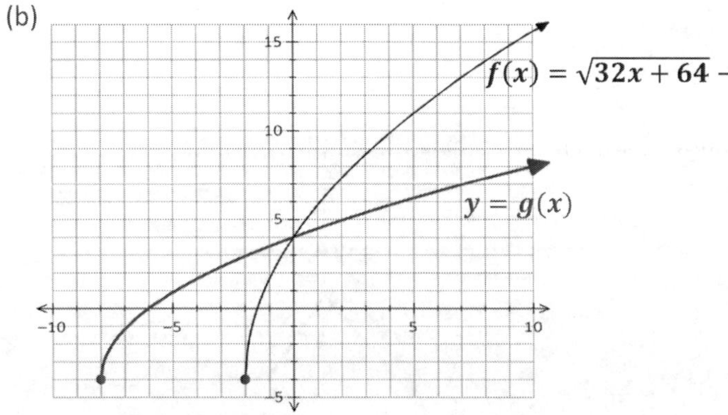

Equation in terms of $f(x)$: ____________

Equation in terms of x: ____________

x*-intercept of* $f(x)$: ____________

(use an algebraic process)

In the previous section we considered problems that involved both a stretch and reflection.

And we found that the order in which we applied the transformations didn't matter.

For example, consider the graph of $f(x) = \sqrt{x+4} - 1$

Suppose we wish to apply a horizontal stretch about the y-axis by a factor of $1/2$ and apply a horizontal reflection about the y-axis.

- We can apply the ***stretch first***, and then the reflection....

Horizontal stretch by a factor of $1/2$

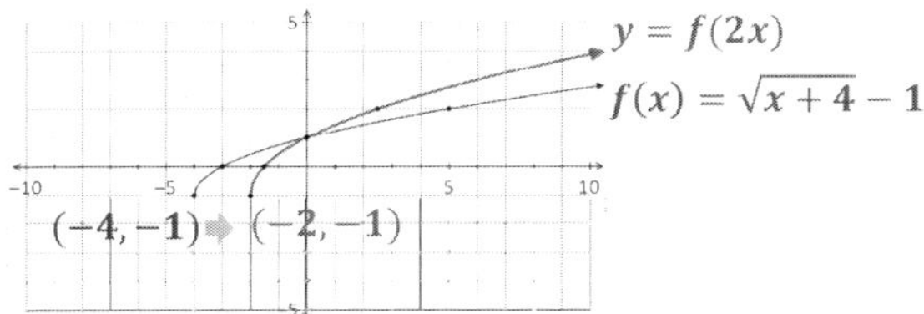

And then horizontal reflection

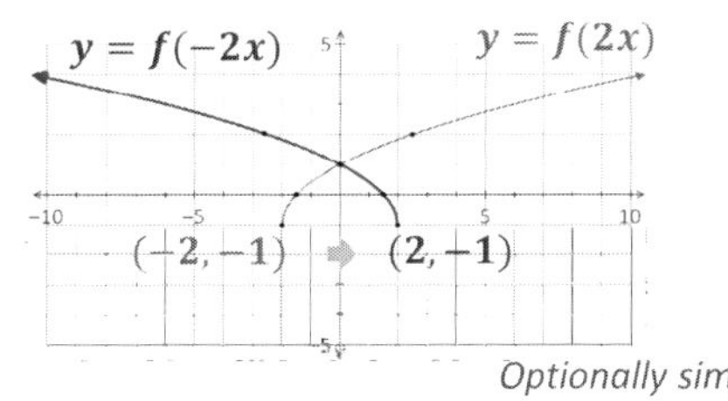

Equation in terms of x: $y = \sqrt{2x+4} - 1$ → $y = \sqrt{-2x+4} - 1$

Optionally simplify $y = \sqrt{-2(x-2)} - 1$

- Or we can apply the ***reflection first***, and then the stretch ....

Horizontal reflection

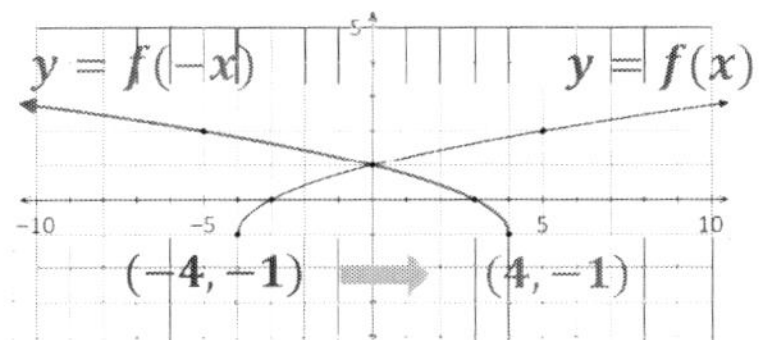

And then the horizontal stretch, factor $1/2$

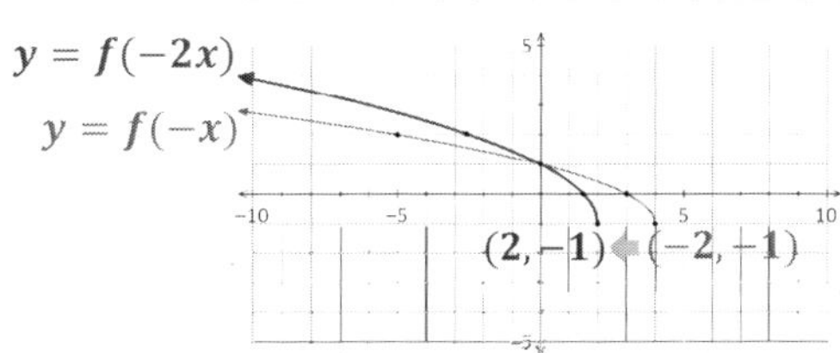

Equation in terms of x: $y = \sqrt{-x+4} - 1$ → $y = \sqrt{-2x+4} - 1$ *Same equation! (and same graph)*

Warm-up **Exploration #1** **Combining a vertical stretch** (or reflection) **with a horizontal translation**

The graph of $f(x) = (x-1)^2 - 4$ is shown below

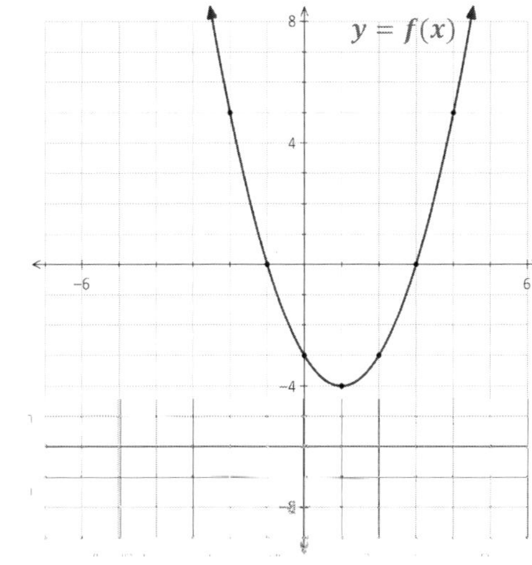

1 On the same grid, construct a new graph of $y = g(x)$ by first applying a vertical stretch about the x-axis by a factor of 2, then applying a horizontal translation 4 units left.

2 Determine an equation of $y = g(x)$, in terms of x, by applying the transformations in opposite order:

- First apply a horizontal translation 4 units left.

- Then apply a vertical stretch by a factor of 2.

3 Does the equation developed in #2 match the graph made in #1? Is the order in which a vertical stretch and horizontal translation are applied relevant?

Exploration #2 Combining a vertical stretch (or reflection) with a vertical translation

The graph of $f(x) = x^2 - 4x + 3$ is on the right

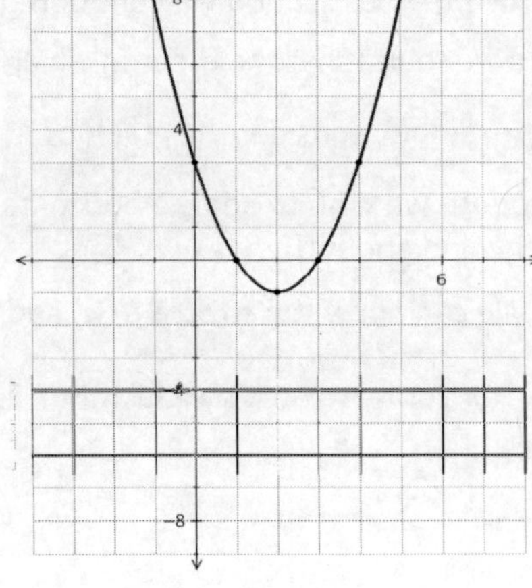

1 ➡ On the same grid, construct a new graph of $y = g(x)$ by:

- First applying a vertical stretch about the x-axis by a factor of 2
- Then applying a vertical translation 3 units down.

Mapping Rule:

2 ➡ Determine an equation for $y = g(x)$,

In terms of $f(x)$: *In terms of x:*

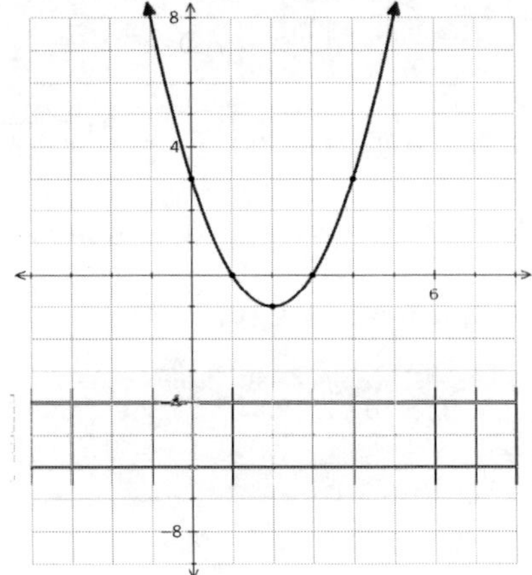

3 ➡ Next, on the same (new) grid construct a new graph of $y = h(x)$ by:

- First applying a vertical translation 3 units down
- Then applying a vertical stretch about the x-axis by a factor of 2

Mapping Rule:

4 ➡ Determine an equation for $y = h(x)$,

In terms of $f(x)$: *In terms of x:*

5 ➡ Compare the graphs and equations above. Is the order in which a vertical stretch and vertical translation are applied relevant?

Exploration #3 Combining a horizontal stretch (or reflection) with a horizontal translation

The graph of $f(x) = \sqrt{x+4} - 1$ is on the right.

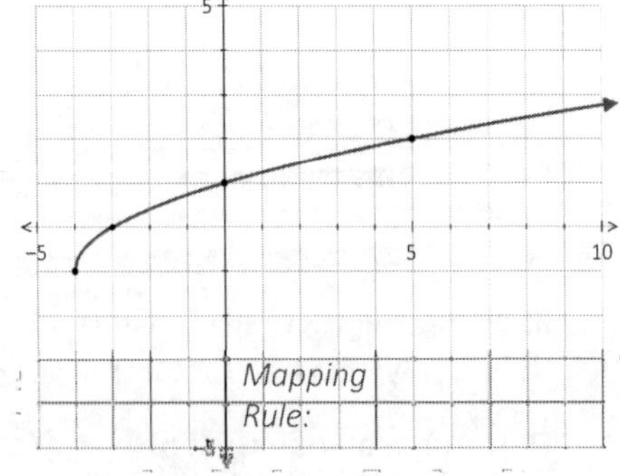

1 ➡ On the same grid, construct a new graph of $y = g(x)$ by:

- First applying a horizontal stretch about the y-axis by a factor of $1/2$
- Then applying a horizontal translation 4 units right.

2 ➡ Determine an equation for $y = g(x)$,

In terms of $f(x)$: *In terms of x:*

3 ➡ Next, on the same grid construct a new graph of $y = h(x)$ by again transforming the graph of $y = f(x)$:

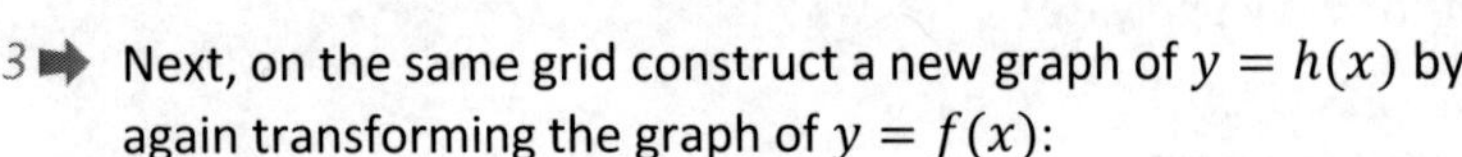

- First applying a horizontal translation 4 units right.
- Then applying a horizontal stretch about the y-axis by a factor of $1/2$

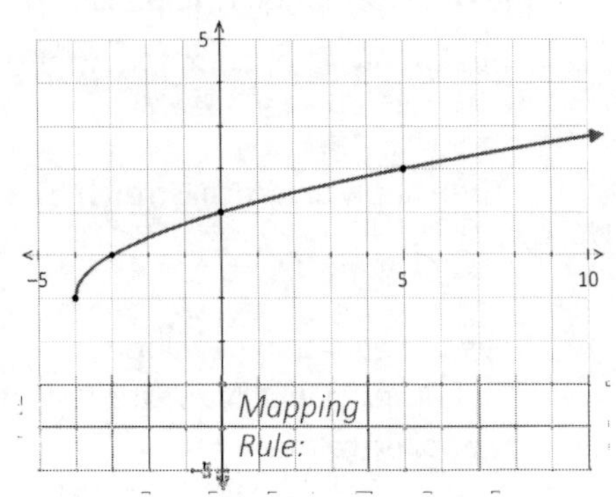

4 ➡ Determine an equation for $y = h(x)$,

In terms of $f(x)$: *In terms of x:*

5 ➡ Compare the graphs and equations above. Is the order in which a horizontal stretch and horizontal translation are applied relevant?

Exploration #4 Analyzing Horizontal Translations

1 ➡ Analyze the following pairs of functions graphed below. Does the horizontal translation match the constant term? *That is, for the group 1,* ***is*** *the horizontal translation from graph ❶ to ❷ "8 units left"?*

Group 1: ❶ $y = \sqrt{x}$

❷ $y = \sqrt{2x+8}$

Group 2: ❶ $y = x^2$

❷ $y = (-x+5)^2$

Group 3: ❶ $y = x^3$

❷ $y = (-\frac{1}{2}x+4)^3$

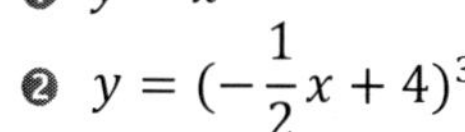

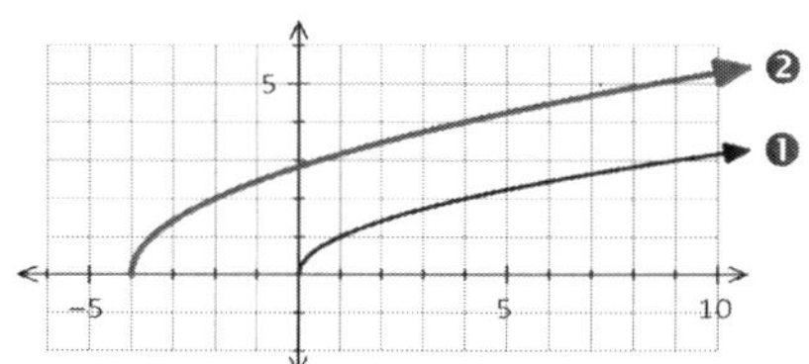

Horizontal translation: ____________

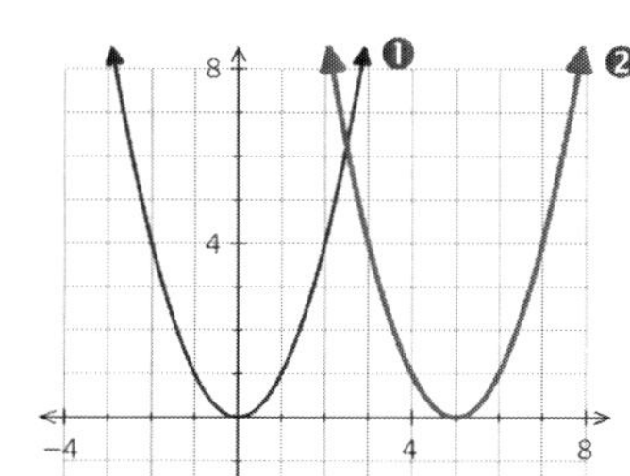

Horizontal translation: ____________

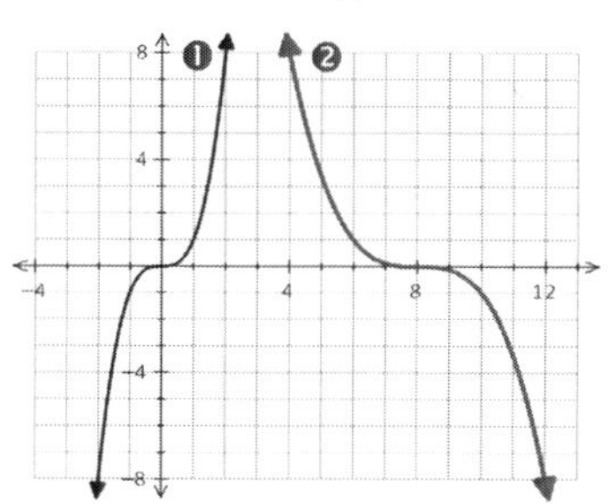

Horizontal translation: ____________

2 ➡ Analyze the corresponding graphs for each pair of functions. In the space provided above, indicate the horizontal translation for each.

3 ➡ Multiple transformations can be seen using the model $y = af[b(x-h)]+k$.
Put each of the y_2 equations above in this form by factoring out the b value. *(The first is done for you!)*

$y = \sqrt{2x+8}$

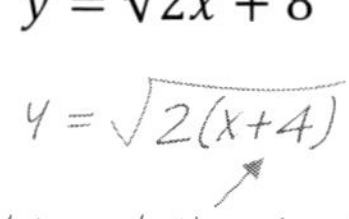

$y = \sqrt{2(x+4)}$

Horizontal translation 4 units left

$y = (-x+5)^2$

$y = (-\frac{1}{2}x+4)^3$

To identify the horizontal translation from a basic graph, first factor out any coefficient.

For example, given $y = (3x-6)^2-7$, the horizontal translation is 2 units right. (and not 6)

$$y = [3(x-2)]^2-7$$

When describing multiple transformations,

First consider... ***Stretches Reflections***

And then... ***Translations***

So given the transformation from $y = f(x)$ to $\boldsymbol{y = af[b(x-h)]+k}$

The mapping rule is: $(x, y) \rightarrow (\frac{1}{b}x+h, ay+k)$

Note that the horizontal translation, "+h", is applied after all x-coordinates are multiplied by the horizontal stretch factor.

...and same goes for the vertical translation "k"!

- The *vertical stretch about the x-axis factor* is $|\boldsymbol{a}|$
- If $a < 0$, there is a *vertical reflection* in the x-axis
- The *horizontal stretch about the y-axis factor* is $\frac{\boldsymbol{1}}{|\boldsymbol{b}|}$
- If $b < 0$, there is a *horizontal reflection* in the y-axis
- The *horizontal translation* is $\boldsymbol{h}$ units, and the *vertical translation* is $\boldsymbol{k}$ units.

Worked Example

Given the graph of $y = f(x)$ on the right, $y = g(x)$ is defined as $g(x) = -f(2x + 6) + 1$.

(a) Construct a mapping rule for the transformation from $y = f(x)$ to $y = g(x)$.

(b) Sketch the graph of $y = g(x)$ on the same grid

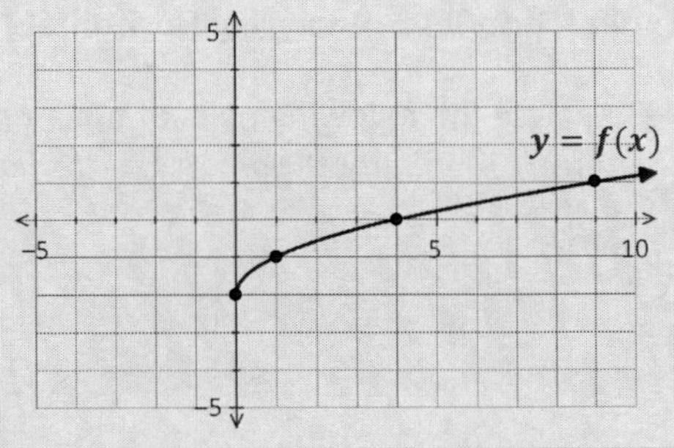

Sol: (a) *First things first – prep the function by factoring out the coefficient of x:*

$$g(x) = -f[(2(x + 3)] + 1$$

Vertical reflection
All y *coords. made negative*

Horizontal translation 3 units left, vertical translation 1 up

Horizontal stretch, factor of 1/2

Multiply all x *coords. by* $\frac{1}{2}$

Mapping Rule: $(x, y) \rightarrow (\frac{1}{2}x - 3, -y + 1)$

(b) *Apply the mapping rule to the indicated (•) points on the graph of* $y = f(x)$*.*

$(0, -2) \rightarrow (\frac{1}{2}(0) - 3, -(-2) + 1) \rightarrow \mathbf{(-3, 3)}$

$(1, -1) \rightarrow (\frac{1}{2}(1) - 3, -(-1) + 1) \rightarrow \mathbf{(-2.5, 2)}$

$(4, 0) \rightarrow (\frac{1}{2}(4) - 3, -(0) + 1) \rightarrow \mathbf{(-1, 1)}$

$(9, 1) \rightarrow (\frac{1}{2}(9) - 3, -(1) + 1) \rightarrow \mathbf{(1.5, 0)}$

Plot each of the points to sketch the graph of $y = g(x)$ →

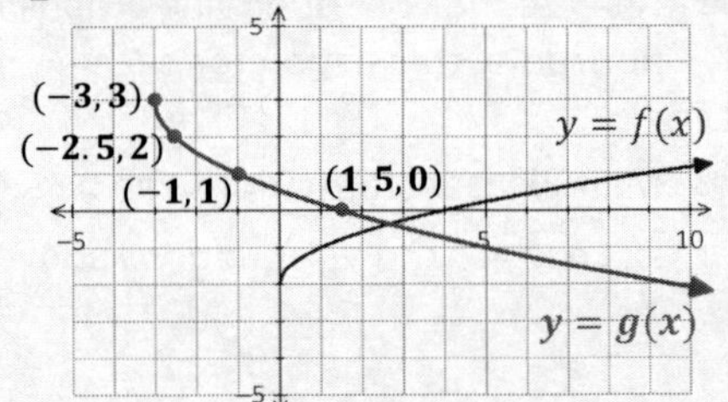

Class Example 1.41 *Applying Multiple Transformations to a Graph*

The graph of $y = f(x)$ is shown on the right.

A function $y = g(x)$ is given by $g(x) = \frac{1}{2}f(-x + 2) + 1$

(a) Construct a mapping rule for the transformation from $y = f(x)$ to $y = g(x)$.

(b) Sketch the graph of $y = g(x)$ on the same grid.

(c) The point $P(0,2)$ is on the graph of $y = f(x)$. Determine the coordinates of the corresponding point on the graph $y = g(x)$

(d) State the domain and range for $y = g(x)$

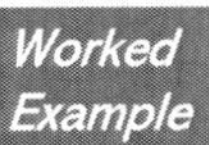

Worked Example A function $y = f(x)$ has a domain of $[-6, 12]$ and a range $[0, 10]$. For the transformed function $y = -3f(2x + 8) - 5$, (i) provide a mapping rule and determine the new (ii) domain and (iii) range.

Solution: First prep the transformed function, factor out the "2" in the brackets:

$$y = -3f[2(x + 4)] - 5$$

Vertical reflection about the x-axis

Vertical sretch about x-axis, factor of 3

Horizontal stretch about y-axis, factor of 1/2

Horiz. translation 4 units left, vertical 5 down

Next, visualize a "possible" graph of $y = f(x)$. (That has the given domain / range):

Possible Graph of $y = f(x)$:

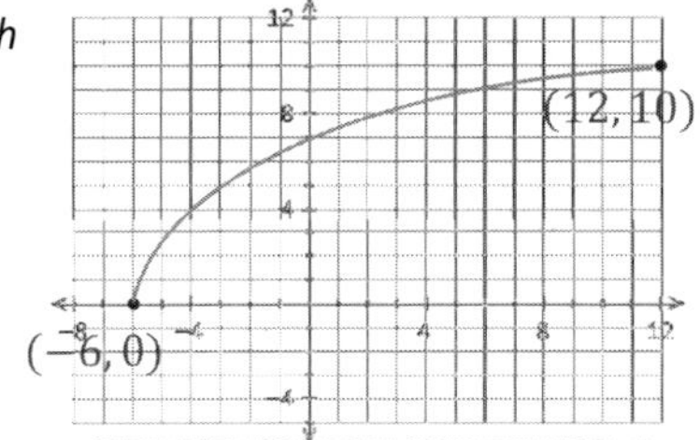

Mapping Rule: $(x, y) \rightarrow (\frac{1}{2}x - 4, -3y - 5)$

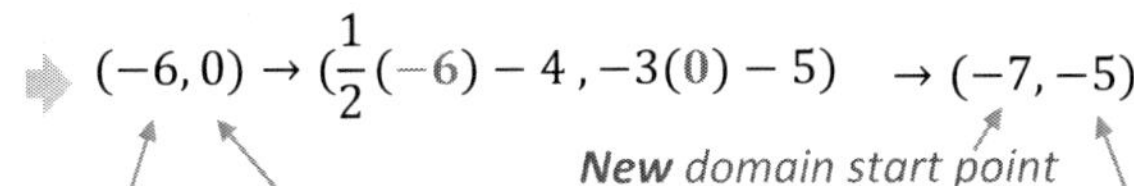

$(-6, 0) \rightarrow (\frac{1}{2}(-6) - 4, -3(0) - 5) \rightarrow (-7, -5)$

Domain start point *Range start* ***New*** *domain start point* ***New*** *Range start point*

$(12, 10) \rightarrow (\frac{1}{2}(12) - 4, -3(10) - 5) \rightarrow (2, -30)$

Domain end point *Range end point* ***New*** *domain end* ***New*** *Range end point*

Transformed Graph:

(−7, −5)

(2, −30)

Note *that drawing "possible graphs" is not essential, but can help visualize!*

New Domain: $[-7, 2]$

New Range: $[-30, -5]$

Class Example 1.42 *Determining the Equation of a Function Given Transformations*

A function $y = f(x)$ has a domain of $[-4, \infty)$ and a range $(-\infty, 9]$. For each of the transformations of $f(x)$ described below, (i) provide a mapping rule and determine the new (ii) domain and (iii) range.

(a) $y = -\frac{1}{3}f(2x)$

(b) $y + 5 = 2f(-x + 1)$

Class Example 1.43 *Applying Multiple Translations to a Function – Graph and Equation*

The graph of $f(x) = |x - 6| - 4$ is shown on the right.

A function $y = g(x)$ is given by $g(x) = -f(3x - 6) + 3$

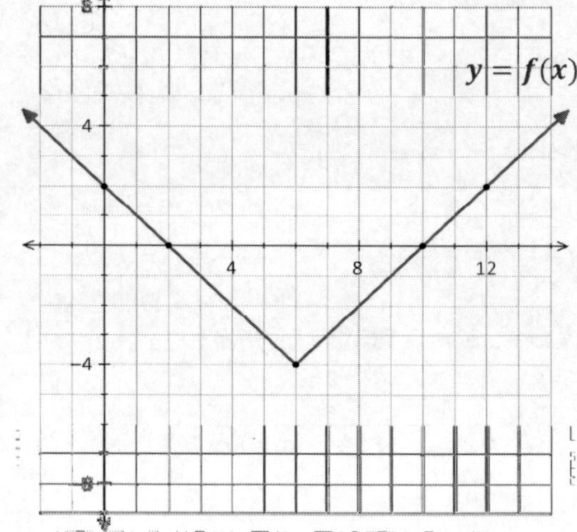

(a) Construct a mapping rule for the transformation from $y = f(x)$ to $y = g(x)$.

(b) Transform each indicated point (•), to sketch the graph of $y = g(x)$ on the same grid.

Class Example 1.44 *Determining the Equation of a Function Given Transformations*

A function $f(x) = 3\sqrt{6x} + 1$ is transformed to $y = g(x)$ and $y = h(x)$, as described below. Construct an equation for each transformed function.

(a) $y = g(x)$ is obtained by vertically stretching the graph of $f(x)$ about the x-axis by a factor of 4, and vertically translating the graph 6 units up.

Equation in terms of $f(x)$:

Equation in terms of x:

(b) $y = h(x)$ is obtained by reflecting the graph of $f(x)$ in the x-axis, stretching the graph about the y-axis by a factor of 1/2, and horizontally translating the graph 3 units right.

Equation in terms of $f(x)$:

Equation in terms of x:

Worked Example

The graph of $y = g(x)$ represents a transformation of $y = f(x)$. Determine an equation for $y = g(x)$, in terms of $f(x)$.

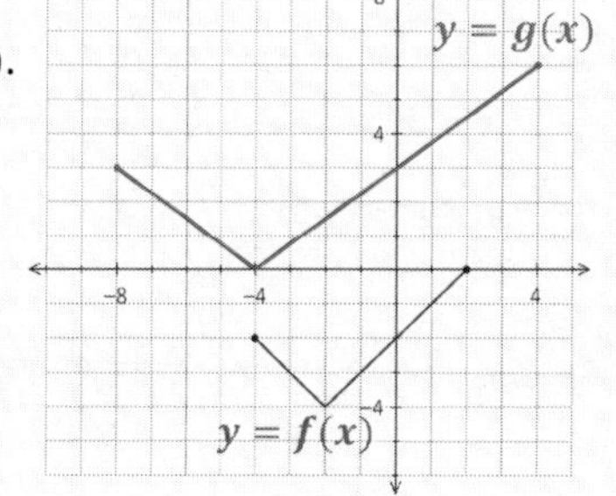

$y = g(x)$

$y = f(x)$

***First** identify the stretches:*

...and $g(x)$ is 6 units tall

Graph of $f(x)$ is 4 units "tall"

➔ **Vertical stretch factor of 1.5**

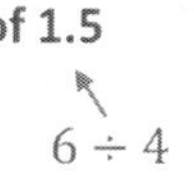

$6 \div 4$

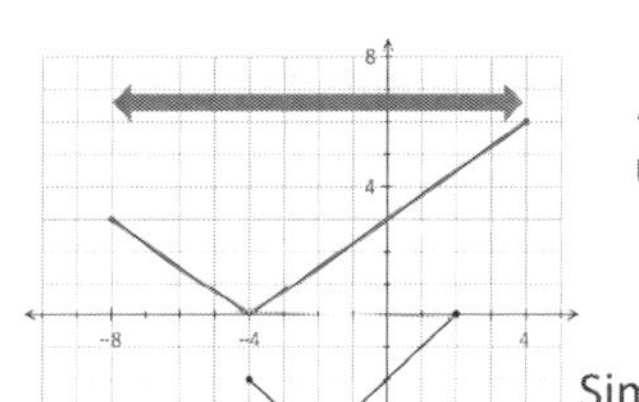

...while $g(x)$ is 12 units wide

Similarly, graph of $f(x)$ is 6 units wide

➔ **Horizontal stretch factor of 2**

$12 \div 6$

After** determining both the horizontal and vertical stretches, **then** identify any **translations:

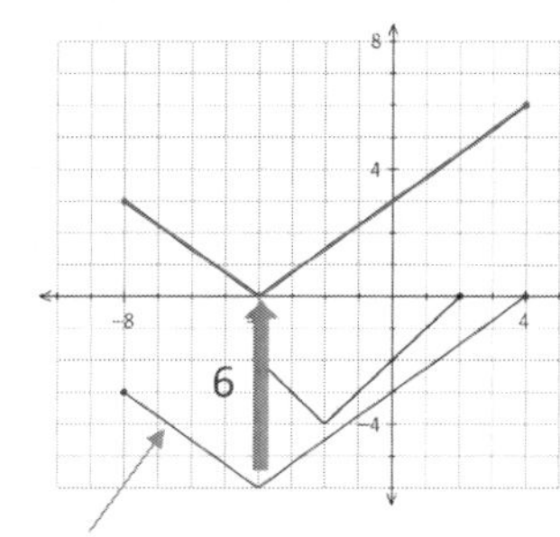

Graph with only stretches applied

Note that after we apply the two stretches to $f(x)$...

...we must shift up 6 units

Equation: $g(x) = 1.5[f(0.5x) + 4]$

Simplifies to: $\mathbf{g(x) = 1.5f(0.5x) + 6}$

Class Example 1.45 *Determining more than one Transformation from a Graph*

For each pair of graphs below, the graph of $y = g(x)$ represents a transformation of of $y = f(x)$ through any of: stretches, reflections, and translations. Determine an equation for $y = g(x)$, in terms of $f(x)$.

(a)

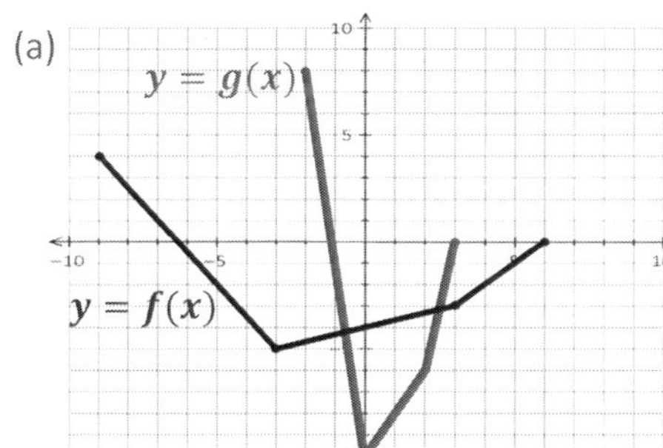

Mapping Rule:

Equation of $g(x)$*:*

(b)

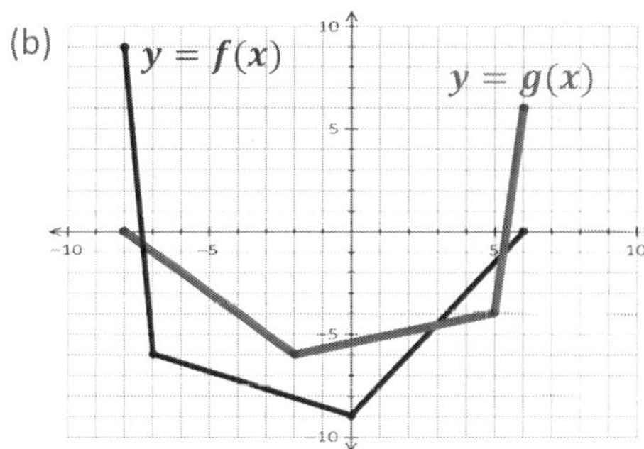

Mapping Rule:

Equation of $g(x)$*:*

1.4 Practice Questions

1. For each graph of $y = f(x)$ below, provide a mapping rule and **sketch** each indicated transformed function.

(a) $g(x) = -f(2x) - 2$

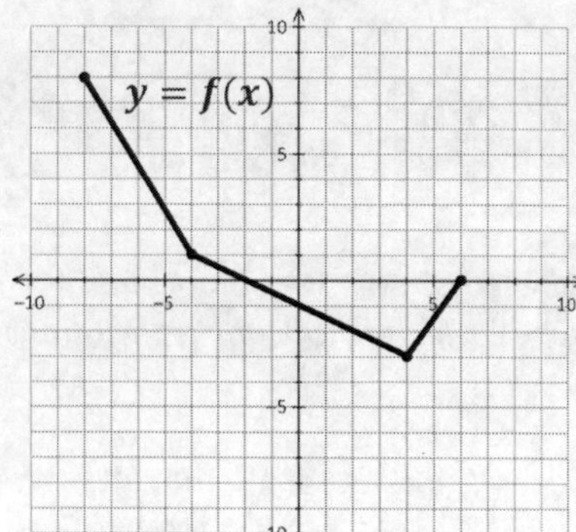

Mapping Rule:

(b) 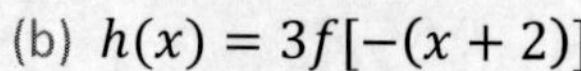 $h(x) = 3f[-(x+2)]$

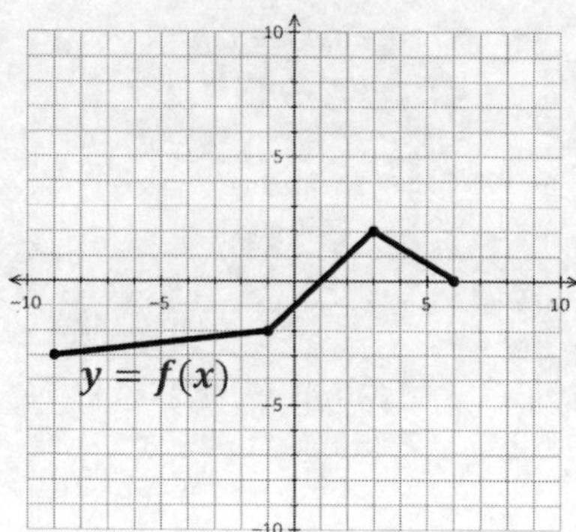

Mapping Rule:

(c) $k(x) = -f(3x - 6) - 1$

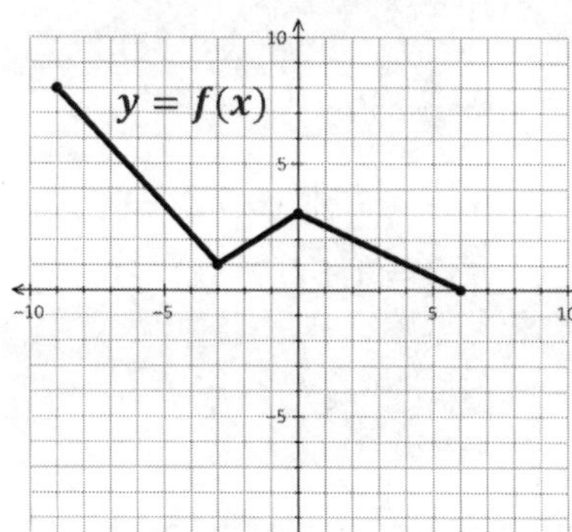

Mapping Rule:

(d) $p(x) = \frac{1}{2}f(-0.5x + 1)$

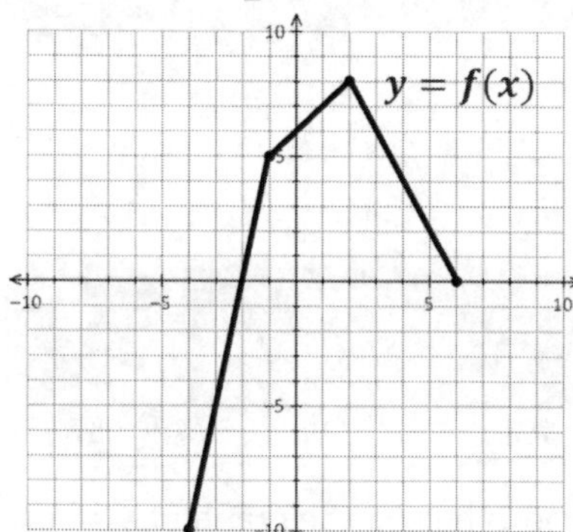

Mapping Rule:

2. Construct a mapping rule for each of the following transformations of a function $y = f(x)$, listed below. Then, determine the new coordinates of a point $P(8, 6)$ on the graph of $y = f(x)$, after the transformation.

(a) $y + 5 = 3f(-x + 2)$

i *Mapping:*

ii *New Point:*

(b) $-2y = f(4x - 8)$

i *Mapping:*

ii *New Point:*

(c) $\frac{1}{4}y = f(-\frac{1}{3}x - 5)$

i *Mapping:*

ii *New Point:*

3. A function $y = f(x)$ has a domain of $[-12, 8]$ and a range $[-6, 10]$. For each of the transformations of $f(x)$ described below, (i) provide a mapping rule and determine the new (ii) domain and (iii) range.

(a) $y = \frac{3}{2}f(\frac{1}{2}x) - 5$ (b) $y + 1 = 3f(-x + 2)$ (c) $-2y = f(4x - 8)$

4. A function $y = f(x)$ has a domain of $\{x \geq -12\,, x \in \mathbb{R}\}$ and a range $\{y \leq 9\,, y \in \mathbb{R}\}$.

For each of the transformations of $f(x)$ described below, (i) provide a mapping rule and determine the new (ii) domain and (iii) range.

(a) $y = 2f(-x) + 3$ (b) $y - 4 = \frac{2}{3}f[\frac{4}{3}(x - 5)]$ (c) $-4y = f(3x + 12)$

Answers from previous page

1. (a) $(x, y) \to (\frac{1}{2}x\,, -y - 2)$ **(b)** $(x, y) \to (-x - 2\,, 3y)$ **(c)** $(x, y) \to (\frac{1}{3}x + 2\,, -y - 1)$ **(d)** $(x, y) \to (-2x + 2\,, \frac{1}{2}y)$

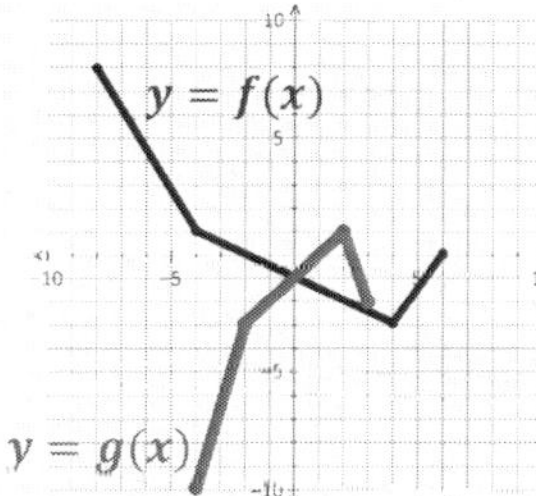

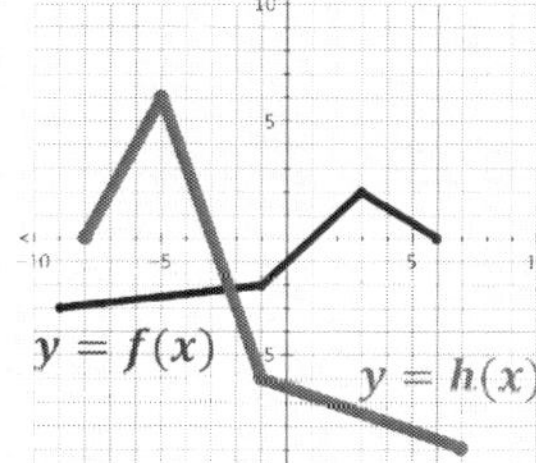

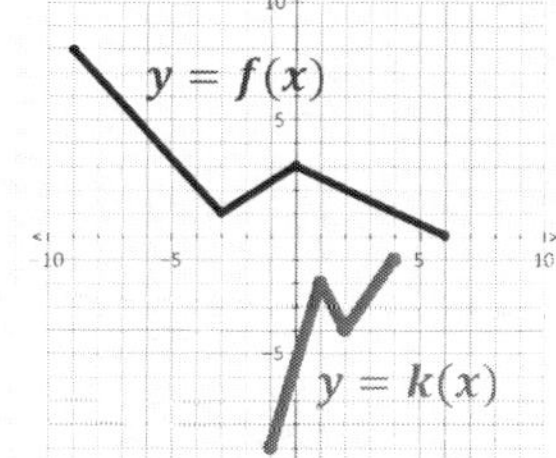

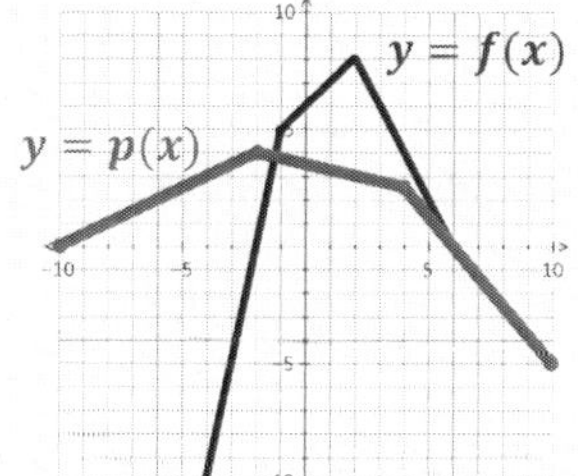

2. (a) i $(x, y) \to (-x + 2\,, 3y - 5)$
ii $(8, 6) \to (-6, 13)$

(b) i $(x, y) \to (\frac{1}{4}x + 2\,, -\frac{1}{2}y)$
ii $(8, 6) \to (4, -3)$

(c) i $(x, y) \to (-3x - 15\,, 4y)$
ii $(8, 6) \to (-39, 24)$

5. The graph of $f(x) = (x-1)^2 - 16$ is shown on the right.

A function $y = g(x)$ is defined in terms of $f(x)$ as:

$$g(x) = -\frac{1}{4}f(x+3) + 5$$

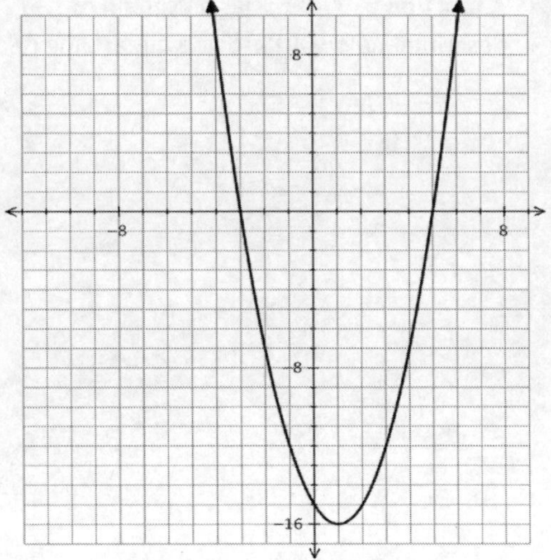

(a) Construct a mapping rule for the transformation from $y = f(x)$ to $y = g(x)$.

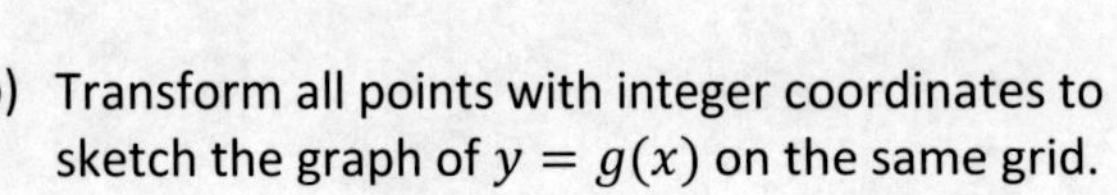

(b) Transform all points with integer coordinates to sketch the graph of $y = g(x)$ on the same grid.

(c) Write an equation for $y = g(x)$, in terms of x.

6. A function $f(x) = (x+4)^2$ is transformed by:

- Reflecting the graph in the line $x = 0$
- Stretching the graph vertically by a factor of 3
- Horizontally translating the graph 1 unit right and vertically translating the graph 2 units down

(a) Construct a mapping rule for the transformation from $f(x)$ to $g(x)$.

(b) Determine an equation for $y = g(x)$

i *in terms of $f(x)$:*

ii *in terms of x:*

Answers from previous page

3. (a) i $(x, y) \rightarrow (2x, \frac{3}{2}y - 5)$
ii *D:* $[-24, 16]$
iii *R:* $[-14, 10]$

(b) i $(x, y) \rightarrow (-x + 2, 3y - 1)$
ii *D:* $[-6, 14]$
iii *R:* $[-19, 29]$

(c) i $(x, y) \rightarrow (\frac{1}{4}x + 2, -\frac{1}{2}y)$
ii *D:* $[-1, 4]$
iii *R:* $[-5, 3]$

4. (a) i $(x, y) \rightarrow (-x, 2y + 3)$
ii *D:* $\{x | x \leq 12, x \in \mathbb{R}\}$
iii *R:* $\{y | y \leq 21, y \in \mathbb{R}\}$

(b) i $(x, y) \rightarrow (\frac{3}{4}x + 5, \frac{2}{3}y + 4)$
ii *D:* $\{x | x \geq -4, x \in \mathbb{R}\}$
iii *R:* $\{y | y \leq 10, y \in \mathbb{R}\}$

(c) i $(x, y) \rightarrow (\frac{1}{3}x - 4, -\frac{1}{4}y)$
ii *D:* $\{x | x \geq -8, x \in \mathbb{R}\}$
iii *R:* $\{y | y \geq -2.25, y \in \mathbb{R}\}$

7. The graph of $f(x) = 2\sqrt{x+4} - 2$ is shown on the right.

A function $y = g(x)$ is given by $g(x) = 3f(-x+2)$

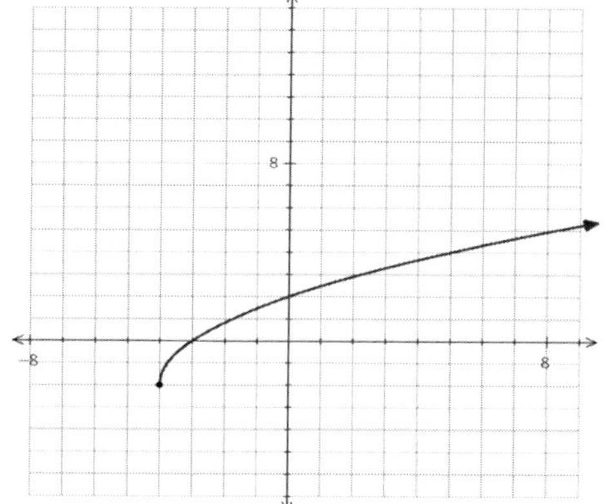

(a) Construct a mapping rule for the transformation from $y = f(x)$ to $y = g(x)$.

(b) Transform all points with integer coordinates to sketch the graph of $y = g(x)$ on the same grid.

(c) Write an equation for $y = g(x)$, in terms of x.

8. The graph of $f(x) = \sqrt{x+9} - 2$ is shown on the right.

A transformed function is given by $y - 1 = -f(0.5x - 2)$

(a) Construct a mapping rule for the transformation.

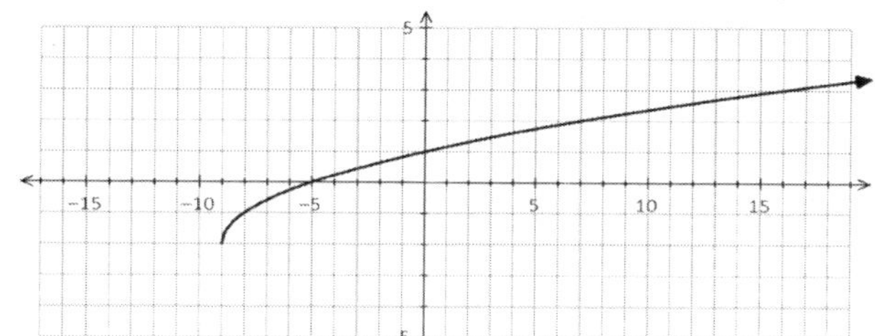

(b) Transform all points with integer coordinates to sketch the graph of the transformed function on the same grid.

(c) Write an equation for the transformed function, in terms of x.

Answers from previous page

5. (a) 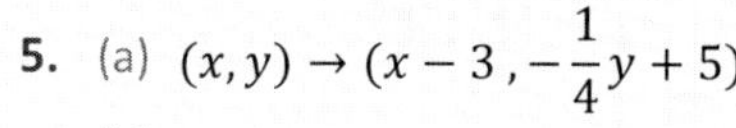
$(x, y) \rightarrow (x-3, -\frac{1}{4}y + 5)$

(b) *Transform vertex:*

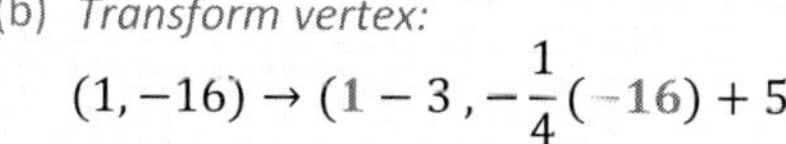
$(1, -16) \rightarrow (1-3, -\frac{1}{4}(-16) + 5)$

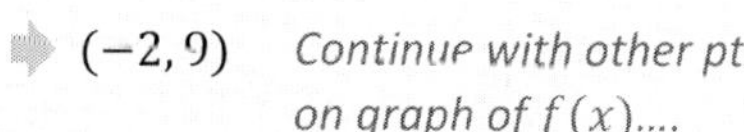
$(-2, 9)$ *Continue with other pts on graph of $f(x)$....*

(c) 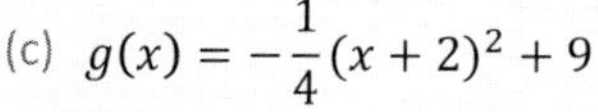
$g(x) = -\frac{1}{4}(x+2)^2 + 9$

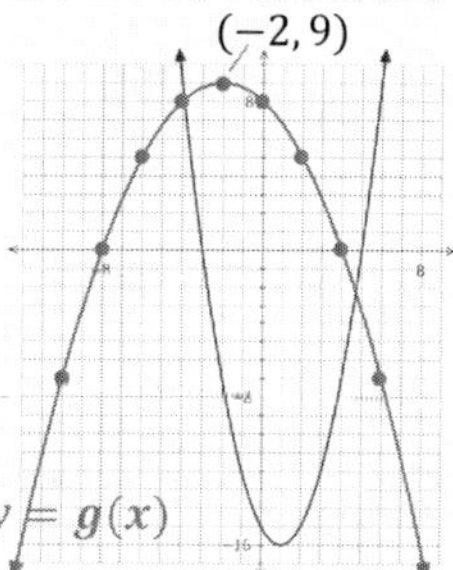

6. (a) $(x, y) \rightarrow (-x+1, 3y-2)$

(b) i $g(x) = 3f[-(x-1)] - 2$

ii $g(x) = 3[-(x-1)+4)]^2 - 2$

$g(x) = 3[-(x-5)]^2 - 2$

Optionally simplify to:

$g(x) = 3(x-5)^2 - 2$

9. The graph of $f(x) = |x - 4| - 5$ is shown on the right.

A function $y = g(x)$ is given by $g(x) = -2f(x + 3) + 4$

(a) Construct a mapping rule for the transformation from $y = f(x)$ to $y = g(x)$.

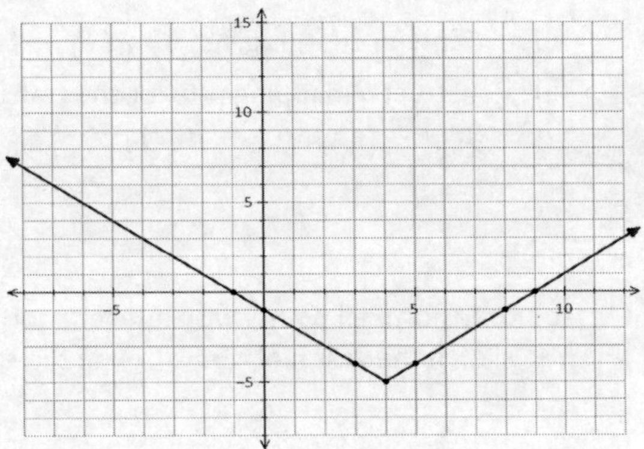

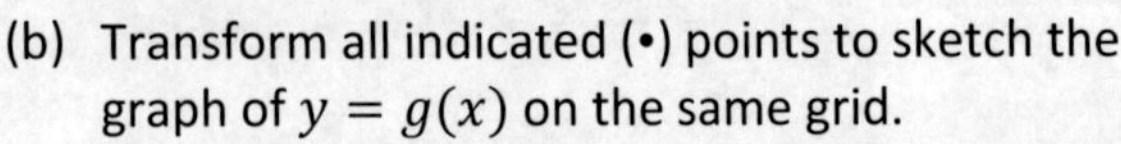

(b) Transform all indicated (•) points to sketch the graph of $y = g(x)$ on the same grid.

(c) Write an equation for $y = g(x)$, in terms of x.

10. The graph of $f(x) = |0.5(x + 4)| - 6$ is shown on the right.

A function $y = g(x)$ is given by $g(x) = f(-4x) - 2$

(a) Construct a mapping rule for the transformation from $y = f(x)$ to $y = g(x)$.

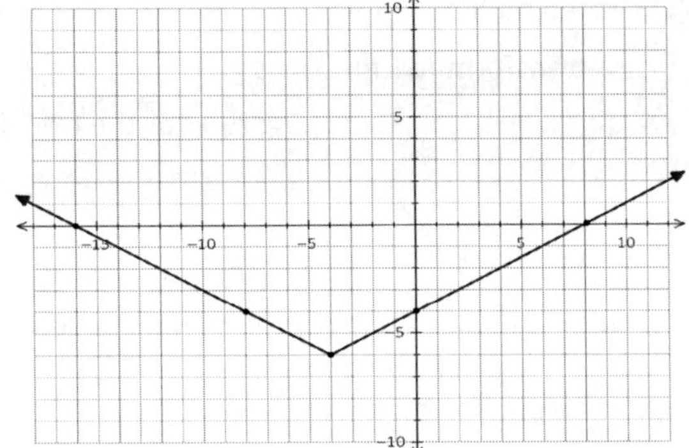

(b) Transform all indicated (•) points to sketch the graph of $y = g(x)$ on the same grid.

(c) Write an equation for $y = g(x)$, in terms of x.

Answers from previous page

7. (a) $(x, y) \to (-x + 2, 3y)$

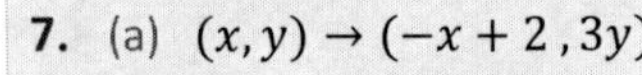

(b) $(-4, -2) \to (6, -6)$

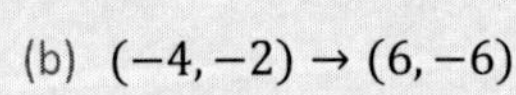

$(-3, 0) \to (5, 0)$

and so on....

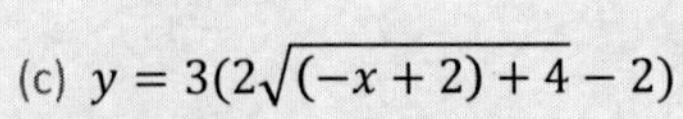

(c) $y = 3(2\sqrt{(-x + 2) + 4} - 2)$

Simplify to:

$\boldsymbol{g(x) = 6\sqrt{-(x - 6)} - 6}$

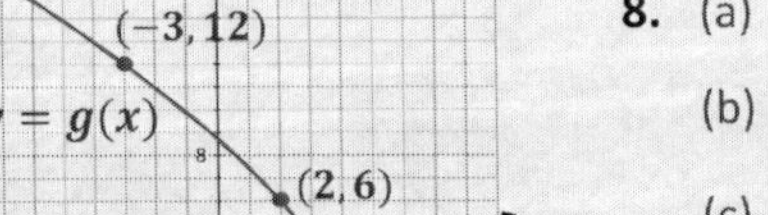

8. (a) $(x, y) \to (2x + 4, -y + 1)$

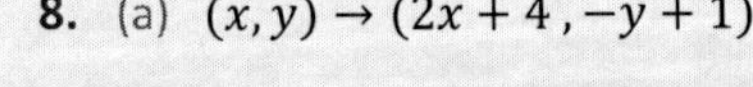

(b) $(-9, -2) \to (-14, 3)$ *and so on....*

(c) $y = -\left(\sqrt{(0.5x - 2) + 9} - 2\right) + 1$ *Simplify to...*

$y = -\sqrt{0.5(x + 14)} + 3$

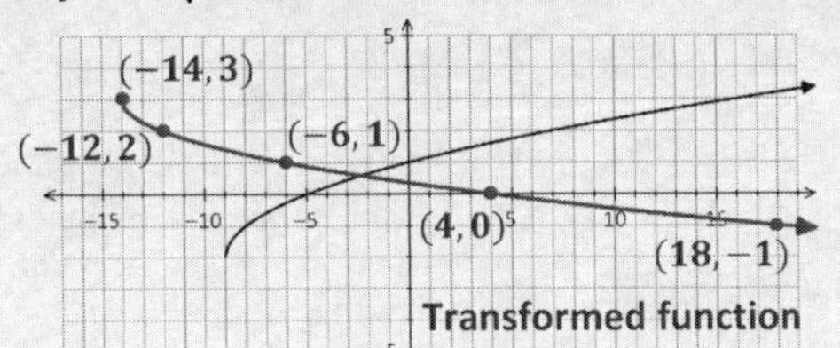

11. A function $f(x) = -\sqrt{x} + 2$ is transformed by:

- Reflecting the graph in the line $y = 0$
- Stretching the graph horizontally about the line $x = 0$ by a factor of 1/4
- Horizontally translating the graph 1 unit right and vertically translating the graph 2 units up

(a) Construct a mapping rule for the transformation from $f(x)$ to $g(x)$.

(b) Determine an equation for $y = g(x)$ i *in terms of $f(x)$:* ii *in terms of x:*

(c) The range of $y = f(x)$ is $(-\infty, 2]$. Determine the range of $y = g(x)$.

(d) The horizontal stretch applied to the graph of $y = f(x)$ can instead be described by a vertical stretch. Determine the magnitude of the vertical stretch.

Answers from previous page

9. (a) $(x, y) \rightarrow (x - 3, -2y + 4)$

(b) $(4, -5) \rightarrow (1, 14)$

$(3, -4) \rightarrow (0, 12)$

and so on....

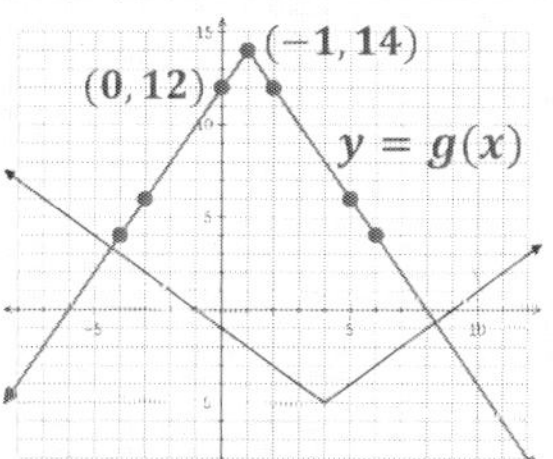

(c) $y = -2(|(x + 3) - 4| - 5) + 4$ *Simplifies to...*

$\mathbf{g(x) = -2|x - 1| + 14}$

10. (a) 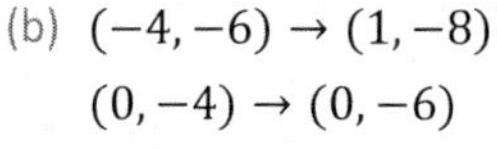
$(x, y) \rightarrow (-\frac{1}{4}x, y - 2)$

(b) $(-4, -6) \rightarrow (1, -8)$

$(0, -4) \rightarrow (0, -6)$

and so on....

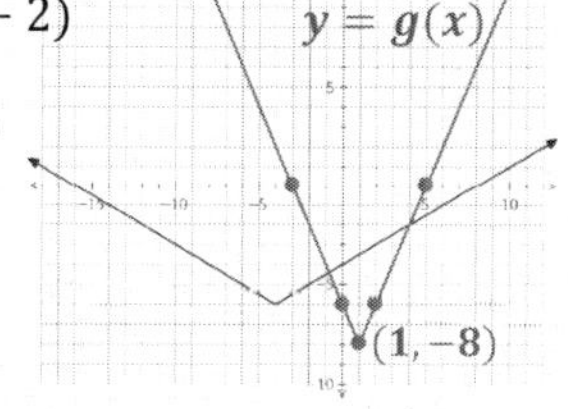

(c) $y = |0.5(-4x) + 2| - 6 - 2$ *Simplifies to...*

$g(x) = |-2x + 2| - 8$ ➧ $\mathbf{g(x) = |-2(x - 1)| - 8}$

12. A function $f(x) = (x + 16)^3$ is transformed by:

- Stretching the graph horizontally by a factor of 1/8
- Horizontally translating the graph 2 units right and vertically translating the graph 3 units down

(a) Construct a mapping rule for the transformation from $f(x)$ to $g(x)$.

(b) Determine an equation for $y = g(x)$ i *in terms of* $f(x)$: ii *in terms of* x:

(c) The horizontal stretch applied to the graph of $y = f(x)$ can instead be described by a vertical stretch and a horizontal translation. Determine the magnitude of the vertical stretch and the magnitude and direction of the horizontal translation.

13. The graph of $f(x) = (x + 7)^2$ is reflected in the y-axis. The resulting graph could also be achieved by horizontally translating the graph of $y = f(x)$. Describe the directing and magnitude of the translation, and verify by applying the transformation to the equation of $f(x)$.

Answers from previous page

11. (a) $(x, y) \to (\frac{1}{4}x + 1, -y + 2)$ (b) i $g(x) = -f[4(x-1)] + 2$

(c) $(-\infty, 0]$ (d) Vert. stretch about x-axis, factor of **2** ii $g(x) = -\left(-\sqrt{4(x-1)} + 2\right) + 2$

simplify.... $\mathbf{g(x) = \sqrt{4(x-1)}}$

14. A function $y = f(x)$ has a domain of $(-\infty, 12]$ and a range of $[-2, \infty)$. The graph is transformed by:

- Reflecting the graph in the line $x = 0$
- Stretching the graph horizontally about the line $x = 0$ by a factor of 2
- Stretching the graph vertically about the line $y = 0$ by a factor of 5
- Horizontally translating the graph 3 units right and vertically translating the graph 4 units up

(a) Construct a mapping rule for the transformation.

(b) The graph of $y = f(x)$ passes through a point $P(-4, 1)$. Determine the coordinates of the corresponding point to P on the transformed graph.

(c) Determine an equation for the transformed function, in terms of $f(x)$.

(d) Determine the domain and range of the transformed function.

15. For each pair of graphs below, the graph of $y = g(x)$ represents a transformation of of $y = f(x)$ through any of stretches, reflections, and translations. Determine an equation for $y = g(x)$, in terms of $f(x)$.

(a)

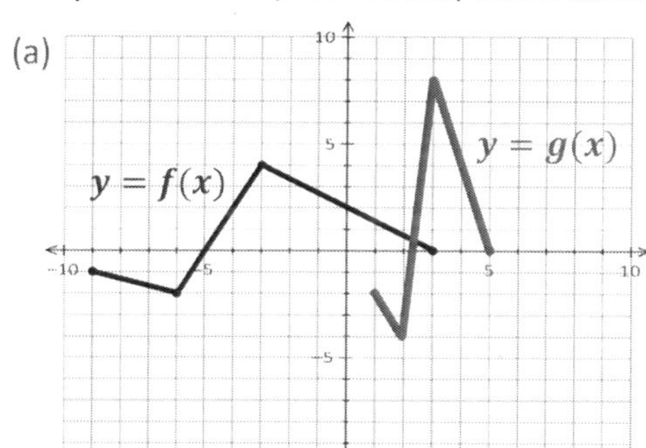

(b)

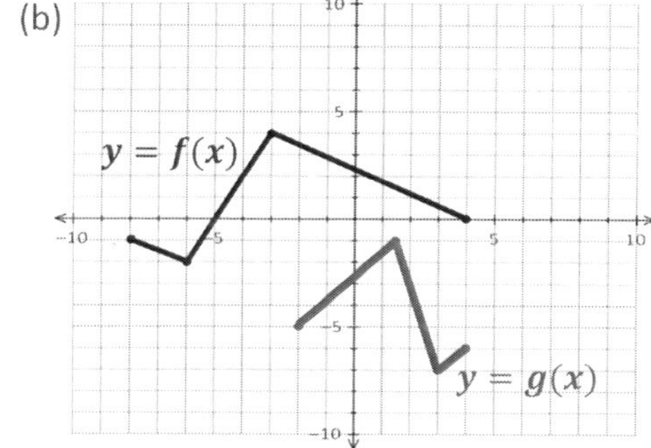

Mapping Rule: ____________

Equation of $g(x)$: ____________

x-intercepts of $g(x)$: ____________

Mapping Rule: ____________

Equation of $g(x)$: ____________

Answers from previous page

12. (a) $(x, y) \to (\frac{1}{8}x + 2, y - 3)$ (b) i $g(x) = f[8(x-2)] - 3$

ii $g(x) = (8(x-2) + 16)^3 - 3$

simplify.... $\mathbf{g(x) = (8x)^3 - 3}$ *further...* $g(x) = 512x^3 - 3$

(c) Vertical stretch about x-axis, factor of 512, plus a horizontal translation 16 units right

13. (a) Horizontal translation 7 units right. To verify, apply horizontal reflection: $y = (-x + 7)^2$, and then simplify: $y = [-1(x-7)]^2 \rightarrow y = (-1)^2(x-7)^2 \rightarrow y = (x-7)^2$ ✓ *checks out!*

16. For each pair of graphs below, the graph of $y = g(x)$ represents a transformation of of $y = f(x)$ through any of stretches, reflections, and translations. Determine an equation for $y = g(x)$, in terms of $f(x)$.

(a)

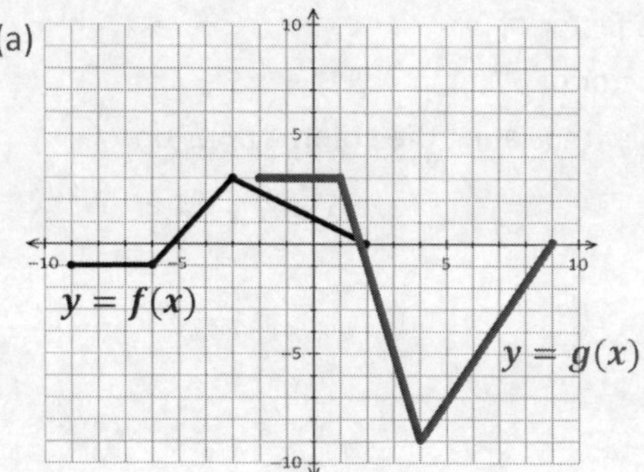

(b)

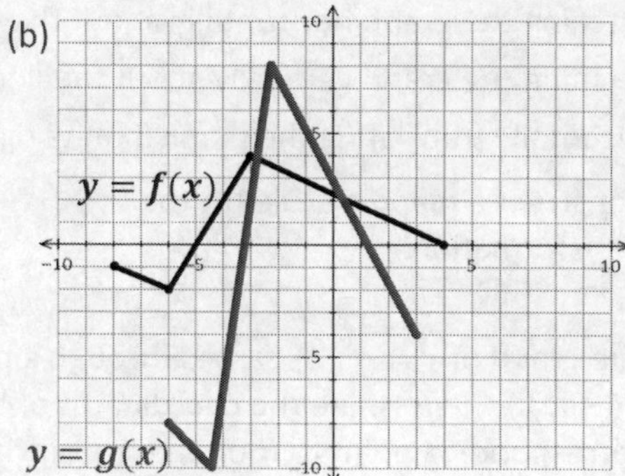

(a) i *Mapping Rule:* ____________

ii *Equation of $g(x)$:* ____________

(b) i *Mapping Rule:* ____________

ii *Equation of $g(x)$:* ____________

17. The graph of $y = f(x)$ is transformed into the graph of $y = g(x)$, as shown below:

An equation for $g(x)$ in terms of $f(x)$ is:

A. $g(x) = f(-2x + 1)$

B. $g(x) = f(2x - 6)$

C. $g(x) = f(\frac{1}{2}x - 3)$

D. $g(x) = f(-\frac{1}{2}x + 1)$

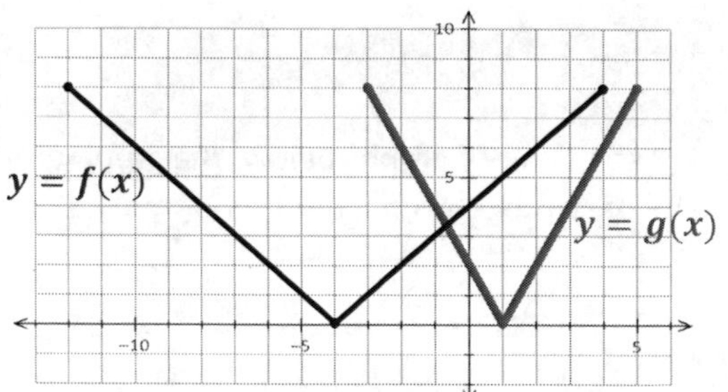

18. The graph of $y = f(x)$ is transformed into the graph of $y = g(x)$, as shown below:

An equation for $g(x)$ in terms of $f(x)$ is:

A. $g(x) = f(\frac{1}{4}x) - 4$

B. $g(x) = \frac{1}{2}f(\frac{1}{4}x)$

C. $g(x) = f(4x) - 4$

D. $g(x) = \frac{1}{2}f(4x)$

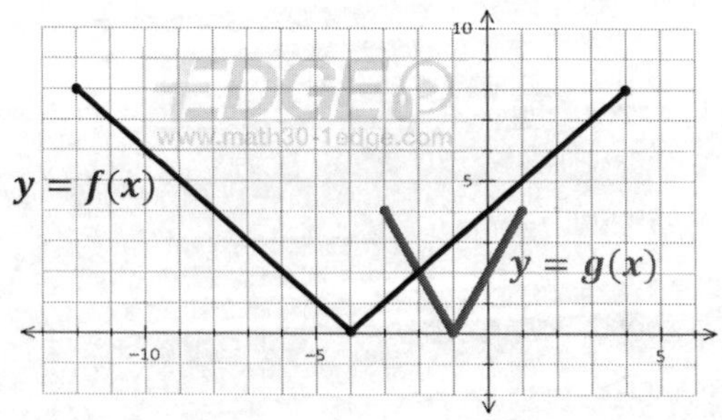

Answers from previous page

14. (a) $(x, y) \rightarrow (-2x + 3, 5y + 4)$ (b) $(11, 9)$

(c) $g(x) = 5f\left[-\frac{1}{2}(x - 3)\right] + 4$

(d) D: $[-21, \infty)$ R: $[-6, \infty)$

15. (a) i $(x, y) \rightarrow (\frac{1}{3}x + 4, 2y)$

ii $g(x) = 2f[3(x - 4)]$

iii $(7/3, 0)$, $(5,0)$

(b) i $(x, y) \rightarrow (-\frac{1}{2}x, y - 5)$

ii $g(x) = f(-2x) - 5$

19. The point $P(3,8)$ is on the graph of $y = f(x)$.

The point corresponding to P on the graph of $y + 2 = 2f(\frac{1}{3}x - 4)$ is:

A. $(21,14)$

B. $(21,12)$

C. $(13,14)$

D. $(13,12)$

20. The mapping rule that describes the transformation from the graph of $y = f(x)$ is $(x,y) \rightarrow (4x+8, 2y)$.

An equation for the transformed function is $y = af[b(x-c)]$, where possible values for a, b, and c are listed on the right.

NR The codes for the values of a, b, and c are, respectively, ____ , ____, and ____.

Codes can be used more than once

Code	Possible values of a, b, and c.
1	1/4
2	2
3	1/2
4	4
5	8
6	16

21. The graph of $y = f(x)$ is shown on the right.

Determine the domain and range of:

$y + 11 = f(-3x - 3)$

i *Domain:* ____________

ii *Range:* ____________

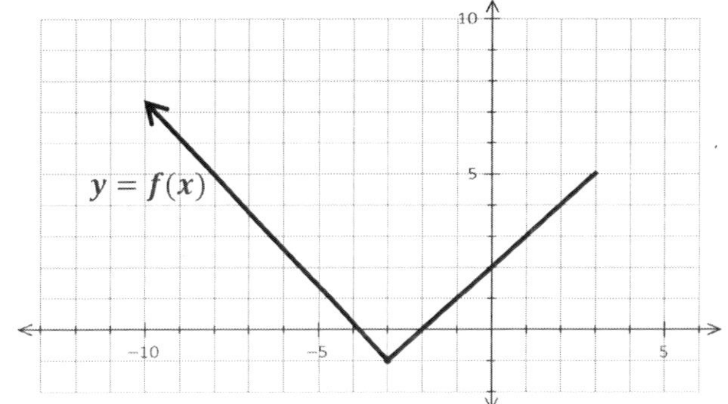

22. A function $y = f(x)$ has a domain of $(-\infty, 12]$ and a range $[-4, 8]$.
A function $g(x) = af(bx)$ has a domain of $(-\infty, 15]$ and a range $[-16, 8]$.

NR The codes for the values of a and b in the equation $g(x) = af(bx)$ are, respectively, ____ ,and ____.

Use the codes on the right. Codes can be used more than once.

Code	Possible values of a and b.
1	1/2
2	2
3	$-1/2$
4	-2
5	5/4
6	4/5

Answers from previous page

16. (a) i $(x,y) \rightarrow (x+7, -3y)$ ii $g(x) = -3f(x-7)$ (b) i $(x,y) \rightarrow (\frac{3}{4}x, 3y-5)$ ii $g(x) = 3f(\frac{4}{3}x) - 5$

17. B **18.** D

23. A function of $y = f(x)$ has a range of $(-\infty, 6]$. The range of $y + 2 = -2f(x)$ is:

A. $[-14, \infty)$

B. $[-10, \infty)$

C. $(-\infty, -14]$

D. $(-\infty, -10]$

24. The graph of the function $y = f(x)$ is shown below. The mapping rule that describes the transformation from $y = f(x)$ to $y = g(x)$ is $(2x - 6, -y + 3)$

Determine the domain and range of $y = f(x)$.

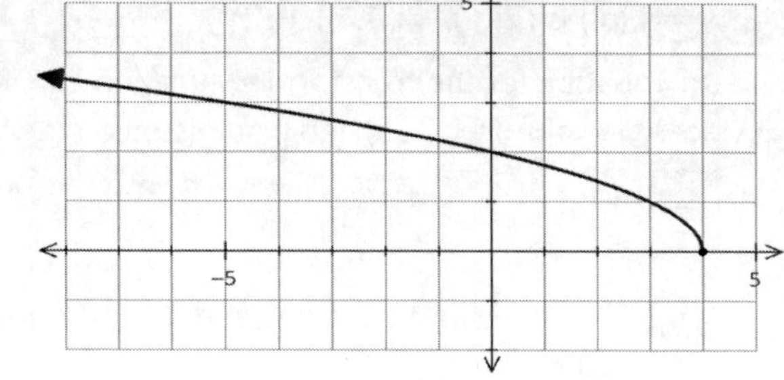

i *Domain:*

ii *Range:*

25. The graph of $y = f(x)$ is shown on the right. The vertex is at point P.

For the graph of $y = -f(\frac{1}{2}x - 2) + 3$, determine the:

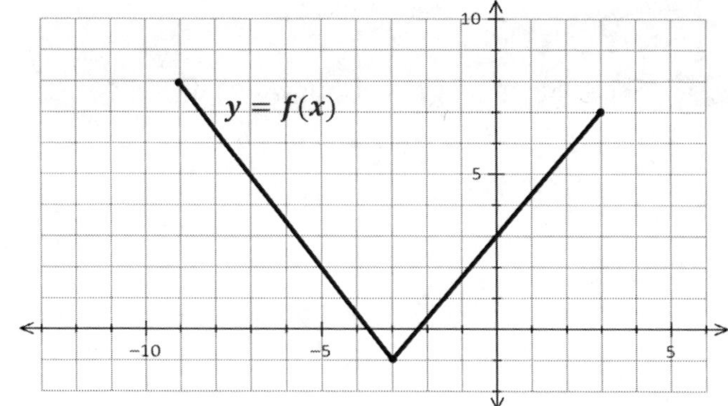

i *Domain:*

ii *Range:* ________

iii *Coordinates of point corresponding to P on graph of $g(x)$:* ________

26. A function of $y = f(x)$ is transformed to $g(x) = 5f[b(x + 3)] - k$. The point $(-6, 8)$ on the graph of $f(x)$ corresponds to the point $(-13, 33)$ on the graph of $y = g(x)$. Determine the values of b and k.

i $b =$ ________

ii $k =$ ________

Answers from previous page and this page

19. A **20.** 215 **21.** i $[-2, \infty)$ ii $[-12, \infty)$ **22.** 46

23. A **24.** i $(-\infty, 2]$ ii $(-\infty, 3]$ **25.** i $[-14, 10]$ ii $[-5, 4]$ iii $(-2, 4)$ **26.** i 3/5 ii 7

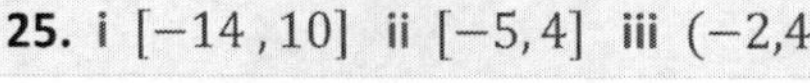

1.5 Inverse of a Relation

You can think of the **inverse** as "undoing", or more specifically – doing the opposite operations in the opposite order.

For example, the inverse of walking into a room and turning on the lights is to turn off the lights and then leave the room.

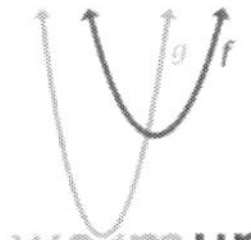

Warm-up **Exploration #1** **Sketching the Graph of an Inverse**

Consider the function $f(x) = \frac{1}{3}x - 2$

1 ➡ **Complete** the first column of the table below by substituting the given values of x into the equation for $f(x)$.

x	$y = f(x)$	$y = g(x)$
-6	$\frac{1}{3}(-6)-2=-4 \Rightarrow (-6,-4)$	
-3		
0		
3		
6		
9		

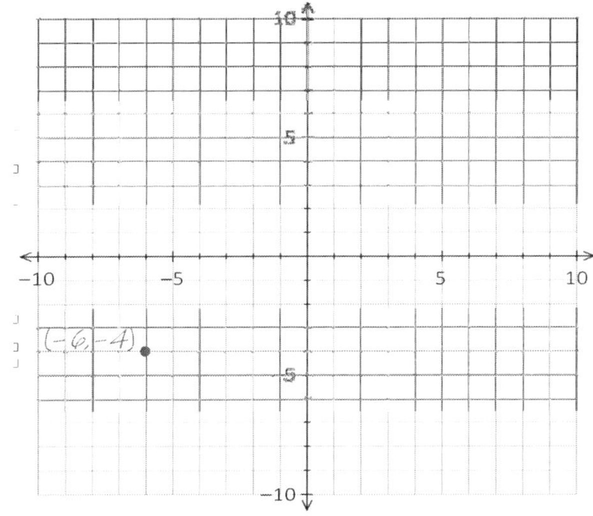

2 ➡ Plot each of the points on the grid on the right to **sketch** the graph of $f(x)$. *The first point is plotted for you.*

3 ➡ **Complete** the $g(x)$ column by interchanging all of the $f(x)$ coordinates. *Note that the first point in the column will be (–4,–6).*

4 ➡ Plot each of the points in the $g(x)$ column to **sketch** the graph of $y = g(x)$ on the same grid.

5 ➡ **Sketch** the graph of $y = x$, also on the same grid.

6 ➡ **Compare** the distances from the line $y = x$ of points on the graph of $f(x)$ and corresponding points on the graph of $g(x)$.

7 ➡ Use terminology from this unit to **describe** the transformation of the graph of $y = f(x)$ to the graph of $y = g(x)$. Where are the **invariant points** in this transformation?

8 ➡ Are the graphs of $y = f(x)$ and $y = g(x)$ functions? Explain.

9 ➡ Determine an **equation** for $y = g(x)$. How does this equation relate to the equation for $y = f(x)$?

The **inverse** of a relation, designated as $x = f(y)$, or $y = f^{-1}(x)$ if the inverse is also a function, is found by interchanging the x and y coordinates.

The *mapping rule* for this transformation is: $(\boldsymbol{x}, \boldsymbol{y}) \rightarrow (\boldsymbol{y}, \boldsymbol{x})$

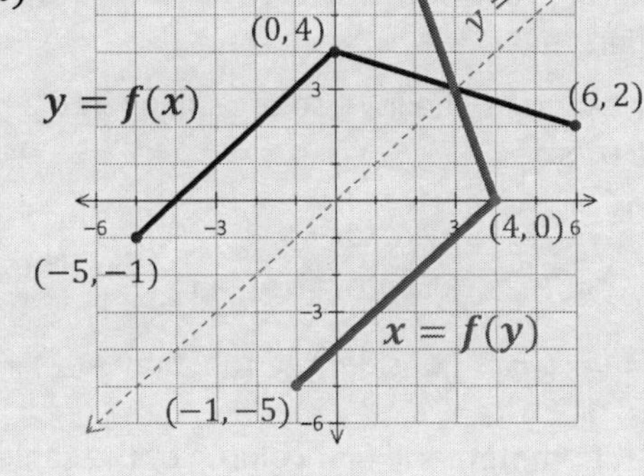

For example, given the graph of $y = f(x)$ on the right...

The inverse is found by interchanging all coordinates so that all points $(x, y) \rightarrow (y, x)$. Note that the inverse is **not a function**.

For example, the point $(-5, -1)$ on the graph of $y = f(x)$ becomes $(-1, -5)$ on the graph of the inverse.

Note that the graph of $y = f(x)$ is *reflected in the line* $y = x$ to become the inverse, that is the graph $x = f(y)$.

So there is an **invariant point** on the line $y = x$, at $(3, 3)$.

➡ The **domain** and **range** also interchange.

$y = f(x)$	$x = f(y)$
D: $[-5, 6]$ *R:* $[-1, 4]$	*D:* $[-1, 4]$ *R:* $[-5, 6]$

Note how this section is called "The Inverse of a **Relation**".

For that you might ask, given your inquisitive nature - *Why is that? What is the difference between a Function and a Relation anyways?*

Once again – great questions! Let's do some review, because yes, you've encountered this before....

A **relation** is a very broad term that describes a set of inputs (think x-coordinates) and outputs (y-coordinates). So any set of ordered pairs, which can be described as an equation, a graph, a domain-range map, is a **relation**.

A **function** meanwhile is a special type of relation where each input (x-coordinate) has exactly one output (y-coordinate). That is, for each x in the domain, there corresponds one (and only one) y.

Relation A – *Not a function*

A: $\{(-2, 5), (0, 4), (1, -2), (1, 3)\}$

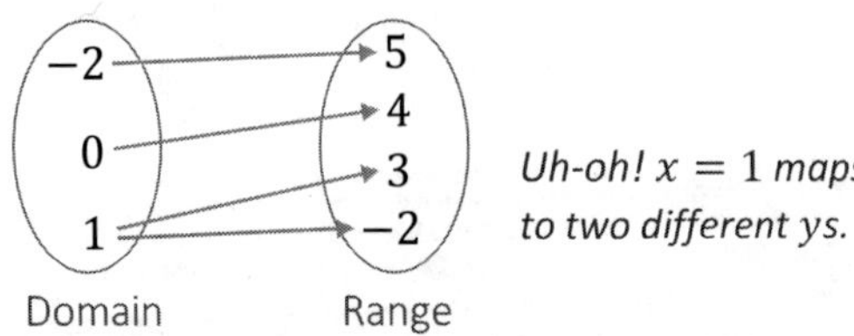

Uh-oh! $x = 1$ *maps to two different ys.*

Note that when $x = 1$ *there are two separate y's. (That's a no-go for functions!)*

Relation B – *Function!*

B: $\{(-2, 5), (0, 4), (1, -2), (2, -2)\}$

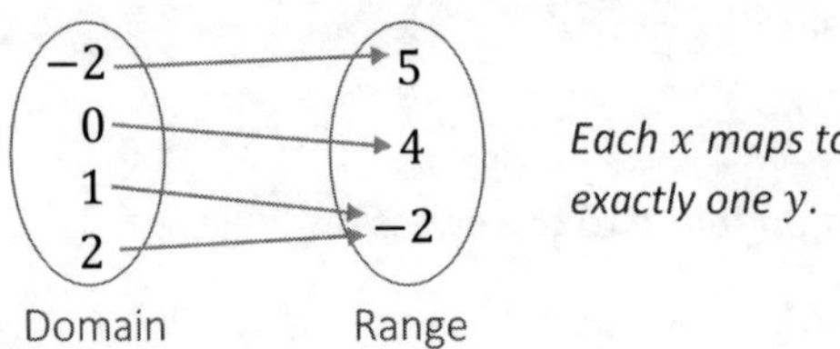

Each x *maps to exactly one* y.

Note that it's OK for two different x's to map to the same y!
(As is the case with $x = 1$ *and* $x = 2$*, which both map to* $y = -2$*)*

The Vertical Line Test – Applying the Definition of Functions to Graphs

Given the graph of a relation, we know it's a function if each x in the domain maps to one and only one y. And an easy way to visualize this is the **vertical line test**.

Given the graph of $y = f(x)$, $f(x)$ is a function if (and only if) **any vertical line touches at most once.**

Function ✓

Any vertical line intersects the graph once

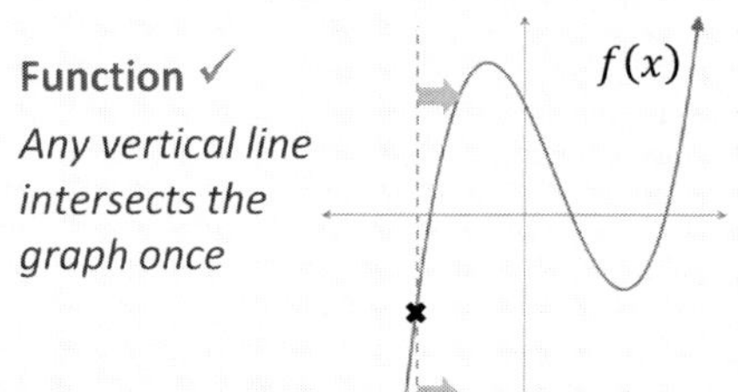

NOT a Function ✖

Vertical line intersects graph in more than one place!

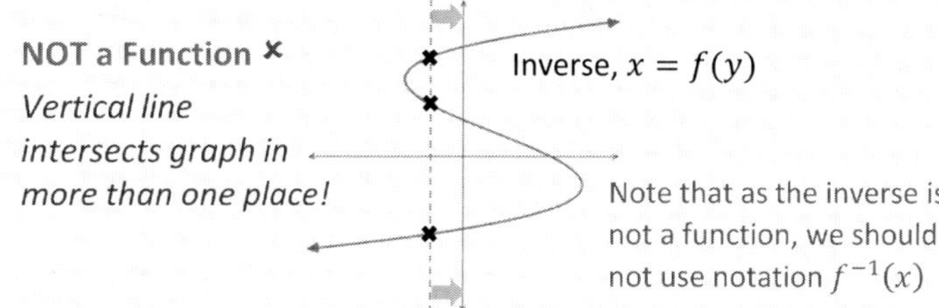

Note that the first graph on the left is the function $f(x) = (x + 2)(x - 1)(x - 3)$.

Whereas the graph on the right is the **inverse**, which has the equation $x = (y + 2)(y - 1)(y - 3)$.

We'd next express the inverse equation in terms of y, which is challenging in this case*! We'll come back to this concept.*

*Switch **x** and **y** in the equation*

The also illustrates, that, given a function $y = f(x)$, its inverse, $x = f(x)$ ***need not be a function.***

So while the inverse can be expressed as either $x = f(y)$ or, *equivalently, $y = f^{-1}(x)$...

We should not use$y = f^{-1}(x)$ when the inverse is not a function.

*Avoid this notation when the inverse is not a function

Also be careful to note: $y = f^{-1}(x)$ **The "-1" here should not be confused with an exponent, which would represent the *reciprocal***

$$f^{-1}(x) \neq \frac{1}{f(x)}$$

The Horizontal Line Test

We saw that the vertical line test can be used to determine if the graph of a relation is a function.

Similarly, we can use the **horizontal line test** to determine if, given the graph of $y = f(x)$, its INVERSE $x = f(y)$ or $y = f^{-1}(x)$ will be a function.

Given the graph of $y = f(x)$, its **inverse** $x = f(y)$ will **be a function** if (and only if) **any horizontal line touches at most once.**

Example 1

$f(x) = (x - 3)^3$

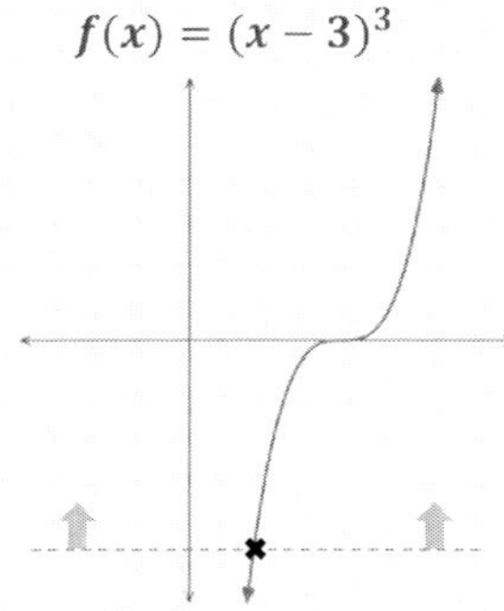

Passes H.L.T.

Any horizontal line will only intersect curve once

➔ *Inverse will pass V.L.T. and* ***will be*** *a function:*

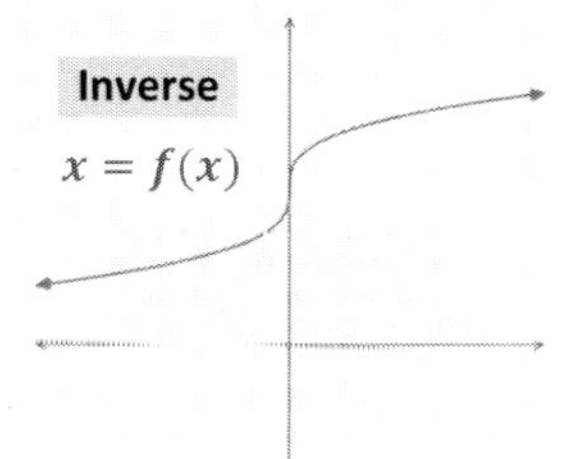

Example 2

$f(x) = (x - 3)^2 - 2$

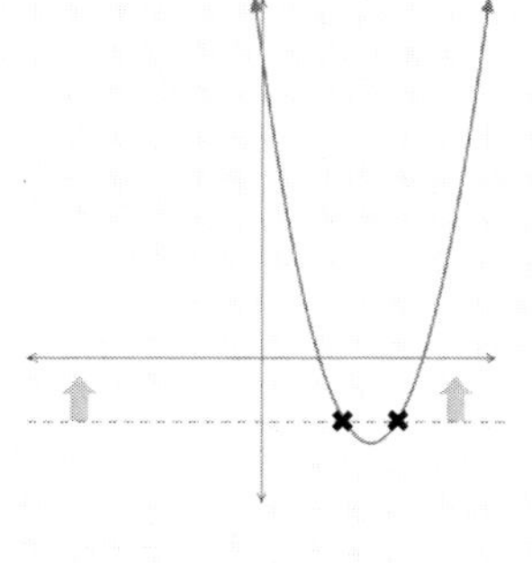

Fails H.L.T.

Horizontal line intersects curve more than once

➔ *Inverse* ***will not be*** *a function:*

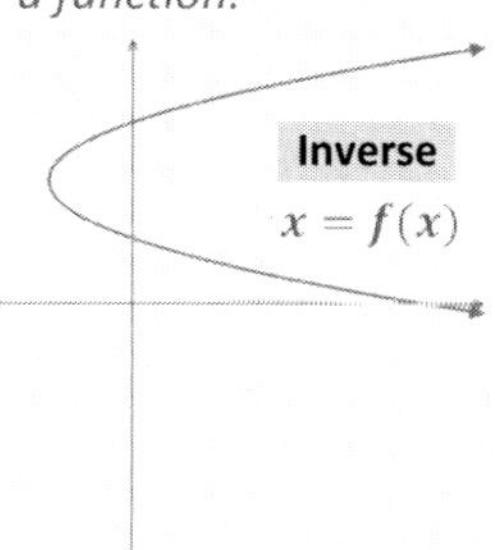

The Inverse as a Reflection

Since the graph of $y = f^{-1}(x)$ is a reflection (in the line $y = x$) of the graph of $y = f(x)$, we can now complete our list of the three types of reflections we consider in this course.

① **Vertical Reflection** in the line $y = 0$

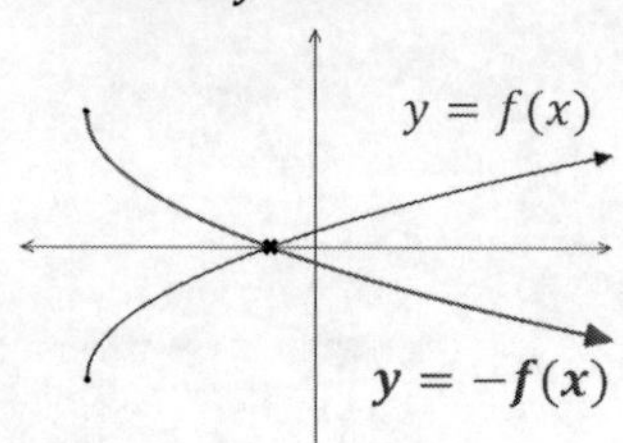

② **Horizontal Reflection** in the line $x = 0$

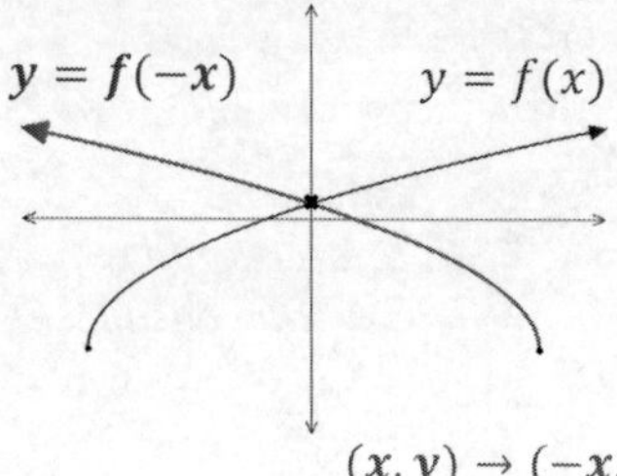

③ **(Inverse) Reflection** in the line $y = x$

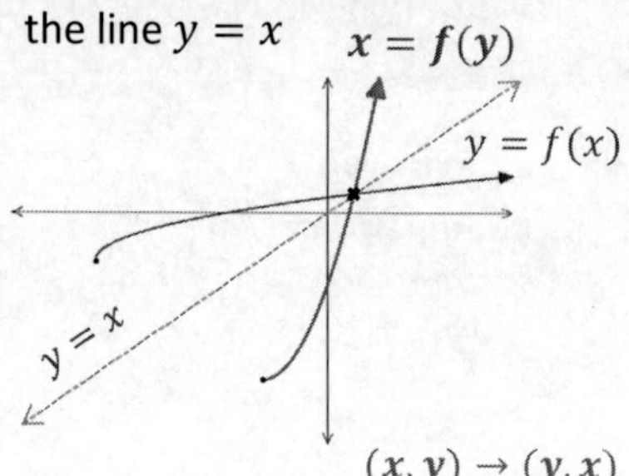

Mapping Rule: $(x, y) \to (x, -y)$ $\quad (x, y) \to (-x, y)$ $\quad (x, y) \to (y, x)$

Worked Example

Given the graph of $y = g(x)$

(a) Sketch the graph of $x = f(y)$ on the same grid

Is the inverse a function?

(b) State the location and coordinates of any invariant point

(c) State the domain and range of $x = f(y)$

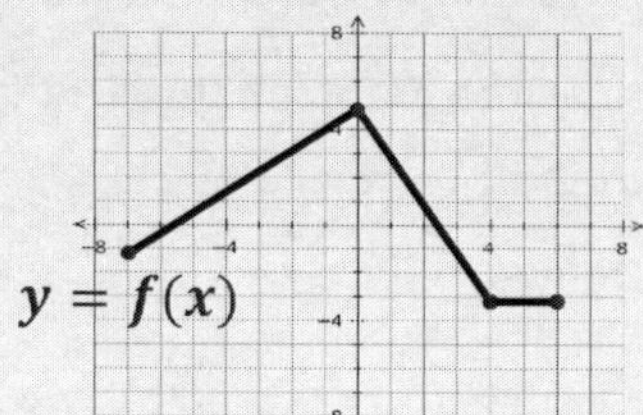

Solution: (a) To **sketch** the graph of the inverse, **transform** all points by:

$$(x, y) \to (y, x)$$

Procced left to right on all points on $f(x)$...

$$(-7, 0) \to (\mathbf{0}, \mathbf{-7})$$
$$(0, 6) \to (\mathbf{6}, \mathbf{0})$$
$$(4, -2) \to (\mathbf{-2}, \mathbf{4})$$
$$(6, -2 \to (\mathbf{-2}, \mathbf{6})$$

Plot points and sketch

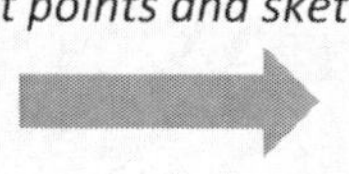

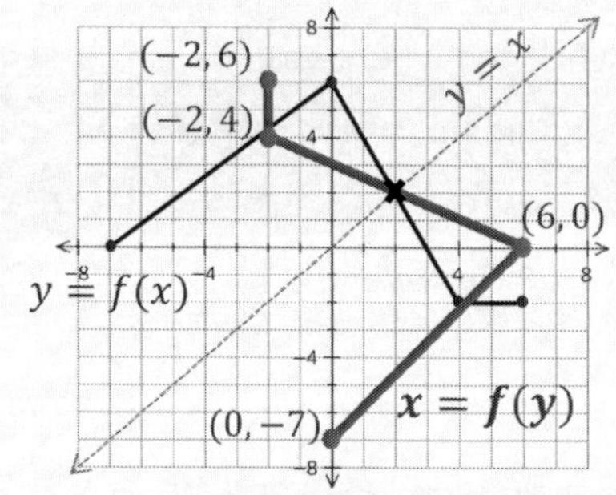

(b) The invariant point occurs where the graph of $y = f(x)$ intersects the line $y = x$. *(That is, where the x and y coordinates are the same, and interchanging has no effect)* Invariant point is at $(\mathbf{2}, \mathbf{2})$ ***on the line*** $\boldsymbol{y = x}$

(c) ***D:*** $\{x | -2 \leq x \leq 6, x \in \mathbb{R}\}$ ***R:*** $\{y | -7 \leq x \leq 6, y \in \mathbb{R}\}$

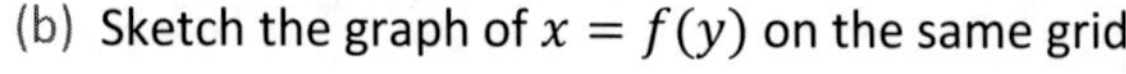

Class Example 1.51 *Sketching the Graph of an Inverse Relation*

The graph of $y = f(x)$ is shown on the right.

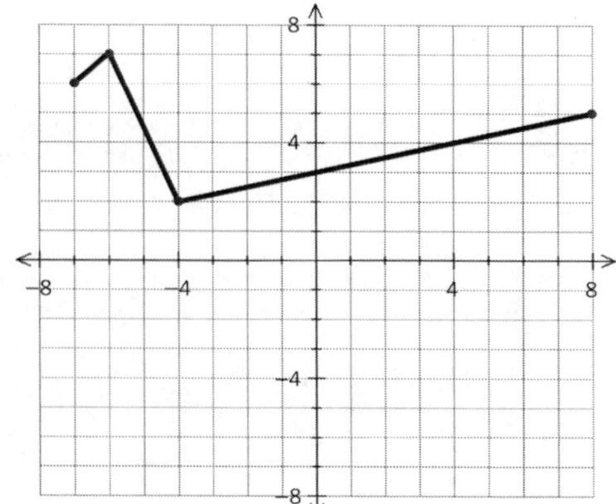

(a) Make a prediction on whether the graph of $x = f(y)$ will be a function.

(b) Sketch the graph of $x = f(y)$ on the same grid

(c) State the coordinates (and location) of any invariant point(s)

(d) State the domain and range of both $y = f(x)$ and $x = f(y)$

$y = f(x)$ *D:*__________ $x = f(y)$ *D:*__________

*R:*__________ *R:*__________

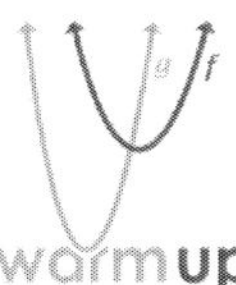

Exploration #2 **Sketching an Inverse Function using the online graphing calculator DESMOS**

To do this exercise, you'll need to go to **desmos.com/calculator**

You can optionally set up a free account, that way you can save your graphs. ☺

Objective: To analyze the graphs of $f(x) = x^2 - 4$ and the inverse, $y = f^{-1}(x)$

Desmos offers some functionality lacking on your graphing calculator, such as graphing in "$x =$" form

1 ➡ Graph the function $\mathbf{y = x^2 - 4}$ and its **inverse**, obtained by *switching x and y in the equation.*

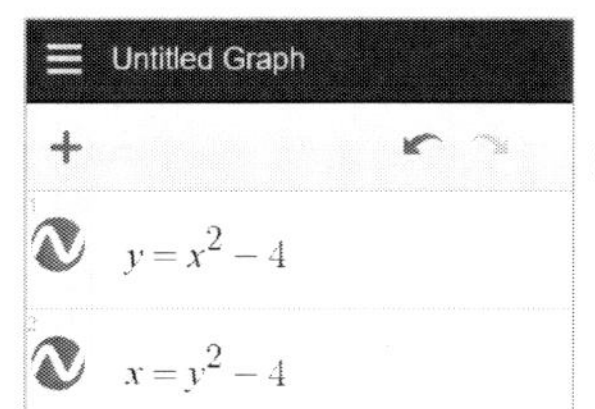

Note: For exponents, use **shift + 4** to access **^**

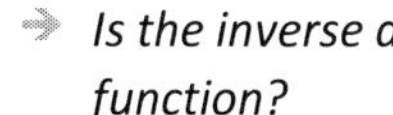

2 ➡ Analyze the graphs, shown below.

➔ *Do the domain and range switch?*

➔ *Can we confirm the graphs are reflections?*

➔ *Is the inverse a function?*

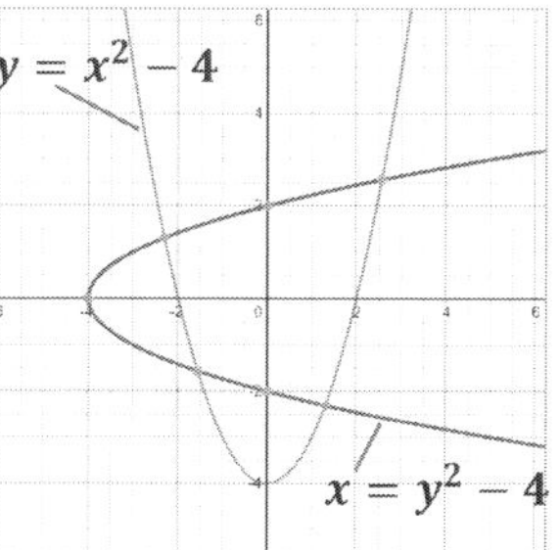

3 ➡ Describe the shape of the graphs of $\mathbf{f(x) = x^2 - 4}$ and its **inverse**.

We can determine "$y =$" form equation of the inverse to the function $y = x^2 - 4$ in one of two ways:

Method 1

By isolating "y" in the inverse shown above. (Where we switched x and y in the equation, to obtain $x = y^2 - 4$

Method 2

By examining the graph of the inverse, and expressing as two radical functions. (One representing the "top" branch, one the bottom)

4 ➡ Determine "$y =$" form equation of the inverse to the function $y = x^2 - 4$.

5 ➡ Use Desmos to graph the function $\mathbf{y = \sqrt{x} + 5}$ and its **inverse**, $\mathbf{x = \sqrt{y} + 5}$.

Sketch here. Describe the shape of each graph.

Can you confirm they are reflections?

Is the inverse a function?

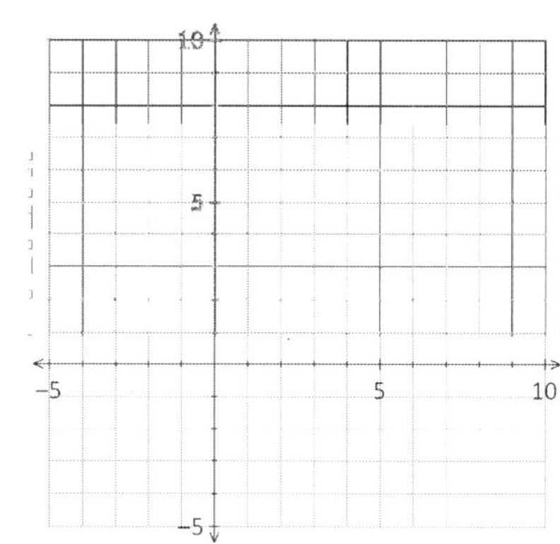

6 ➡ Determine the "$y =$" form equation of the inverse of the function $y = \sqrt{x} + 5$.

Worked Example

Given the function $f(x) = \sqrt{x+2}$, graphed on the right,

(a) Sketch the graph of $y = f^{-1}(x)$ on the same grid
Transform all indicated points (•). Is the inverse a function?

(b) State the domain and range of $y = f^{-1}(x)$

(c) Determine an equation, in terms of x, for $y = f^{-1}(x)$

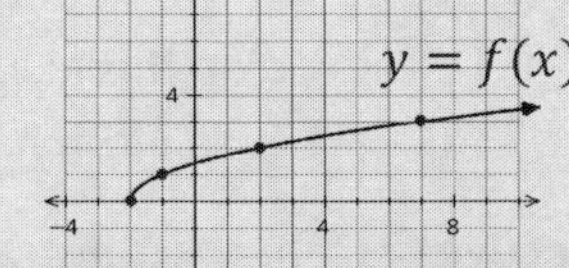

Solution: (a) To **sketch** the graph of the inverse, **transform** all points by:

$$(x, y) \to (y, x)$$

Procced left to right on all points on $f(x)$...

$$(-2, 0) \to (\mathbf{0}, \mathbf{-2})$$
$$(1, -1) \to (\mathbf{-1}, \mathbf{1})$$
$$(2, 2) \to (\mathbf{2}, \mathbf{2})$$
$$(3, 7) \to (\mathbf{7}, \mathbf{3})$$

Plot points and sketch

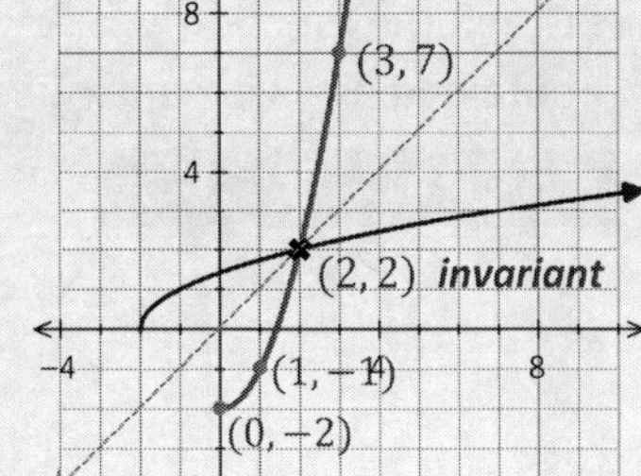

(b) For $y = f^{-1}(x)$, *D:* $[-2, \infty)$ *R:* $[0, \infty)$

The range of $f(x)$ *The domain of $f(x)$* **➔ *Domain and Range interchange!***

(c) For the equation, start with $y = \sqrt{x+2}$ (use "$y =$" instead of "$f(x)$") and interchange x and y.

$x = \sqrt{y+2}$ *Square both sides to isolate y*

$(x)^2 = (\sqrt{y+2})^2$

$y + 2 = x^2$ ➡ $\mathbf{f^{-1}(x) = x^2 - 2}$; $\boxed{x \geq 0}$

↙ *We must* ***restrict the domain****, which is the range of $y = f(x)$.*

Class Example 1.52 *Determining the Graph and Equation of an Inverse*

The graph of $f(x) = x^2 - 9$ is shown on the right.

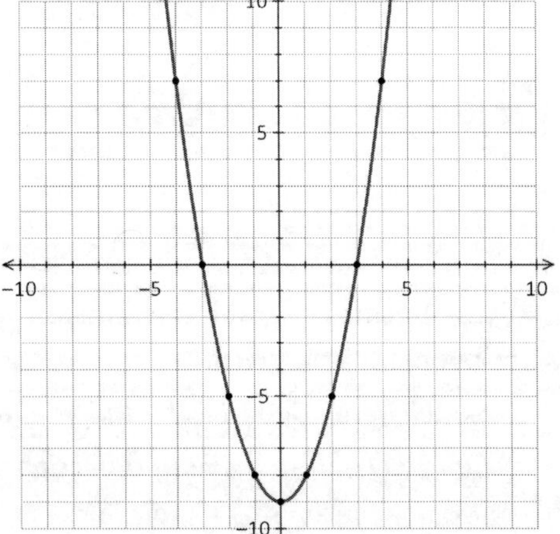

(a) Sketch the graph of $x = f(y)$ on the same grid by transforming each indicated (•) point.

(b) Describe the transformation from the graph of $y = f(x)$ to that of $x = f(y)$.

(c) Describe where any invariant points can be found, and how many there are.

(d) State the domain and range of both $y = f(x)$ and $x = f(y)$.

$y = f(x)$ *D:* ________ $x = f(y)$ *D:* ________

R: ________ *R:* ________

(e) Determine an equation for $x = f(y)$, in terms of x.

Let's revisit the worked example on the previous page. A common mistake is to *forget to restrict the domain,* and stating the equation of the inverse as just $f^{-1}(x) = x^2 - 2$. *(graphed below)*

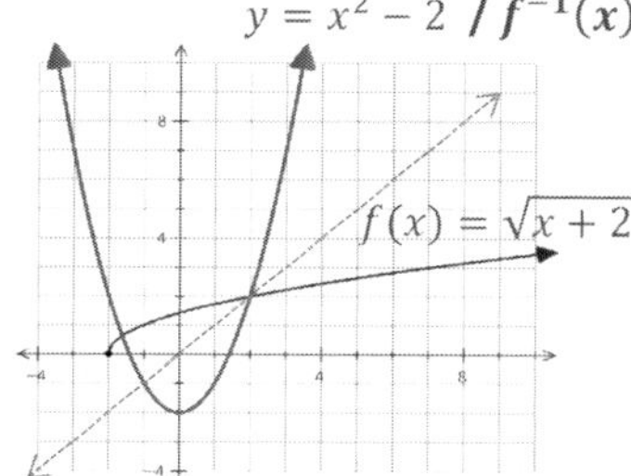

← But it's important to see how – ***this graph makes no sense!***

- *The graph of $f(x)$* ***is a half-parabola (sideways)***
- *So graph of inverse, which recall is a* ***reflection*** *about $y = x$, must also be a* ***half-parabola!***
- *Therefore we must restrict the domain of the inverse!*

  Domain of $y = f^{-1}(x)$ is equal to the range of $y = f(x)$

You can sketch the graph of $y = f^{-1}(x)$ on your graphing calculator!

(But it's a bit cumbersome. Ready? Let's go!)

Suppose we wish to sketch the inverse of $f(x) = x^2 - 4$.

❶ Input the equation for $y = f(x)$ into y_1 and graph.

```
Plot1  Plot2  Plot3
■\Y1■X²-4
■\Y2=
```

❷ Access the **draw inverse** function in your calculator by keying in "2nd" + "Program" [2nd] [prgm]

❸ Select #8, "DrawInv" →

```
NORMAL FLOAT AUTO REAL
DRAW POINTS ST
1:ClrDraw
2:Line(
3:Horizontal
4:Vertical
5:Tangent(
6:DrawF
7:Shade(
8:DrawInv
9↓Circle(
```

❹ We want to instruct the calculator to draw the inverse of what we inputted into y_1. With your cursor set after "DrawInv", key in "VARS", then scroll to "Y-VARS", then "Function", finally select "Y1". *Once you have this on your screen, hit ENTER!* →

```
NORMAL FLOAT AUTO REAL
DrawInv Y1
```

Follow these steps to get "Y1" after "DrawInv"

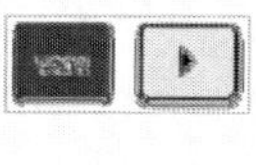

```
VARS Y-VARS COLOR
1:Function...
2:Parametric...
3:Polar...
4:On/Off...
```

```
FUNCTION
1:Y1
2:Y2
3:Y3
4:Y4
```

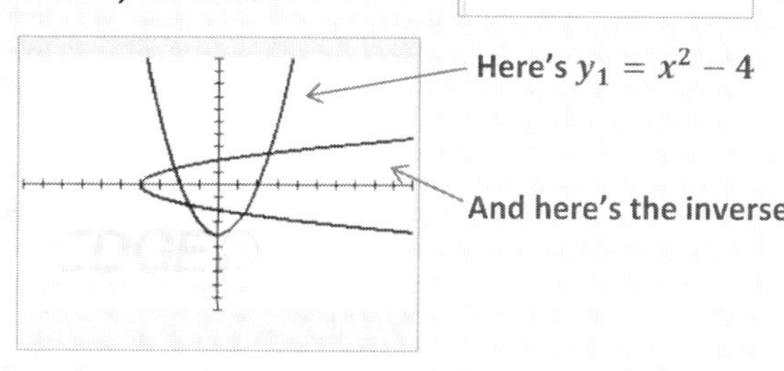

Class Example 1.53 *Determining the Inverse of a Linear Function*

The graph of $f(x) = 4x - 3$ is shown on the right.

(a) Sketch the graph of the reflection of $f(x)$ in the line $y = x$ by transforming the four indicated (•) points.

(b) State the coordinates of any invariant points.

(c) Determine an equation for the reflected graph, in terms of x.

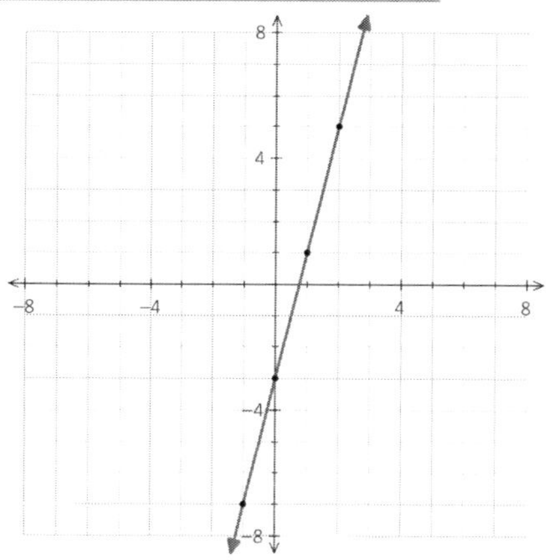

Equations of Inverses – *Opposite Operations in the Opposite Order!*

Consider the following pairs of function and inverse:

$f(x) = 2x + 6$ → *Multiply x by 2, then add 6*

$f^{-1}(x) = \dfrac{x-6}{2}$ → *Subtract 6 from x, then divide by 2*

$f(x) = x^2 - 4$ → *Square x, then subtract 4*

$y = \pm\sqrt{x+4}$ *(inverse)* → *Add 4 to x, then square root*

$f(x) = \sqrt{x+4} - 1$ → *Add 4 to x, then square root, then subtract 1*

$f^{-1}(x) = (x+1)^2 - 4$ → *Add 1 to x, then square, then subtract 4*

Can you see the pattern?

For each pair the equation of the inverse represents the ***opposite operations in the opposite order***!

In previous examples we've practiced a method for obtaining the equation of an inverse to a function. However it could be useful to remember this pattern as a double-check of our results.

Class Example 1.54 *Determining the Inverse of a Radical Function*

The graph of $f(x) = \sqrt{x+4} - 1$ is shown on the right.

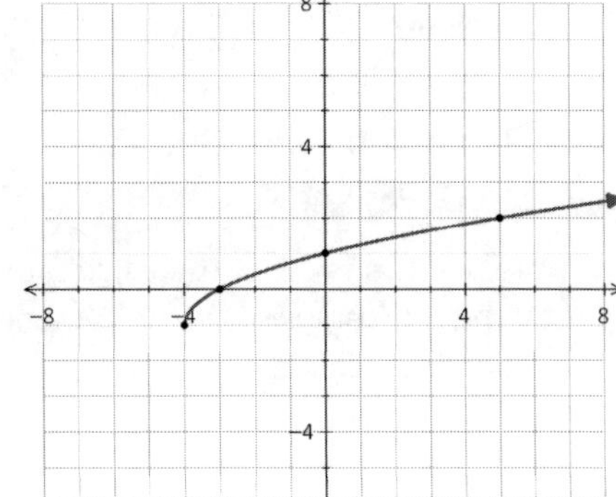

(a) Sketch the graph of $y = f^{-1}(x)$ on the same grid by transforming each of the four indicated (•) points.

(b) Describe where any invariant points can be found, and how many there are.

(c) State the domain and range of both $y = f(x)$ and $y = f^{-1}(x)$.

$y = f(x)$ D:__________ $y = f^{-1}(x)$ D:__________

R:__________ R:__________

(d) Determine an equation for $y = f^{-1}(x)$, in terms of x.

Class Example 1.55 *Restricting the Domain so that the Inverse is a Function*

For each of the functions below, provide a restriction on the domain so that the inverse would be a function.

(a) $f(x) = (x-3)^2 - 2$

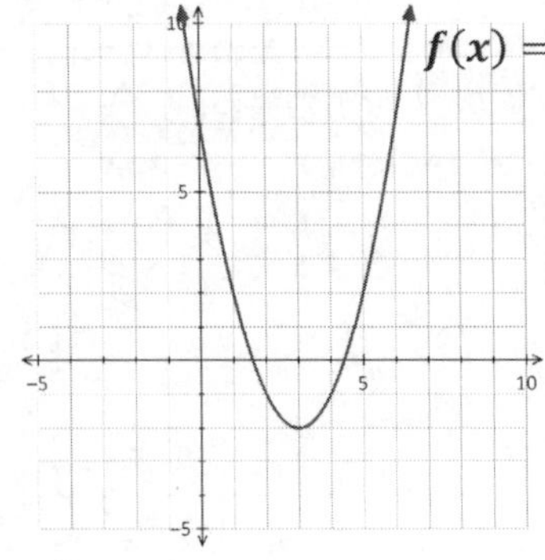

(b)

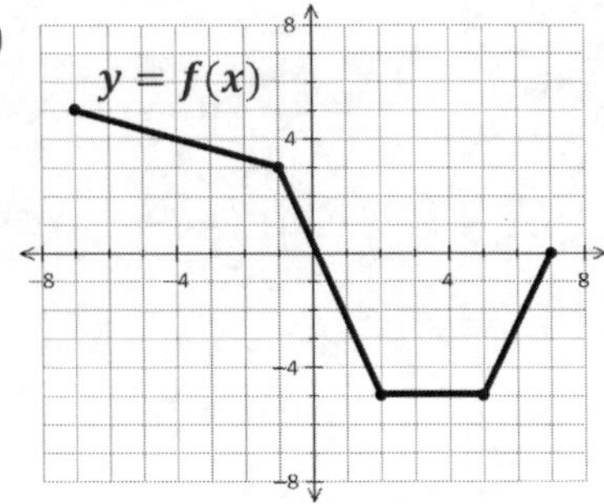

Chapter 2 POLYNOMIAL FUNCTIONS

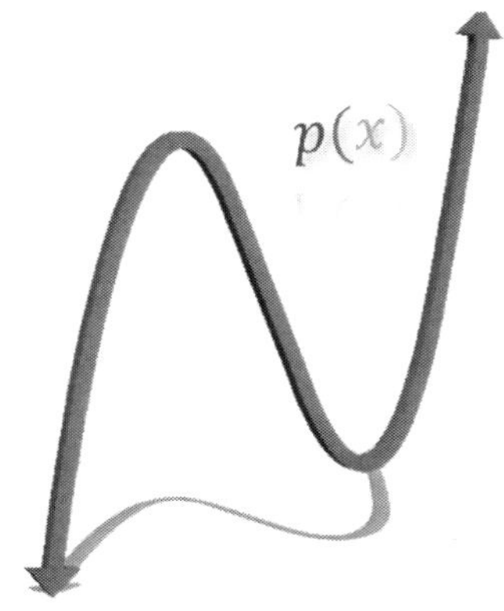

2.1 Characteristics of Polynomial Functions

You've already studied Polynomial functions! *Remember these?*

In Math 10C you studied **Linear Functions**

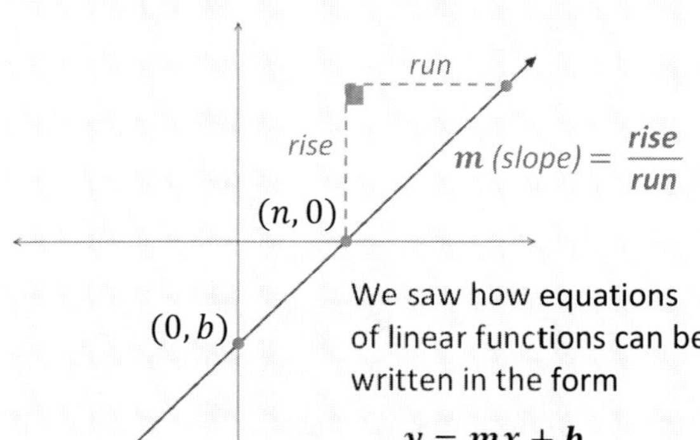

We saw how equations of linear functions can be written in the form

$$y = mx + b$$

*Where m is the **slope** of the line, and b is the **y-intercept***

Note that the linear functions can also be written in the form $y = m(x - n)$

*Where n is the **x-intercept***

➡ These are **degree 1** Polynomial Functions

And in Math 20-1 you studied **Quadratic Functions**

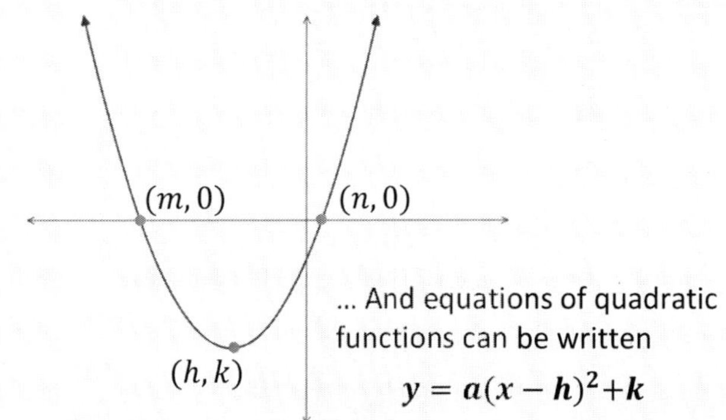

... And equations of quadratic functions can be written

$$y = a(x - h)^2 + k$$

*Where a is the **vertical stretch**, and the coordinates of the **vertex** are (h, k).*

Note that the quadratic functions can also be written in the form $y = a(x - m)(x - n)$

*Where m, n are **x-intercepts***

➡ These are **degree 2** Polynomial Functions

Determine an equation for each of the following functions:

1 ➡

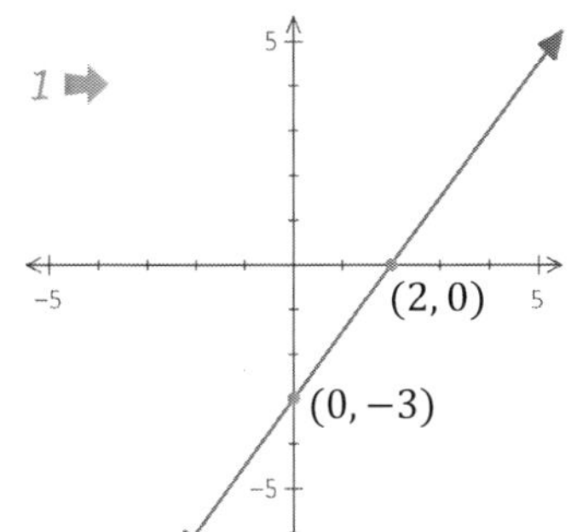

2 ➡

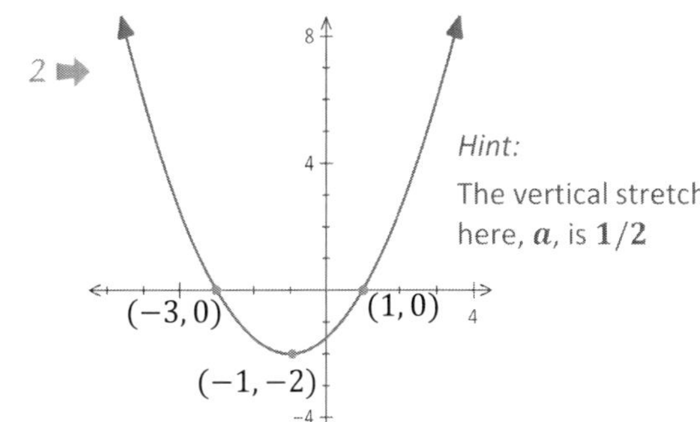

Hint:
The vertical stretch here, a, is $1/2$

Defining Polynomial Functions

A **polynomial function**, such as a linear function, quadratic, or cubic, involves only non-negative integer powers of x.

Any polynomial function can be written in the form

$$p(x) = a_n x^n + a_{n-1}x^{n-1} + a_{n-2}x^{n-2} + \ ... + a_2x^2 + a_1x^1 + a_0$$

Where:
- *n is a whole number, representing the degree of the function*
- *a_n to a_0 are real numbers, representing the coefficients, with a_n designated the **leading coefficient** (coeff. of the highest degree term)*

This *general formula* may look complicated, but a few **polynomial function examples** should show its simplicity:

- $y = 8$ — This is **degree 0** (constant function)
- $y = -5x + 4$ — This is **degree 1** (linear), with a leading coefficient of -5
- $f(x) = 2x^2 - 4x + 3$ — This is **degree 2** (quadratic), with a leading coefficient of 2
- $y = x^3 + \frac{1}{2}x^2 - 7x - 1$ — This is **degree 3** (quadratic), with a leading coefficient of 1
- $p(x) = 3x^4 - \sqrt{2}x^3 + 5x - 8$ — This is **degree 4** (quartic), with a leading coefficient of 3
- $p(x) = -2x^5 + x^4 - 3x^2 - 6$ — This is **degree 5** (quintic), with a leading coefficient of -2

These examples are all written in **descending order of degree**, where terms are arranged starting with the highest degree term, starting with the **leading coefficient**. (The coefficient of the highest degree term)

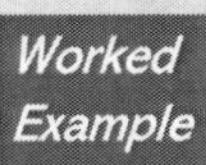

Worked Example

Identify which of the following are polynomial functions. For each that is a polynomial function; state the degree and leading coefficient:

(a) $y = -6x^2 + \frac{3}{x} - 8$ (b) $y = 3x^4 - 2x^5 - 3x^2 - 1$ (c) $y = 5x^2 - 3\sqrt[3]{x} + 1$

Solution: (a) The middle term can be written $3x^{-1}$, which is **NOT POLYNOMIAL** as exponent of x is not a whole number

(b) All exponents are whole numbers, and all coefficients are real numbers. Hello, you **POLYNOMIAL FUNCTION**. Degree is **5** *(degree of entire poly function is that of highest degree term).* Leading coefficient is $\mathbf{-2}$. *(the terms are not in descending order of degree – the 2nd term should be re-arranged to the "front"!)*

(c) The middle term can be written $3x^{1/3}$, which is **NOT POLYNOMIAL** as the exponent is not a whole number

Class Example 2.11 *Identifying Polynomial Functions*

Identify which of the following are polynomial functions. For each that is a polynomial function; state the degree and leading coefficient:

(a) $y = 4x^5 - 3x^2$ (b) $y = 2\sqrt{x} + 5x$ (c) $y = \sqrt{3}\,x + 7$ (d) $y = -2x^3 + 5x^{-1}$

Polynomial functions can be of any whole number degree n – but for this course we'll only deal with functions where $n \leq 5$.

And while the coefficients ***can*** be any real number – we'll mainly stick with **integer** coefficients.

Classifying Polynomial Functions by Degree

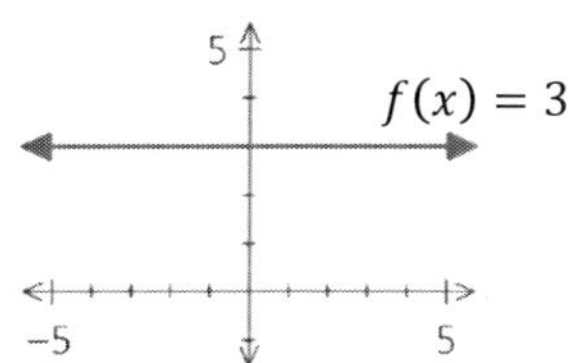

A **polynomial function** can be of any whole number degree, including zero!
The graph on the right is of the constant function $f(x) = 3$, which is a polynomial function of degree zero.

Let's now acquaint ourselves with **some examples of polynomial functions**, degree 1 through 5.

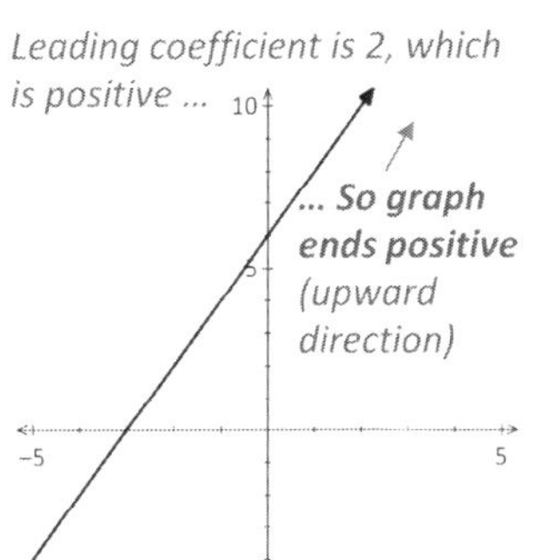

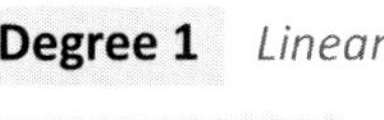

Degree 1 *Linear*

$f(x) = 2x - 3$

Domain: $\{x \in \mathbb{R}\}$

Range: $\{y \in \mathbb{R}\}$

End Behavior:
starts negative in quad III,
ends positive in quad I

# of intercepts: 1

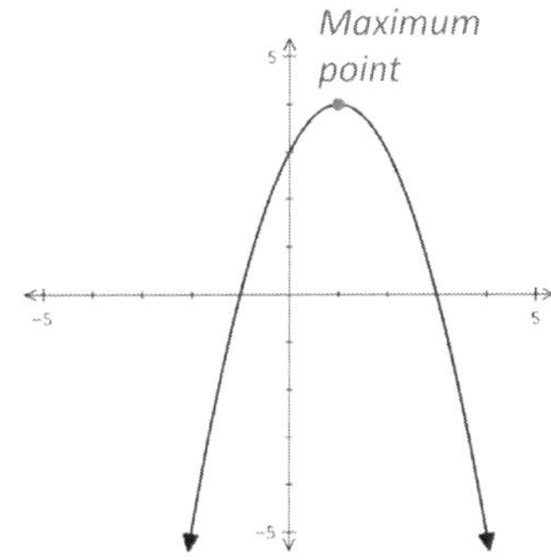

Degree 2 *Quadratic*

$g(x) = -x^2 + 2x + 3$

Domain: $\{x \in \mathbb{R}\}$

Range: $\{y | y \leq 4, y \in \mathbb{R}\}$

End Behavior:
starts negative in quad III,
ends negative in quad IV

# of intercepts: 2

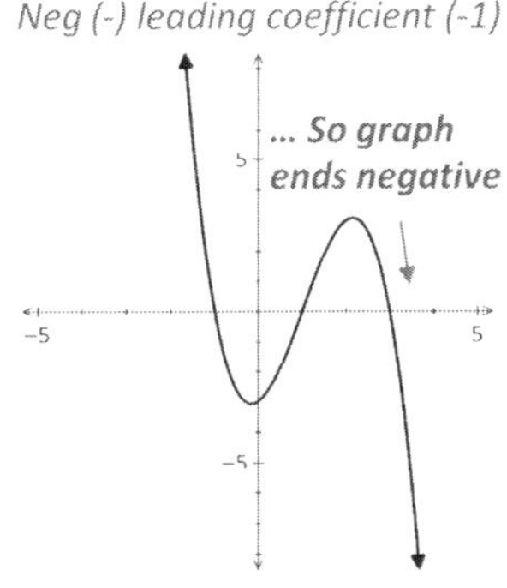

Degree 3 *Cubic*

$k(x) = -x^3 + 3x^2 + x - 3$

Domain: $\{x \in \mathbb{R}\}$

Range: $\{y \in \mathbb{R}\}$

End Behavior:
starts positive in quad II,
ends negative in quad IV

# of intercepts: 3

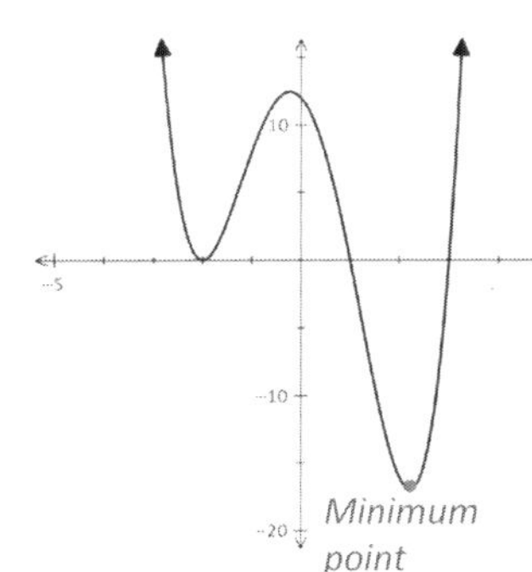

Degree 4 *Quartic*

$p(x) = x^4 - 9x^2 - 4x + 12$

Domain: $\{x \in \mathbb{R}\}$

Range: $\{y | y \geq -16.9, y \in \mathbb{R}\}$

End Behavior:
starts positive in quad II,
ends positive in quad I

# of intercepts: 3

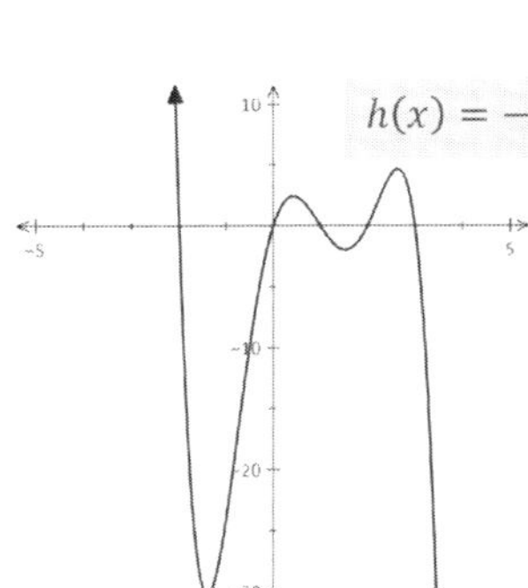

Degree 5 *Quintic*

$h(x) = -x^5 + 4x^4 + x^3 - 16x^2 + 12x$

Domain: $\{x \in \mathbb{R}\}$

Range: $\{y \in \mathbb{R}\}$

End Behavior:
starts positive in quad II,
ends negative in quad IV

# of intercepts: 5

→ The functions on the left are **odd degree** – and the graphs start and end in the opposite direction. *For example, the degree 3 and 5 functions start positive and end negative.*

Odd functions have no max or min point, must have at least one x-intercept, and have a **range** $\{y \in \mathbb{R}\}$**.**

→ The functions above / on the right are **even degree**. As such, the graphs start and end in the same direction. *For example, the degree 4 function starts positive and ends positive.*

Even functions have either a maximum or minimum point, and the **range** is restricted accordingly.

If the **sign of the leading coefficient is positive** (see the degree 1 and 4 examples above), the graph "ends positive", or heading upward in quadrant I.

And if the leading coefficient is positive and the **degree is even** (as with the degree 4 example above), the graph will have a **minimum point**.

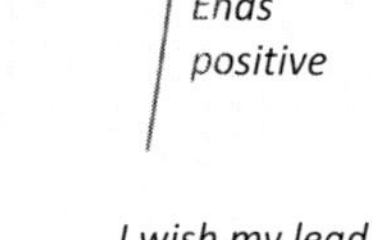

If the sign of the leading coefficient is **negative** (see the degree 2, 3 and 5 examples), the graph "**ends negative**", or downward, in quad IV.

And if the leading coefficient is negative and the function degree is even (as with the degree 2 function), the graph will have a **maximum point**.

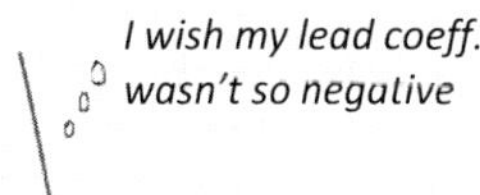

There is a relationship between the degree of a polynomial function and the number of x-intercepts on the graph.

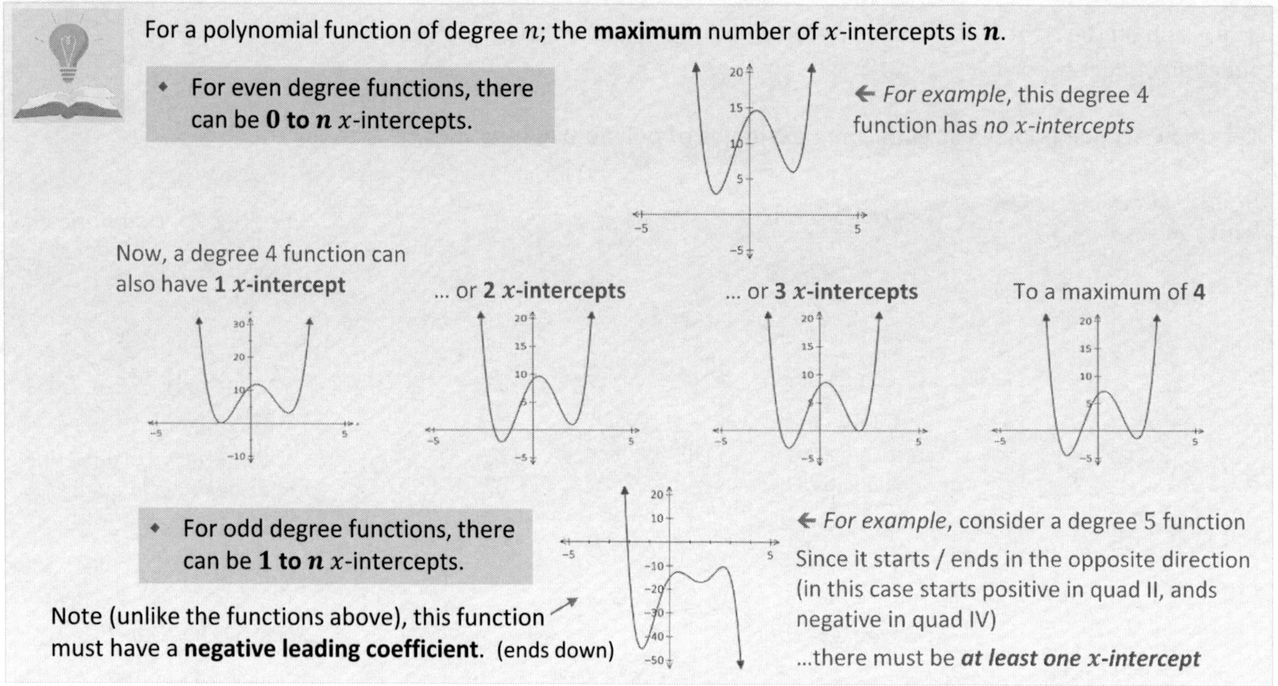

And we should know how to spot a Polynomial Function Graph!

On the previous page we saw the relationship between the degree of a polynomial function and certain characteristics of the graph. You might next ask – how can we immediately tell that a graph is of a polynomial function, and not some other function we study in Math 30?

And once again – great question! Cheers to your inquisitive nature.
Let's dive into that, with a couple of key distinguishing points:

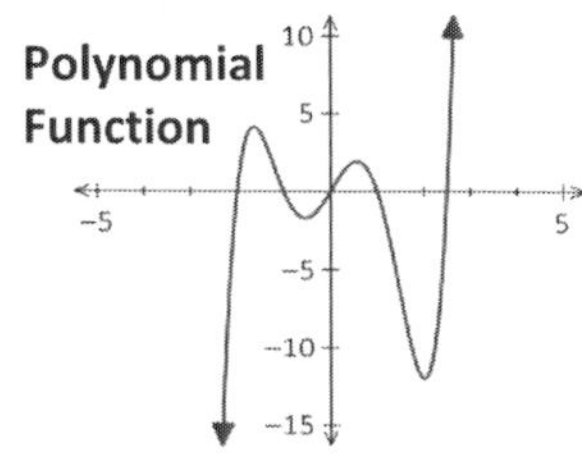

Graph can be drawn without lifting your pencil.

- The first key point is that all polynomial functions have a domain $\{x \in \mathbb{R}\}$. That means graphs of polynomial functions:

  - *Have no start or end points, like, for example, radical function graphs.*
  - *Have no vertical asymptotes or any other type of discontinuity, as with rational function graphs.*

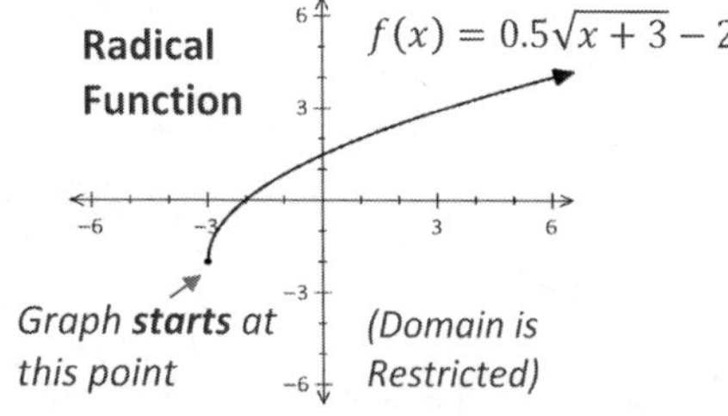

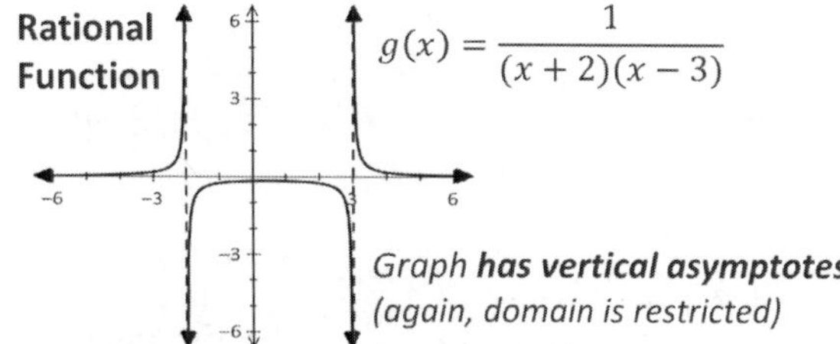

- The second point is polynomial function graphs have no horizontal asymptotes (like exponential functions) and there is no periodic pattern (as with some trig graphs).

  So graphs will always both start and end in either an upward or downward position.

  For example, this polynomial function graph....

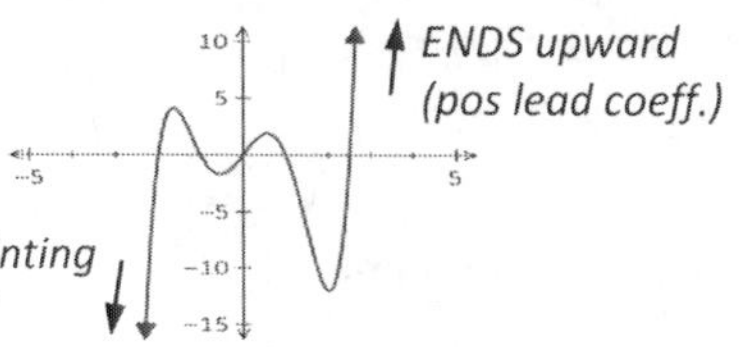

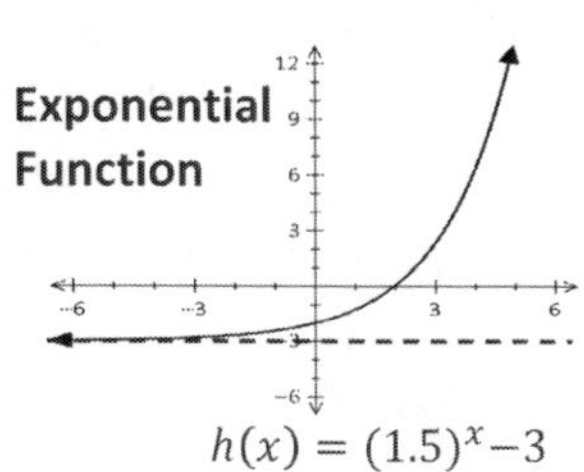

Worked Example

For the polynomial function $p(x) = -x^4 + 3x^3 + 7x^2 - 15x - 18$;

Without using your graphing calculator, state:

i - The start and end behavior of the graph ii - The number of possible x-intercepts

iii - Whether or not the graph will have a minimum or maximum point

iv - The domain of the function and the y-intercept

Use your graphing calculator to determine:

v - The x-intercepts of the graph vi - The range of the function

Sol.:

i - The degree of the function is 4; since it's even, the graph will start and end in the same direction. And since the leading coefficient is negative (that is, "-1"), the graph will end negative / heading downward. So.... The graph **starts negative in quadrant III**, and **ends negative in quadrant IV**.

ii - Degree 4 (even), so there can be **between 0 and 4 x-intercepts**.

iii - Even degree, graph starts / ends in the same direction. Negative leading coefficient, so which means the graph ends negative. Therefore the graph will **have a maximum point,** which can be found graphically.

iv - All polynomial functions have domain $\{x \in \mathbb{R}\}$. The y-intercept is the same as the constant value, so $(0, -18)$

v - x-intercepts are the same as the zeros of the function. The zero function is in CALC menu, found be entering 2nd trace

vi - For the range, find the **MAXIMUM**, which is also in the CALC menu.

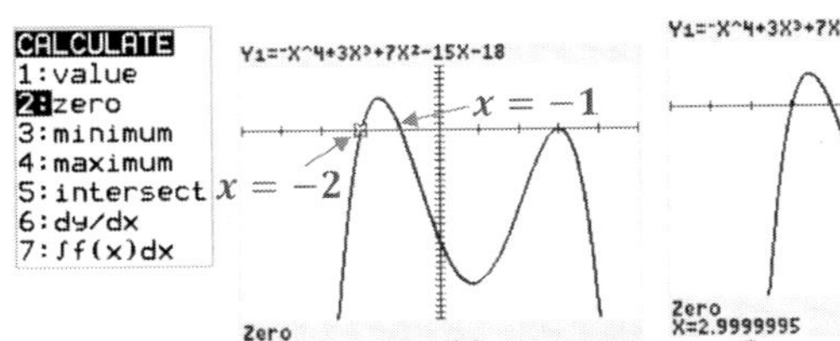

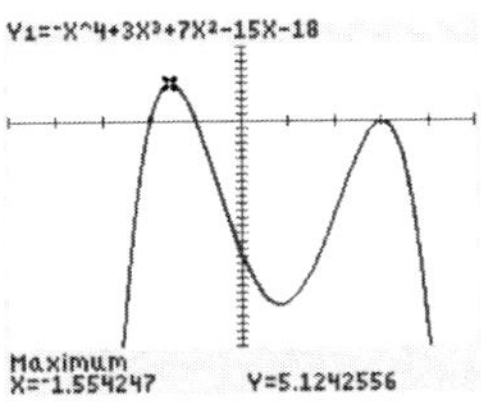

Find the zeros one at a time...

Note: sometimes the calc *adds decimals.* Here, the actual value is just 3.

So, **x-intercepts** are $(-2, 0)$, $(-1, 0)$, and $(3, 0)$

The **range** is: $\{y \leq 5.12, y \in \mathbb{R}\}$

Note that the maximum is provided as an approximate value, to the nearest hundredth.

Class Example 2.12 *Characteristics of Polynomial Functions*

For each of the following polynomial functions, without using your graphing calculator, state:

i - The start and end behavior of the graph ii - The number of possible x-intercepts

iii - Whether or not the graph will have a minimum or maximum point

iv - The domain of the function and the y-intercept

Use your graphing calculator to determine:

v - The x-intercepts of the graph vi - The range of the function

(a) $y = x^3 - 2x^2 - 5x + 6$

i - Start / end

ii - # of x-ints

iii - Max or min?

iv - Domain:

y-intercept:

v - Coords of x-ints:

vi - Range:

(b) $y = x^5 + x^4 - 7x^3 - 13x^2 - 6x$

i - Start / end

ii - # of x-ints

iii - Max or min?

iv - Domain:

y-intercept:

v - Coords of x-ints:

vi Range:

Class Example 2.13 *More Characteristics of Polynomial Functions*

For the function $p(x) = -x^4 - 3x^3 + 7x^2 + 15x - 18$, without using your graphing calculator, state:

i - The start and end behavior of the graph

ii - The number of possible x-intercepts

iii - Whether or not the graph will have a minimum or maximum point

iv - The domain of the function and the y-intercept

Lastly, **sketch** the graph on the grid below.
Label any intercepts and max / min points

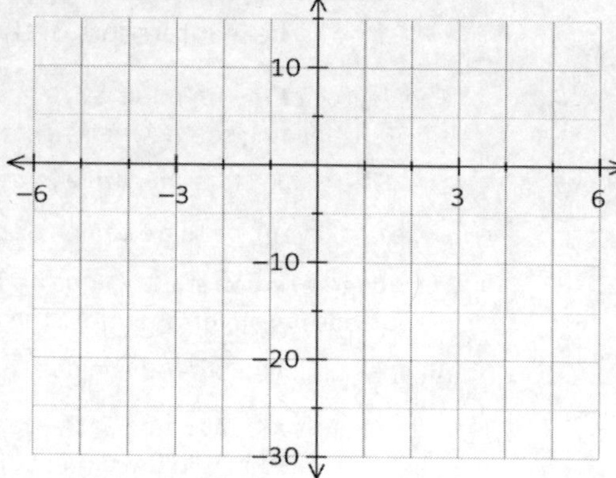

Then, use your graphing calculator to determine:

v - The x-intercepts of the graph

vi - The range of the function *nearest hundredth*

Class Example 2.14 *Identifying Polynomial Functions*

For each of the polynomial functions listed below indicate the graph number that matches.
(Use reasoning – try without using your graphing calculator)

(a) $y = x^4 - 2x^3 - 7x^2 + 8x + 12$ ________

(b) $y = -x^5 + 11x^3 + 6x^2 - 28x - 24$ ________

(c) $y = x^5 - x^4 - 9x^3 + 13x^2 + 8x - 12$ ________

(d) $y = x^4 - 4x^3 - x^2 + 16x - 12$ ________

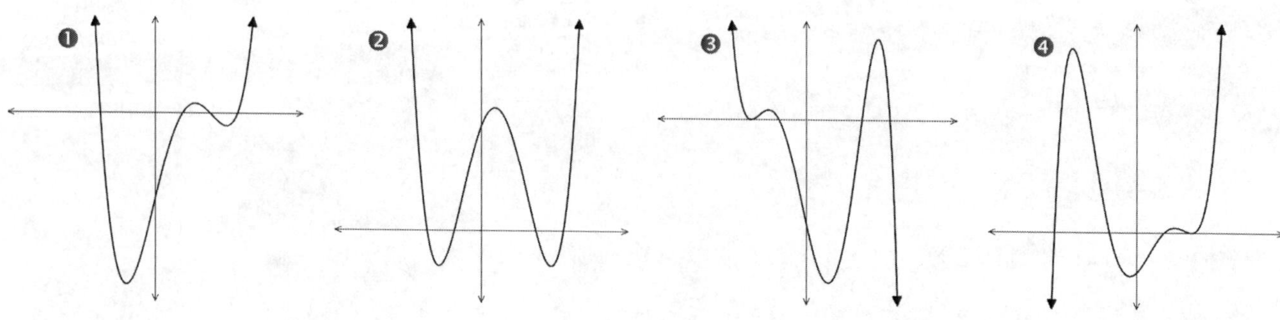

Applications of Polynomial Functions

In Math 20, we saw how a certain type of polynomial function, the quadratic (degree 2) function, has applications in parabolic motion and finance. (To name just two!)

Remember finding the ***maximum height of a ball***? →

Height of Golf Ball
$y = -16x^2 + 100x$

Height (in feet)

Time (in seconds)

Above – classic math 20 question, do you remember how to find the **max height**?

First – match the window to what's given. Graph $y_1 = -16x^2 + 100x$

```
WINDOW
 Xmin=0
 Xmax=7
 Xscl=1
 Ymin=0
 Ymax=200
 Yscl=1
 Xres=1
```

Then – from the CALC menu, select #4, "MAXIMUM"

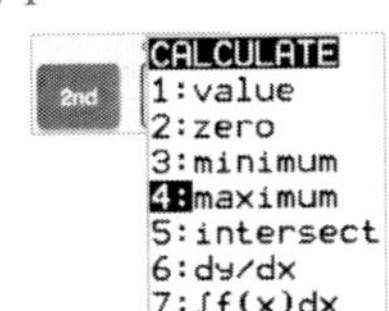

Next – For "left bound", hit enter anywhere to the left of the max. Do the same for "right bound", anywhere to the right.

The MAX should be between these arrows.

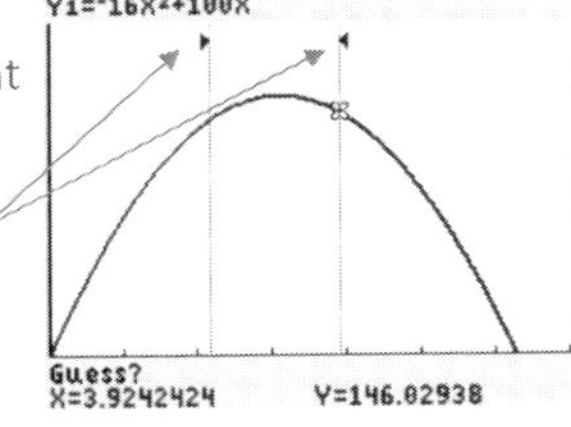

Finaly – Hit enter for "Guess". The max value is the y-**coord.**

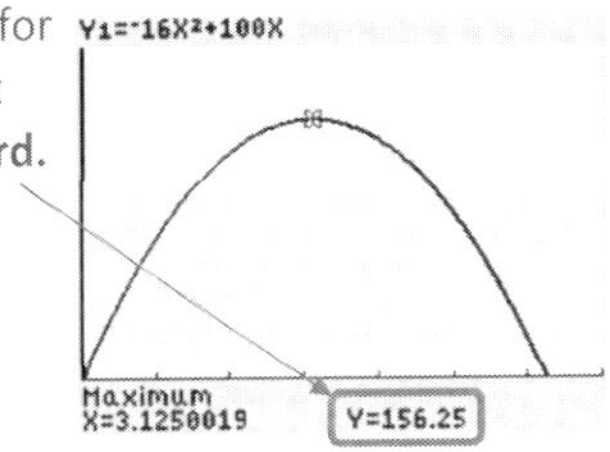

So, the **maximum height** of the ball is **156.24 feet**, after **3.1 seconds.**

Class Example 2.15 *Constructing and Analyzing a Polynomial Equation*

A box is with no lid is made by cutting four squares (each with a side length "x" from each corner of a 24 cm by 12 cm rectangular piece of cardboard.

(a) Determine a function that models the volume of the box.

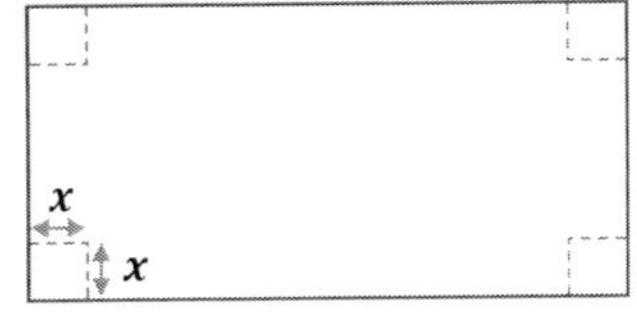

(b) Use technology to graph the function, and sketch below. Label each axis, provide a scale, and indicate any intercepts or max / min points.

*Use your **graphing calculator** to obtain these... you'll need to "trial-and-error" a suitable viewing window, indicate in your sketch below.*

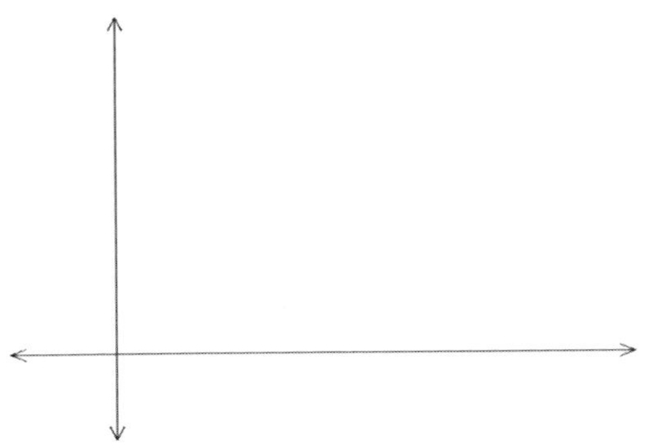

(c) State the domain of the function, with respect to the "real-world" constraints of the problem.

(d) State the value of "x" that gives the maximum volume. (Round to the nearest hundredth)

(e) State the maximum volume of the box, (Round to the nearest cm^3)

Class Example 2.16 *Constructing and Analyzing a Polynomial Equation*

A box with a lid can be created by removing two congruent squares from one end of a rectangular 8.5 inch by 11 inch piece of cardboard. The congruent rectangles removed from the other end as shown. (The shaded rectangles represent the waste, or removed portions that will not be used in the box)

(a) In the diagram below there are two congruent rectangles; one that will form the **base** of the box, and one that will be the **top**. Complete the diagram by providing the missing dimensions (indicated with ➔/🡻) for the base and top.

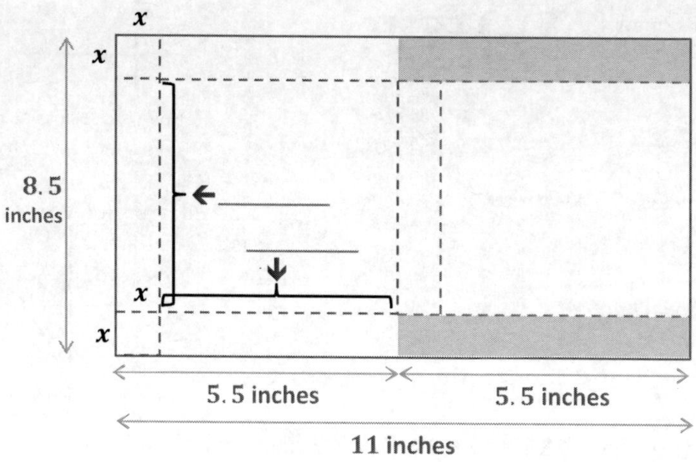

(b) Determine a function that models the volume of the box.

(c) Use technology to graph the function, and sketch on the grid provided.

Label each axis, provide a scale, and indicate any intercepts or max / min points.
Use your ***graphing calculator*** *to obtain these... you'll need to "trial-and-error" a suitable viewing window, indicate on the grid.*

(d) State the domain of the function, with respect to the "real-world" constraints of the problem.

(e) State the value of "x" that gives the maximum volume. (Round to the nearest thousandth)

(f) State the maximum volume of the box, (Round to the nearest thousandth)

(g) State the dimensions that yield the maximum volume. (Round to the nearest thousandth)

2.2 Dividing Polynomials and the Remainder Theorem

Zeros of a Polynomial Functions

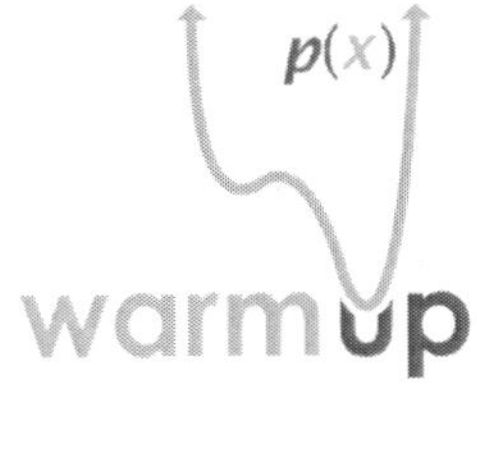

Warm-Up #1 Each of the polynomial functions below is written in both expanded and factored form.

Expanded Form	Factored Form
$f(x) = 6x - 12$	$f(x) = 6(x - 2)$
$g(x) = x^2 - 2x - 8$	$g(x) = (x + 2)(x - 4)$
$p(x) = x^3 - 7x + 6$	$p(x) = (x + 3)(x - 1)(x - 2)$

1 ➡ State the **zeros** (values of x for which the function equals zero) beside the factored form of each function above.

2 ➡ State the relationship between the factored form of a polynomial equation, and the zeros / x-intercepts of its graph.

3 ➡ Convert the following 2nd degree polynomial functions to factored form, to determine the zeros / x-intercepts of the graphs: (a) $g(x) = x^2 - x - 6$ (b) $f(x) = 2x^2 + 7x - 4$

4 ➡ Suppose we wish to factor a 3rd degree function $p(x) = x^3 - 6x^2 + 11x - 6$.
We are told an equivalent; partially factored form is: $p(x) = (x - 1)(x^2 - 5x - 6)$
Determine the fully factored form of $p(x)$,
and state the **zeros** of the function.

The **zero** of a function is the value of x for which the function equals zero.
This is represented graphically by the **x-intercepts**.
So if a function $f(x)$ has zeros of $x = -3$, 1 and 4, then the graph will have x-intercepts at $(-3, 0)$, $(1, 0)$ and $(4, 0)$.

The zeros of a function

The x-intercepts on the graph

The zeros of $f(x)$ are $-3, 1$ and 4

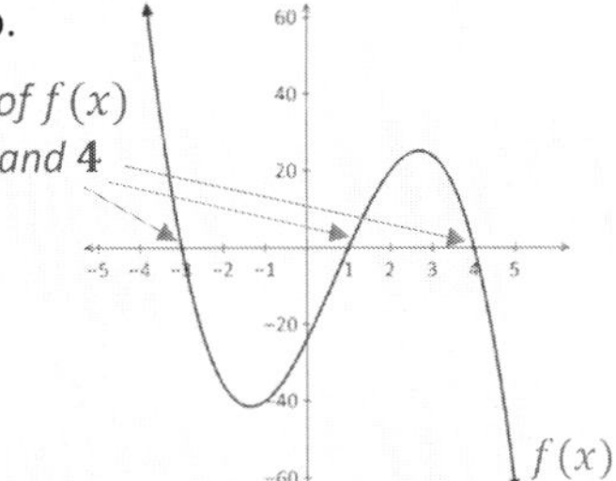

When a polynomial function is in **factored form**, the zeros of the function can be easily identified.

For example, our function here is $f(x) = -2(x + 3)(x - 1)(x - 4)$.

Zeros are: -3 1 4

Bringing it all together.....

A useful tool in determining characteristics of polynomial functions (in this case, zeros) – is **factoring**.

In prior courses we factored a lot of second degree (quadratic) polynomial functions, such as $g(x) = x^2 - x - 6$ and $f(x) = 2x^2 + 7x - 4$

So, what we now need is a method to factor 3rd (or higher) degree polynomials, so that we can algebraically determine the zeros of a function such as $p(x) = x^3 - 6x^2 + 11x - 6$.

And this method involves **dividing polynomials**, which is what we'll look at in this section.

Dividing Polynomials

A method we can use to fully factor a polynomial is to divide a known factor.

For example, if we wish to factor the number 35, we can divide one of its factors, 7. ➡ $35 \div 7 = 5$

So the factors of 35 are 5 and 7

Similarly, if we wish to factor the polynomial $4x^5 + 6x^3 - 10x^2$, we can divide one of its factors, $2x^2$. ➡ $\frac{4x^5 + 6x^3 - 10x^2}{2x^2} = 2x^3 + 3x - 5$

So the factors of $4x^5 + 6x^3 - 10x^2$ are $2x^2$ and $2x^3 + 3x - 5$

And if the first factor is a binomial, we can use **long division**.

Warm-Up #2 Consider the following two long-division calculations:

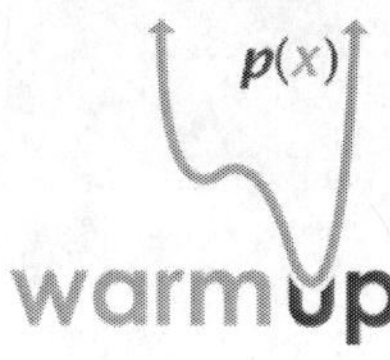

(a) $15 \overline{)348}$, quotient 23

30

48

45

3

(b) $x + 2 \overline{)x^3 - 2x^2 - 9x + 18}$, quotient $x^2 - 4x - 1$

$x^3 + 2x^2$

$-4x^2 - 9x$

$-4x^2 - 8x$

$-x + 18$

$-x - 2$

20

1 ➡ For each calculation above, identify the divisor, dividend, quotient, and remainder

2 ➡ Study the steps for the calculation of (b), shown below.

$x + 2 \overline{)x^3 - 2x^2 - 9x + 18}$, quotient x^2

$x^3 + 2x^2$

$-4x^2$

Step 1 *Divide the first term in the dividend, x^3, by the first term in the divisor, x. Indicate the result as the first term of the quotient.*

Step 2 *Multiply the result from step 1, x^2, through the divisor, $x + 2$.*

Step 3 *Subtract the result from step 2, $x^3 + 2x^2$, from the first two terms in the dividend, $x^3 - 2x^2$.*

Continued...

$x + 2 \overline{)x^3 - 2x^2 - 9x + 18}$, quotient $x^2 - 4x$

$x^3 + 2x^2$

$-4x^2 - 9x$

$-4x^2 - 8x$

$-x$

Step 4 *"Bring down" the next term in the dividend, $-9x$*

Step 5 *Repeat step 1 - divide the first term of the resulting line from step 4, $-4x^2$, by the first term in the divisor. Indicate the result as the second term of the quotient.*

Step 6 *Repeat step 2 ... multiply the result from the previous step, "$-4x$", through the divisor, $x + 2$.*

Steps 7, 8, 9... *Repeat step 3 ... then bring down the last term (18), then repeat steps 1, 2, and 3 again! (Full solution is **above**)*

3 ➡ Follow the steps above to divide: $(x^3 - 2x^2 - 5x + 6) \div (x + 1)$

$x + 1 \overline{)x^3 - 2x^2 - 5x + 6}$

State your result in the form:

$$\text{Quotient} + \frac{\text{Remainder}}{\text{Divisor}}$$

Solution (using an alternate method) is on the next page

Recall on the last page we saw that $348 \div 15$ equaled 23, remainder 3.

This result can be shown two ways: $348 \div 15 = 23 + \dfrac{3}{15}$ (Quotient: 23; Remainder: 3; Divisor: 15) } $Q + \dfrac{R}{D}$ form **or...** $348 = 23 \times 15 + 3$ — $Q \times D + R$ form

And we saw that $(x^3 - 2x^2 - 9x + 18) \div (x + 2)$ came to $x^2 - 4x - 1$, remainder 20.
(Polynomial: $x^3 - 2x^2 - 9x + 18$; Divisor: $x + 2$; Quotient: $x^2 - 4x - 1$; Remainder: 20)

This can similarly be shown two ways:

$$\frac{x^3 - 2x^2 - 9x + 18}{x + 2} = x^2 - 4x - 1 + \frac{20}{x + 2} \quad \Big\} \ \frac{P}{D} = Q + \frac{R}{D} \text{ form}$$

$$x^3 - 2x^2 - 9x + 18 = (x^2 - 4x - 1)(x + 2) + 20 \quad \} \ P = Q \times D + R \text{ form}$$

Synthetic Division

On the previous page we saw a method of long division for polynomials, which is somewhat cumbersome.

Luckily, there's a shorter method that makes use of just the coefficients. Let's look at how synthetic division can be used for the problem we faced on the bottom of the previous page: $(\boldsymbol{x^3 - 2x^2 - 5x + 6}) \div (\boldsymbol{x + 1})$

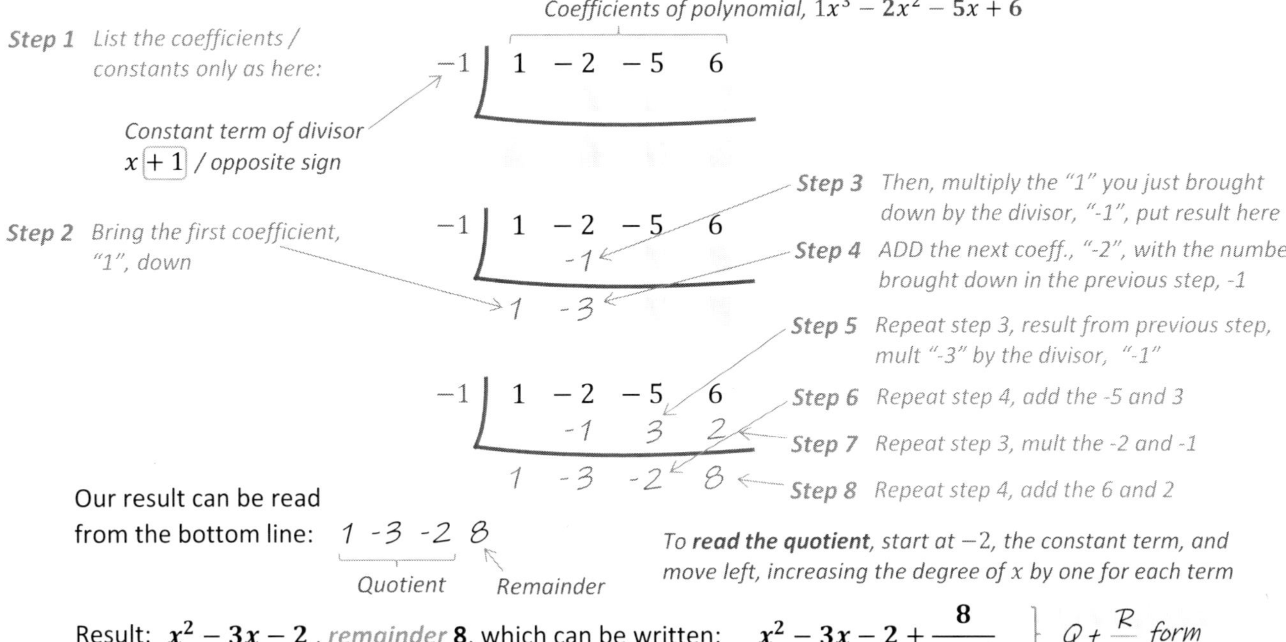

Result: $\boldsymbol{x^2 - 3x - 2}$, *remainder* **8**, which can be written: $\boldsymbol{x^2 - 3x - 2 + \dfrac{8}{x + 1}}$ } $Q + \dfrac{R}{D}$ form

Class Example 2.21 *Dividing Polynomials - Long Division and Synthetic*

Use **both long division** and **synthetic division** to find $(2x^3 - 3x^2 - 5x + 10) \div (x - 3)$

Express your answer in the form

$Q + \dfrac{R}{D}$ and $P = Q \times D + R$.

Class Example 2.22 *Dividing Polynomials*

Use synthetic division to find $(\mathbf{x^3 - 6x + 20}) \div (\mathbf{x + 2})$, expressing your answer in the form $P = Q \times D + R$.

Verify your answer by also using long division.

Hint: As there is no x^2 term, indicate with a "0" in your first step / setup.

EXTENSION: Synthetic Division where the Divisor is in the Form $ax - b$

Suppose we wish to find the result of: $(2x^3 + 3x^2 - 17x + 12) \div (2x + 1)$. *(The divisor has a coefficient $\neq 1$)*

If we're using long division – this is no sweat! There is no change to our procedure. *So that's always an option for this type of problem.* But if we wish to use **Synthetic Division**, we do need to be mindful in our approach.

Worked Example Use synthetic division to find the result for $(2x^3 + 3x^2 - 17x + 12) \div (2x + 1)$. Express your answer in the form $P = Q \times D + R$.

Solution: First note that $2x + 1 = 2\left(x + \frac{1}{2}\right)$ So our first step is to divide by $x + \frac{1}{2}$

$-\frac{1}{2}$	2	3	−17	12
	↓	−1	−1	9
	2	2	−18	21

So we have: $P(x) = (2x^2 + 2x - 18)\left(x + \frac{1}{2}\right) + 21$ *Now, this isn't the nicest form! So we tweak things a bit by first factoring out a "2" from the quotient.*

$= 2(x^2 + x - 9)\left(x + \frac{1}{2}\right) + 21$ *And then multiplying that "2" through the other factor, $x + \frac{1}{2}$*

$= (x^2 + x - 9)\left((2)x + \frac{1}{2}(2)\right) + 21$

$= \mathbf{(x^2 + x - 9)(2x + 1) + 21}$ } $P = Q \times D + R$ form

Class Example 2.23 *Synthetic Division – when the Divisor Zero is a Fraction*

Use synthetic division divide:
$(9x^3 + 18x^2 - x + 2) \div (3x + 1)$.

Express your answer in the form
$P = Q \times D + R$.

Division can be used to determine the factored form of a Polynomial.

Worked Example One of the factors of $P(x) = x^3 - 3x^2 - 4x + 12$ is $x - 2$. Find the remaining factors by dividing $P(x) \div (x - 2)$. Then, state the zeros of the function.

Solution: We want to find $(x^3 - 3x^2 - 4x + 12) \div (x - 2)$ ... use synthetic division.

2	1	−3	−4	12

➡

2	1	−3	−4	12
		2	−2	−12
	1	−1	−6	0

⬅ Quotient is $\boldsymbol{x^2 - x - 6}$, with no remainder

So, $P(x) = (x^2 - x - 6)(x - 2)$ } $P = Q \times D + R$ form *(remainder is 0, which is essential!)*

Factor the quotient to get $\boldsymbol{P(x) = (x + 2)(x - 3)(x - 2)}$

The ZEROS of the function are: $\boldsymbol{-2}$, $\boldsymbol{2}$ and $\boldsymbol{3}$

Class Example 2.24 *Using Synthetic Division to Factor a Polynomial*

One of the factors of $\boldsymbol{P(x) = x^3 + 2x^2 - 5x - 6}$ is given as $\boldsymbol{x + 3}$.

(a) Find the remaining factors by dividing $P(x) \div (x + 3)$.
(Use synthetic division, express in the form $P = Q \times D + R$, then factor the quadratic quotient)

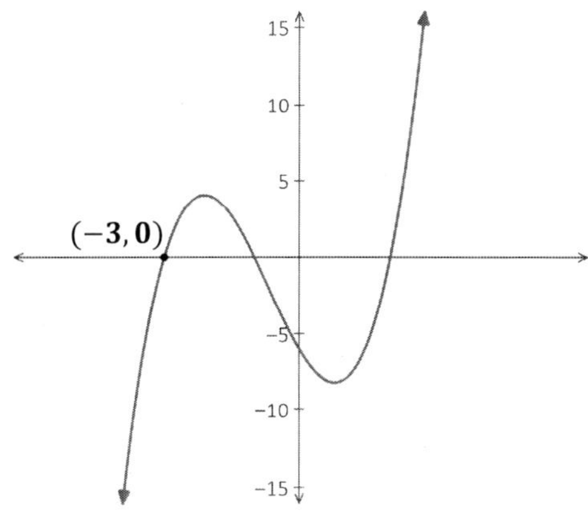

(b) Use your result from (a) to state the zeros of $P(x)$.

(c) Label the coordinates of the x-intercepts on the graph of $P(x)$ above.

The Remainder Theorem

Sometimes it is useful to know the remainder in polynomial division, without having to do the entire division. Lucky for us – we can find just that using the **remainder theorem.**

The **remainder theorem** states that when a polynomial, $P(x)$, is divided by a binomial in the form $x - a$, the remainder is $\boldsymbol{P(a)}$.

Worked Example Use the remainder theorem to determine the remainder in each division:

(a) $\boldsymbol{f(x) = x^4 - x^3 - 7x^2 + x + 6}$ is divided by $x - 2$.

(b) $\boldsymbol{P(x) = 3x^3 + 4x^2 - 5x - 2}$ is divided by $2x - 1$.

Sol.: (a) Remainder is $f(2)$

(evaluate f at the *zero* of the divisor, $x - 2$)

$f(2) = (2)^4 - (2)^3 - 7(2)^2 + (2) + 6$

$= 16 - 8 - 28 + 2 + 6$

$= \mathbf{-12}$ *The remainder is -12*

(b) Zero of the divisor is $\frac{1}{2}$, so find $P(\frac{1}{2})$

$P\left(\frac{1}{2}\right) = 3\left(\frac{1}{2}\right)^3 + 4\left(\frac{1}{2}\right)^2 - 5\left(\frac{1}{2}\right) - 2$

$= \frac{3}{8} + 1 - \frac{5}{2} - 2 \Rightarrow = -\frac{\mathbf{25}}{\mathbf{8}}$ *Remainder is* $-\frac{25}{8}$

Class Example 2.25 *Finding Remainders using the Remainder Theorem*

Use the remainder theorem to determine the remainder for each division:

(a) $\boldsymbol{P(x) = 2x^4 + 5x^3 - 12x^2 + x + 4}$ is divided by $\boldsymbol{x + 1}$.

(b) $\boldsymbol{P(x) = 4x^3 - 15x^2 + 12x + 4}$ is divided by $\boldsymbol{4x + 1}$.

Worked Example When $P(x) = x^3 + 2x^2 + ax - 8$ is divided by $x + 3$, the remainder is -5. Use the remainder theorem to determine the value of a.

Solution: The remainder is given as -5. However the remainder can also be found by evaluating $P(-3)$.

$(-3)^3 + 2(-3)^2 + a(-3) - 8 = -5 \leftarrow$ *... equals what we were given as the remainder*

The remainder, $P(3)$ $\quad -27 + 18 - 3a - 8 = -5 \Rightarrow -3a = 12 \Rightarrow \boldsymbol{a = -4}$

Class Example 2.26 *Applying the Remainder Theorem*

When $\boldsymbol{P(x) = x^3 + ax^2 - 16x + 48}$ is divided by $\boldsymbol{x + 2}$, the remainder is 60. Use the remainder theorem to determine the value of a.

Worked Example

A polynomial function $y = f(x)$, shown on the right, is degree 3, has a leading coefficient of 1, and has integral zeros.

(a) Given the corresponding points on the graph, state the remainder when $y = f(x)$ is divided by:

i $(x + 3)$ ii $(x - 1)$

(b) Determine the factored form equation for $y = f(x)$.

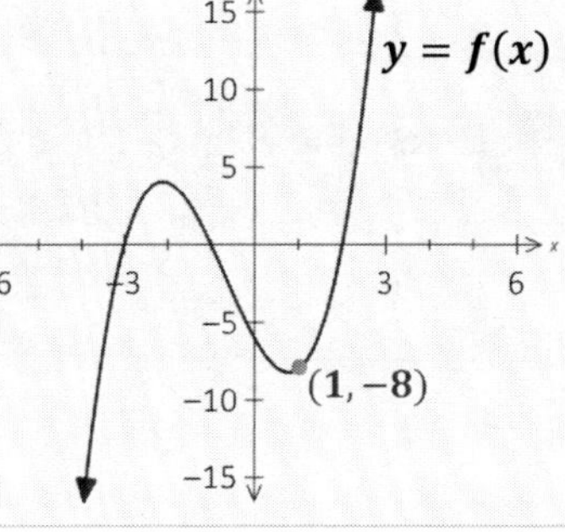

Solution: **(a)** We know that when $f(x)$ is divided by $(x - a)$, the remainder is $f(a)$.

So when $f(x)$ is divided by $(x + 3)$, the remainder is $f(-3)$. *Which is the y-coord on the graph at* $x = -3$.

→ **i** – The remainder when divided by $(x + 3)$ is $\mathbf{0}$ **ii** – The remainder when divided by $(x - 1)$ is $\mathbf{-8}$

(b) Each zero (x-intercept) on the graph of $y = f(x)$ corresponds to a factor in the equation.

So as we are given the leading coefficient is 1, we know the factored form equation is:

$$\boldsymbol{f(x) = (x + 3)(x + 1)(x - 2)}$$ *As there are zeros at* $-3, -1,$ *and* 2

Class Example 2.27 *Relating the Graph with Remainders and Factors*

A polynomial function $y = P(x)$, shown on the right, is degree 3, has a leading coefficient of 1, and has integral zeros.

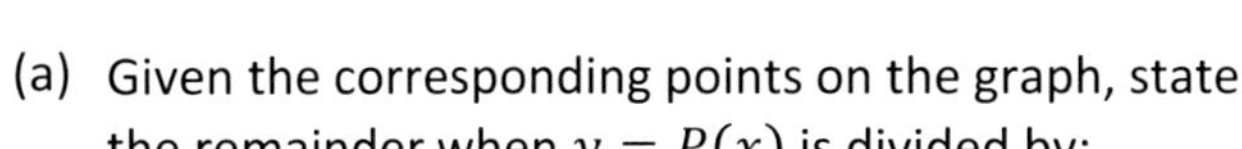

(a) Given the corresponding points on the graph, state the remainder when $y = P(x)$ is divided by:

i $(x + 2)$

ii $(x - 4)$

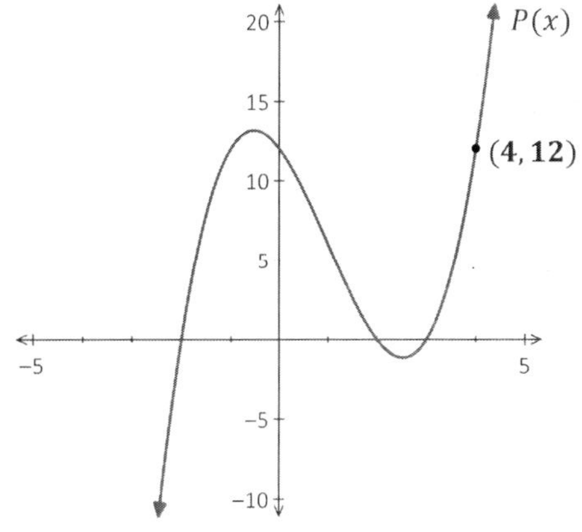

(b) State the factored form equation for $y = P(x)$.

2.2 Practice Questions

1. Determine each quotient, Q, using **long division**. Express in the form $\dfrac{P(x)}{x-a} = Q + \dfrac{R}{D}$.

(a) $(x^3 + 4x^2 - 4x - 16) \div (x - 1)$

(b) $(2m^4 - 3m^2 + 1) \div (m + 2)$

(c) $(2x^3 - 9x^2 - 2x + 24) \div (x - 3)$

(d) $(12x^3 - 5x^2 + x) \div (4x - 3)$

(e) $(25x^2 - 5) \div (5x + 1)$

(f) $(8x^3 + 27) \div (2x + 3)$

2. Determine each quotient, Q, using **synthetic division**. Express in the form $\boldsymbol{P(x) = D \times Q + R}$.

(a) $(x^3 - 2x^2 - 9x + 18) \div (x - 1)$

(b) $(m^5 + 8m^3 + 2m - 15) \div (m + 2)$

(c) $(3x^4 + 5x^3 + x - 2) \div (x + 1)$

(d) $(3x^3 - x - 3) \div (x + 2)$

(e) $(2x^3 + x^2 - 7x - 6) \div (2x - 1)$

(f) $(8x^3 - 27) \div (2x - 3)$

Answers to Practice Questions on the previous page

1. (a) $x^2 + 5x + 1 - \dfrac{15}{x - 1}$ (b) $2m^3 - 4m^2 + 5m - 10 + \dfrac{21}{m + 2}$ (c) $2x^2 - 3x - 11 - \dfrac{9}{x - 3}$

(d) $3x^2 + x + 1 + \dfrac{3}{4x - 3}$ (e) $5x - 1 - \dfrac{4}{5x + 1}$ (f) $4x^2 - 6x + 9$ *no remainder*

3. A stainless steel holding tank is built in the shape of a rectangular prism. The height of the tank is $x + 5$ feet, as shown, and the volume of the tank is $V = 2x^3 + 13x^2 + 13x - 10$.

The area of the base can be found by dividing the volume by the height.

(a) Determine an expression for the area of the base.

(b) By factoring the expression developed in (a), determine the factored-form of the expression representing the volume.

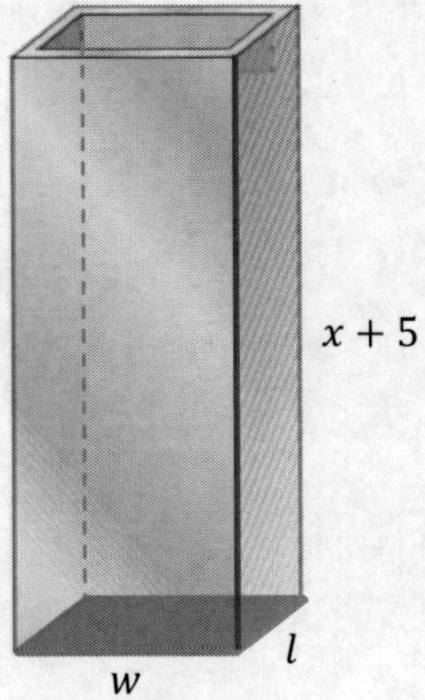

4. A rectangular prism has a volume which can be expressed:

$$\mathbf{V = 9x^3 - 3x^2 - 5x - 1.}$$

Determine the possible measures for w and h, in terms of x, if the length is $3x + 1$ as shown.

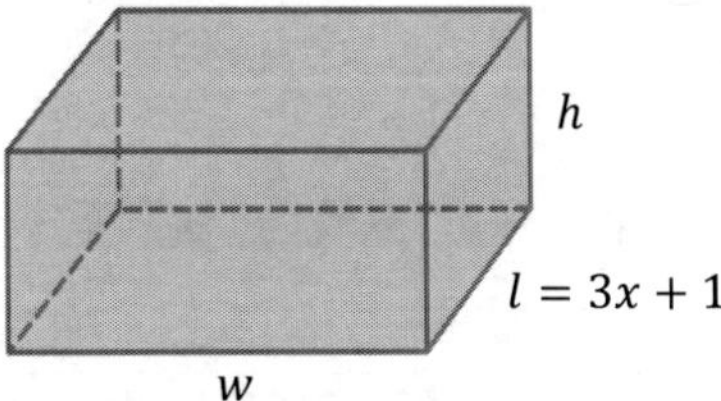

Answers to Practice Questions on the previous page

2. (a) $x^3 - 2x^2 - 9x + 18 = \underbrace{(x^2 - x - 10)}_{\text{Quotient}}\underbrace{(x-1)}_{\text{Divisor}} + \underbrace{8}_{\text{Remainder}}$ (b) $P(x) = \underbrace{(m^4 - 2m^3 + 12m^2 - 24m + 50)}_{\text{Quotient}}\underbrace{(m+2)}_{\text{Divisor}} - \underbrace{115}_{\text{R}}$

(c) $3x^4 + 5x^3 + x - 2 = (3x^3 + 2x^2 - 2x + 3)(x + 1) - 5$ (d) $3x^3 - x - 3 = (3x^2 - 6x + 11)(x + 2) - 25$

(e) $2x^3 + x^2 - 7x - 6 = (x^2 + x - 3)(2x - 1) - 9$ (f) $8x^3 - 27 = (4x^2 + 6x + 9)(2x - 3)$

5. Use the remainder theorem to find the remainder for each division below:

(a) $(x^3 - 3x + 4) \div (x - 2)$

(b) $(x^4 - 2x^3 - 7x^2 + 8x + 12) \div (x + 3)$

(c) $(2x^4 + x^3 - 13x^2 - 21x - 9) \div (x + 2)$

(d) $(8x^3 + 125) \div (2x + 5)$

(e) $(2x^3 - x^2 - 2x + 1) \div (2x + 1)$

(f) $(9x^3 - 3x^2 + 6x + 1) \div (3x - 1)$

6. When $P(x) = x^3 - x^2 + ax - 8$ is divided by $x + 1$, the remainder is -16. Use the remainder theorem to determine the value of a.

Answers to Practice Questions on the previous page

3. (a) base $= 2x^2 + 3x - 2$ (b) $V = (\boldsymbol{x} + \mathbf{5})(\mathbf{2}\boldsymbol{x} - \mathbf{1})(\boldsymbol{x} + \mathbf{2})$ **4.** $V = (x - 1)(3x + 1)^2$ $\boldsymbol{w} = \mathbf{3}\boldsymbol{x} + \mathbf{1}, \boldsymbol{h} = \boldsymbol{x} - \mathbf{1}$

7. When $P(x) = 16x^3 + kx^2 - x + 1$ is divided by $4x - 1$, the remainder is 0. Use the remainder theorem to determine the value of k.

8. When $P(x) = x^3 + ax^2 + bx - 6$ is divided by $x - 1$, the remainder is -2. When $P(x)$ is divided by $x + 3$ the remainder is -66. Determine the values of a and b. Need a hint? *See the bottom of the next page.*

9. When $\boldsymbol{P(x) = x^4 + ax^2 - 4x + b}$ is divided by $x + 2$, the remainder is 34. When $P(x)$ is divided by $x - 1$ the remainder is -2. Determine the values of a and b.

Answers to Practice Questions on the previous page

5. (a) **6** (b) **60** (c) **5** (d) **0** (e) **3/2** (f) **3**

6. $a = 6$

10. One of the factors of a function $\boldsymbol{P(x) = x^3 + 5x^2 - 2x - 24}$ is $x + 4$.

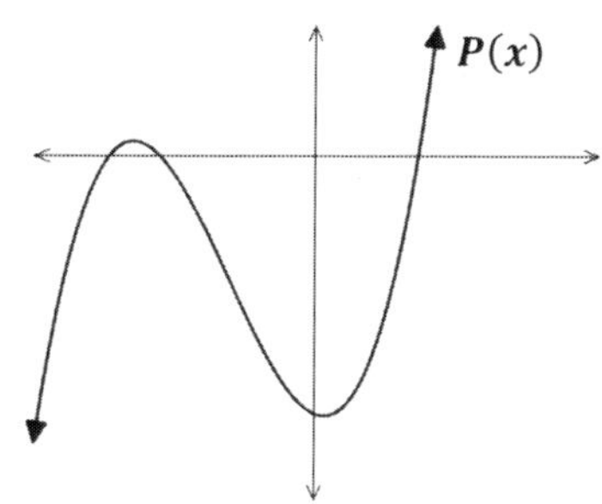

(a) Use synthetic division to determine the quotient when $P(x)$ is divided by $x + 4$.

(b) Factor the quadratic quotient obtained in (a), to state the fully factored form of $P(x)$.

(c) Each factor relates to an x-intercept. Label the coordinates of the three x-intercepts on the graph of $P(x)$ on the right.

11. A polynomial function $y = f(x)$, shown on the right, is degree 3, has a leading coefficient of 1, and has integral zeros.

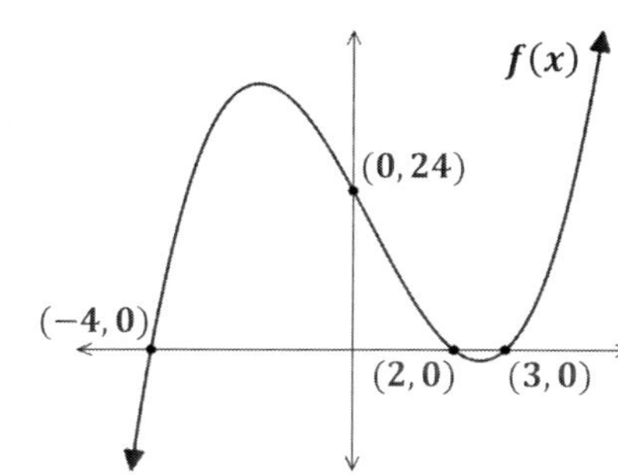

(a) Given the corresponding points on the graph, state the remainder when $y = f(x)$ is divided by:

i x ii $(x + 4)$

(b) State the factored form equation of $y = f(x)$.

Answers to Practice Questions on the previous page

7. $k = -16$

8. **HINT:** The solution involves setting up and solving a system of two equations.

① $(1)^3 + a(1)^2 + b(1) - 6 = -2$
$a + b = 3$

② $(-3)^3 + a(-3)^2 + b(-3) - 6 = -66$
$9a - 3b = -33$

Solve by substitution: ① $a = 3 - b$ → substitute "$3 - b$" for "a" in equation ②: $9(\boldsymbol{3 - b}) - 3b = -33$

Solve ① for "a"

$27 - 12b = -33$

$\boxed{b = 5}$

Now substitute into ① to find "a"

① $a = 3 - (5)$

$\boxed{a = -2}$

9. $a = 3, b = -2$

12. Exam Style **NR** When $p(x) = x^3 + 4x^2 + x - 6$ is divided by $x - 2$ the remainder is _____.

13. When the function $p(x) = x^4 + 2x^3 - 7x^2 - 8x + 12$ is divided by $x + 2$, the quotient is:

Exam Style

A. $x^3 - 7x + 6$

B. $x^3 + 4x^2 + x - 6$

C. $x^3 + 3x^2 - 4x - 12$

D. $x^3 - x^2 - 4x + 4$

14. When the function $p(x) = x^3 + 4x^2 - 3x - 18$ is divided by $2x - 1$, the remainder is:

Exam Style

A. $-\frac{147}{8}$

B. -16

C. $-\frac{125}{8}$

D. -12

Answers to Practice Questions on the previous page and this page

10. (a) Quotient: $\mathbf{x^2 + x - 6}$ *(remainder 0)* (b) $\mathbf{P(x) = (x+4)(x+3)(x-2)}$
(c) *Label coords:* $\mathbf{(-4, 0), (-3, 0)}$*, and* $\mathbf{(2, 0)}$

11. (a) i The y-coord at $x = 0$ is **24** → Remainder is **24**
ii The y-coord at $x = -4$ is 0 → Remainder is **0**
(b) $\mathbf{P(x) = (x+4)(x-2)(x-3)}$
Zero at $x = -4$ ...and $x = 2$ ...and $x = 3$

12. 20 **13.** A **14.** A

2.3 The Factor Theorem

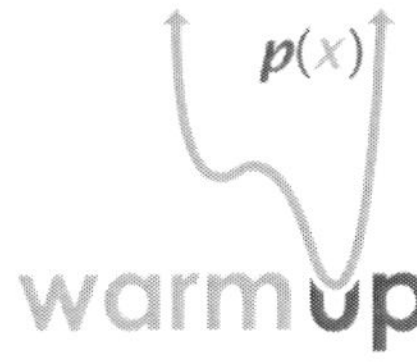

The graph of $P(x) = x^3 - 2x^2 - 11x + 12$ is shown on the right. The graph has integer x-intercepts.

The factored form of $P(x)$ is
$P(x) = (x - a_1)(x - a_2)(x - a_3)$;
where a_1, a_2, and a_3 are zeros of the function.

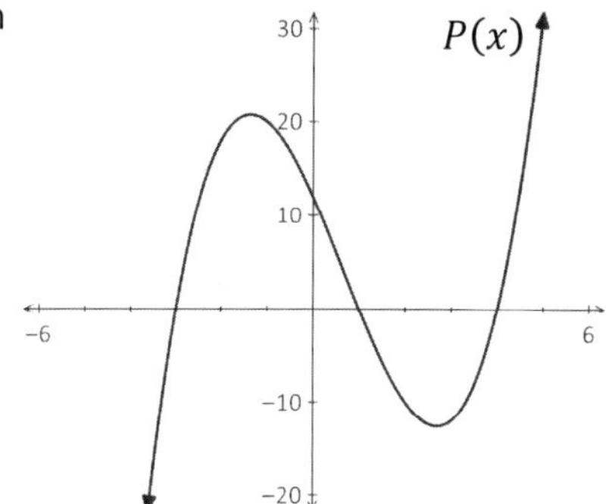

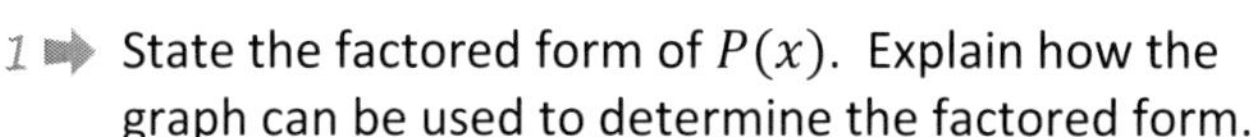

1 ➡ State the factored form of $P(x)$. Explain how the graph can be used to determine the factored form.

2 ➡ Find 3 remainders; for when $P(x)$ is divided by each of its three factors as stated above.

3 ➡ State the relationship between the factors of a polynomial expression, the zeros of the corresponding polynomial function, and the remainder theorem.

Recall that the **remainder theorem** states that when a polynomial, $P(x)$, is divided by a binomial in the form $x - a$, the remainder is $\boldsymbol{P(a)}$.

The **factor theorem** states that $x - a$ is a factor of a polynomial function $P(x)$, if $\boldsymbol{P(a) = 0}$. (That is, if dividing by a factor gives no remainder)

Worked Example Use the factor theorem to show that the polynomial function $P(x) = x^3 + 5x^2 + 3x - 9$ has factors of $(x + 3)$ and $(x - 1)$.

Solution: The factor theorem states that if $(x - a)$ is a factor of $P(x)$, then $P(a) = 0$. *(That is, there is no remainder)*

- Test $(x + 3)$ ... *is it a factor?*

  $(x + 3)$ is a factor if $P(-3) = 0$

  $P(-3) = (-3)^3 + 5(-3)^2 + 3(-3) - 9$

  $= -27 + 45 - 9 - 9$

  $= \mathbf{0}$ ⬅ *No remainder ... $(x + 3)$ **is a factor***

- Test $(x - 1)$ ... *is it a factor?*

  $(x - 1)$ is a factor if $P(1) = 0$

  $P(1) = (1)^3 + 5(1)^2 + 3(1) - 9$

  $= 1 + 5 + 3 - 9$

  $= \mathbf{0}$ ⬅ *No remainder ... $(x - 1)$ **is a factor***

Note: *Here's the graph of $P(x)$. The x-intercepts are $(-3, 0)$ and $(1, 0)$.*

15
10
5
−5
−10
−15
−6
6

Class Example 2.31 *Using the Factor Theorem to test given Binomials*

A polynomial function is given as $\boldsymbol{P(x) = 3x^4 - 4x^3 - 19x^2 + 8x + 12}$. Use the factor theorem to show that:

(a) $(x - 1)$ is a factor of $P(x)$ (b) $(3x + 2)$ is a factor of $P(x)$ (c) $(x + 1)$ is a NOT a factor

Class Example 2.32 *Applying the Factor Theorem*

The binomial $\boldsymbol{x+2}$ is a factor of $\boldsymbol{P(x)=3x^3+kx^2-2x-8}$. Determine the value of k.

The Integral Zero Theorem

Let's take stock of what we've seen so far:

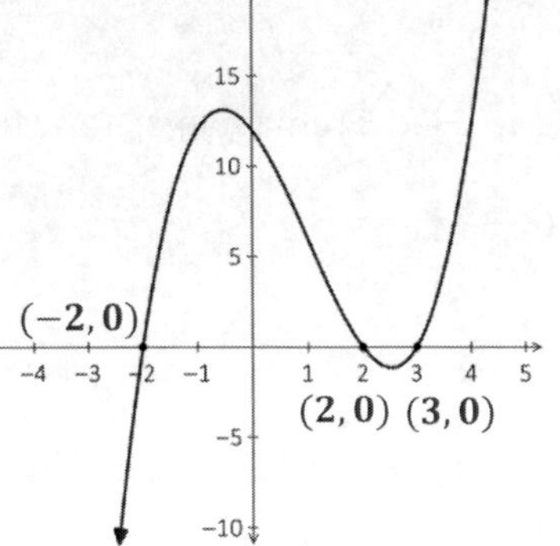

- When a polynomial is in factored form, it's zeros (that is, the x-intercepts of its graph) are easily discernible.
  - *For example*, the function $P(x)=x^3-3x^2-4x+12$ is shown ➔
  - *Written in factored form*, it's: $P(x)=(x+2)(x-2)(x-3)$

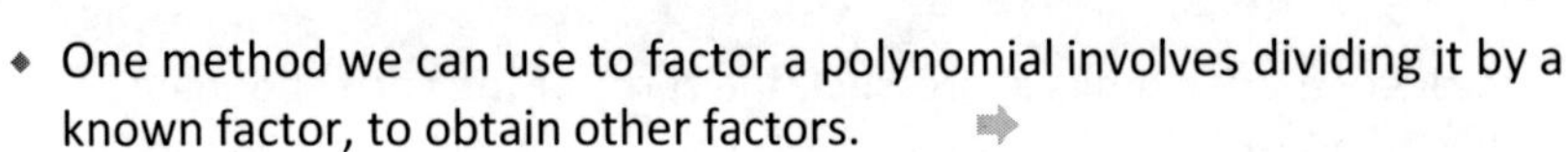

- One method we can use to factor a polynomial involves dividing it by a known factor, to obtain other factors. ➡
  - *For example*, suppose we know that the function $P(x)=x^3-7x-6$ has a zero at $x=3$. That means one of the factors is $(x-3)$...

    And the other factor(s) can be found by finding $(x^3-7x-6)\div(x-3)$

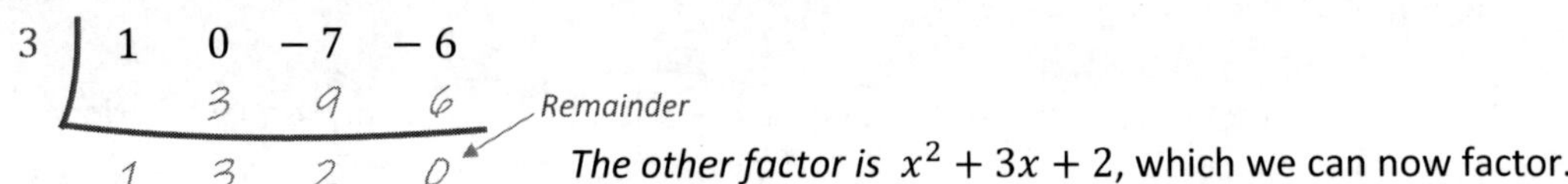

The other factor is x^2+3x+2, which we can now factor.

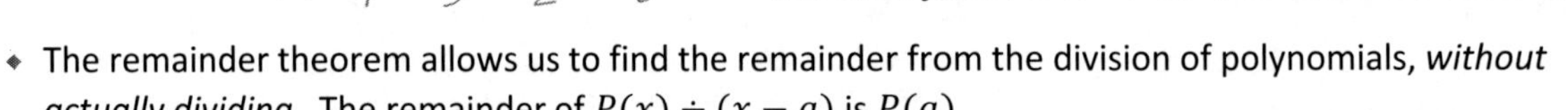

- The remainder theorem allows us to find the remainder from the division of polynomials, *without actually dividing*. The remainder of $P(x)\div(x-a)$ is $P(a)$.
- The factor theorem states that $(x-a)$ is a factor of $P(x)$ if $P(a)=0$. (That is, if dividing gives no remainder, as we saw in the synthetic division above)

  *Consider what happens when a **number** is divided by one of its factors.*

  For example, the factors of 35 are 7 and 5. ➔ $35\div 7$ is 5, with no remainder.
  ➔ $35\div 5$ is 7, with no remainder.

The question now is – *how do we find that first factor?*

The **integral zero theorem** identifies the relationship between the factors of a polynomial and the constant term of the polynomial.
It states that if $(x-a)$ is a factor of $P(x)$, then a is a factor of the constant term of $P(x)$.

For example, consider the polynomial x^3-7x-6.

The integral zero theorem asserts that $(x+3)$ ***could be*** a factor.
Spoiler alert – it isn't – but $(x-3)$ is!

But we know that $(x+4)$ ***will not be*** a factor, as "4" is not a factor of the constant term of the polynomial, "6"

Think – if $(x+4)$ was a factor... $(x+4)(x\ \square)(x\ \square)$
... there's no way these would multiply to 6

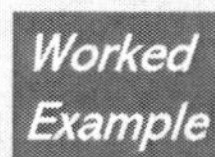

Worked Example

Given the polynomial function $P(x) = 2x^3 - 5x^2 - 11x - 4$,

(a) Use an algebraic process to fully factor $P(x)$. Verify graphically.

(b) State the zeros of $P(x)$.

(a)  **Step 1** Potential zeros: $\pm 4, \pm 2, \pm 1$

By Integral Zero Theorem, any zero must be a factor of 4. (There are six potential zeros, the positive and negative of the factors of 4)

Step 2 Test $(x-1)$: $P(1) = 2(1)^3 - 5(1)^2 - 11(1) - 4 \rightarrow = -18$ Not 0.... so, $(x-1)$ is <u>not</u> a factor!

Test $(x+1)$: $P(-1) = 2(-1)^3 - 5(-1)^2 - 11(-1) - 4 \rightarrow = \mathbf{0}$ So $(x+1)$ <u>IS</u> a factor!

Find the first factor through "guess and test". Use factor theorem to test each potential zero. Remember: If $P(a) = 0$ (that is, if division gives no remainder), then $x - a$ is a factor.

Step 3 Divide $(2x^3 + 7x^2 - 5x - 4) \div (x+1)$:

-1	2	-5	-11	-4
	$\downarrow$	-2	7	4
	2	-7	-4	$0 \Leftarrow R$

quotient

Result: $2x^2 - 7x - 4$

Divide $P(x)$ by the found factor from step 2

The remaining factors are given by the quotient

Step 4 $P(x) = (x+1)(2x^2 - 7x - 4)$ *Express your division result in the form $P(x) = D \times Q + R$.*

Step 5 $P(x) = (x+1)(2x+1)(x-4)$ *Factor the quadratic quotient – **and we're done!***

(b) Zeros of $P(x)$ are $x = -1, -1/2$, and 4

*Each factor corresponds to a **zero**:*

$P(x) = (x+1)(2x+1)(x-4)$

$x + 1 = 0$ $\quad 2x + 1 = 0$ $\quad x - 4 = 0$

Class Example 2.33 *Applying Factor and Integral Zero Theorems to Fully Factor*

For the polynomial function $P(x) = x^3 - x^2 - 10x - 8$,

(a) Use the steps outlined above to fully factor $P(x)$.

(b) State the zeroes of $P(x)$.

(c) The graph of $P(x)$ is below. Use the results above to label the coordinates of the x-intercepts.

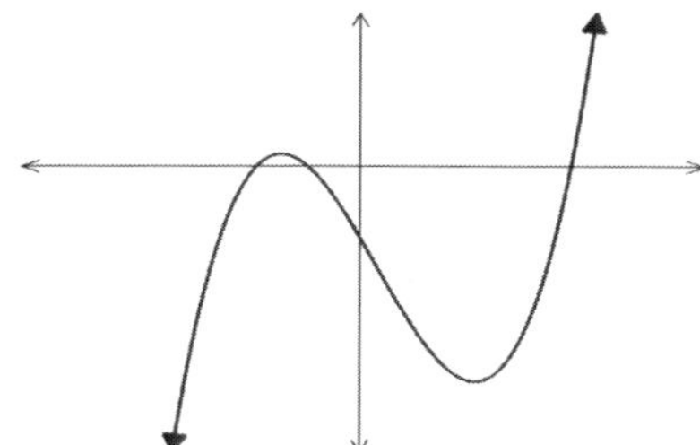

(d) State the roots of the equation of $x^3 - x^2 - 10x - 8 = 0$.

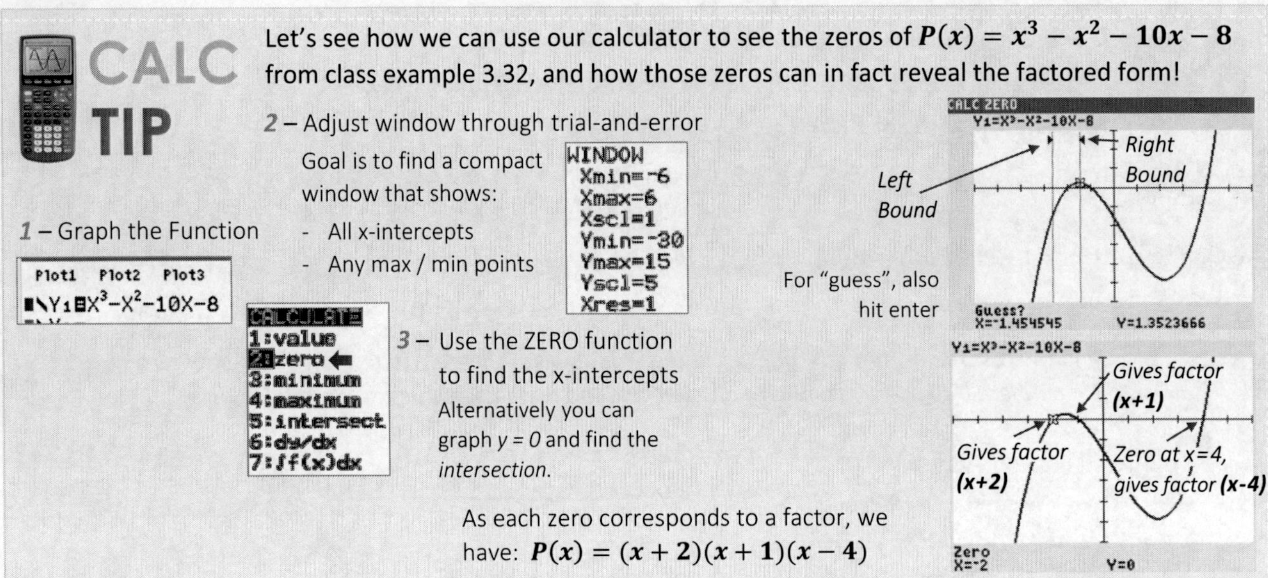

As you may have noticed, your graphing calculator is like an answer key for these sort of factoring questions! *(Given the graph of a polynomial function, each zero corresponds to a factor, so remember to check your work!)*

Class Example 2.34 *Factoring Polynomials with Irrational Zeros*

For the polynomial function $f(x) = 2x^3 - 13x + 10$:

(a) Use the steps outlined on the previous page to fully factor $P(x)$.

(b) State the **exact** zeroes of $P(x)$.
Hint: quadradic formula will be required!

(c) State the **exact** roots of the equation $2x^3 - 13x + 10 = 0$.

Class Example 2.35 *Factoring Degree 4 (or higher) Polynomials*

For the polynomial function $\boldsymbol{p(x) = 3x^4 - 2x^3 - 21x^2 + 32x - 12}$:

(a) Use the steps previously outlined to fully factor $P(x)$.

Hint: As its degree 4, steps 1 through 3 must be performed twice!

(b) State the zeroes of $P(x)$.

Class Example 2.36 *Problem Solving with the Factor Theorem*

For the polynomial function $\boldsymbol{P(x) = x^5 - 3x^4 - 6x^3 + 10x^2 + kx}$, one of the zeros is -1.

It can be algebraically determined that the largest zero can be written in the form $a + \sqrt{b}$. Determine the values of a and b, respectively.

2.3 Practice Questions (Sample, first page)

Access them all at math30-1edge.com

1. State a possible factored form equation for each function described below, where each is a degree 3 polynomial function with all zeros listed.

(a) $P(-2) = 0, P(1) = 0$, and $P(5) = 0$

(b) $p(-2) = 0, p(-3/2) = 0$, and $p(1) = 0$

(c) $g(-4) = 0, g(5/2) = 0$, and $g(0) = 0$

2. Use the factor theorem to determine whether $x - 1$ is a factor of each of the following polynomial functions:

(a) $y = x^3 + 9x^2 + 15x - 25$

(b) $y = x^4 + x^3 - 8x - 8$

(c) $y = 3x^3 - 2x^2 - 5x + 4$

(d) $y = 3x^4 - 5x^3 - 12x^2 + 12x + 16$

(e) $y = x^3 - 3x^2 + 3x - 1$

3. Use the factor theorem to determine whether $x + 3$ is a factor of each of the following polynomial functions:

(a) $P(x) = x^3 - x^2 - 6x$

Answers to Practice Questions visit math30-1edge.com for many, many more practice questions, and step-by-step solutions!!

1. (a) $\mathbf{P(x) = (x + 2)(x - 1)(x - 5)}$ (b) $\mathbf{p(x) = (2x + 3)(x + 2)(x - 1)}$ (c) $\mathbf{g(x) = x(2x - 5)(x + 4)}$

2. (a) $x - 1$ is a factor if $(x^3+9x^2 + 15x - 25) \div (x - 1)$ gives no remainder.
Evaluate $(1)^3+9(1)^2 + 15(1) - 25$ .... which is 0. As such, we have shown that **YES**, $x - 1$ is a factor!
Use a similar process for the (b), (c), (d), and (e), substitute "1" to see if you get no remainder (that is, ZERO).
(b) *No* (c) Yes (d) *No* (e) Yes

3. (a) Evaluate $P(-3)$ to get $(-3)^3 - (-3)^2 - 6(-3)$, which is -18 (NOT "0"), so $x - 3$ **is NOT a factor**.

2.4 Further Analysis of Polynomial Function Graphs

Warm-Up #1 Labeled A to E below are five separate functions, provided in both expanded equivalent and factored form.

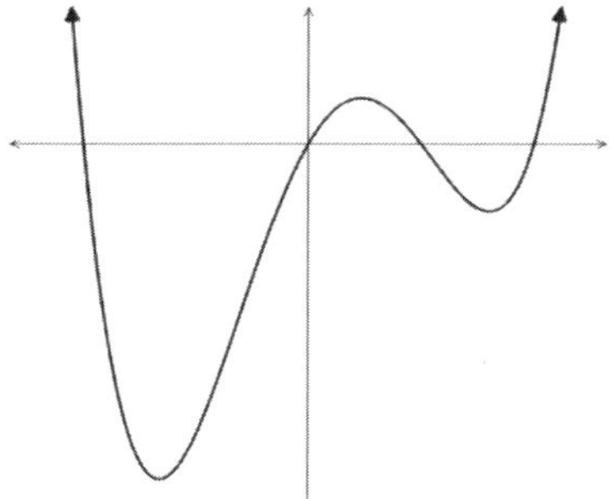

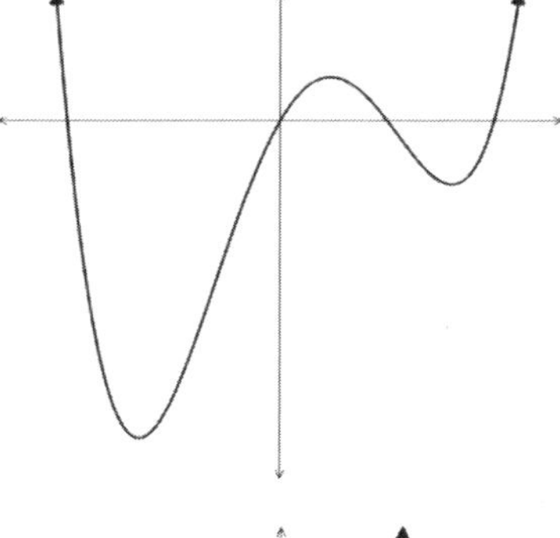

1 ➡ For each function below, *without using your graphing calculator,* match with the appropriate graph on the right, and determine each of the indicated graph characteristics.

A
Expanded form: $y = -2x^3 - 6x^2 + 12x + 16$
Factored form: $y = -2(x+1)(x-2)(x+4)$

Degree: _____ y-intercept: _______ Max / min?: _______

x-intercepts: ______________________

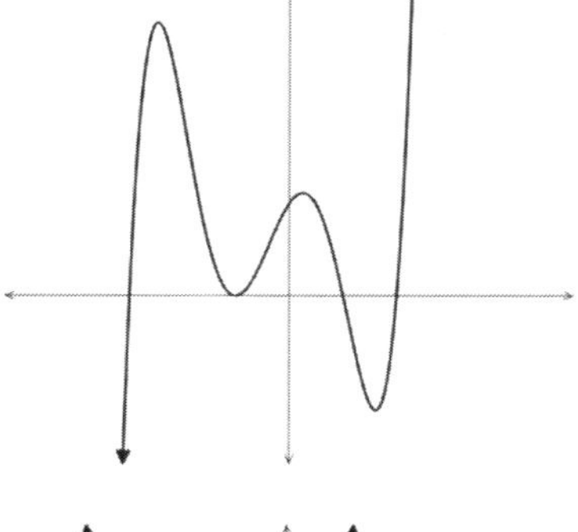

B
Expanded form: $y = x^4 + 6x^3 + 9x^2 - 4x - 12$
Factored form: $y = (x+2)^2(x-1)(x+3)$

Degree: _____ y-intercept: _______ Max / min?: _______

x-intercepts: ______________________

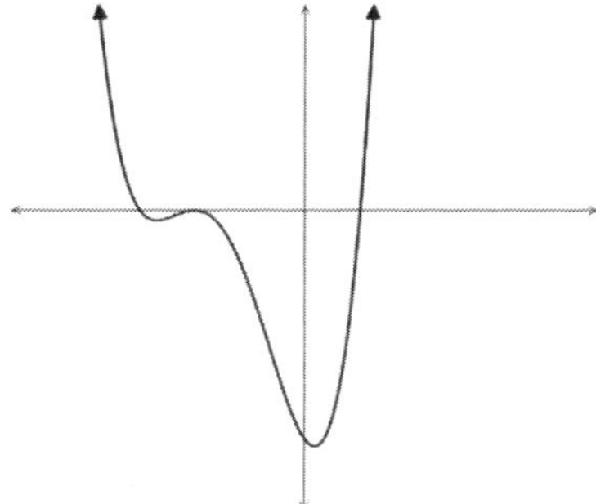

C
Expanded form: $y = \frac{1}{2}x^4 - x^3 - 8x^2 + 16x$
Factored form: $y = \frac{1}{2}x(x+4)(x-2)(x-4)$

Degree: _____ y-intercept: _______ Max / min?: _______

x-intercepts: ______________________

D
Expanded form: $y = x^5 - 8x^3 + 6x^2 + 7x - 6$
Factored form: $y = (x+3)\,(x+1)(x-1)^2(x-2)$

Degree: _____ y-intercept: _______ Max / min?: _______

x-intercepts: ______________________

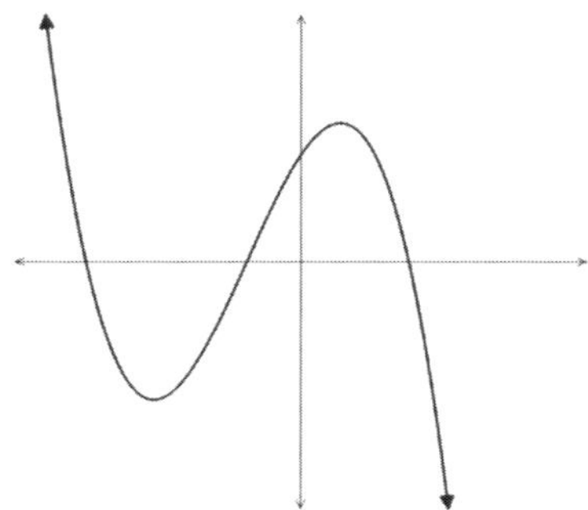

E
Expanded form: $y = x^5 + 2x^4 - 6x^3 - 8x^2 + 5x + 6$
Factored form: $y = (x+3)\,(x+1)^2(x-1)(x-2)$

Degree: _____ y-intercept: _______ Max / min?: _______

x-intercepts: ______________________

2 ➡ Functions B, D, and E each have a degree 2 factor, such as such as $(x+2)^2$ for function B.
Analyze and explain the effect on the graph, when a factor is squared.

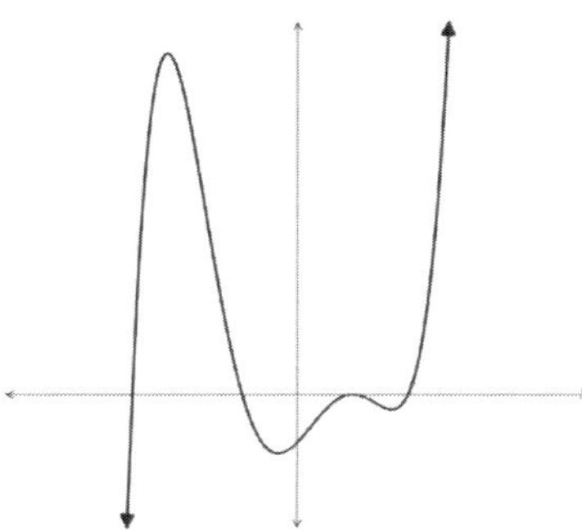

Warm-Up #2 Below are the graphs of the basic linear, quadratic, cubic, quartic, and quintic functions.

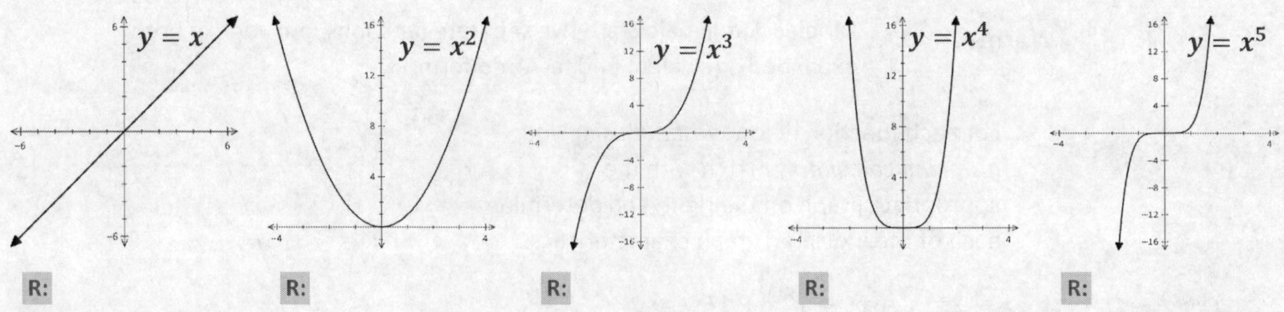

R: **R:** **R:** **R:** **R:**

1 ➡ State the domain for each function and indicate the range (" **R:**") for each above.
Predict the range of the functions $y = x^6$ and $y = x^7$.

2 ➡ Describe the effect on the graph at the x-intercept of $y = x^n$; when the degree n is even.

The graph of $y = x^n$ where $n \in N$ follows a pattern based on whether n is ODD or EVEN.

If n is **even**, the graph will have a min point at $(0, 0)$ and therefore the range is $[0, \infty)$.

The graph will "bounce" at the x-intercept

At $n = 4$, 6, etc... the bounce becomes more U-shaped

If n is **odd**, the graph has no max or min point and therefore the range is $(-\infty, \infty)$.
And the graph crosses the x-axis at the x-intercept. If n is 3 or greater, we get this shape: ➔

Point of inflection

Enrichment: A **point of inflection** occurs where a curve changes from *concave down* to *concave up* (or vice-versa).
Note that this terminology, however interesting, is not in the curriculum!

Concave down

Concave up

For example, the curves of $y = x^3$ and $y = x^5$ are *concave down* on the interval $x < 0$ and *concave up* on $x > 0$.
There is a **point of inflection** at $x = 0$. Also, note how at $n = 5$ (and 7, 9, etc) the graph is *flatter* at the inflection pt.

Warm-Up #3 The functions below all have the form $y = (x + 2)(x - 3)^n$, when $n = 2, 3, 4$, and 5.

$y = (x + 2)(x - 3)^2$

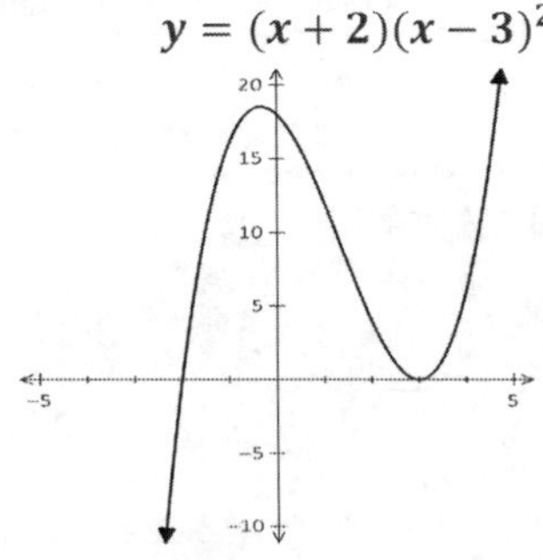

$y = (x + 2)(x - 3)^3$

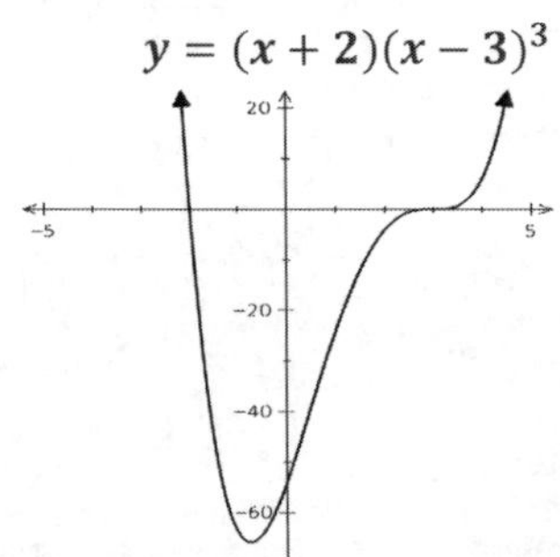

$y = (x + 2)(x - 3)^4$

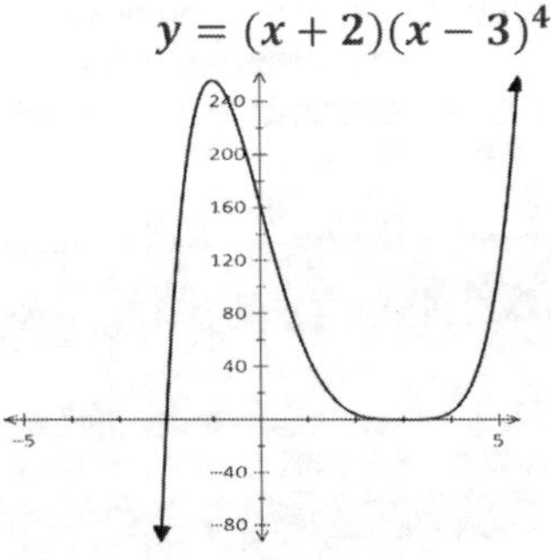

$y = (x + 2)(x - 3)^5$

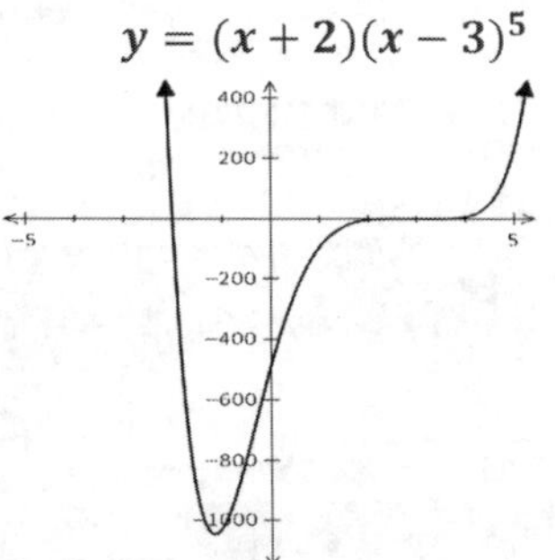

1 ➡ Describe the shape of the graph of $y = (x + 2)(x - 3)^n$ around the zero $x = 3$, when:

(a) $n = 2$ or 4

(b) $n = 3$ or 5

2 ➡ Sketch the graphs of $y = (x + 2)(x - 3)^6$ and $y = (x + 2)(x - 3)^7$ in your calculator. (You'll have to make your y *min* and y *max* very large!) Predict the effect on the graph at a zero associated with:

An even degree factor ________________ An odd degree $(n \geq 3)$ factor ________________

Effect of Factors on the Graph of a Polynomial Function

As we've previously seen, each x-intercept (**zero**) on the graph of a polynomial function relates to a factor in the equation.

For example, consider the function $P(x)$ on the right, which has an equation:

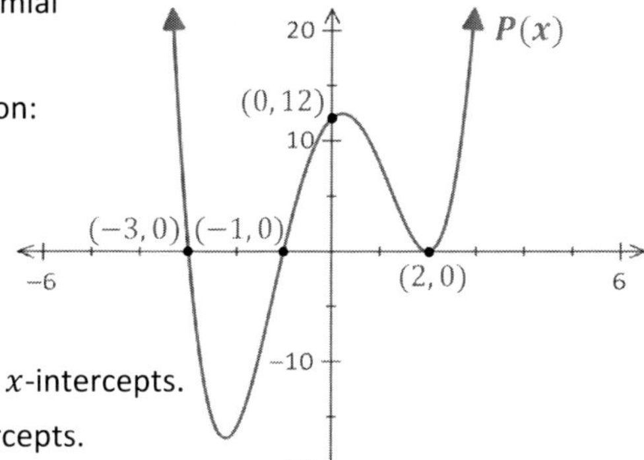

$$P(x) = (x+3)(x+1)(x-2)^2$$

*The **zeros** of $P(x)$ are:* $-\mathbf{3}$, $-\mathbf{1}$, *and* $\mathbf{2}$

Notice that:

- $P(x)$ is degree 4, meaning its graph could theoretically have at most **4** x-intercepts.
  However as one of the factors is degree 2, the graph has only 3 x-intercepts.

$$P(x) = (x+3)(x+1)(x-2)^2$$

Graph "bounces" at the x-intercept provided by this factor

- The leading coefficient of $P(x)$ is 1 (positive), meaning the graph ends positive in quadrant I.

  If we changed it to say -2, giving $f(x) = -2(x+3)(x+1)(x-2)^2$

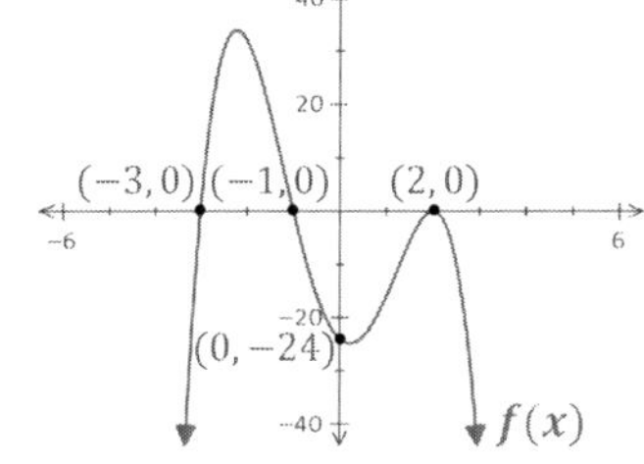

  ...the graph would end negative in quadrant IV, but the x-intercepts would remain the same. (they are invariant to this vertical stretch & reflection)

  Notice that the y-intercept (as with all y-coordinates) is multiplied by -2.

- If we changed the degree of the last factor from 2 to 3, the degree of the function would now be odd. (That is, 5 instead of 4)

$$g(x) = (x+3)(x+1)(x-2)^3$$

Graph has a "point of inflection" at the x-intercept provided by this factor

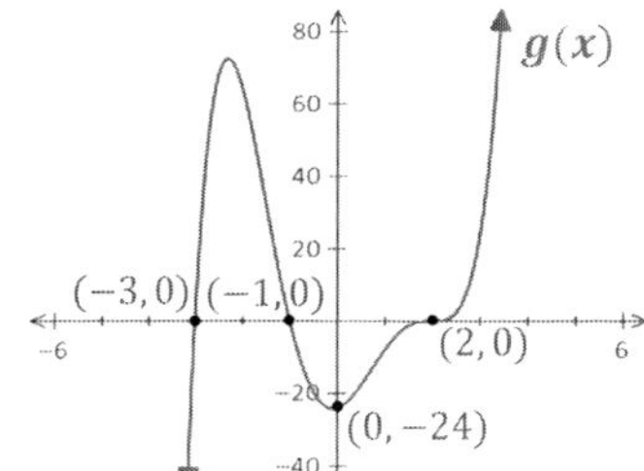

Consider a factor of a polynomial function $y = (x+a)(x-b)^n$;

We say that the factor repeats n times, and the zero $x = a$ has a **multiplicity** of $\boldsymbol{n}$.

As we saw on the previous page, the shape of the graph around a zero relates to its multiplicity.

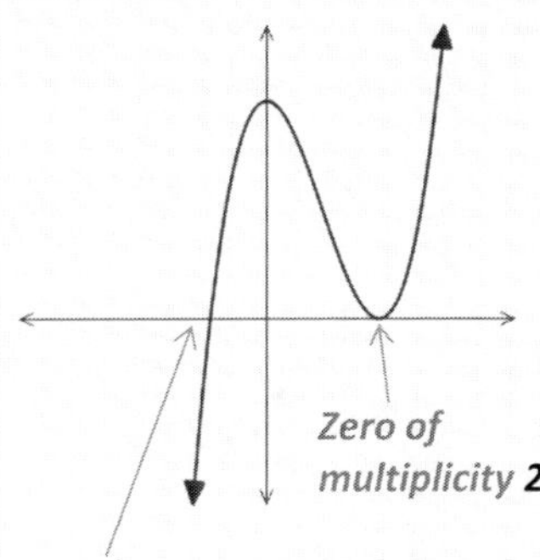

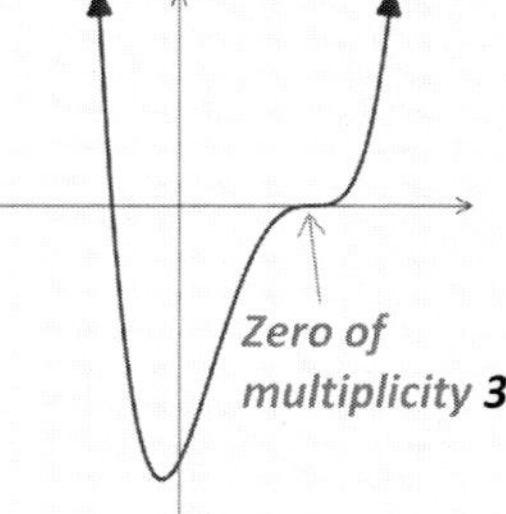

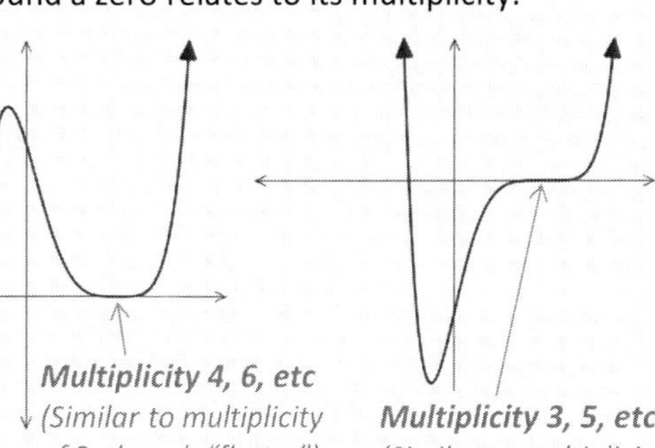

Note that the **zeros** of a function correspond to the **roots** of the related equation.

For example,

$P(x) = 2(x+1)(x-2)^3$ has zeros at $x = -1$ and 2.

Its graph has x-intercepts $(-1,0)$ and $(2,0)$.

$2(x+1)(x-2)^3 = 0$ has roots of $x = -1$ and 2.

Class Example 2.41 *Zeros and Multiplicity – Determining an Function Equations*

Each of the graphs below are given by a polynomial function with a leading coefficient of either 1 or -1, and integer zeros. Determine an equation, in factored form and of minimum degree, for each function.

(a)

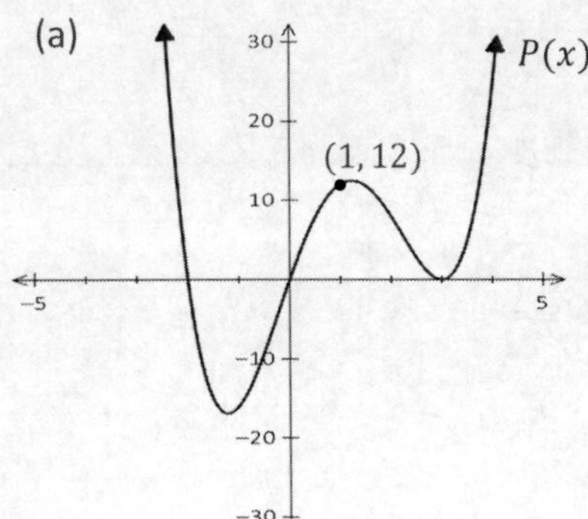

(b)

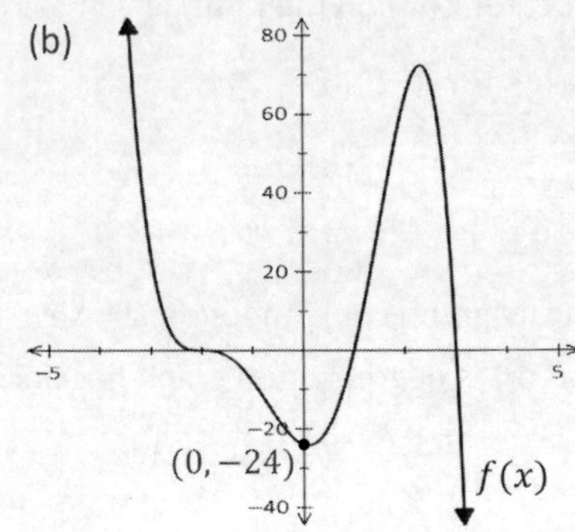

Class Example 2.42 *Analyzing Polynomial Functions*

For each of the following polynomial functions, determine (without a graphing calculator, if possible) the:

i - The start and end behavior of the graph

ii - The coordinates of the x and y-intercepts

Then, use your calculator to determine the:

iii - Coordinates of any absolute maximum / minimum points (rounded to the nearest hundredth if necessary), and the range of the function.

***Sketch** each graph* →

(a) $P(x) = 2(x + 2)(x - 1)^2(x - 3)$

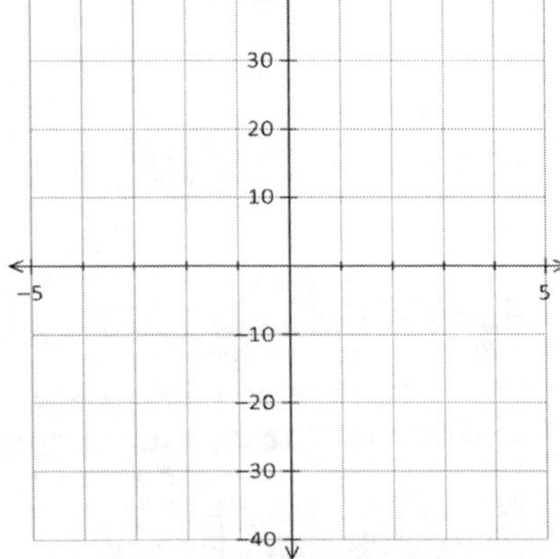

(b) $P(x) = -(x + 2)^3(x - 2)$

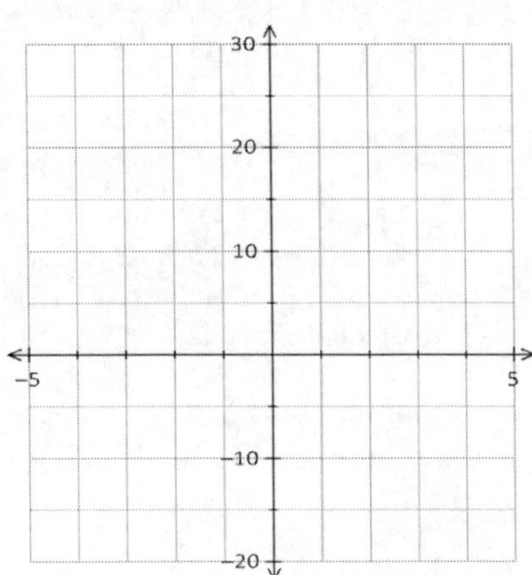

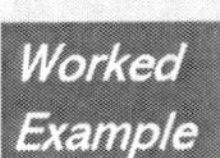

The graph of the polynomial function $P(x)$, shown on the right, has a y-intercept of $(0, 36)$.

Determine a factored form equation for $P(x)$, of minimum degree.

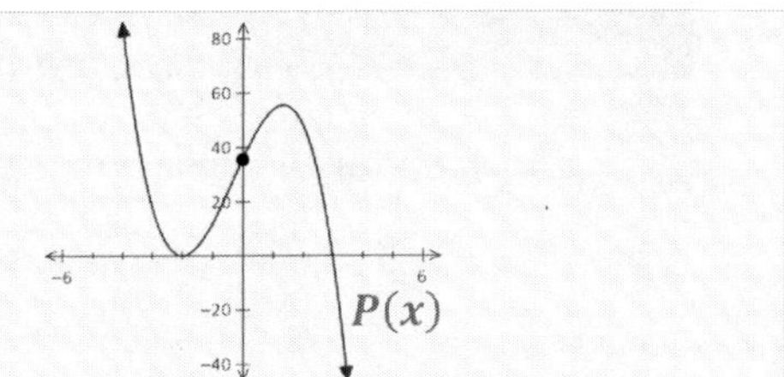

Solution: $P(x)$ has two zeros, each of which corresponds to a factor.

Always include "$\mathbf{a}$", representing the ***vertical stretch***

$$P(x) = a(x+2)^2(x-3)$$

Zero of $x = -2$, graph "bounces", so multiplicity is 2.

Zero of $x = 3$, multiplicity is 1.

Use any point on the graph, here we're given $(0, 36)$, to solve for "a".

Substitute $\mathbf{0}$ for x and $\mathbf{36}$ for y. *That is, for $P(x)$.*

$$36 = a(0+2)^2(0-3)$$

$$36 = -12a$$

$$a = -3 \qquad \boxed{P(x) = -3(x+2)^2(x-3)}$$

Check on your graphing calculator:

Match the window to the graph above.

Graph, y-int., matches

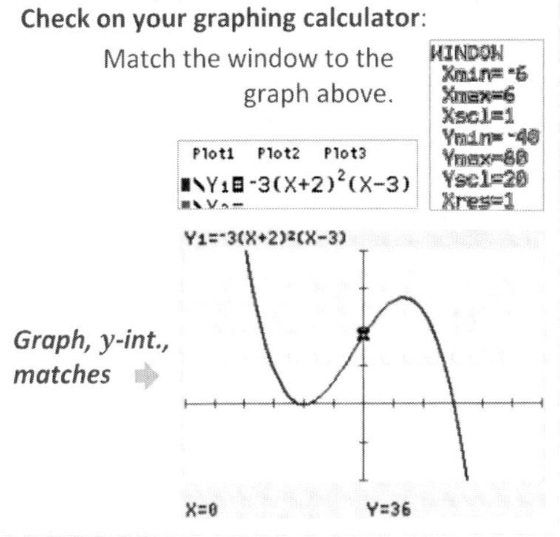

Class Example 2.43 *Bringing it all Together –Polynomial Function Equations from Graphs*

Each of the graphs below are given by a polynomial function with integer zeros. For each, i – state the minimum degree of the function, and ii - determine a factored form equation of minimum degree.

(a)

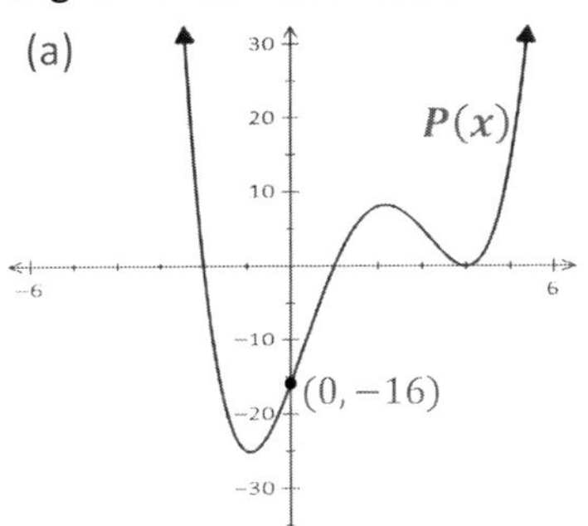

(b)

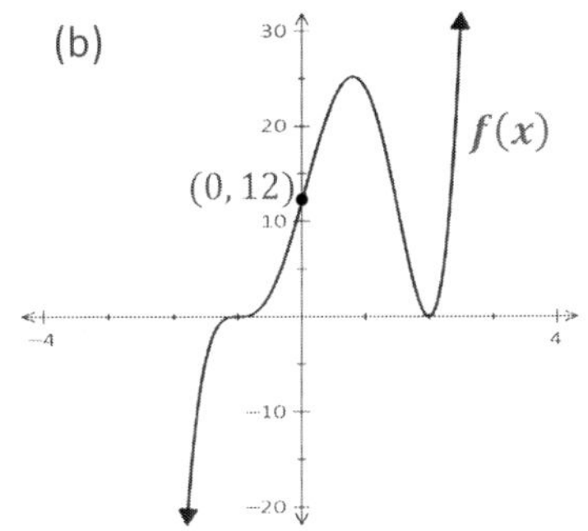

Ch. 2 Polynomial Functions REVIEW (sample)

First three pages only provided here,
Access them all at math30-1edge.com

1. Which of the following are polynomial functions (include equations and graphs)

(a) $y = 2x^5 - 14x^2 + \sqrt{2}$ (b) $y = -5x^4 - 14x^{-1} + 1$ (c) $y = 2(5)^x - 2$ (d) $y = 2x^2 - 5\sqrt{x}$

(e)

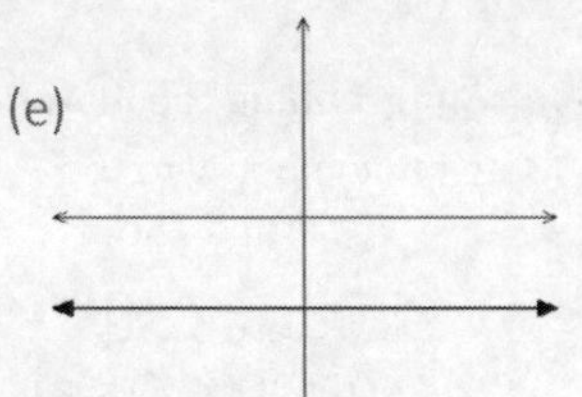

(f)

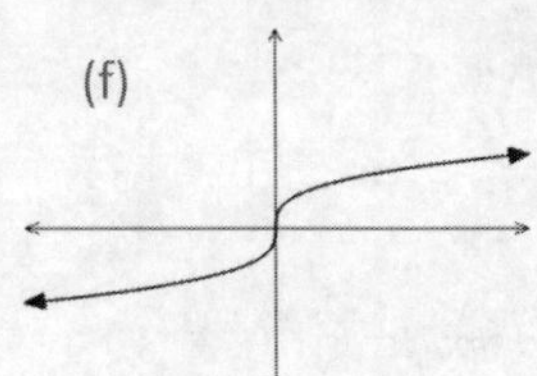

(g)

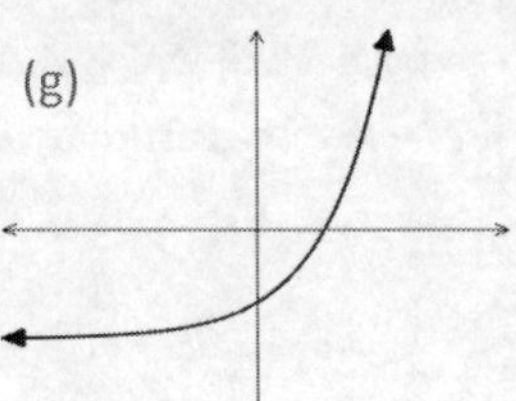

(h)

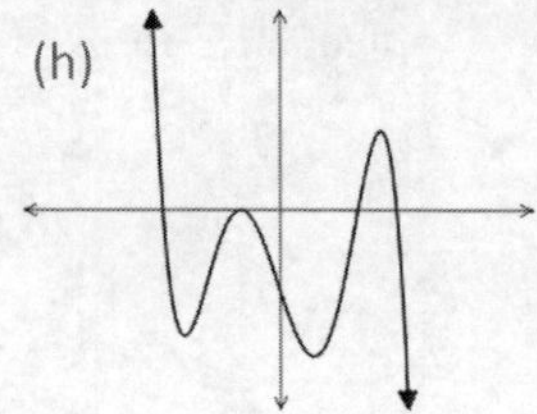

2. For each of the following polynomial functions, determine the following characteristics without using your calculator and without factoring.

i - The start and end behavior of the graph
ii - Whether or not the graph will have a minimum or maximum point
iii - The domain of the function and the y-intercept

Use your graphing calculator to determine:

iv - The range of the function (round to the nearest hundredth)

(a) $\mathbf{y = -4x^4 - 40x^3 - 77x^2 + 109x - 30}$

i - Start / end

ii - Max or min?

iii - Domain:
y-intercept:

iv - Range:

(b) $\mathbf{y = 3x^5 + 14x^4 - x^3 - 60x^2 - 36x}$

i - Start / end

ii - Max or min?

iii - Domain:
y-intercept:

iv - Range:

3. Divide each of the following polynomials, using either long division or its synthetic counterpart. Express results in the form $P = Q \times D + R$

(a) $(2x^3 + 9x^2 - 6x - 40) \div (x + 1)$

(b) $(2x^4 - 11x^3 + 6x^2 + 45x - 54) \div (2x - 3)$

4. A package for mailing must have dimensions where sum of its height and the perimeter of the base is no more than 96 cm. Find the dimensions of the box of maximum volume that can be sent if the base is a square.

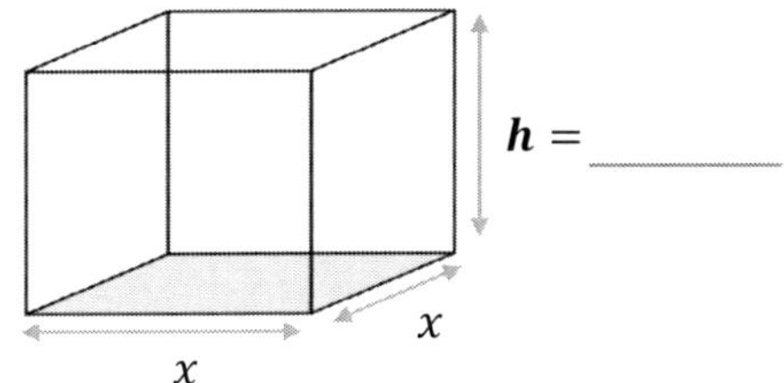

(a) Determine a function that represents the Volume of the box.

(b) Use technology to graph the function obtained in (b) with a suitable viewing window.

Provide your sketch below, labeling any max/mins and intercepts..

(c) Provide a domain and range for your function obtained in (b), with respect to the "real world" constraints of the problem.

Domain: ______ Range: ______

(d) State the maximum volume of the box that can be sent.

(e) State the dimensions for the box that provides the maximum volume.

5. Use the remainder theorem to find the remainder when:

(a) $2x^3 + 9x^2 - 6x - 40$ is divided by $(x - 2)$

(b) $8x^3 + 27$ is divided by $(2x - 3)$

6. Exam Style **NR** The same remainder is obtained when a polynomial function $P(x) = x^3 - bx^2 - 4x + 12$ is divided by $(x + 1)$ as when it's divided by $(x - 4)$.

The value of b is _____.

Answers to Review Questions from Previous Page

1. (a), (e), and (h) are polynomial functions

2. (a) i - Starts negative in quad III, ends neg in quad IV
ii - Maximum point iii - $\{x \in \mathbb{R}\}$ iv - $\{y | y \leq 36.25, y \in \mathbb{R}\}$

(b) i - Starts negative in quad III, ends pos in quad I
ii - No max / min iii - $\{x \in \mathbb{R}\}$ iv - $\{y \in \mathbb{R}\}$

3. (a) $2x^3 + 9x^2 - 6x - 40 = \mathbf{(2x^2 + 7x - 13) \times (x + 1) - 27}$

(b) $2x^4 - 11x^3 + 6x^2 + 45x - 54 = \mathbf{(x^3 - 4x^2 - 3x + 18) \times (2x - 3) + 0}$

7. Use an algebraic process to **(i)** fully factor each of the following polynomial functions, showing all steps and processes. Then, **(ii)** state the zeros of each function. and **(iii)** provide a sketch of the graph, labelling all x and y intercepts.

(a) $y = x^3 - 2x^2 - 13x - 10$

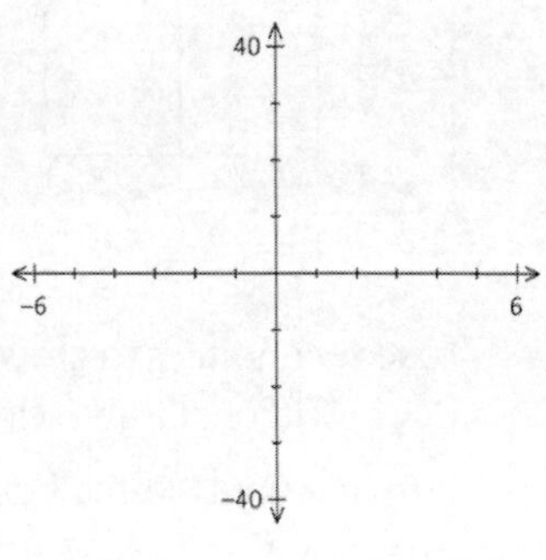

(b) $y = 2x^4 + 13x^3 + 21x^2 - 9x - 27$

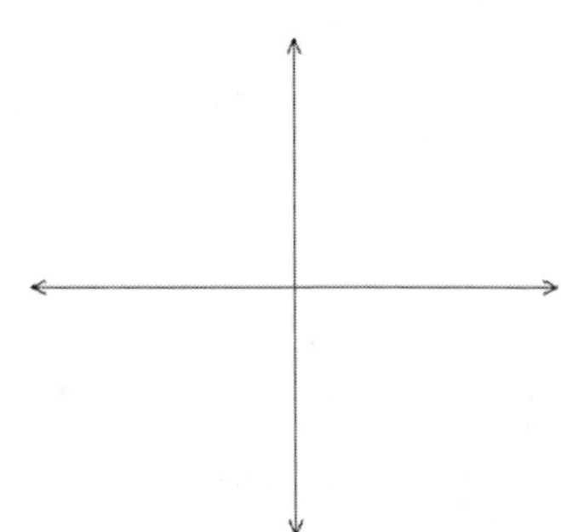

Determine a suitable window on your own! Provide the scale you used here.

Answers to Review Questions from Previous Page and this page

4. (a) $V = x^2(96 - 4x)$

(c) Domain: $[0, 24]$
Range: $[0, 8192]$

(d) Max V: $8192\ cm^3$

(e) 16 cm x 16 cm *base*,
32 cm *height*

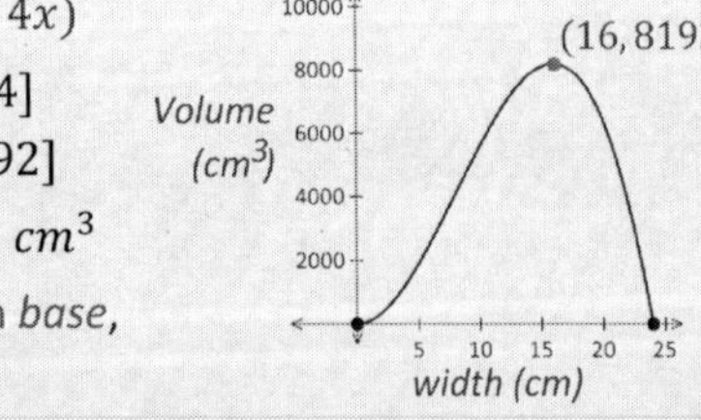

5. (a) **0** (b) **54**

6. **3**

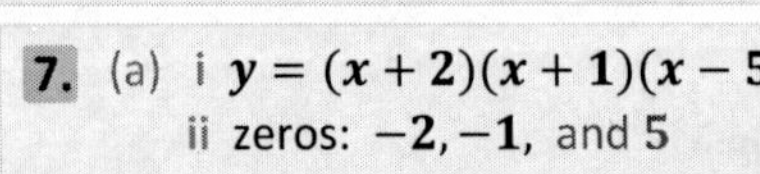

7. (a) i $\boldsymbol{y = (x+2)(x+1)(x-5)}$
ii zeros: $\boldsymbol{-2, -1}$, and $\boldsymbol{5}$
iii

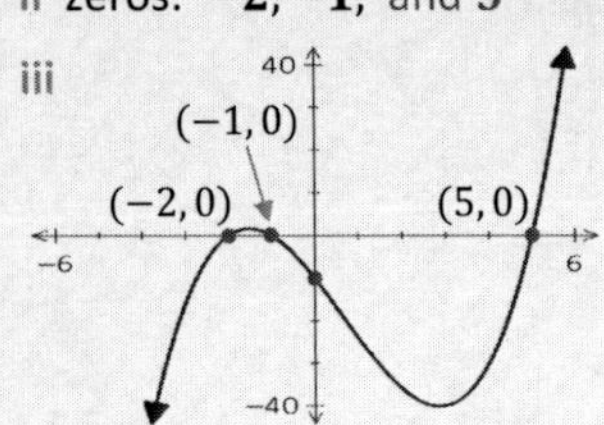

(b) i $\boldsymbol{y = (x-1)(2x+3)(x+3)^2}$
ii zeros: $\boldsymbol{-3, -3/2}$, and $\boldsymbol{1}$
iii

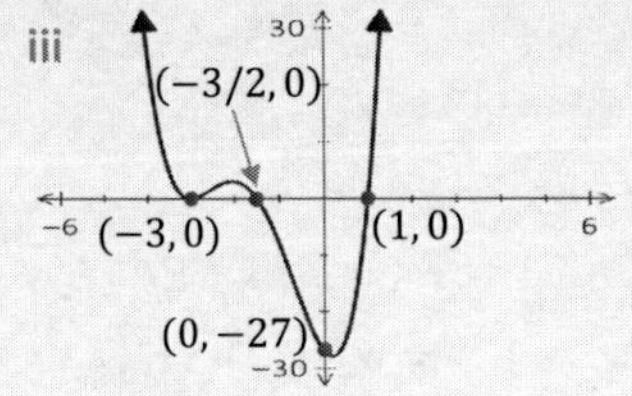

Chapter 3 EXPONENTIAL and LOGARITHMIC FUNCTIONS

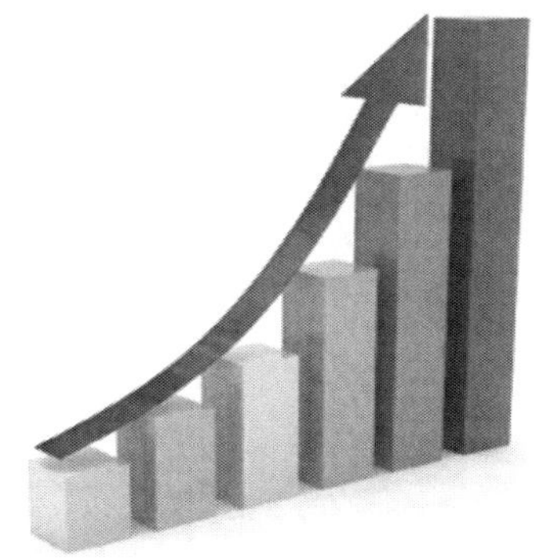

3.1 Exponential Expressions and Equations

Background Skills – Exponent Rules

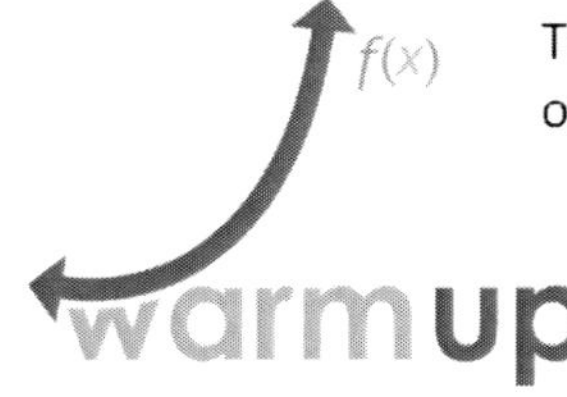

To successfully complete this unit, and even enjoy it, we must first brush up our sills on a concept last thoroughly visited in Math 10C – **exponents**.

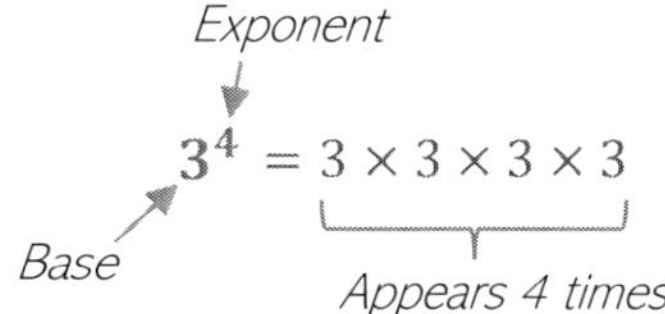

Exponent Rules *(Remember these?)*

Visit math30-1edge.com for solutions to all warm-ups and class examples

Name	Rule	Example (simplify each)
Product of Powers	$(b^m)(b^n) = b^{m+n}$	$x^3 \times x^4 =$
Quotient of Powers	$\frac{b^m}{b^n} = b^{m-n}$	$\frac{x^5}{x^3} =$
Power of a Power	$(b^m)^n = b^{mn}$	$(y^3)^2 =$
Power of a Product	$(ab)^m = a^m b^m$	$(4x^4)^2 =$
Power of a Quotient	$\left(\frac{a}{b}\right)^m = \frac{a^m}{b^m}$	$\left(\frac{2x}{y^2}\right)^3 =$
Zero Exponent	$b^0 = 1$	$(7y^2 \times 2y^6)^0 =$
Negative Exponents	$b^{-n} = \frac{1}{b^n}$ or $\frac{1}{b^{-n}} = b^n$	$\frac{1}{2^{-4}} =$
Neg. Exp. *fraction base*	$\left(\frac{a}{b}\right)^{-n} = \left(\frac{b}{a}\right)^n$	$\left(\frac{2}{3}\right)^{-2} =$
Rational Exponents	$b^{\frac{m}{n}} = \sqrt[n]{b^m}$ or $(\sqrt[n]{b})^m$	$8^{\frac{2}{3}} =$

Answers are at the bottom of the next page

← Try Each First!

Don't peek!

Evaluate each of these three, try w/o using your calc!

Worked Example Simplify Each of the Following Expressions

(a) $(2x^3y^2)(-3x^4y^3)^2$ (b) $\left(\dfrac{2x^3y^4}{10x^2y^6}\right)^3$ (c) $\left(\dfrac{12m^{-3}n^{-1}}{8m^2n^{-2}}\right)^{-2}$

(a) Start with 2nd bracket, apply exp. "2"

$= (2x^3y^2)(9x^8y^6)$

$(-3)^2$ $(x^4)^2$ $(y^3)^2$

Square everything inside, one at a time

Answer:

$$= \mathbf{18x^{11}y^8}$$

2×9 $x^3\times x^8$ $y^2\times y^6$ *add exponents*

note Final answers should not have *negative exponents*

(b) Start by simplifying "inside"

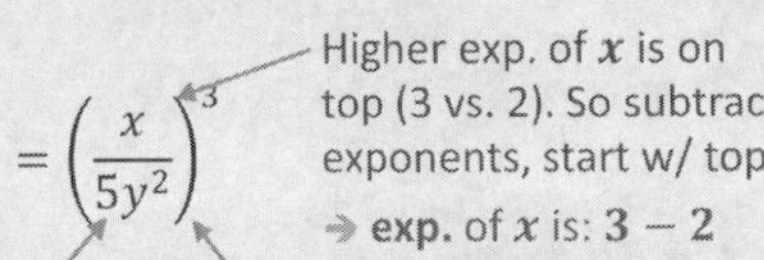

$= \left(\dfrac{x}{5y^2}\right)^3$

$10 \div 2$

Higher exp. of x is on top (3 vs. 2). So subtract exponents, start w/ top

→ **exp.** of x is: $3-2$

Higher exp. of y is on bottom to start. So subtract exponents, start w/ bottom

→ **exp. of y is: $6-4$**

Then, apply outside exp., "**3**", to both the numerator and denom.:

$= \dfrac{(x)^3}{(5y^2)^3}$

$$= \mathbf{\dfrac{x^3}{125y^6}}$$

(c) Start by simplifying "inside"

Exp. of m is *higher on bottom.* So start w/ n term, which is *higher on top.*

→ **exp. of n is: $-1-(-2)$**

$= \left(\dfrac{3n^1}{2m^5}\right)^{-2}$

12/8 reduces

Exp. of m is higher on bottom.

→ **exp. of m is: $2-(-3)$**

Next, apply the rule: $\left(\dfrac{a}{b}\right)^{-n} = \left(\dfrac{b}{a}\right)^{n}$

$= \left(\dfrac{2m^5}{3n}\right)^2 = \dfrac{(2m^5)^2}{(3n)^2}$

$$= \mathbf{\dfrac{4m^{10}}{9n^2}}$$

Class Example 3.11 *Simplifying Expressions with Exponents*

Visit **math30-1edge.com** for solutions to all warm-ups and class examples

Simply Each of the following:

(a) $(5x^5y^3)^2$ (b) $(-4m^3n^2)^2(10mn^3)$ (c) $\left(\dfrac{12a^2b^3}{6b^4}\right)^3$ (d) $\left(\dfrac{x^{-2}y^3}{2x^{-1}y^{-2}}\right)^{-3}$

Class Example 3.12 *Simplifying Expressions with Exponents*

Evaluate Each of these: (Try without a calculator!)

(a) 8^{-2} (b) $\left(\dfrac{2}{3}\right)^{-3}$ (c) $8^{\frac{5}{3}}$ (d) $9^{-0.5}$ (e) $25^{-\frac{3}{2}}$

Solving Exponential Equations

An **exponential equation** has the variable in the exponent, for example: $2^{3x-4} = 32$

One method to solve involves writing all terms in the same base. Let's see how that works here...

$2^{3x-4} = 2^{\square}$

We need to re-write 32 as a power of 2. (Since the left side is a power of 2)

→ **32** can be written as 2^5

$2^{3x-4} = 2^5$

Now our equation consists of two power terms of the same base.

$3x - 4 = 5$

Since the expressions are equal, and the bases are equal, then **the exponents must also be equal.**

$3x = 9$

$x = 3$

To solve an **exponential equation**:

1. Re-write all terms with the *same base*
2. If necessary, simplify using Exponent Rules, so there is a single term on each side
3. Set the exponents equal and solve

Worked Example Solve the Exponential Equation: $9^{x-7} = 27^{2x-9}$

Algebraic Solution:

Re-write 9 and 27 using the **same base**

$\square^{x-7} = \square^{2x-9}$

$(\mathbf{3})^2 \quad (\mathbf{3})^3$

So equation becomes:

$(\mathbf{3^2})^{x-7} = (\mathbf{3^3})^{2x-9}$

First simplify each side....

$3^{2x-14} = 3^{6x-27}$ *multiply exponents*

Once both sides are fully simplified, ***set the exponents equal***

$2x - 14 = 6x - 27$

$-4x = -13$

$x = \dfrac{13}{4}$

Verify numerically on your calc:

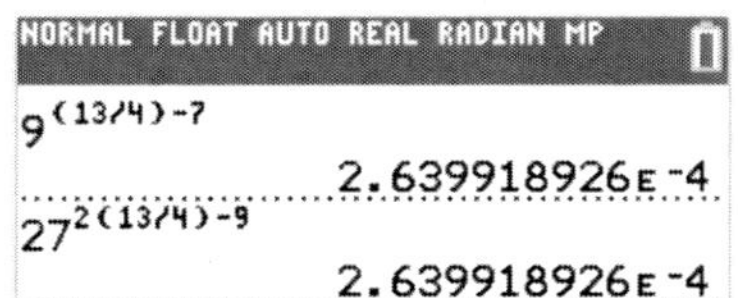

Substitute $x = 13/4$ into **both sides** of orig. equation

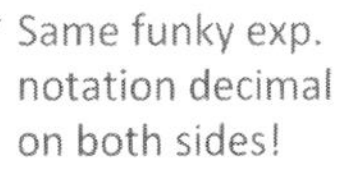

✓ Same funky exp. notation decimal on both sides!

→ *So our solution is **verified***

Solve graphically on your calculator:

One method is to:

1) Set Equation to zero: $9^{x-7} - 27^{2x-9} = 0$

2) Graph y_1 = *left side* , y_2 = *right side*

3) Sol. is the x-coord. of pt. of intersect

```
Plot1  Plot2  Plot3
\Y1=9^(X-7)-27^(2X-9)
\Y2=0
```

2nd + TRACE to get CALC menu

```
CALCULATE
1:value
2:zero
3:minimum
4:maximum
5:intersect
6:dy/dx
7:∫f(x)dx
```

Window Note:

We don't need a "perfect" window setting – it's not crucial to even see the graphs!

```
WINDOW
Xmin=-10
Xmax=10
Xscl=1
Ymin=-10
Ymax=10
```

So long as the solution is within these values, we're good. (If not, adjust)

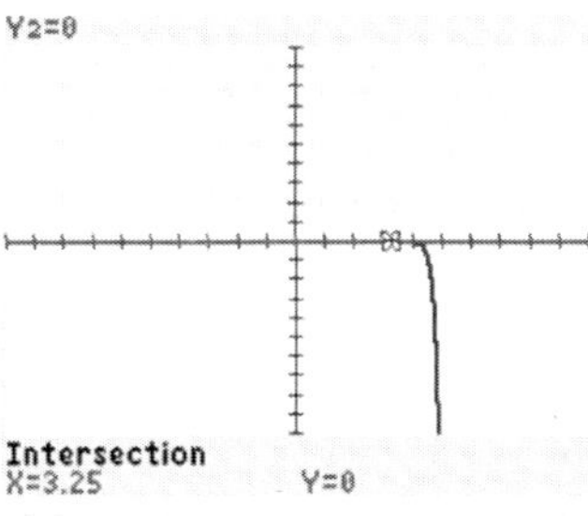

Class Example 3.13 *Solving Exponential Equations*

For the equation: $8^{x+2} = \left(\dfrac{1}{4}\right)^{x+3}$

(a) Solve algebraically

(b) Verify numerically and solve graphically on your calc.

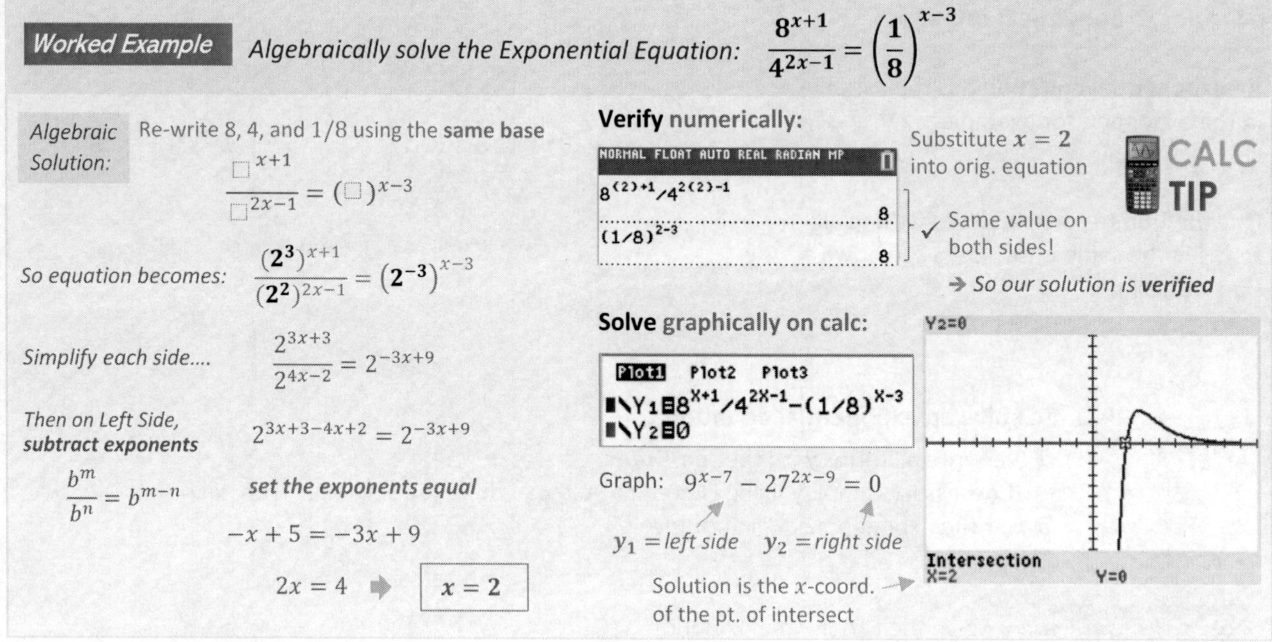

Class Example 3.14 *Solving Exponential Equations of three terms*

For the equation: $4^{x-1} = 8^{(2x+2)}16^{(x-3)}$

(a) Solve algebraically

(b) Verify numerically on your calc.

(c) Solve graphically on your calc.

Class Example 3.15 *Solving Equations where the base is unknown*

Algebraically determine the value of x in the exponential equation $\left(\frac{a}{b}\right)^{(4x-1)} = \left(\frac{b^3}{a^3}\right)^{(x+2)}$, where $a \neq b$, $a \neq 0$, and $b \neq 0$.

Class Example **3.16** *Solving Equations where the variable is in the base*

Algebraically solve the equation $2(3x-1)^{-\frac{3}{4}} = 16$, and verify your solution.

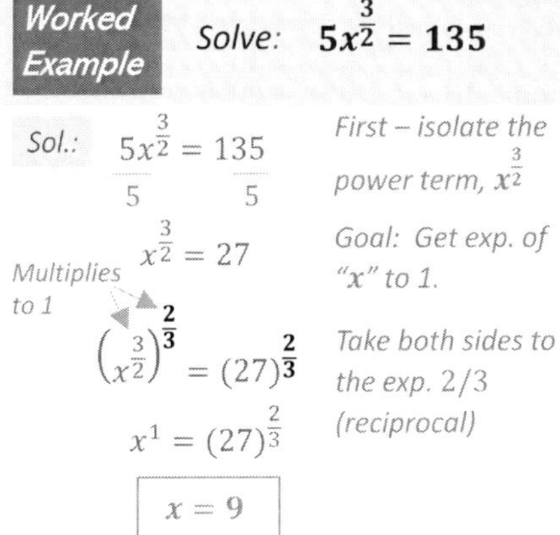

Applications of Exponential Equations

Finally, we explore real world situations that can be modeled using exponential equations – where some initial value ($\boldsymbol{a}$) has a multiplication factor ($\boldsymbol{b}$) applied every certain period of time ($\boldsymbol{p}$).

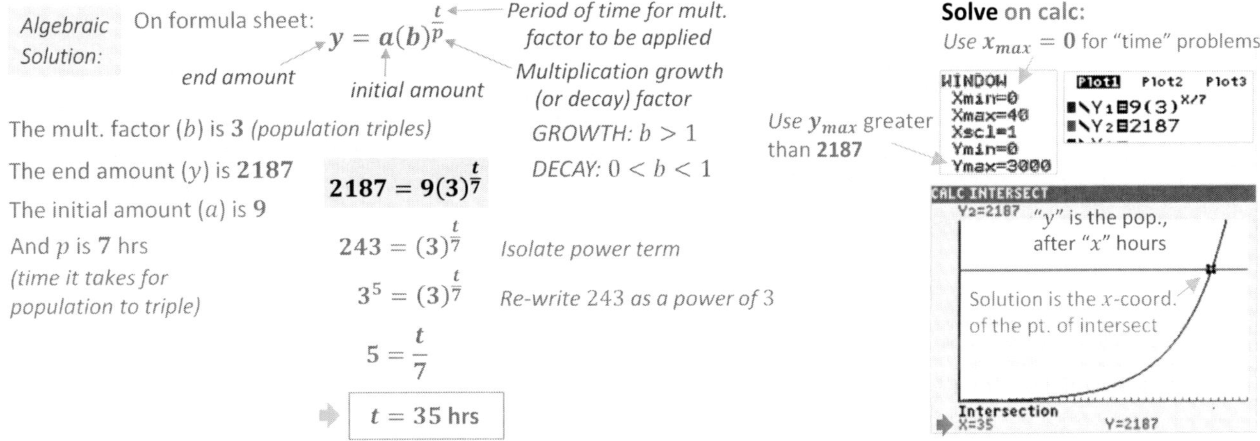

Class Example **3.17** *Applications of Exponential Equations*

A particularly strong investment fund has doubled in value over the past 5 years. Assuming that the fund continues this performance, setup and algebraically solve an equation to determine how long it would take for a \$5 000 investment to grow to \$80 000.

3.1 Practice Questions (Sample)

SAMPLE of our high-quality, diploma-exam based practice questions. To access all questions, including REVIEW section, visit **Math30-1edge.com** *for EDGE ONLINE access*

1. Fully simply each of the following expressions:

(a) $(5x^2y^3)(3xy^{-1})^2$

(b) $\dfrac{(5x^2y^3)(8x^3y^2)}{-10x^5y^2}$

(c) $\left(\dfrac{4m^2n^4}{-6m^5n^2}\right)^3$

(d) $\left(\dfrac{27x^{-2}y^2}{3x^{-3}y^5}\right)^2$

(e) $\left(\dfrac{2x^2y^{-4}}{x^{-1}y^2}\right)^{-2}$

2. Evaluate each, showing simplification steps: *Try first without a calculator, use your calc to verify!*

(a) $\left(-\dfrac{3}{4}\right)^{-2}$

(b) $81^{\frac{3}{4}}$

(c) $\left(\dfrac{25}{16}\right)^{-\frac{3}{2}}$

(d) $\dfrac{2^{-3}}{\left(\dfrac{8}{50}\right)^{-\frac{1}{2}}}$

Answers to Practice Questions – get all the rest at math30.1.com!

1. (a) $45x^4y$ (b) $-4y^3$ (c) $-\dfrac{8n^6}{27m^9}$ (d) $\dfrac{81x^2}{y^6}$ (e) $\dfrac{y^{12}}{4x^6}$ **2.** (a) $\dfrac{16}{9}$ (b) 27 (c) $\dfrac{64}{125}$ (d) $\dfrac{1}{20}$

3.2 Graphs of Exponential Functions

We saw in the previous section how exponential equations involved terms where the variable is in the exponent. **Exponential functions** are can be used to model many real-world situations.

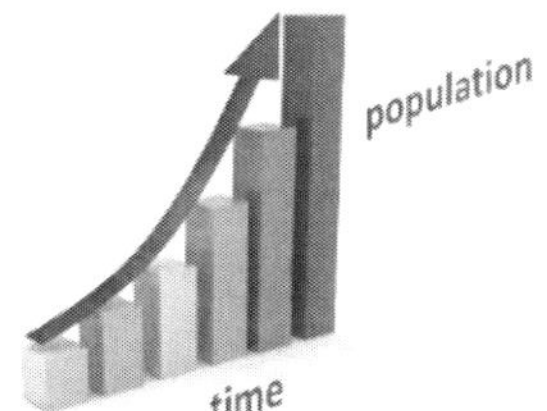

- *The world population*
- *The value of an investment earning positive interest*
- *The measured amount of a decaying radioactive isotope*
- *The value of a used car*
- *The temperature of a cup of hot chocolate as it cools*

Exponential functions can model any of these, given certain parameters.

We'll dive further into applications in section 4.7, after learning a bit about logarithms. For now we'll focus on the basic graphs.

*These three involve **negative exponential growth**, or **exponential decay*** →

Exploration #1 The Graph of $y = 2^x$

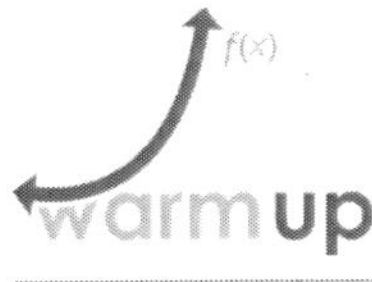

1 ➡ Complete the table of values below, and plot the remaining points on the graph. Then, sketch the smooth curve that goes through each of your points. Then – you will have graphed your first exponential function.

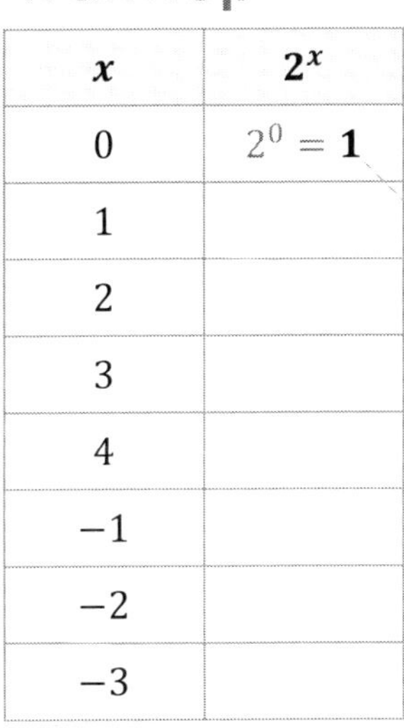

x	2^x
0	$2^0 = \mathbf{1}$
1	
2	
3	
4	
-1	
-2	
-3	

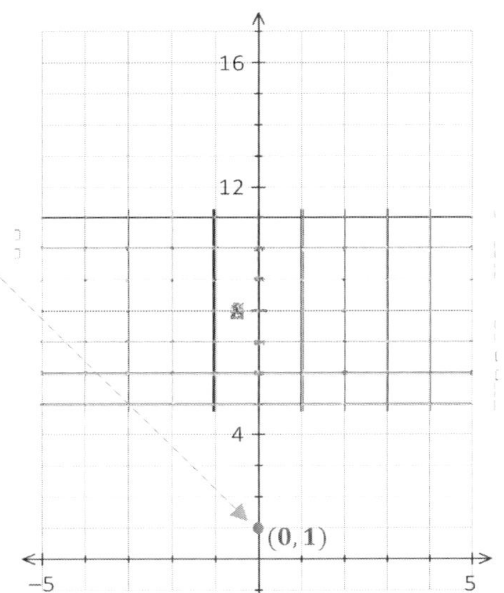

2 ➡ ***Function Essentials:***

State each of the following

Domain

Range

Asymptote

x-intercept

y-intercept

The Graph of $y = (\frac{1}{2})^x$

Next, we sketch the graph of the function obtained by **horizontally reflecting** the graph of $y = 2^x$, about the line $x = 0$.

3 ➡ Use transformations to show that the resulting function equation is: $y = \left(\frac{1}{2}\right)^x$

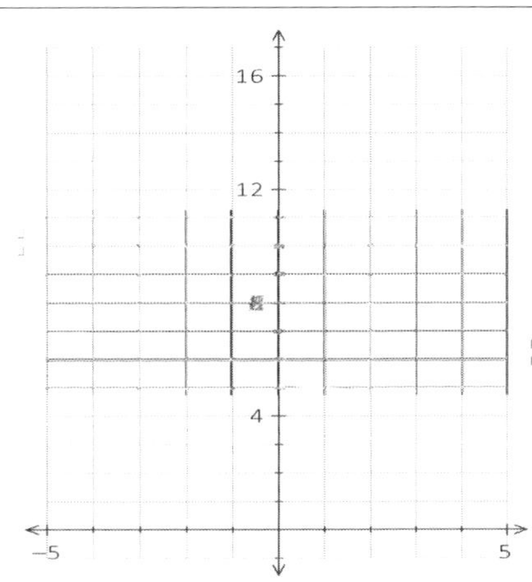

4 ➡ ***Function Essentials:***

Are there any differences from the graph of $y = 2^x$?

We introduce the *basic exponential function* as: $y = 2^x$

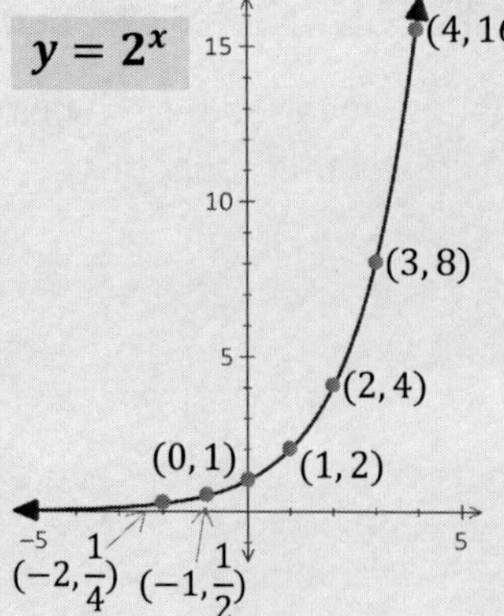

Applying a *horizontal reflection* gives: $y = (\frac{1}{2})^x$

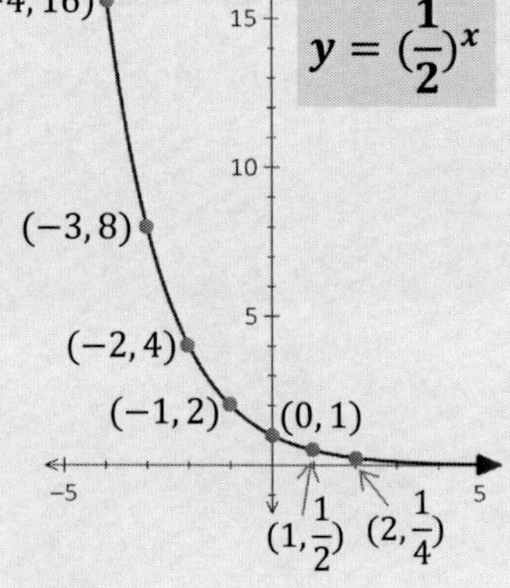

For any graph, $y = b^x$ $\quad b > 0, b \neq 1$:

Domain: $\{x \in \mathbb{R}\}$	*y-intercept:* $(0, 1)$	*Asymptote:* **Horizontal Asymptote at** $y = 0$
Range: $\{y \mid y > 0, y \in \mathbb{R}\}$	*x-intercept:* n/a	↑ *As* x *gets larger and larger the graph approaches, but never touches, the x-axis.*

Note how the characteristics; domain, range, intercepts and asymptote – are the same for any base of $y = b^x$.

And about that base, **b***....*

- ☞ It can't be "0" *Or else y would just be 0 for every x ....* $y = 0^x$
- ☞ It can't be "1" *Or else y would just be 1 for every x ....* $y = 1^x$
- ☞ It can't be negative *Note that f we allowed negative bases, any even value of x would return a positive value*

So we define the base to be: $b > 0, b \neq 1$:

Now with that, let's explore the effect of changing the base:

Exploration #2 Comparing Graphs in the form $y = b^x$, $b > 1$, for different b values

Analyze the graphs on the right. All points with integer coordinates are shown.

1 ➡ For each equation listed to the right, indicate the number of the matching graph.

2 ➡ Describe the effect on the graph of $y = b^x$, $b > 1$, as b gets larger.

3 ➡ Describe the effect on the graph of $y = b^x$, $b > 1$, as b gets *closer to 1*.

$y = \left(\frac{3}{2}\right)^x$

$y = \left(\frac{10}{9}\right)^x$

$y = 4^x$

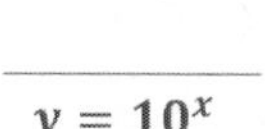

$y = 10^x$

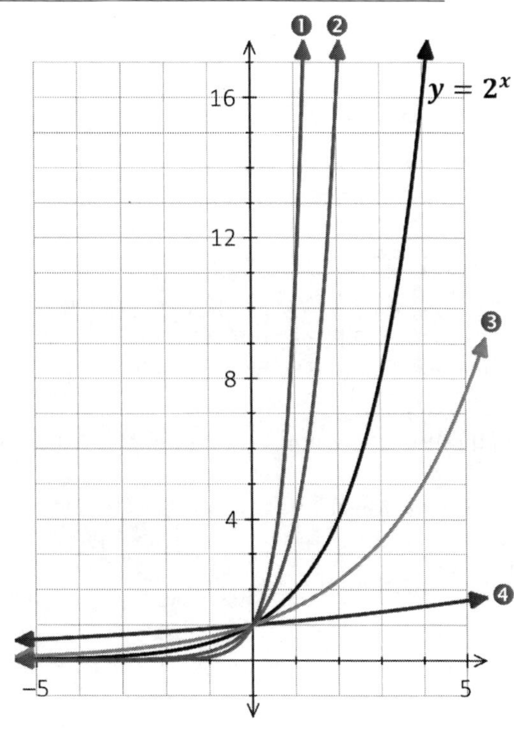

Exploration #3 Comparing Graphs in the form $y = b^x$, where b is between 0 and 1

Analyze the graphs on the right. All points with integer coordinates are shown.

1 ➡ For each equation on the right, indicate the number of the graph that matches.

2 ➡ Describe the effect on the graph of $y = b^x, 0 < b < 1$, as b gets *closer to 0.*

3 ➡ Describe the effect on the graph of $y = b^x, 0 < b < 1$, as b gets *closer to 1.*

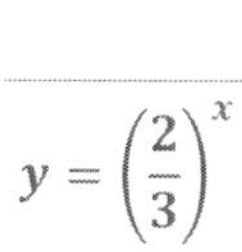

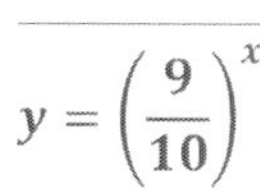

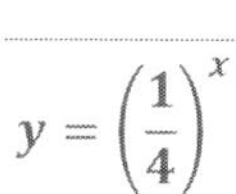

$y = \left(\frac{1}{10}\right)^x$

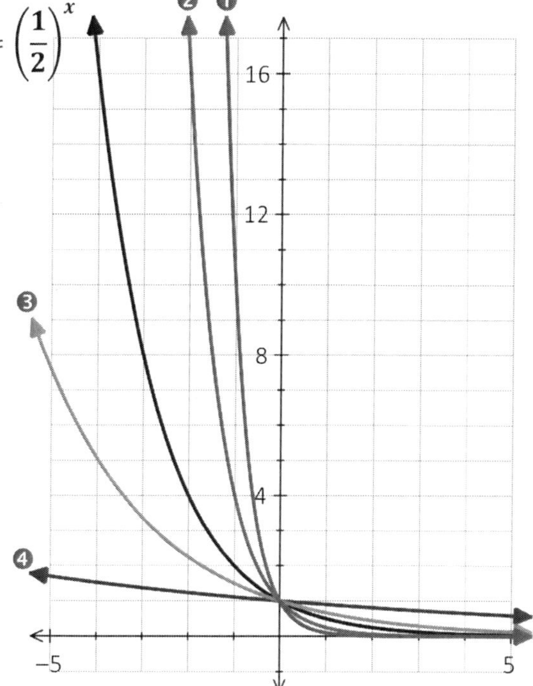

Note that for all of these graphs::

$\{x \in \mathbb{R}\}$	$\{y > 0, y \in \mathbb{R}\}$	H.A. at $y = 0$	$y = 1$	n/a
Domain	Range	Asymptote	y-intercept	x-intercept

Given the graph of $y = b^x$ ...

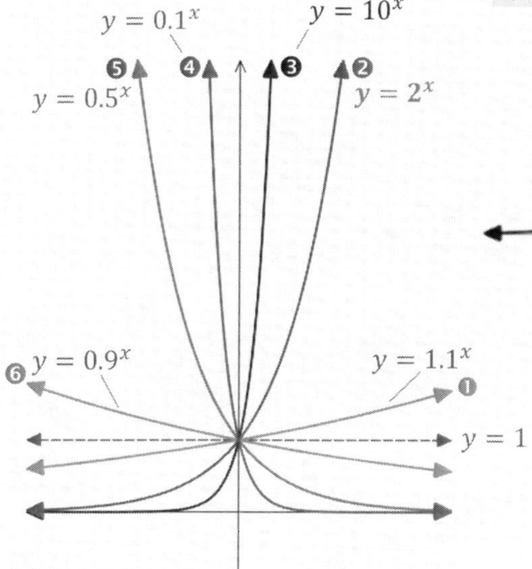

If $b > 1$, the graph ***increases***

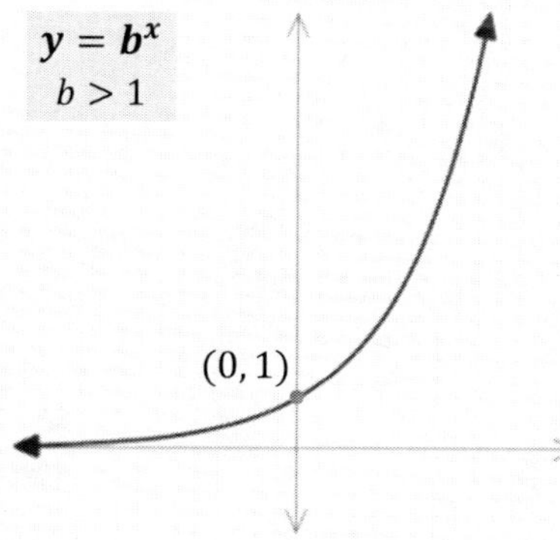

If $0 < b < 1$, the graph ***decreases***

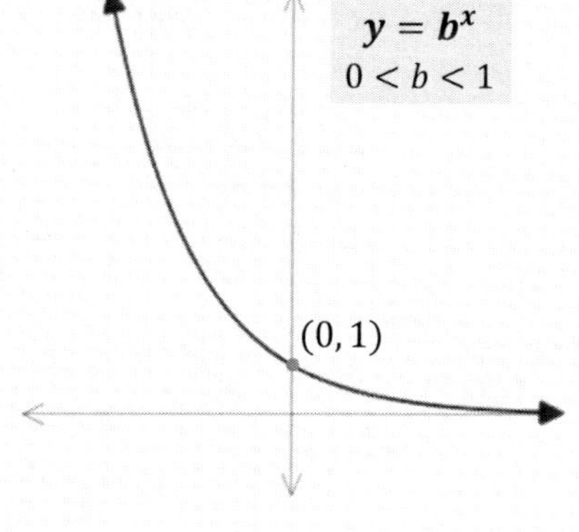

If $b > 1$, the graph *bends-up* from the horizontal line $y = 1$

The larger the base (the further from "1"), the greater the increase... compare graphs ❶,❷, and ❸ on the left.

The smaller the base (the further from "1"), the greater the decrease... compare graphs ❹, ❺, and ❻ on the left.

We'll next explore the effect of adding a vertical stretch, a, and vertical translation, k, to the graph of $y = b^x$....

Exploration #4

Comparing Graphs of $\boldsymbol{y = a(b)^x}$, $a \neq 0, b > 1$ for different a values

Analyze the graphs on the right. All points with integer coordinates are shown. The corresponding functions for each of graphs ❶, ❷, and ❸ can be written in the form $y = a(2)^x$.

Hint: The value of "a" can be solved for, or obtained through reasoning, or through exploration with your graphing calculator.

1 ➡ **State the equation** of the function for:

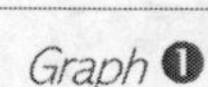

_______ *Graph* ❶

_______ *Graph* ❷

_______ *Graph* ❸

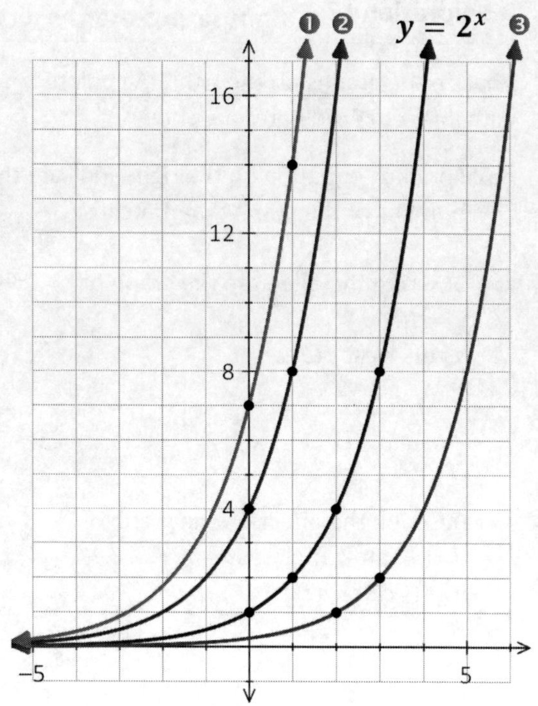

2 ➡ **Describe** the effect of $\boldsymbol{a}$ on the graph of $y = a(b)^x$.

Exploration #5

Effect of Parameter "$\boldsymbol{k}$" in $\boldsymbol{y = a(b)^x + k}$, $a \neq 0, b \neq 1$

Graphs ❶ and ❷ on the right are obtained by applying a vertical translation to the graph of $y = 3(2)^x$.

The horizontal asymptote (HA) for graph ❶ is shown.

1 ➡ **Explain** how the value $\boldsymbol{k}$, where $a > 0$, affects whether the graph has an x-intercept.

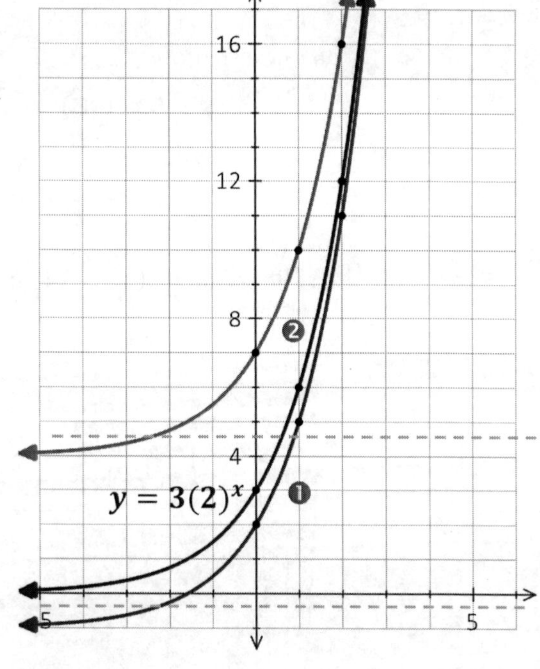

2 ➡ **State the indicated characteristics** for each graph:

	Range	Asymptote	y-intercept
Graph ❶			
Graph ❷			

Given the graph of $y = a(b)^x$; $a > 0$ ...

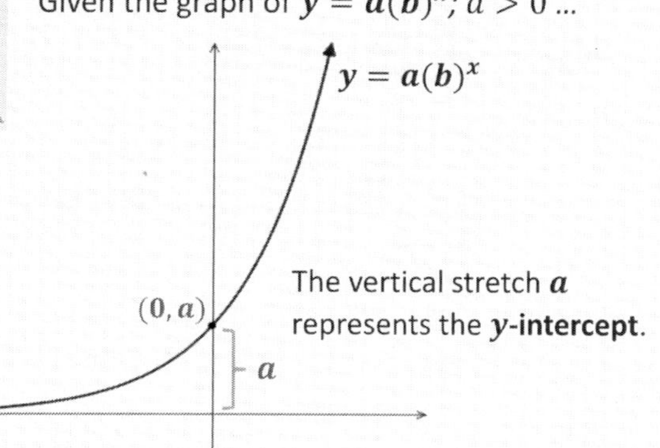

Given the graph of $y = a(b)^x$; $a > 0$ ...

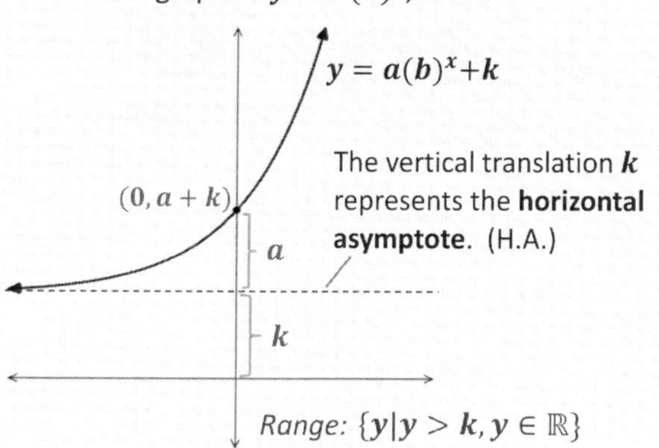

Class Example 3.21 *Determining Graph Characteristics from an Equation*

Given the function $y = 3(2)^x + 4$,

(a) Without graphing, state the range, asymptote, and y-intercept.

(b) Use reasoning to state whether the graph will have an x-intercept. Explain.

(c) Sketch and label all characteristics.

Exploration #6 The graph of $y = a(b)^x$, $a < 0, b \neq 1$

The graph on the right can each be written in the form $f(x) = a(3)^x$.

1 ➡ Determine the value of a to state an equation for $f(x)$.

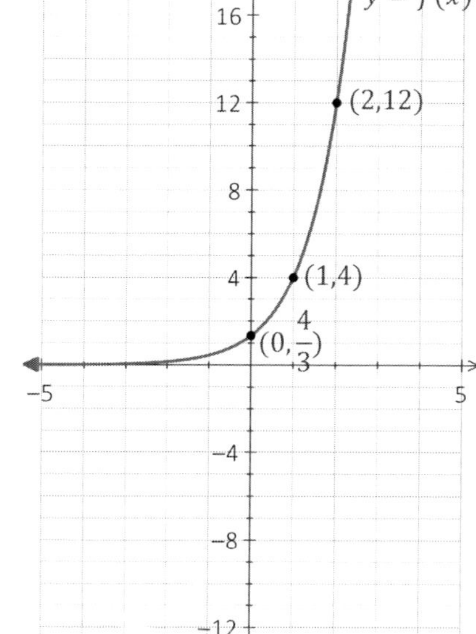

2 ➡ On the same grid, sketch the graph of $y = g(x)$, obtained by reflecting the graph of $f(x)$ about the line $y = 0$. Label it graph ❷.

3 ➡ State the equation of $y = g(x)$, both in terms of $f(x)$ and in terms of x.

An **exponential function** can be written $\boldsymbol{y = a(b)^x + k}$.

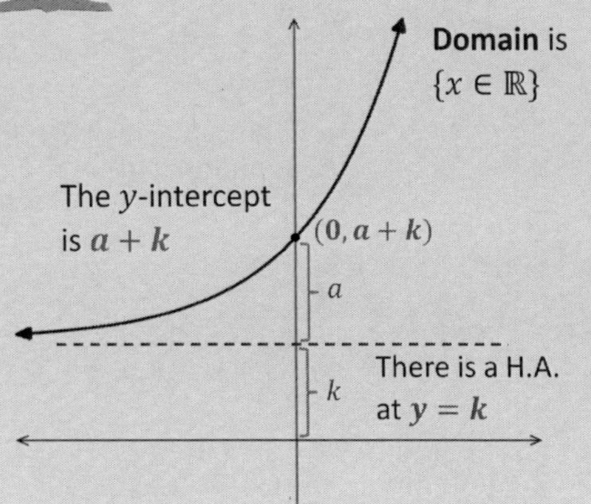

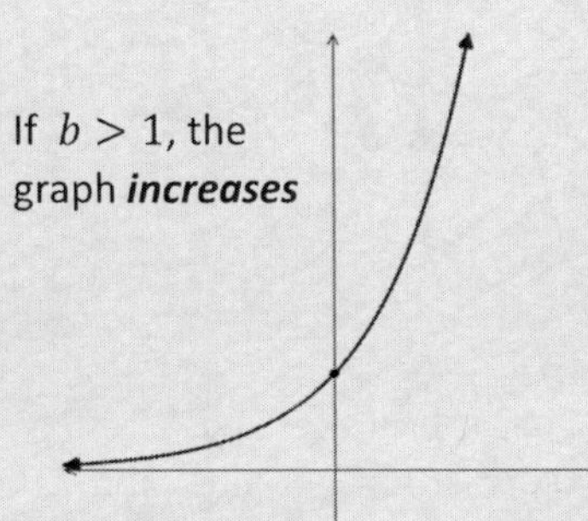

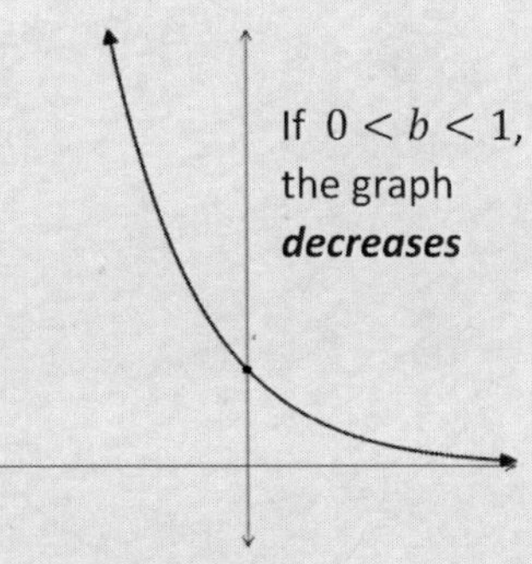

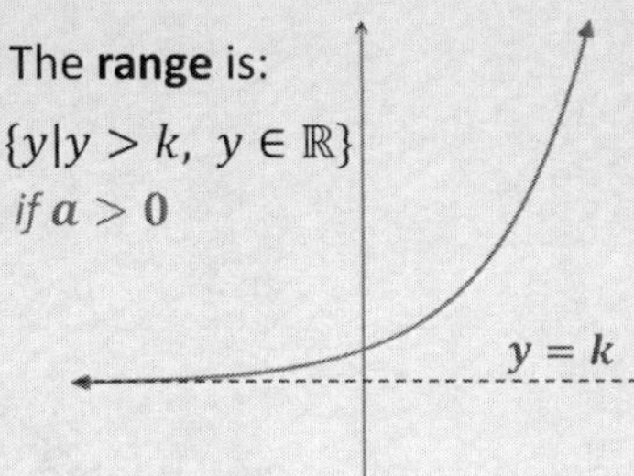

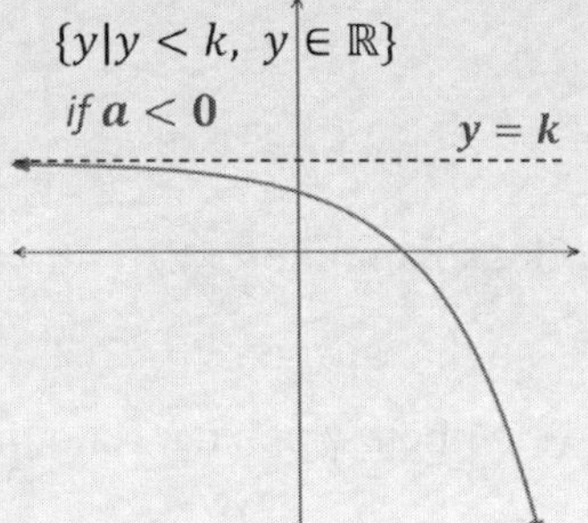

- For the y-intercept, set $x = 0$ and evaluate.

$$y = a(b)^0 + k$$
$$= a(1) + k \quad \textit{since } b^0 = 1 \textit{ for any } b$$
$$= \boldsymbol{a + k}$$

- For the x-intercept, set $y = 0$ and solve.

$$\boldsymbol{0} = a(b)^x + k$$

Solve the resulting equation for the x-intercept, if it exits.

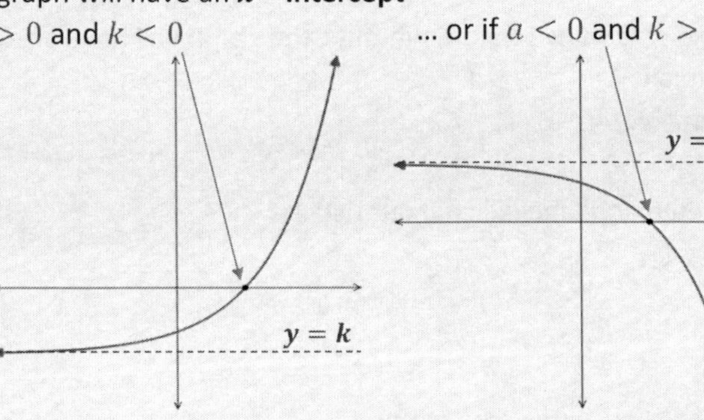

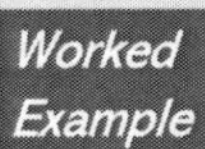

Given the function $\boldsymbol{y = -4\left(\frac{1}{2}\right)^x + 16}$, state the domain, range, asymptote, and any intercepts. Sketch and label all characteristics.

Solution: For all exponential functions (unless restricted by some application), the **DOMAIN** is $\{\boldsymbol{x} \in \mathbb{R}\}$

For **range consider two things:**

- The **H.A. is at** $\boldsymbol{y = 16}$ (the vertical translation). So, the graph is either entirely above or entirely below $\boldsymbol{y = 16}$.
- Since $a < 0$ (there is a negative in front of the function), the graph opens down, and is **entirely below** the line $y = 16$.

So, the **RANGE** is $\{\boldsymbol{y} | \boldsymbol{y < 16},\ \boldsymbol{y} \in \mathbb{R}\}$

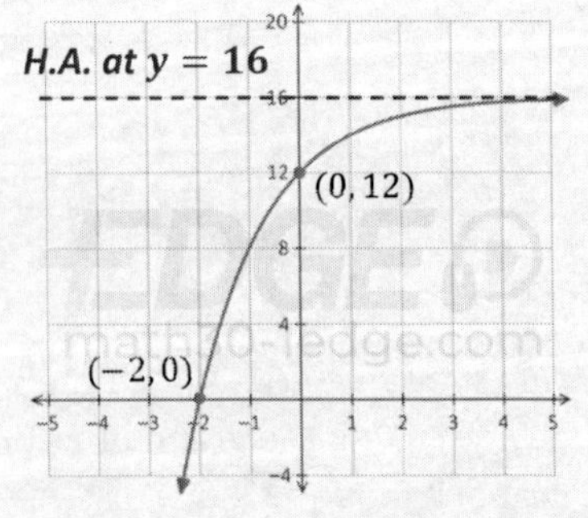

For **y-intercept**, substitute $x = 0$ and evaluate:

$$y = -4\left(\tfrac{1}{2}\right)^0 + 16$$
$$= -4(1) + 16 \Rightarrow = 12$$

$\Rightarrow$ $\boxed{(\mathbf{0, 12})}$

For **x-intercept**, substitute $y = 0$ and solve:

$$\boldsymbol{0} = -4(\tfrac{1}{2})^x + 16$$
$$4(\tfrac{1}{2})^x = 16$$
$$(\tfrac{1}{2})^x = 4$$
$$(2^{-1})^x = 2^2$$
$$2^{-x} = 2^2$$
$$-x = 2$$
$$x = -2$$

$\boxed{(\mathbf{-2, 0})}$

$$\boldsymbol{y = -4\left(\frac{1}{2}\right)^x + 16}$$

Horiz. Asymptote

Base is less than 1, so graph "falls right".
*But $a = -4$ is negative, so **vertically reflect***

Copyright © RTD Learning 2020

Class Example 3.22 *Determining an Exponential Function Equation from a Graph*

Given the function $y = -3(4)^x + 48$,

(a) State the domain, and analyze the characteristics of equation to determine the range.

(b) Use an algebraic process to determine the x and y intercepts. *Verify using your graphing calculator.*

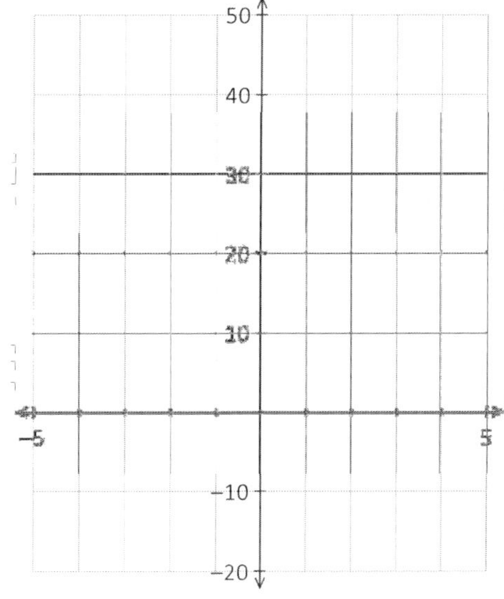

Class Example 3.23 *Identifying graphs in the form* $y = ab^x$

Each graph on the right represents an exponential function that can be written in the form $y = a(b)^x$; $b > 1$

Use reasoning to match each equation with a graph number.

(a) $y = b^x$ ______

(b) $y = \frac{2}{5}(b)^x$ ______

(c) $y = 5b^x$ ______

(d) $y = 4b^x$ ______

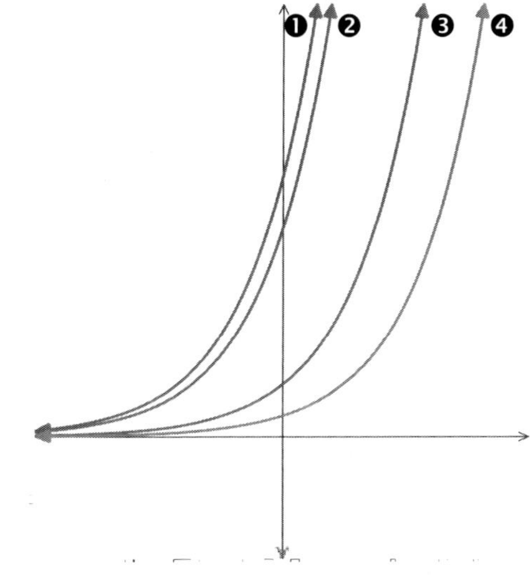

Class Example 3.24 *Identifying graphs in the form* $y = b^x$

Each graph on the right represents an exponential function that can be written in the form $y = (b)^x$; $b \neq 1$

Use reasoning to match each equation with a graph number.

(a) $y = \left(\frac{1}{2}\right)^x$ ______

(b) $y = \left(\frac{1}{8}\right)^x$ ______

(c) $y = \left(\frac{1}{4}\right)^x$ ______

(d) $y = \left(\frac{2}{3}\right)^x$ ______

(e) $y = \left(\frac{3}{4}\right)^x$ ______

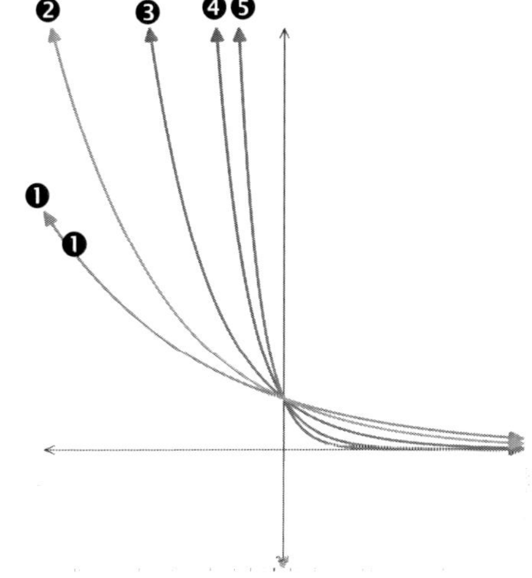

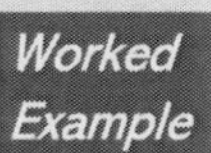

On the graph shown the horizontal asymptote and points (•) shown have integer values. Determine an equation for the corresponding are exponential function, in the form $y = a(b)^x + k$.

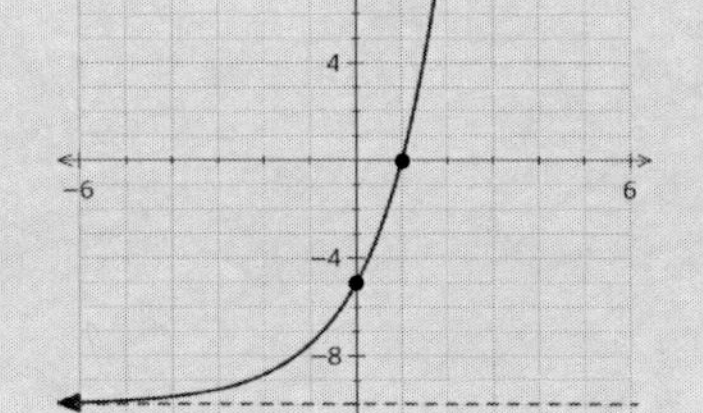

Solution: *Start by identifying the H.A. (horizontal asymptote), which defines the* $\mathbf{k}$ *value.* → $\mathbf{k = -10}$

So we have $\mathbf{y = a(b)^x - 10}$

Next, the value of $\mathbf{a}$ *(vertical stretch) can be counted as the distance from the* ***H.A.*** *and the* ***y-intercept.*** → $\mathbf{a = 5}$

So now we have $\mathbf{y = 5(b)^x - 10}$

Finally, use any other point on the graph, such as $(1, 0)$*, to solve for* $\mathbf{b}$*.*

$$0 = 5(b)^1 - 10$$

$$10 = 5(b)$$

$$b = 2$$

So now that we have k*,* a*, and* b*, we can state the equation:*

$$\mathbf{y = 5(2)^x - 10}$$

For reference...

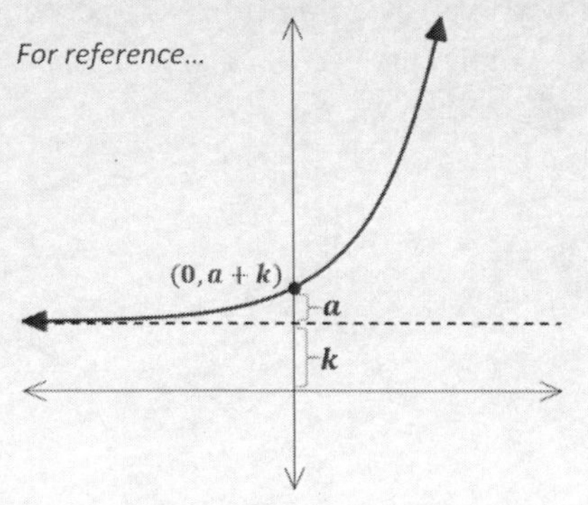

Class Example 3.25 *Determining an Exponential Function Equation from a Graph*

For the graph below, the horizontal asymptote and points indicated (•) are all integer values. Determine an equation for the corresponding exponential function, in the form $y = a(b)^x + k$.

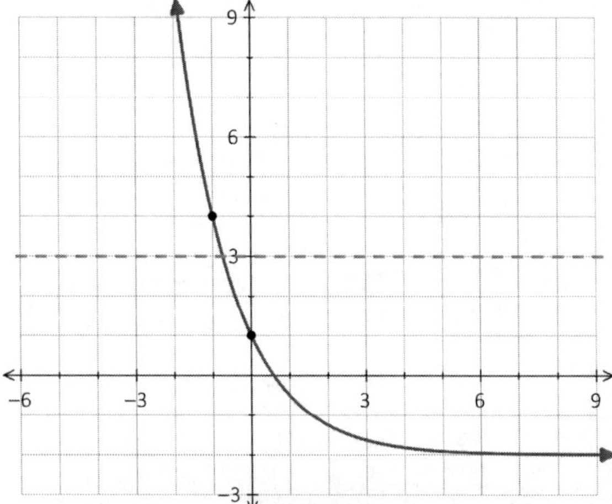

3.3 The Logarithmic Function

What exactly are *logarithms*?

Logarithms can be helpful in putting very large (or very small) numbers on a *human-friendly scale.*

For example, on the pH scale →

Ammonia, which has a pH of 12, is ***11 higher*** than battery acid, which has a pH of 1...

But Ammonia is ***one hundred billion*** times as alkaline!

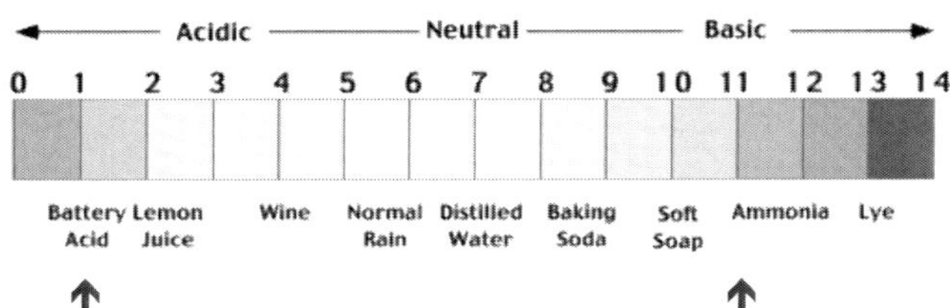

Every increase of 1 on the pH scale means a tenfold increase in the alkalinity of a substance. (Similarly, every decrease of one means a tenfold increase in the acidity) The pH scale (and others – such as the Richter scale, or Decibel scale), are examples of ***logarithmic scales***. *Teaser – more on those in section 4.8!*

Orders of magnitude

The difference between large numbers can be hard to comprehend.

A "trick" we can use is to write these numbers in terms of **inputs**, that, is, as a power base 10.

10 000	*Ten Thousand*	→	4	*Since $10^4 = 10\,000$*
1 000 000	*One Million*	→	6	*Since 10^6 = 1 million*
1 000 000 000 000	*One Trillion*	→	12	*Since 10^{12} = 1 trillion*
1 000 000 000 000 000	*One Quadrillion*	→	15	*Since 10^{15} = 1 Quadrillion*

These inputs are called logarithms.

IE *- The logarithm, base 10, of one million is 6*

Defining the Logarithmic Function as an Inverse

Pt. on graph of $y = 2^x$	Corresponding pt. on graph of inverse
$(-3, \quad)$	
$(-2, \quad)$	
$(-1, \quad)$	
$(0, \quad)$	
$(1, \quad)$	
$(2, \quad)$	
$(3, \quad)$	
$(4, \quad)$	

Exploration #1

The graph of $f(x) = 2^x$ is shown:

1 ➡ Use the graph to complete the ← table on the left.

2 ➡ Sketch the graph of the inverse on the same grid on the right →

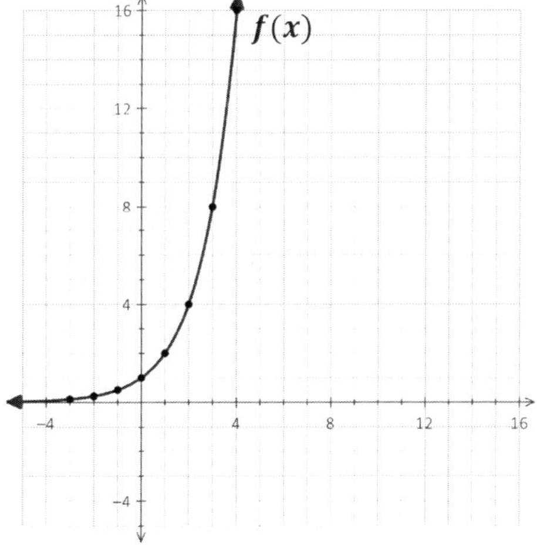

4 ➡ Determine the equation of the inverse function

3 ➡ Indicate the domain and range of each function:

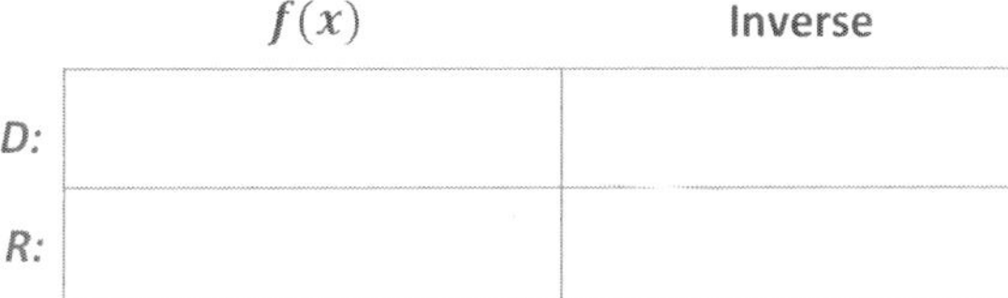

	$f(x)$	Inverse
D:		
R:		

Converting Between Exponential and Logarithmic Form

The **logarithmic function** is defined as the inverse of the exponential function.

In an **exponential function**, the input (x) is the exponent, and the output (y) is the value of some base b to the power of x.

$y = b^x$ — *output (value of the power term)*: y; *input (exponent)*: x; *base*: b

In a **logarithmic function** the input (x) is value of the power, and the output (y) is exponent.

$x = b^y \iff y = log_b x$ — *output (the exponent)*: y; *input (value of the power term)*: x; *base*: b

Exploration #2

Converting between Logarithmic and Exponential form

1 ➡ Complete the table on the right by converting each logarithmic form to exponential form

2 ➡ Re-write each of the following in logarithmic form

(a) $5^3 = 125$ (b) $\left(\frac{1}{8}\right)^{-2} = 64$

Logarithmic Form	Exponential Form	Value of y
$y = log_2 8$	$2^y = 8$	**3** Since $2^3 = 8$
$y = log_3 9$		
$y = log_{10} 1000$		
$y = log_2\left(\frac{1}{16}\right)$		
$y = log(0.01)$		
$y = log_5 \sqrt{5}$		

Worked Example *Convert each to Logarithmic form:* (a) $m^3 = 125$ (b) $4^x = -3$

Solution: *When converting to log form...* $log_{\square}(\square) = \square$ ← The log is equal to the **exponent**

base (subscript); What the power term is equal to (in brackets)

(a) Start with the **base**: $log_m(\ \ ) = \_\_$

☞ Next, **think:** *The logarithm is an **exponent***

$log_m(\ \ ) = 3$

So, the "125" *must go here*

➡ $\boldsymbol{log_m(125) = 3}$

(b) Start with the **base**: $log_4(\ \ ) = \_\_$

The ***exponent*** goes here

$\boldsymbol{log_4(-3) = x}$

Class Example 3.31 *Converting from Exponential to Logarithmic Form*

Convert each of the following to logarithmic form:

(a) $4^x = \frac{1}{64}$ (b) $y = \left(\frac{1}{2}\right)^{-3}$ (c) $10^{-2} = 0.01$

A logarithm that is base 10 is called the **common log**. With common logs, we do not need to indicate the base.

$log_{10}100$ *is the same as* $log100$ — *No base specified – that means base* ***10****!*

ENRICHMENT: A logarithm that is base "*e*" is called the **natural log**.

$log_e(x)$ *is the same as* $ln(x)$ — *The natural log is written* ln, *and the base is "e"*

"e**" is an irrational mathematical constant, equal to approximately **2.718**, that has many applications in exponential growth. The base "e" and the natural log are studied in Math 31.*

To get "e" on your calculator, key in: 2nd ÷

*So, your calculator actually has **two** log buttons:* log ← base **10**; ln ← base "*e*"

To evaluate a logarithm or solve for an unknown value in a log expression, we **convert to exponential form.**

Worked Examples *Solve each for x:* (a) $log_3 243 = x$ (b) $log_{1/2} x = -4$

Solution: *Convert each to exponential form...*

$(\ )^{\square} = \square$ — base; What the log is equal to (*logs are exponents*); What the power term is equal to

(a) Start with the **base**: $3^x = 243$

Next, write the constant in base 3, and *set exps equal* $3^x = 3^5$ ➡ $x = 5$

(b) $\left(\frac{1}{2}\right)^{-4} = x$

$(2)^4 = x$ ➡ $x = 16$

CALC TIP

The default base for logarithms on your calculator is base 10, the common logarithm

$10^2 = 100$

For other bases, your calculator may have a "LOGBASE" function.

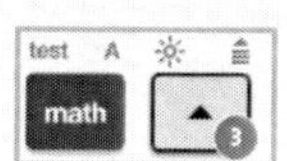

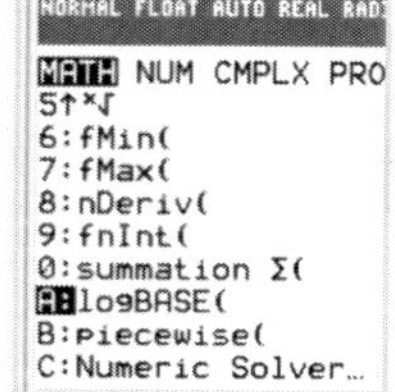

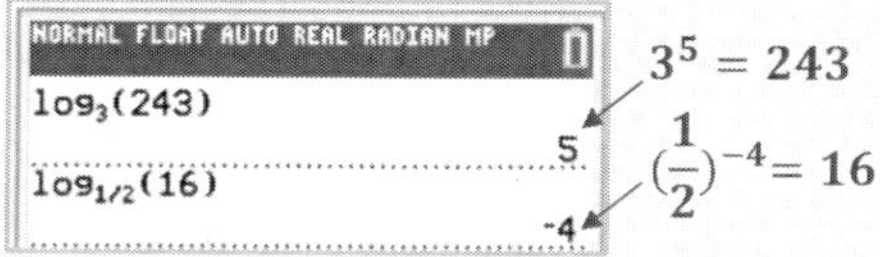

$3^5 = 243$

$\left(\frac{1}{2}\right)^{-4} = 16$

If your calculator does not have LOGBASE, you'll have to use the change of base identity.

Jump ahead to page 179 to see how to do this!

Class Example 3.32 *Converting to Exponential Form*

Determine the unknown value for each *(Try without using the LOG function of your calculator)*:

(a) $log_5\left(\frac{1}{125}\right) = x$ (b) $log_2(x) = 5$ (c) $log_b 81 = 4$ (d) $log_{64} 512 = x$

Class Example 3.33 *Evaluating Logarithms*

Without using the LOG function of your calculator - evaluate each of the following logarithms.

(a) $log_2\left(\frac{1}{32}\right)$ (b) $log_{\frac{1}{3}}(9)$ (c) $log(0.001)$ (d) $log_3(9\sqrt{27})$

(e) $log_b(1)$ (f) $log_b(b)$ (g) $log_2 2^5$

The example above included three noteworthy logarithms:

$log_b(1) = \mathbf{0}$ — Since $b^{\mathbf{0}} = 1$

$log_b(b) = \mathbf{1}$ — Since $b^{\mathbf{1}} = b$

$log_b(b^n) = \boldsymbol{n}$ — Since $b^{\boldsymbol{n}} = b^n$

Class Example 3.34 *Evaluating Logarithms – Exam Style question*

A student is using an algebraic approach to solve the equation $log_{16}8\sqrt[3]{4} = x$.

One of his steps is to write the linear equation $ax = \frac{bc}{3}$, where bc represents the two-digit numerator and $a, b, c, \in I$.

NR The values of a, b, and c, are, respectively, ____, ____, and ____.

The Change of Base Identity

Exploration #3

1 ➡ Fill in the blanks – evaluate each of the following logarithms by converting to exponential form:

(a) i $log_{16}(256) =$ ____ *(Think: $16^{\square} = 256$?)*

ii $\dfrac{log_4(256)}{log_4(16)} = \dfrac{\_\_\_\_}{\_\_\_\_} =$ ____ *(Think: $4^{\square} = 256$? $log_4(256)$; $log_4(16)$ Think: $4^{\square} = 16$?)*

iii $\dfrac{log_2(256)}{log_2(16)} = \dfrac{\_\_\_\_}{\_\_\_\_} =$ ____ *($2^{\square} = 256$? $log_2(256)$; $log_2(16)$ $2^{\square} = 16$?)*

iv Compare your results from i, ii, and iii above

(b) i $log_9(6561) =$ ____ *(Think: $9^{\square} = 6561$?)*

ii $\dfrac{log_3(6561)}{log_3(9)} = \dfrac{\_\_\_\_}{\_\_\_\_} =$ ____ *($log_3(6561)$; $log_3(9)$)*

iv Compare your results from i and ii above

2 ➡ Fill in the blanks: i $log_5(125) =$ ____ *($5^{\square} = 125$?)*

Use your **calculator** to determine the value of: ii $\dfrac{log(125)}{log(5)} =$ ____ iii $\dfrac{ln(125)}{ln(5)} =$ ____ *(Evaluate using your calculator)*

A reminder that natural logs, which are heavily used in Math 31, are not part of the curriculum for this course.

iv Compare your results from i, ii, and iii above

3 ➡ Fill in the blanks: i $log_2(32) =$ ____

Use your **calculator** to determine the value of: ii $\dfrac{log(32)}{log(2)} =$ ____ iii $\dfrac{ln(32)}{ln(2)} =$ ____ *(Evaluate using your calculator)*

iv Compare your results from i, ii, and iii above

The **Change of Base Identity** shows how any logarithm of any base b, can be written using ***any other base*** $\boldsymbol{a}$.

$$log_b(c) = \frac{log_a c}{log_a b}$$

One consequence of this is that we can evaluate any logarithm on our calculator, by expressing as common logs. (base 10) *This is particularly handy if your calculator does not have LOGBASE!*

For example, $log_2(32)$ can be found by converting to base 10:

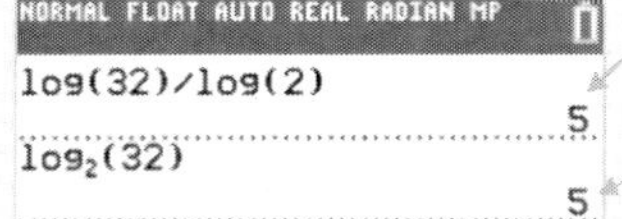

Evaluate in base 10, using change of base. *Note you could also use $ln(32) \div ln(2)$.*

You get the same result using LOGBASE!

Graphs of Logarithmic Functions

We already saw how the graph of logarithmic function is *the inverse* of an exponential function.

→ *The inverse of the exponential function* $y = b^x$ *is* $\boldsymbol{x = b^y}$, *which can be written* $\boldsymbol{y = log_b x}$

> *note* *The domain of* $\boldsymbol{y = log_b x}$ *is restricted, as you can't take the log of zero or negatives.*
>
> *Consider trying to evaluate* $\boldsymbol{log_2(0)}$, *or* $\boldsymbol{log_2(-4)}$
>
> *This is asking:* $2^{\square} = 0$? *And this is:* $2^{\square} = -4$?
>
> ***Neither is possible – domain is*** $\{x|x > 0, x \in \mathbb{R}\}$

We should be familiar with the graph of $\boldsymbol{y = log_b x}$, when $b > 1$ and when $0 < b < 1$.

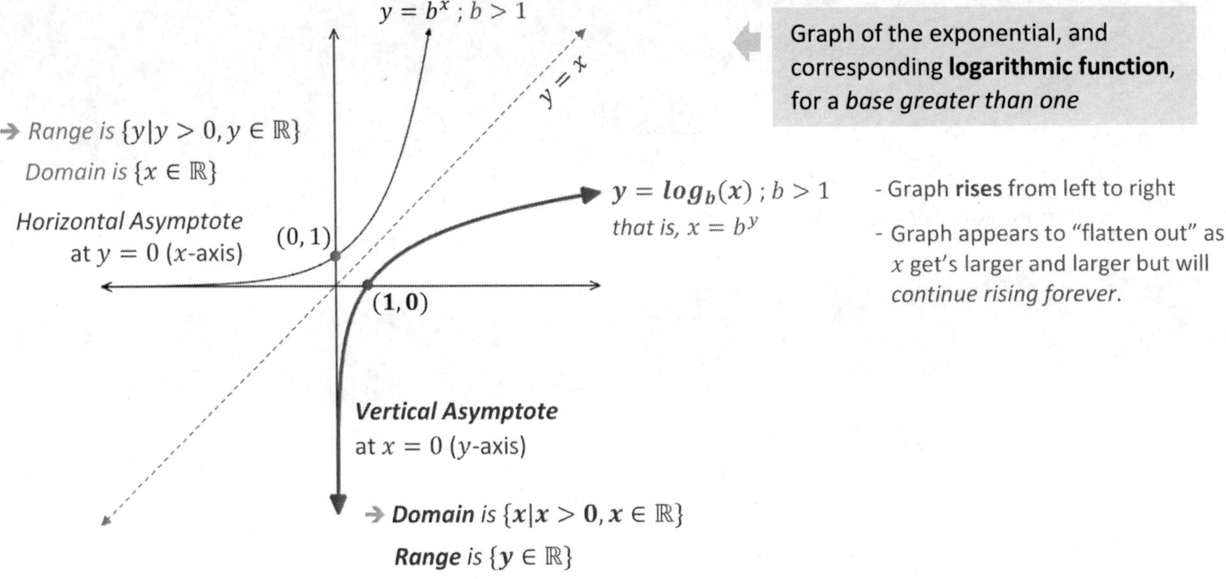

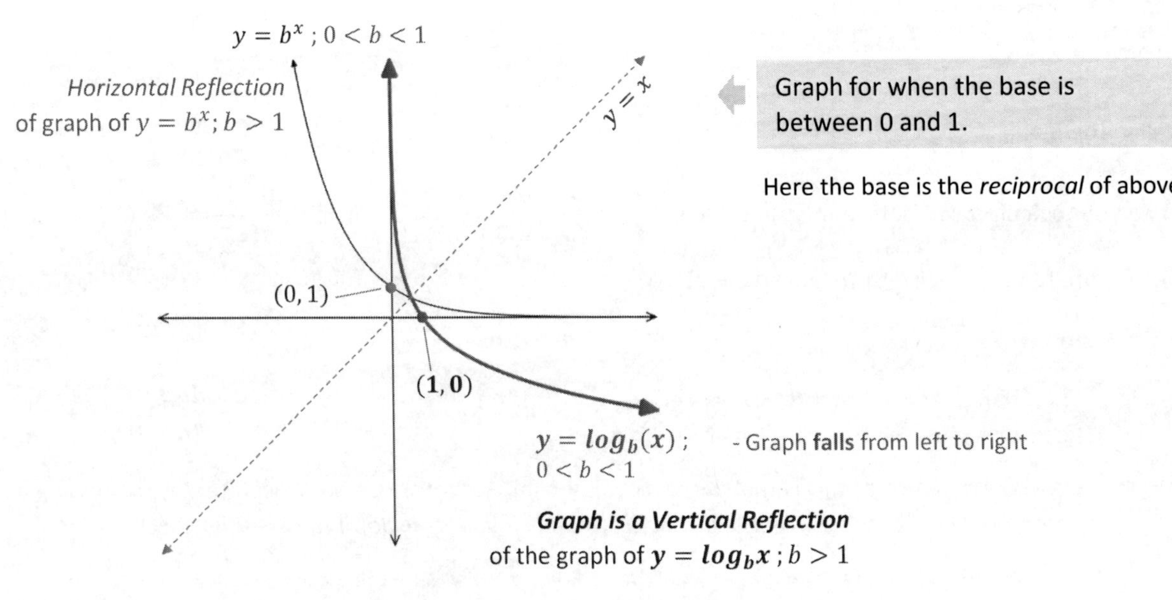

We can explore the graphs described above using our **graphing calculators**, with bases of 2 and ½:

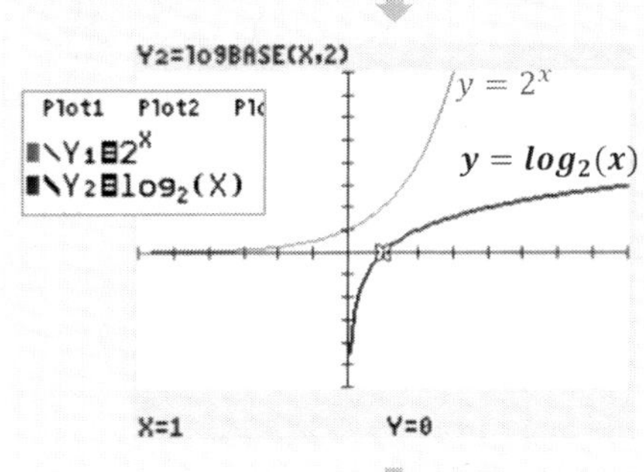

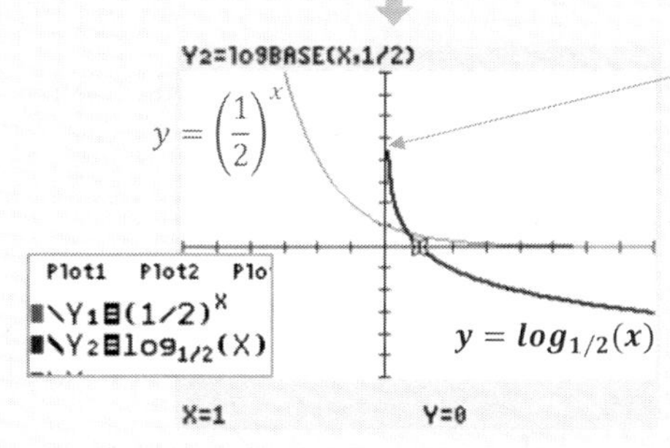

Calculator lies: *It appears that the log graph "starts" here, in fact it's a resolution issue – there is a vertical asymptote at* $x = 0$. (So graph continues "up" forever!)

Range is $\{y \in \mathbb{R}\}$

Recall you can enter this using the change of base identity, $y = \frac{log(x)}{log(2)}$ *... If you don't have LOGBASE on your calc!*

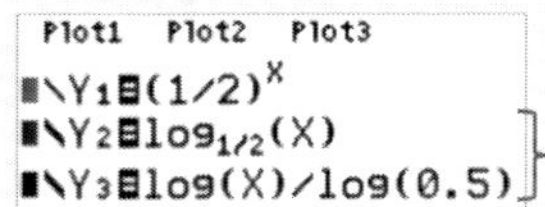

These are identical

Note: We can also graph these on **desmos.com**:

(There you can enter function in "x=" form)

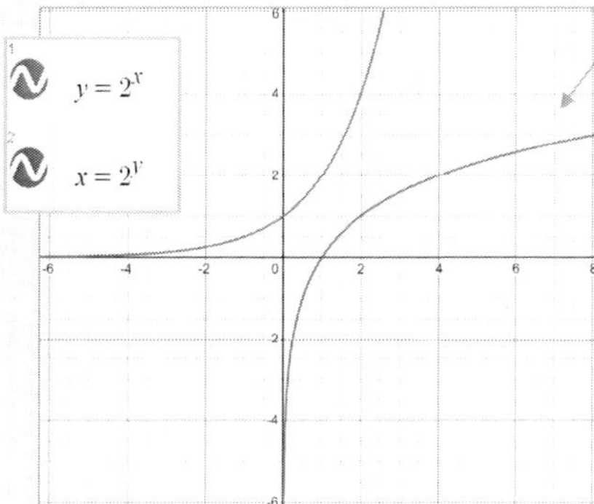

This is the graph of $\boldsymbol{y = log_2(x)}$.

(On DESMOS we can enter the simpler form of the INVERSE, $\boldsymbol{x = 2^y}$)

We'd get the *same graphs* inputting this:

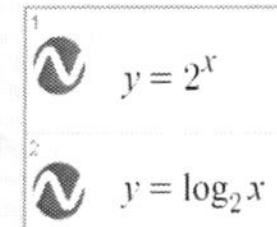

Note: Key in "shift" ... "dash" to enter the base 2

Worked Example For the function $f(x) = log_3(2x + 3)$, state the domain, range, and coordinates of any intercepts, and asymptote. Sketch the graph.

Solution:

For $\boldsymbol{y}$-intercept, substitute $x = 0$:

$y = log_3(2(\mathbf{0}) + 3)$

$y = log_3(3)$ ← *Think:* $3^{\square} = 3$ *?*

$y = 1$

$(0, 1)$

For $\boldsymbol{x}$-intercept, substitute $y = 0$:

$\mathbf{0} = log_3(2x + 3)$

$3^0 = 2x + 3$ *Convert to exponential form*

$1 = 2x + 3$

$-2 = 2x$

$x = -1$

$(-1, 0)$

For ***domain***:

$f(x) = log_3(\mathbf{2x + 3})$

What's being logged must be greater than zero

$2x + 3 > 0$

$2x > -3$

$\{\boldsymbol{x | x > -3/2, x \in \mathbb{R}}\}$

The domain also defines the V.A:

V.A at $\boldsymbol{x = -3/2}$

Range is $\{\boldsymbol{y} \in \mathbb{R}\}$

(for all log functions)

V.A. at $x = -\frac{3}{2}$

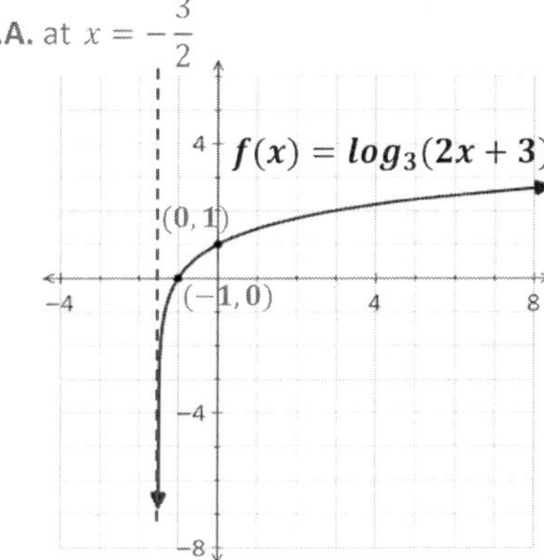

Class Example 3.35 *Analyzing a Logarithmic Function*

Consider the function $\boldsymbol{g(x) = log_2(5 - 2x)}$. Use an algebraic process to determine the:

(i) domain, (ii) range, (iii) asymptote, (iv) coordinates of any intercepts
(exact values, for the y-intercept express as a logarithm)

Sketch the graph, labeling all relevant characteristics.

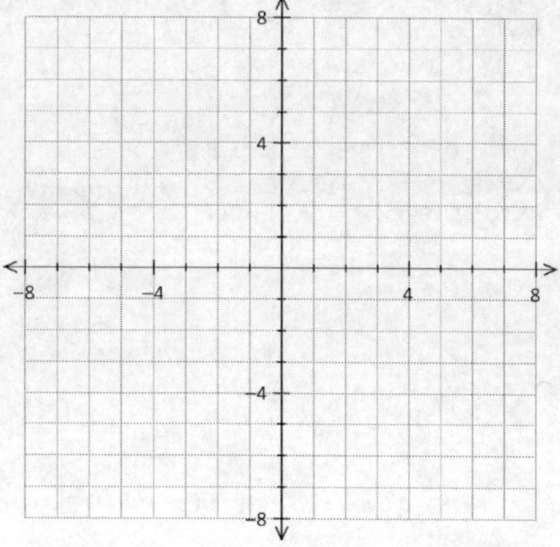

Class Example 3.36 *Analyzing a Logarithmic Function*

A function is defined by $\boldsymbol{f(x) = 2(3)^x - 6}$.

(a) Determine the equation of the simplified inverse function, $f^{-1}(x)$

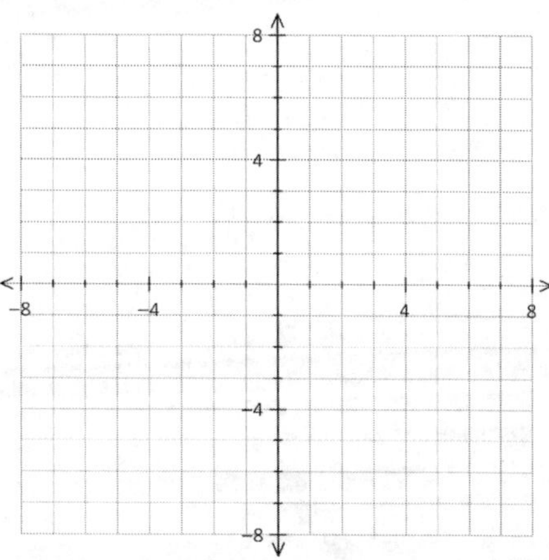

(b) Determine the domain, asymptote, x-intercept, and y-intercept of $y = f^{-1}(x)$. Sketch.

3.4 Logarithm Laws

In section 4.1 we reviewed the exponent rules. A corresponding concept is the Laws of Logarithms. *Remember – the value of a logarithm is an exponent, after all!*

We start by considering the sum of two logarithms of the same base, such as: $log_2 8 + log_2 4$

Is there a simple way to evaluate this? See if you can spot the patterns below.

1 ➡ $\mathbf{log_2 8 + log_2 4}$

$2^{\square} = 8$? $2^{\square} = 4$?

$= \underline{3} + \underline{2}$

$= \underline{5}$ *Copy answer here*

Now find x: $log_2 x = \underline{5}$

$x = \underline{32}$

So.....

$\mathbf{log_2 8 + log_2 4} = \underline{log_2 32}$

2 ➡ $\mathbf{log_3 9 + log_3 27}$

$3^{\square} = 9$? $3^{\square} = 27$?

$= \_\_\_\_ + \_\_\_\_$

$= \_\_\_\_$ *Copy answer here*

Now find x: $log_3 x = \_\_\_\_$

$x = \_\_\_\_$

So.....

$\mathbf{log_3 9 + log_3 27} = \_\_\_\_\_\_\_\_$

3 ➡ $\mathbf{log 100\,000 - log 1000}$

$10^{\square} = 100\,000$? $10^{\square} = 1000$?

$= \_\_\_\_ - \_\_\_\_$

$= \_\_\_\_$ *Copy answer here*

Now find x: $log x = \_\_\_\_$

$x = \_\_\_\_$

So.....

$\mathbf{log 100\,000 - log 1000} = \_\_\_\_\_\_\_\_$

---

4 ➡ $\mathbf{log_2 8^2} = log_2(\_\_\_)$

$= \_\_\_\_$

Now evaluate this:

$= 2 log_2 8$

$= 2(\_\_\_)$ ➡ $= \_\_\_\_$ *So.....* $\mathbf{log_2 8^2} = \_\_\_\_\_\_\_\_$

The LOGARITHM LAWS:

❶ $log_b(M \times N) = log_b M + log_b N$

❷ $log_b\left(\frac{M}{N}\right) = log_b M - log_b N$

❸ $log_b(M)^n = n log_b M$

❹ $log_b(c) = \frac{log_a c}{log_a b}$

Change of base identity, introduced in 3.3

Example:

$\underbrace{log(1\,000\,000)}_{=6} = \underbrace{log_b 1000}_{=3} + \underbrace{log_b 1000}_{3}$

$\underbrace{log_2(4)}_{=2} = log_2\left(\frac{64}{16}\right) = \underbrace{log_2 64}_{=6} - \underbrace{log_2 16}_{4}$

$\underbrace{log_3 729}_{=6} = log_3(9^3) = 3\underbrace{log_3 9}_{=2} = 3(2) = \mathbf{6}$

$\underbrace{log_{16} 256}_{=2} = \frac{log_4 256}{\underbrace{log_4 16}_{=\frac{4}{2}}} = \frac{log_2 256}{\underbrace{log_2 16}_{=\frac{8}{4}}} = \frac{log 256}{\underbrace{log 16}_{}}$ *Verify on calc!*

Worked Examples Evaluate each expression (a) $\log 50 + \log 20$ (b) $\log_3 108 + \log_3 10 - \log_3 120$

Solution:

(a) Write as single log:

$= log(50 \times 20)$

$= log(1000)$ ➡ $\boxed{= 3}$

(b) *Pos (+) logs go up here*

$= log_3\left(\frac{\quad}{\quad}\right)$ ➡ $= log_3\left(\frac{108 * 10}{120}\right)$ ➡ $= log_3(9)$

Neg (-) logs down here

➡ $\boxed{= 2}$

Class Example 3.41 *Evaluating a Logarithmic Expression using Log Laws*

Evaluate each of the following:

(a) $log_4 8 + log_4 32$

(b) $log_6 288 - log_6 8$

(c) $log_5 10 + log_5 100 - log_5 8$

Worked Examples Evaluate each using the Logarithm of a Power Law (a) $\log_5 25^6$ (b) $\log_2 \sqrt[4]{8}$

Solution:

(a) $= 6\underbrace{log_5 25}_{2}$ *Since* $5^2 = 25$

$= 6(2)$ ➡ $\boxed{= 12}$

(b) $= log_2 8^{\frac{1}{4}}$ *Re-write radicals using rational exponents*

$= \frac{1}{4} log_2 8$ *Use log law:* $log_b(M)^n = n log_b M$

$= \frac{1}{4}(3)$ ➡ $\boxed{= 3/4}$

Class Example 3.42 *Evaluating a Logarithmic Expression using Log Laws*

Simplify using logs laws, to evaluate each of the following:

(a) $log_3 81^7$

(b) $log_4 \sqrt[3]{32}$

(c) $3log50 - 2log15 + log\left(\frac{9}{50}\right)$

Worked Example Simplify the following expression using log laws $\mathbf{3logA - 2logB - \left(\frac{1}{3}logC^6 - \frac{1}{2}logA\right)}$

Sol.: **Step 1** *#'s in front of log expressions become exponents*

$= logA^3 - logB^2 - log(C^6)^{1/3} + logA^{1/2}$ ← *While we're at it, distribute the (-) sign through the brackets*

pos (+) neg(-) neg (-) pos (+)

$= logA^3 - logB^2 - log\,C^{6\times1/3} + logA^{1/2}$

Step 2 *Prepare to write as a single log... pos. logs, ship term to the numerator, neg. logs, ship to denominator*

Pos (+) logs, terms go up here

$= log_3\left(\frac{\quad}{\quad}\right)$

Neg (-) logs down here

Step 3 *Simplify terms in the brackets using exponent rules*

$= log_3\left(\frac{A^3 \times A^{1/2}}{B^2 \times C^2}\right)$ ➡ $\boxed{= log_3\left(\frac{A^{7/2}}{B^2C^2}\right)}$ *Simplified from $A^{3+1/2}$*

Class Example 3.43 *Simplifying a Logarithmic Expression using Log Laws*

Express each as a single log:

(a) $2log_2x^3 - 4log_2\sqrt{x} - \frac{1}{3}log_2x^9$

(b) $2log_3(xy^3) + 3log_3(x^2z) - [4log_3(\sqrt{y}) + \frac{1}{3}log_3(z)]$

Worked Examples Expand and simplify using log laws: $\mathbf{log_2\left(\frac{\sqrt[4]{x^3}}{8}\right)}$

Solution:

$= log_2x^{3/4} - log_28$ *Expand using the log law: $log_b\left(\frac{M}{N}\right) = log_bM - log_bN$*

$= \frac{3}{4}log_2x - log_28$ *Isolate log term using the log law: $log_b(M)^n = nlog_bM$*

➡ $\boxed{= \frac{3}{4}log_2x - 3}$ *Evaluate log_28 ... **think** $2^{\square} = 8$?*

Class Example 3.44 *Evaluating a Logarithmic Expression using Log Laws*

Expand each of the following using log laws. Evaluate and simplify where possible.

(a) $log_3(81x^7)$

(b) $log_5\left(\frac{\sqrt{A}}{125B^2}\right)$

(c) $log\left(\frac{\sqrt{1000}}{\sqrt[3]{x^2}}\right)$

Worked Examples Working with log laws

If $log_3 2 = x$ and $log_3 5 = y$, express each in terms of x and y.

(a) $\boldsymbol{log_3 1800}$ (b) $\boldsymbol{log_3 \frac{50}{27}}$

Sol.: (a) *Express 1800 as multiples of 2 and 5*

Use your calc to determine how many times "2" and "5" divide into 1800...

```
1800/2        900
Ans/2         450   three factors of 2
Ans/2         225
Ans/5          45   two factors of 5
Ans/5           9
```

... and 9 left over

So, re-write as:

$= log_3 (2^3 \times 5^2 \times 9)$ *This is 1800*

$= log_3 2^3 + log_3 5^2 + log_3 9$

Split up using log laws... goal is to isolate $\boldsymbol{log_3 2}$ *and* $\boldsymbol{log_3 5}$

$= 3log_3 2 + 2log_3 5 + log_3 9$

Given - this is "x" *And this is "y"* *And this log is "2"*

$\Rightarrow$ $\boxed{= 3x + 2y + 2}$

(b) *Similarly break down "50"*

$= log_3 \left(\frac{2 * 5 * 5}{27}\right)$ *this is "3"*

$= log_3 (2 \times 5^2) - log_3 27$

$= log_3 2 + log_3 5^2 - 3$

we must isolate "$\boldsymbol{log_3 5}$*", as we're given that's "y"*

$= log_3 2 + 2log_3 5 - 3$

$\Rightarrow$ $\boxed{= x + 2y - 3}$

Class Example 3.45 *Evaluating a Logarithmic Expression using Log Laws*

If $log_2 3 = A$ and $log_2 5 = B$, express each of the following in terms of A and B.

(a) $log_2 720$ (b) $log_2 \left(\frac{75}{\sqrt{8}}\right)$

Worked Examples Evaluating expressions using log laws

If $log_5 x = 7$ and $log_5 y = 4$, evaluate: $\boldsymbol{log_5 \left(\frac{25x^2}{\sqrt{y}}\right)}$

Sol.: *Split up using log laws:* $= log_5 (25x^2) - log_5 \sqrt{y}$

$= log_5 25 + log_5 x^2 - log_5 y^{1/2}$

this is "2" *Isolate* $\boldsymbol{log_5 x}$ *and* $\boldsymbol{log_5 y}$

$= 2 + 2log_5 x - \frac{1}{2} log_5 y$

given, this is 7 *...and this is 4*

$= 2 + 2(7) - \frac{1}{2}(4)$ $\Rightarrow$ $\boxed{= 14}$

Class Example 3.46 *Evaluating a Logarithmic Expression using Log Laws*

If $log_3 x = 5$ and $log_3 y = 8$, evaluate each of the following expressions.

(a) $log_3 (9x^2 y^3)$ (b) $log_3 \left(\frac{81\sqrt{y}}{x^3}\right)$

Class Example 3.47 *Evaluating a Logarithmic Expression using Log Laws*

The expression $\mathbf{3log_2 4\sqrt[3]{y} = x + 2log_2 z}$, written in terms of y, is:

A. $\frac{2^x z^2}{64}$

B. $\frac{2^x z^2}{4}$

C. $\frac{2^z x^2}{64}$

D. $\frac{2^z x^2}{4}$

Class Example 3.48 *Logarithmic Identities*

Use the laws of logarithms to show that each equation below is true for all values x, where $x, b > 0,\ x, b \neq 1$.

(a) $(log_b x)(log_x b) = 1$

(b) $log_{\frac{1}{b}} x = -log_b x$

3.4 Practice Questions (Sample)

SAMPLE of our high-quality, diploma-exam based practice questions. Visit ***Math30-1edge.com*** *for full access.*

1. Express each as a single logarithm, and then evaluate:

(a) $log20 + log5$ (b) $log_5 10 + log_5 15 + log_5 \frac{25}{6}$ (c) $log_2 80 + log_2 50 - log_2 125$

(d) $3log_3 12 - 3log_3 4$ (e) $2log_2 5 + \frac{1}{2}log_2 36 - log_2 \frac{75}{2}$ (f) $\frac{1}{2}log_5 250 - \frac{1}{2}log_5 2$

2. Use the power logarithm law to evaluate each without a calculator:

(a) $log_2 8^{10}$ (b) $log\sqrt[3]{100}$ (c) $log_5 25^{1.5}$

3. Express each as a single logarithm in simplest form:

(a) $2logA - 3logB$ *For (a) only, also state any restrictions on the variable*

(b) $2log_5(4m) + \frac{1}{2}log_5 n - 3log_5 m$

(c) $2\,log(x^3y) - 3\,log(xy^3) - (logx^2 + \frac{1}{3}logy)$ (d) $2log_b a + 3log_b(2bc) - log_b(2c^3) + 2$

Practice Questions answers (we also provide step-by-step solutions

1. (a) 2 (b) 4 (c) 5 (d) 3 (e) 2 (f) 1.5 **2.** (a) 30 (b) 2/3 (c) 3

3. (a) $log\frac{A^2}{B^3}$ (b) $log_5\frac{16\sqrt{n}}{m}$ (c) $log\frac{x}{y^{22/3}}$ (d) $log_b(4a^2b^5)$

3.5 Solving Exponential Equations using Logs, and Applications

Earlier we saw how some exponential equations can be solved by *re-writing terms in the same base*.

This method works great - when terms *can be* written in the same base: → $2(5)^{x-1} = 50$

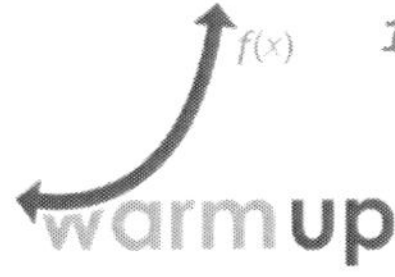

1 ➡ **Algebraically solve** the following equation, by re-writing in the same base.

$$2(5)^{x-1} = 50$$

2 ➡ **Show** that we cannot as easily solve the following equation using the same method

$$2(5)^{x-1} = 60$$

Solving any Exponential Equation

Only some equations can be solved using the method we learned in 4.1, namely re-writing in the same base.
In order to solve any exponential equation we need a broader method – and that method involves **logarithms**.
(bet you saw that coming)

To solve any exponential equation (and not just those special cases where we can equate the bases), we can either:

solve: $\mathbf{2(5)^{x-1} = 60}$

Method 1:

Convert to logarithmic form

$\frac{2(5)^{x-1}}{2} = \frac{60}{2}$ — *First, isolate the power term*

$(5)^{x-1} = 30$ — *Convert to log form*

$log_5(30) = x - 1$ — *Isolate x*

$x = log_5(30) + 1$ — *Exact solution*

$x \approx 3.11$ — *Approx. solution*

Use calculator for approx. solutions → $\log(5)^{x-1} = \log 30$

Method 2:

Or, take the logarithm of both sides

$(5)^{x-1} = 30$ — *Isolate power term*

$log(5)^{x-1} = log30$ — *"log" both sides*

... and isolate x using the power log law

$(x-1)log5 = log30$

$x - 1 = \frac{log30}{log5}$ ➡ $x = \frac{log30}{log5} + 1$

Exact solution ↑
(same as $\boldsymbol{log_5(30) + 1}$)

Class Example 3.51 *Solving an exponential equation using logarithms*

Use an algebraic process to solve each of the following equations. *State solutions correct to the nearest hundredth.*

(a) $(3)2^{5-4x} = 300$

(b) $2(5)^{x-2} = 152$

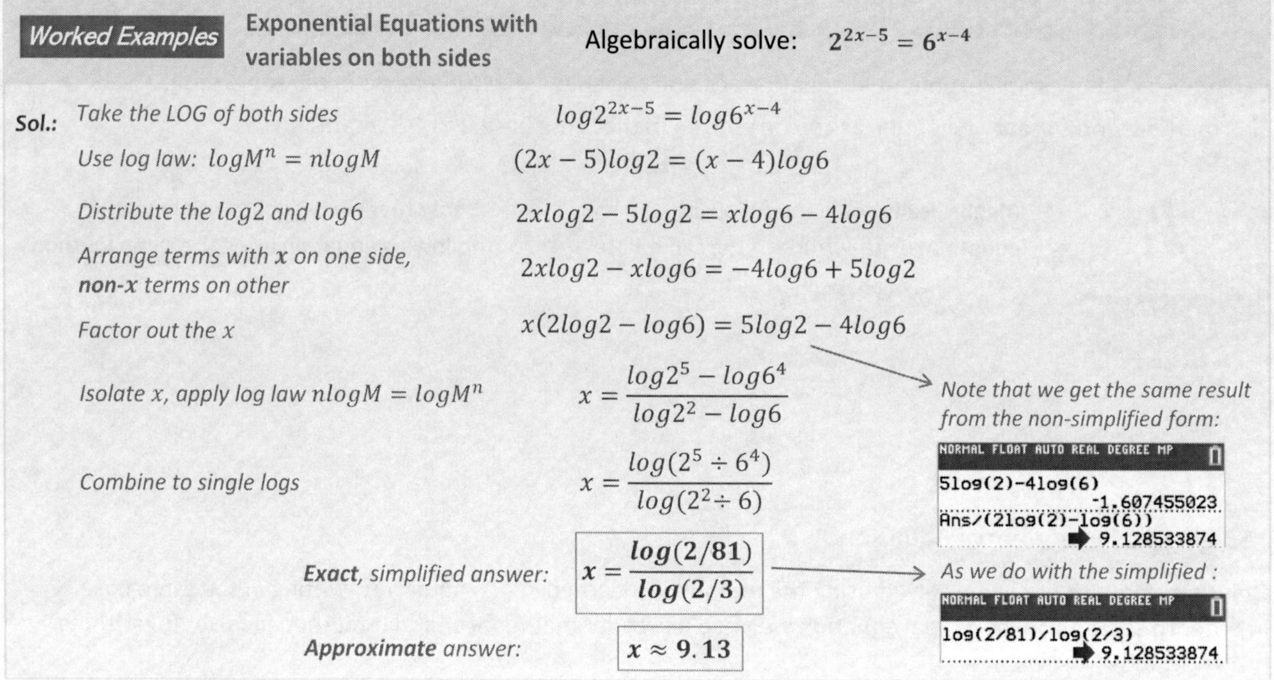

Worked Examples **Exponential Equations with variables on both sides** Algebraically solve: $2^{2x-5} = 6^{x-4}$

Sol.:

Take the LOG of both sides	$log2^{2x-5} = log6^{x-4}$
Use log law: $logM^n = nlogM$	$(2x-5)log2 = (x-4)log6$
Distribute the log2 and log6	$2xlog2 - 5log2 = xlog6 - 4log6$
*Arrange terms with **x** on one side, **non-x** terms on other*	$2xlog2 - xlog6 = -4log6 + 5log2$
Factor out the x	$x(2log2 - log6) = 5log2 - 4log6$
Isolate x, apply log law $nlogM = logM^n$	$x = \dfrac{log2^5 - log6^4}{log2^2 - log6}$
Combine to single logs	$x = \dfrac{log(2^5 \div 6^4)}{log(2^2 \div 6)}$
***Exact**, simplified answer:*	$\boldsymbol{x = \dfrac{log(2/81)}{log(2/3)}}$
***Approximate** answer:*	$\boldsymbol{x \approx 9.13}$

Note that we get the same result from the non-simplified form:

```
NORMAL FLOAT AUTO REAL DEGREE MP
5log(2)-4log(6)
                -1.607455023
Ans/(2log(2)-log(6))
                 9.128533874
```

As we do with the simplified :

```
NORMAL FLOAT AUTO REAL DEGREE MP
log(2/81)/log(2/3)
                 9.128533874
```

Class Example 3.52 *Solving an exponential equation using logarithms*

Use an algebraic process to solve each of the following equations. *State solutions as both simplified exact and approximate values.*

(a) $2^{x+3} = 5^{2x-1}$

(b) $2(6)^{x+2} = 3^{2x-3}$

Applications of Exponential Equations

The world is replete with applications of exponential growth (or, when the values are decreasing over time, decay). And success in this unit includes being able to set up (and solve) exponential equations based on real world problems. *Consider each of the following scenarios to provide a response for each. Answers are at the bottom of the next page.*

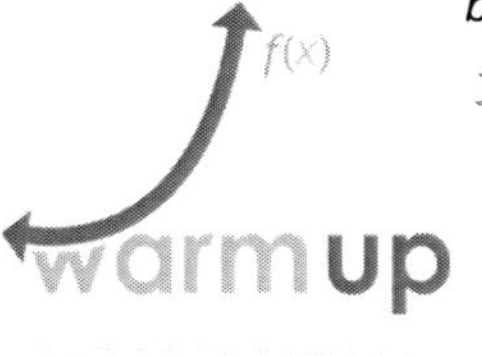

Warm-up #2

1 ➡ The amount of bacteria in colony is increasing in an exponential pattern. Initially there are 100 bacteria present, and that amount is doubling every day.

(a) Determine the amount of bacteria that will be present after 4 days. Show your calculations.

(b) Construct an equation that models the amount of bacteria present, y, after t days.

2 ➡ The amount of bacteria in colony is rapidly increasing in an exponential pattern. Initially there are 100 bacteria present, and that amount is doubling every two hours.

(a) Determine the amount of bacteria that will be present after 6 hours. Show your calculations.

(b) Construct an equation that models the amount of bacteria present, y, after t hours.

3 ➡ The amount of bacteria in colony is rapidly increasing in an exponential pattern. Initially there are 1000 bacteria present. The amount is increasing by 15% per hour.

(a) Determine the amount of bacteria that will be present after 3 hours. Show your calculations.

(b) Construct an equation that models the amount of bacteria present, y, after t hours.

For modelling exponential growth (or decay) applications, we can use: $y = a(b)^{\frac{t}{p}}$

Where some initial amount a grows to y after some amount of time t.
And b is the *multiplication factor of growth*, which is applied *every p units of time.*

$y = a(b)^{\frac{t}{p}}$

Initial amount — *Period in which b is applied* — *growth factor*

if $b > 1$ → **growth*
*$0 < b < 1$ → **decay***

***Type 1** – Values are doubling (or tripling, or "half-ing", etc)*

b *is the factor of growth (or decay). Values are doubling, $b = 2$ Tripling, $b = 3$ Half-life problems, $b = 1/2$. (and so on)*

p *is the period of time which the growth occurs. example) In #9 above $p = 3$, for #10, $p = 0.5$*

Type 2** – Values are increasing (or decreasing) by some **percentage

b *here $b = 1 +$ **growth rate*** (as a decimal) *if decreasing, growth rate is negative*

p *for these "percentage growth" type problems is often (but not always) **1**, such as #8 above - decrease 12% per "1" year*

Class Example 3.53 *Setting up exponential growth equations*

For each of the scenarios below, **construct an equation** that models the amount of some quantity ($\mathbf{y}$) after "$\mathbf{t}$" units of time.

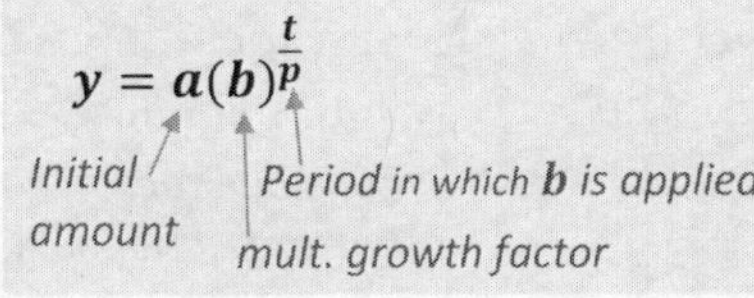

1. 100 bacteria are present in a culture, which is doubling in size every day.

1b. Use your equation to determine the amount of bacteria after 5 days.

2. 100 bacteria are present in a culture, which is tripling in size every day.

3. 100 bacteria are present in a culture, which is doubling in size every three days.

4. 100 bacteria are present in a culture that is decreasing in size. The half-life for the amount bacteria is 5 days.

5. 150 bacteria are present in a culture. Every four days, 75% of the bacteria remain.

6. A \$1000 investment is made, that will earn 4% interest per year.

6b. Use your equation to determine the amount of money after 5 years.

7. The population of a country is 25 million and is forecast to grow by 1.4% per year.

8. A \$40 000 new car will depreciate in value by 12% per year.

9. A \$40 000 new car will depreciate in value by 30% every three years.

10. A \$1000 investment is made, earns 5% annual interest, compounded semi-annually. Determine an equation that models the value after t years. Note: Semi-annually means "twice per year", or once every half-year

Worked Example

The population Stonewall, Manitoba, is 4800. Provincial forecasts predict that the population will increase by 2.1% per year. According to this model, determine how long it will take, correct to the nearest tenth of a year, for the population to reach 6000.

Solution:

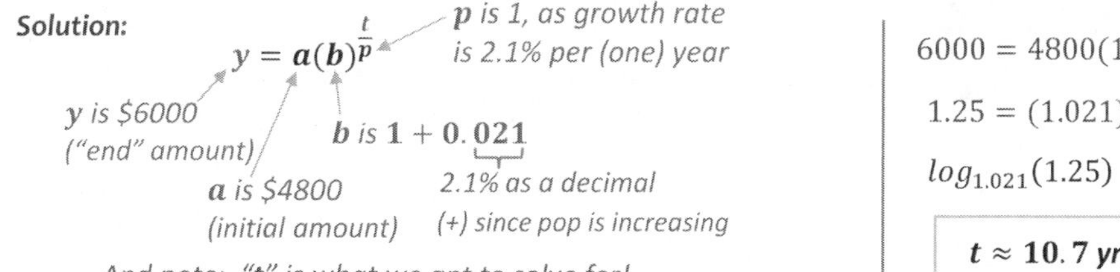

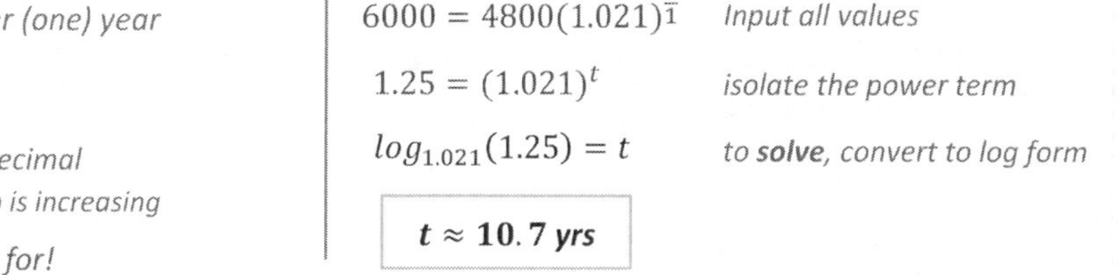

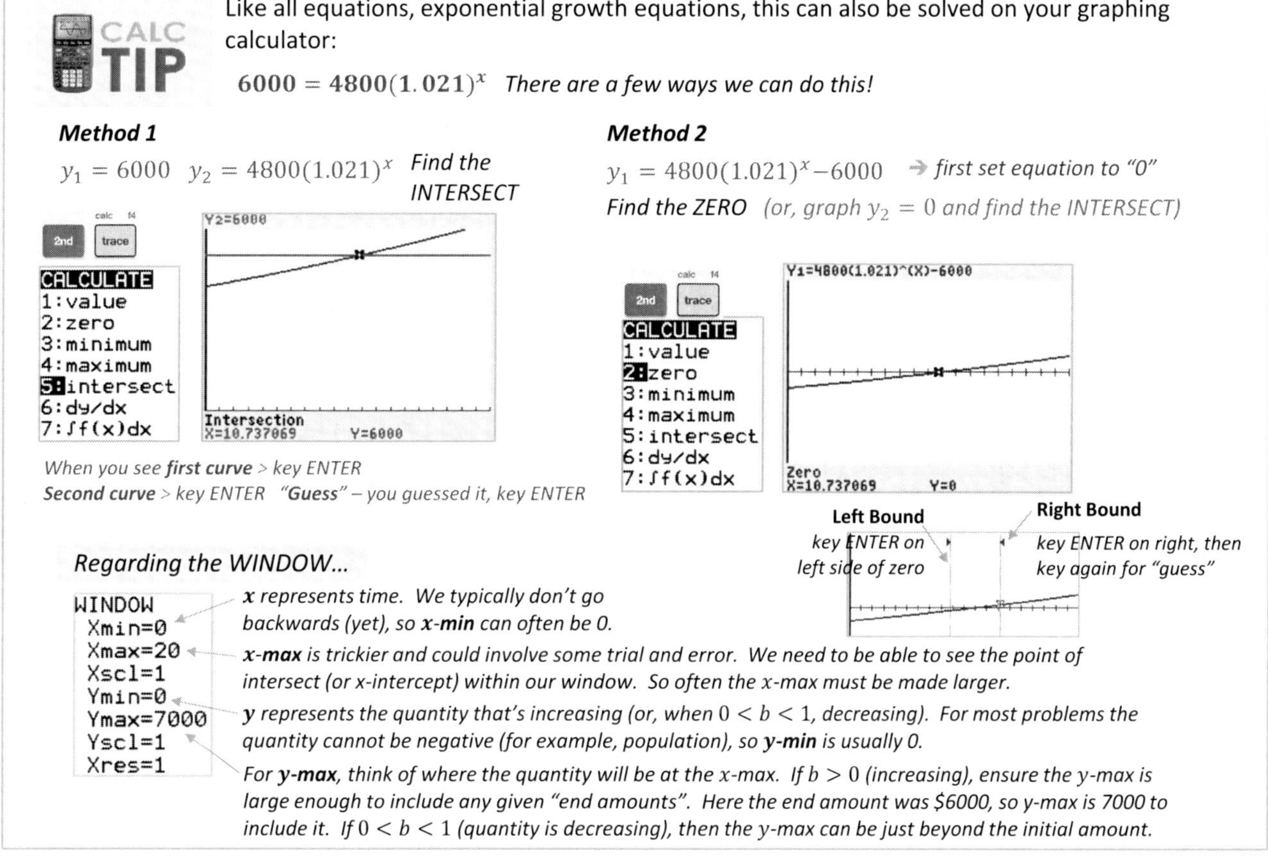

Class Example 3.54 *Modeling exponential growth problems*

A particular strong investment fund promises investors an annual return of 12%. Assuming this growth rate is obtained, determine how long it would take for a $10 000 investment to grow to $25 000, correct to the nearest tenth of a year.

A Closer Look at Half-Life

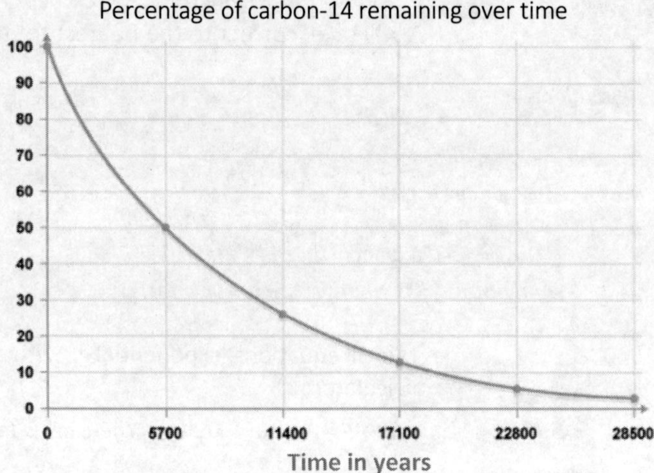

Have you ever heard of *carbon-dating*? Scientists use this method to determine the age of materials that originated from living (carbon-based) organisms.

Carbon-14 is a radioactive isotope present in all organic materials. It has a **half-life** of 5730 years, which means every 5730 years, the amount present is cut in half.

The graph on the right shows the percentage of carbon-14 present over time. Notice that after 5700 years (rounded), the amount is 50% (half). Then after 11400 years, it's half of that, or 25%. (and so-on)

A function that models the percentage of carbon-14 present as a function of time in years is:

$$P = 100\left(\frac{1}{2}\right)^{\frac{t}{5730}}$$

Notice that *in t = 5730 years, the exponent will be "1", so the initial percentage (100) will be multiplied by ½ once.*

And in t = 11460 years, the exponent will be "2", so the initial percentage (100) will be multiplied by ½ twice. (It will be 1/4th)

Worked Example

5 mg of a radioactive substance is slowly decaying, such that its half-life is 211 years. Determine how long it would take for the amount of the substance to decay to 1 mg.

Solution:

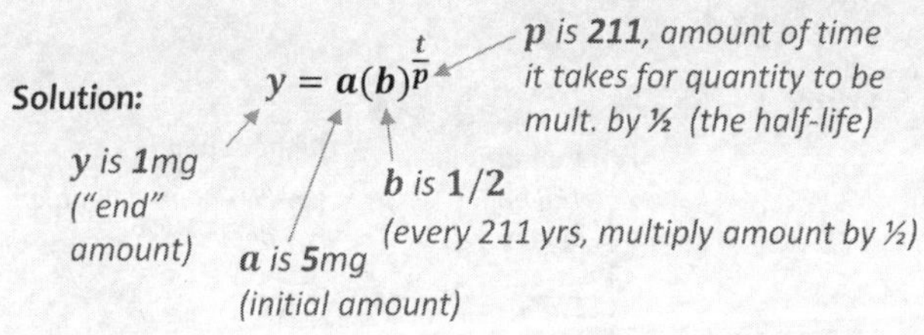

And ***note****: For "doubling period / half-life" problems, we often need to decide:*

Are we GIVEN "p", or do we WANT "p"? Here – we are given "p", that is given the half-life.

$$1\,mg = 5\,mg\left(\frac{1}{2}\right)^{\frac{t}{211}}$$ *Input all values*

$$\frac{1}{5} = \left(\frac{1}{2}\right)^{\frac{t}{211}}$$ *isolate the power term*

$$log_{1/2}(1/5) = \frac{t}{211}$$ *to* ***solve****, convert to log form*

$$t = 211log_{1/2}(1/5)$$

$$\boxed{t \approx 490\ yrs}$$

Class Example 3.55 *Application of Half-life*

A gallbladder scan involved the injection of 0.65 cc's of Technetium-99m, a radioactive isotope. After 10 hours there was 0.20 cc's measured in the body.

(a) Determine the half-life of Technetium-99m.
Correct to the nearest tenth

(b) Determine the rate of loss per hour of the substance. *Correct to the nearest tenth of a percent.*

Class Example 3.56 *Modelling exponential growth – equation forms*

On January 1st, 1998, the population of Ufa, Russia, was 941 500. By January 1st, 2020, the population had gown to 1 136 000. Assuming the rate of growth for the population remains constant,

(a) Determine how long it would take for the population of Ufa to double, correct to the nearest year.

(b) Determine the average annual growth rate for Ufa, correct to the nearest hundredth of a percent.

Class Example 3.57 *Exponential decay*

In a particular body of water, the intensity of light decreases by 15% every 3 metres of depth. Assuming that at the surface the light intensity is at 100%, determine:

(a) The percentage, correct the nearest whole number, of light that would be present at a depth of 10 m, correct to the nearest whole number.

(b) The depth at which the light intensity would be just 10% of that on the surface, correct to the nearest whole number.

While being able to provide algebraic solutions is essential - these problems can also be solved using your graphing calc.

For example in 4.56 (a) we must solve: $1\ 136\ 000 = 941\ 500(2)^{22/x}$

Graph y_2 = left side *y_1 = right side*

*... and find the point of **intersect***

*Note that we must adjust the window! The x-min and y-min should again be 0, while the x-max **must be large enough to show the point of intersect.***

The y-max must be in the millions as y represents the population.

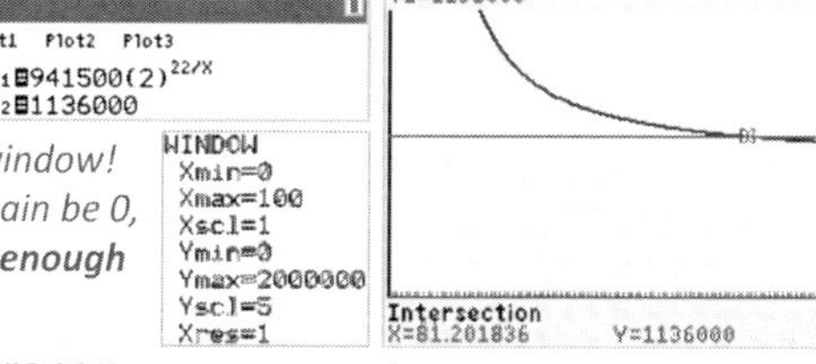

***Solution** is the x-coord. of the pt. of intersection.*

Compound Interest

With investment problems interest is typically applied once per year, as interest rates are always "per year". However annual interest rates can be compounded over any period, such as semi-annually (twice per year), quarterly (four times per year), or monthly.

For example – consider a $1000 investment earning 5% annual interest. How much would it grow to after 4 years, if interest is compounded:

- Annually:

$$A = 1000(1 + 0.05)^4$$

$$= \mathbf{\$1215.51}$$

- Semi-Annually:

$$A = 1000\left(1 + \frac{0.05}{2}\right)^8$$

$$= \mathbf{\$1218.40}$$

Compounded 2 times per yr for 4 years

Interest rate is annual, so divide by # of compounds per yr.

- Daily:

$$A = 1000\left(1 + \frac{0.05}{365}\right)^{1460}$$

$$= \mathbf{\$1221.39}$$

365×4

Interest rate divided by 365 compounds /yr.

It's not an earth-shattering difference, but with more compounding periods per year comes faster growth!

Class Example 3.58 *Compound Interest*

Jeremiah invests $10 000 into a GIC (a guaranteed investment certificate) that earns 5.8% interest compounded quarterly. (4 times per year)

(a) Assuming he makes no additional deposits or withdrawals, predict the value of the investment in 15 years. Round to the nearest whole dollar.

(b) Determine how long it would take for the investment to double. Round to the nearest year.

(c) Construct an alternative function that models the value of the investment after t years, using the doubling period found in (b). Show that this new function is equivalent by determining the value in 15 years, as in (a).

3.6 Solving Logarithmic Equations and Log Scales

1 – Solving Logarithmic Equations

You may recall we solved some logarithmic equations already in section 4.3. Let's see what you remember, try solving each of the following:

1 ➡ Solve: $log_{\frac{1}{3}}(x) = -2$

2 ➡ Solve: $log_{16}(x - 1) = \frac{3}{4}$

For each of these simple equations, finding the solution involves *converting to exponential form*. This is a common method for solving logarithmic equation, so let's call it a ***type 1 logarithmic equation***.

Now, as we're want to do in Math 30-1, we're going to kick things up a notch. Let's consider logarithmic equations that involve first applying laws of logarithms or some other simplification.

3 ➡ Simplify the left side of the following equation using log laws. Then, algebraically solve: $log_2 x - log_2 5 = 3$

4 ➡ Simplify the left side of the following equation using log laws. Then, algebraically solve: $log_3 x + log_3 4 = log_3 24$

5 ➡ **Verify** your solution to #4 by substituting your answer back into the original equation.

6 ➡ **Graphically solve** the equation from #4 using your calculator.

To solve a **logarithmic equation:**

Step 1 On one or both sides, use laws of logarithms to write separate terms as a *single logarithm of coefficient 1.*

Step 2 If there is only a logarithmic term on one side, then convert to exponential form.

If there are log terms on both sides, then set what's being logged on both sides equal to one another. (That is, drop the logarithms)

Step 3 Isolate for x, and numerically verify any answers.

Note: Be suspect of your solutions! Sometimes solutions that appear valid would result in ***logging a negative*** when substituting back into the original equation. That – is not allowed! In such cases we label the offending solution as **EXTRANEOUS and** reject it.

Worked Example Algebraically solve the equation $\boldsymbol{log_6 x + 2 = 3 - log_6(x-1)}$, and numerically verify any solutions.

Solution:

$log_6 x + log_6(x-1) = 1$	*Arrange log terms on one side, non-log term on the other*
$log_6[x(x-1)] = 1$	*Use log laws to combine to a single log*
$x(x-1) = 6^1$	***Convert to exponential form***
$x^2 - x - 6 = 0$	*Expand and set equation to zero*
$(x-3)(x+2) = 0$	*FACTOR to solve*

$x = 3$ *or* ~~$x = -2$~~

$x = -2$ *is* ***EXTRANEOUS***

$$\boxed{x = 3}$$

TEST / verify each solution by substituting back into the original equation:

$log_6(\mathbf{3}) + 2 = 3 - log_6(\mathbf{3} - 1)$

$log_6(3) + 2 = 3 - log_6(2)$

$log_6(\mathbf{-2}) + 2 = 3 - log_6(\mathbf{-2} - 1)$

$log_6(-2) + 2 = 3 - log_6(-3)$

$\boldsymbol{x = -2}$ *is rejected as it would have us taking the log of negatives (not allowed!)*

Solve Graphically:

```
Plot1 Plot2 Plot3
\Y1=log6(X)+log6(X-1)-1
\Y2=0
```

→ *Either graph* $y_1 =$ ***left side***, $y_2 =$ ***right side*** *of equation "as is", or, as in this case, first set the equation to 0.*

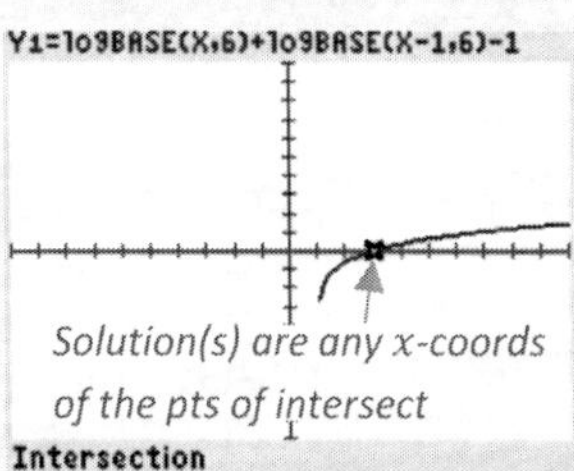

CALC TIP

→ *For LOG equations, or any other similar, EXTRANEOUS solutions do not show up on your graphing calc. This is a useful double-check!*

Class Example 3.61 *Solving Logarithmic Equations that involve a constant term*

Use an algebraic process to solve the following equation. Verify both numerically and graphically.

$$log_9(x-5) + log_9(x+3) = 1$$

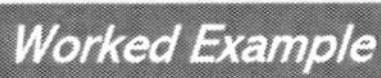

Algebraically solve the equation $\boldsymbol{log_3(3-x) = log_3(x+5) + log_3 3}$, and numerically verify any solutions.

Sol: $log_3(3-x) - log_3(x+5) = log_3 3$ *Arrange the "x" log terms on one side, "non-x" log terms on the other*

$$log_3\left(\frac{3-x}{x+5}\right) = log_3 3$$ *Use log laws to simplify each side into a single logarithm with a coefficient of 1*

$$\frac{3-x}{x+5} = 3$$ *Since the logs are the same, the arguments (what's "being logged") must equal one another. Set the arguments equal.* ***(drop the logarithms)***

Cross-multiply to solve

$$3(x+5) = 3-x$$

$$3x + 15 = 3 - x$$

$$4x = -12$$

$$\boldsymbol{x = -3}$$

Test / verify solution by substituting back into the original equation:

$$log_3(3-(\mathbf{-3})) - log_3(\mathbf{-3}+5) = log_3 3$$

$$log_3(6) - log_3(2) = log_3 3$$

$$log_3\left(\frac{6}{2}\right) = log_3 3 \checkmark$$

NOTE that even though the solution is negative, substituting into the original equation does not result in "logging negatives". So we do not reject this solution!

Class Example 3.62 *Solving Logarithmic Equations where all terms involve logs*

Use an algebraic process to solve each of the following equations. Verify both numerically and graphically.

(a) $log(9-2x) - log(x-2) = logx$

(b) $log_2(2-x) = log_2 6 - log_2(1-x)$

Class Example 3.63 *Solving Logarithmic Equations that involve a constant term*

Use an algebraic process to solve the following equation. Verify numerically.

$$log_4(x+2) - log_4(x-4) = \frac{1}{2}$$

2 – Applications Involving Logarithmic Scales

The Richter Scale

The Richter scale (developed by Charles Richter in 1935) is used to compare the relative size of earthquakes.

The Richter scale is *logarithmic*, an increase of one on the scale represents a tenfold increase in earthquake intensity. So, an earthquake measuring 5.3 is ten times as intense as one measuring 4.3, and a hundred times as intense as one measuring 3.3.

Richter Scale of Earthquake Magnitude

Magnitude level	Category	Effects	Occurrence / year
Less than 1.0 to 2.9	micro	Generally, not felt without special instruments	over 100 000
3.0 to 3.9	minor	Felt by many people; no damage	12000 to 100 000
4.0 to 4.9	light	Felt by all; minor breakage of objects	2000 to 12 000
5.0 to 5.9	moderate	Some damage to weak structures	200 to 2000
6.0 to 6.9	strong	Moderate damage in populated areas	20 to 200
7.0 to 7.9	major	Serious damage over large areas; loss of life	3 to 20
8.0 and higher	great	Severe destruction and loss of life over large areas	fewer than 3

Richter values are determined by taking the logarithm of the amplitude (height) of the largest seismic wave. However, the problems we encounter in Math 30-1 involve comparing Richter scale (and other log scale) values.

Earthquake intensity is given by: $\boldsymbol{I = I_0 \times 10^M}$

Where:

I is the earthquake intensity

M is the magnitude, or **Richter** value

I_0 is a *reference intensity*, or intensity on a standard day

An earthquake in New York in 1884 had a magnitude measured at 5.5 on the Richter Scale. 22 years later an earthquake in San Francisco had a magnitude of 7.9. How many times as intense was the San Francisco earthquake, correct to the nearest whole number?

Solution: *We want to find:* $\frac{I_{SF}}{I_{NY}}$ *That is, the intensity of the San Francisco earthquake* ***divided by*** *the intensity of the one in New York. For each earthquake, use:* $I = I_0 \times 10^M$

So that gives us: $\frac{I_{SF} = I_0 \times 10^{7.9}}{I_{NY} = I_0 \times 10^{5.5}} \Rightarrow = \frac{\cancel{I_0} \times 10^{7.9}}{\cancel{I_0} \times 10^{5.5}} \Rightarrow = 10^{7.9-5.5} \Rightarrow =$ 251 *times as intense*

To find how many times as intense one earthquake, measuring M_1 on the Richter scale is compared to another (smaller) earthquake, measuring M_2 on the Richter scale, use:

$$\frac{I_1}{I_2} = \frac{\cancel{I_0} \times 10^{M1}}{\cancel{I_0} \times 10^{M2}} \Rightarrow \frac{I_1}{I_2} = \frac{10^{M1}}{10^{M2}} \Rightarrow \boxed{\frac{I_1}{I_2} = 10^{M1-M2}}$$

That is, $\boldsymbol{10^{M1-M2}} =$ *"times as intense"*

Class Example 3.64 *Log Scales – determining the order of magnitude*

A 2009 earthquake in Italy measured 6.3 on the Richter scale. How many times more intense was a 1950 earthquake in India, which measured 8.7? *Answer to the nearest whole number.*

Worked Example A 1963 earthquake in Macedonia measured 6.9 on the Richter scale. One year later an earthquake in Alaska had 200 times the intensity. Determine the magnitude (Richter scale value) of the Alaska earthquake.

Sol: *This time we are given:* $\frac{I_A}{I_M} = \mathbf{200}$ *For each earthquake, intensity is given by:* $\boldsymbol{I = I_0 \times 10^M}$ ← *Richter value*

$200 = \frac{I_0 \times 10^x}{I_0 \times 10^{6.9}}$ ⇒ $200 = 10^{x-6.9}$ ⇒ $log_{10}200 = x - 6.9$ ⇒ $x = log_{10}200 + 6.9$ ⇒ $\boxed{\boldsymbol{x = 9.2}}$

Convert to log form to solve

Class Example 3.65 *Determining a Richter scale value*

An earthquake in Loma Prieta, California in 1989 measured 7.1 on the Richter scale, and collapsed a section of the San Francisco – Oakland Bay Bridge. Determine the Richter scale value of an earthquake that had $1/2000$th the intensity. *Answer to the nearest tenth.*

The Ph Scale

Like the Richter scale, the pH scale (0 to 14) is logarithmic. A change of one on the **Ph scale** results in a tenfold change in the hydrogen ion (H^+) concentration.

The pH of pure water is 7. (It's neutral).
Solutions lower than 7 are **acidic**. Every decrease one down the pH scale represents a tenfold increase in H^+ concentration relative to pH 7.

Solutions greater than 7 are **basic**. Every increase one up on the pH scale represents a tenfold decrease in H^+ concentration relative to pH 7. An alkali is a water-soluble base, so all **alkaline** solutions are also basic.

pH is determined by taking the negative logarithm of the hydrogen ion concentration: $\boldsymbol{pH = -log[H^+]}$

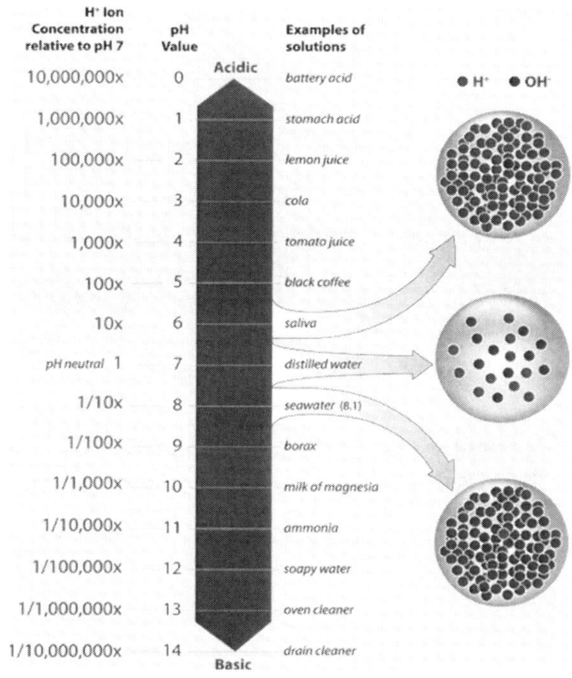

To find how many times as acidic or alkaline one solution is to another, use:
$\mathbf{10^{pH_1 - pH_2}} =$ *"times as acidic / alkaline"*

Note *this is the same approach as Richter Scale problems!*

Class Example 3.66 *Log Scales – Using the pH formula*

The pH of black coffee was measured to be 5.2.
Find the hydrogen ion concentration, in moles/L.
Answer in scientific notation, to one decimal place

Class Example 3.67 *Log Scales – comparing pH levels*

Sea water has a pH of 8.4, and acid rain a pH of 4.6. Determine how many times as alkaline seawater is to pure water, and how many times as acidic acid rain is to pure water.

The Decibel Scale

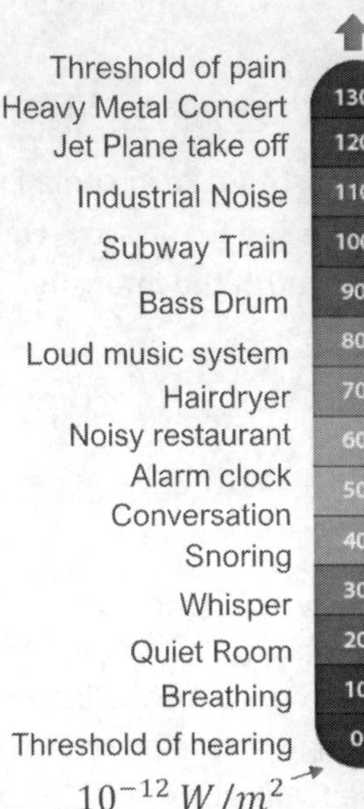

Finally, we come to the Decibel scale. Like the Richter and pH scales, it's logarithmic.

Unlike those scales however, this one has a *scale of 10* associated to it.

Decibel ← Ahh yes, this means ten! So **1** Decibel = 10 "Bels"

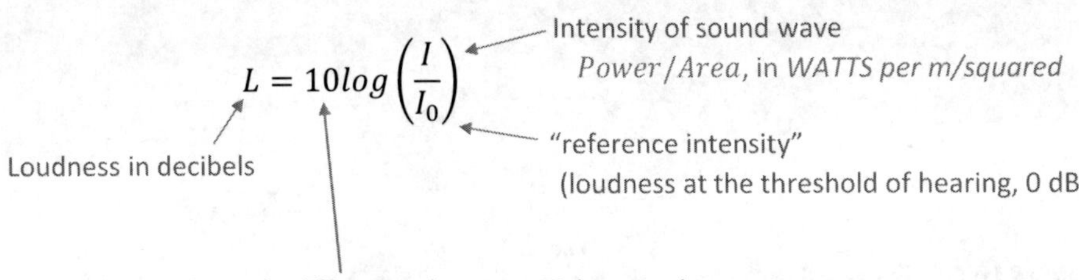

As with Richter and pH scales, the Decibel scale problems we'll encounter primarily involve ***comparing intensities***.

To compare the loudness intensity of two sounds measured in Decibels:
- Convert the Decibel measures to **Bels**, by dividing by 10
- As with pH and Richter scale problems, express as powers of 10

$\mathbf{10^{B_1 - B_2}} =$ *"times as intense sound level"*

Worked Example A conversation between two people in a park measures 55 decibels. A jackhammer nearby measures 104 decibels. How many times as intense is the sound of the jackhammer?

Solution: *First, convert each measure to Bels.* Conversation: **5.5 Bels** Jackhammer: **10.4 Bels**

Then, express each as powers of 10. $\mathbf{10^{10.4-5.5}}$ (10.4: *Jackhammer (louder)*; 5.5: *Conversation*) ➡ = **79 433** *times as intense (louder)*

Class Example 3.68 *Log Scales – comparing decibel levels*

(a) Determine how many times as intense the sound of a lawnmower is, at 82 dB, compared to leaves rustling in the wind, at 48 dB. *Round to the nearest whole number.*

(b) Determine the decibel level of a power tool that has 375 times the intensity of sound of a lawnmower.

Chapter 4 RADICAL FUNCTIONS, RATIONAL FUNCTIONS

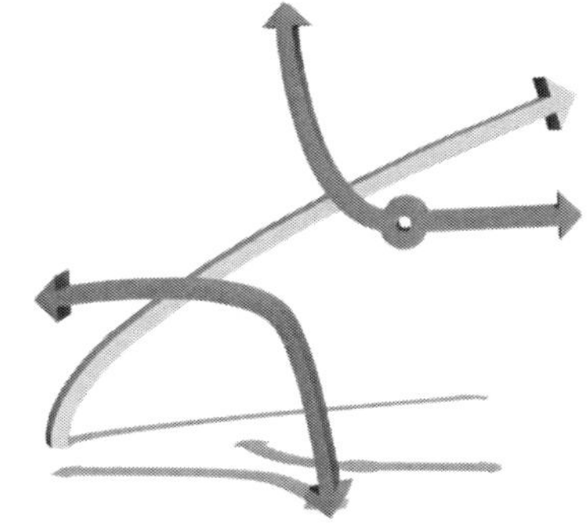

4.1 The Radical Function

As we encountered in Chapter 1, the basic radical function is $f(x) = \sqrt{x}$.

In the first part of this chapter, we'll get re-acquainted with this function, and further examine some graph characteristics.

1 ➡ Complete the **table of values** below

Visit math30-1edge.com for solutions to all warm-ups and class examples

x	$f(x) = \sqrt{x}$
0	
1	
4	
9	
16	
−1	

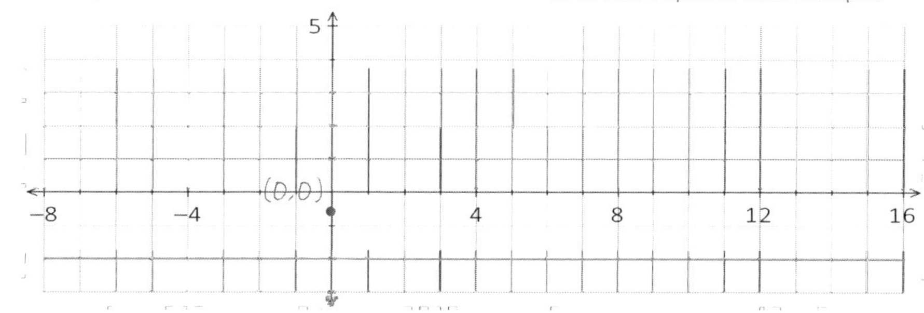

2 ➡ Plot the points above to **sketch** the graph of $f(x) = \sqrt{x}$. Label the graph ①.

3 ➡ **State** the **domain** and **range** of the function. ______ Domain ______ Range

4 ➡ Use transformations to **sketch** the graph of $g(x) = \sqrt{-(x-9)}$ on the same grid above. Label it ②.

$(0, 0) \rightarrow$

$(1, 1) \rightarrow$

$(4, 2) \rightarrow$

$(9, 3) \rightarrow$

$(16, 4) \rightarrow$

5 ➡ Algebraically determine the **domain** of $y = g(x)$, by setting what's under the square root sign *greater than or equal to zero*.

6 ➡ Algebraically confirm the **y-intercept** of $y = g(x)$, by setting the value of x to zero and evaluating.

7 ➡ Algebraically confirm the **x-intercept** of $y = g(x)$, by setting the value of y to zero and solving.

The basic radical function is $\boldsymbol{y = \sqrt{x}}$:

The shape is a half-parabola (sideways)

Recall – it's the inverse of $y = x^2\ ; x \geq 0$

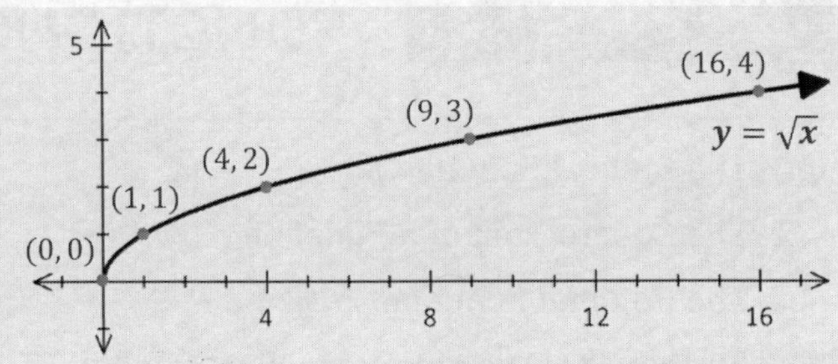

The **domain** is restricted as we cannot square root negative numbers. *D:* $\{x|x \geq 0\ ; x \in \mathbb{R}\}$

The **range** is defined by the starting point (which is the minimum). *R:* $\{y|y \geq 0\ ; y \in \mathbb{R}\}$

Using **transformations,** we can graph any radical function in the form $\boldsymbol{y = a\sqrt{b(x-h)} + k}$

We can also find graph characteristics (**domain**, **range**, $\boldsymbol{x}$ and $\boldsymbol{y}$**-intercepts**) of functions in this form using familiar methods shown in the example below.

Worked Example

Use transformations to sketch the graph of $y = -2\sqrt{x+9} + 2$. Then, use the equation of the function to determine the domain, range, and x and y-intercepts.

Solution: ◆ ***For the GRAPH*** - Construct a mapping rule for transforming the basic graph of $y = \sqrt{x}$:

$(x, y) \rightarrow (x - 9, -2y + 2)$

$(0, 0) \rightarrow (0 - 9, -2(0) + 2)$
$\rightarrow (-9, 2)$

Then transform *key points** on $y = \sqrt{x}$

$(1, 1) \rightarrow (-8, 0)$

$(4, 2) \rightarrow (-5, -2)$

Plot each of these and connect as a smooth "half-parabola" curve!

$(9, 3) \rightarrow (0, -4)$

$(16, 4) \rightarrow (7, -6)$

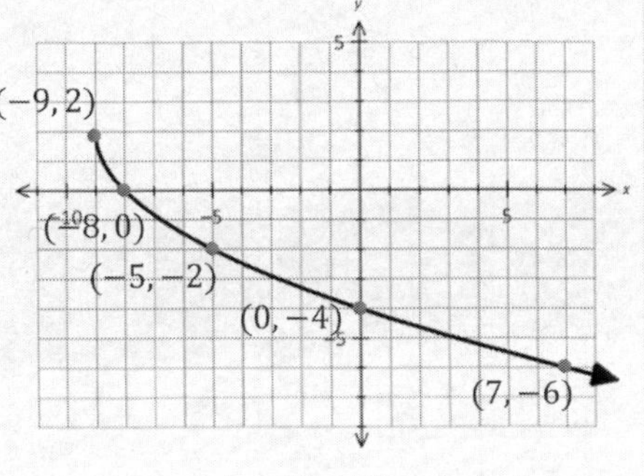

* Use points on $y = \sqrt{x}$ with integer coordinates ... to do so chose *perfect square* x-coordinates.
(Use x = 0, 1, 4, 9, and 16 as these all have whole number square roots.)

◆ ***For the DOMAIN*** – We can't square root negatives! So set whatever's under the square root sign greater than or equal to zero.

$$y = -2\underbrace{\sqrt{x+9}} + 2$$

$$x + 9 \geq 0$$

$$x \geq -9$$

Domain is: $\{\boldsymbol{x}|\boldsymbol{x} \geq \boldsymbol{-9}, \boldsymbol{x} \in \mathbb{R}\}$

◆ ***For the RANGE*** – Unlike with domain, we must consider the graph.

Specially, we note that compared to the basic graph:

– There's been a vertical reflection
So graph opens down, range will have the form $\boldsymbol{y \leq y}$***-coord of "start point"***

– There's been a vertical translation 2 units up
So the y-coord of start point is **2**

Range is: $\{\boldsymbol{y}|\boldsymbol{y} \leq \boldsymbol{2}, \boldsymbol{y} \in \mathbb{R}\}$

◆ ***For the y-INTERCEPT*** – Set the x to zero in the equation and evaluate.

$$y = -2\sqrt{(0) + 9} + 2$$

$$= -2\sqrt{9} + 2$$

$$= -2(3) + 2$$

$$= -4$$

$\boldsymbol{y}$**-intercept is:** $(\boldsymbol{0}, \boldsymbol{-4})$

◆ ***For the x-INTERCEPT*** – Set the y to zero in the equation and solve the resulting equation.

$$0 = -2\sqrt{x+9} + 2$$

$$-2 = -2\sqrt{x+9}$$

Now *square both sides*

$$1 = \sqrt{x+9} \Rightarrow (1)^2 = \left(\sqrt{x+9}\right)^2 \Rightarrow 1 = x + 9$$

$$\Rightarrow x = -8$$

$\boldsymbol{x}$**-intercept is:** $(\boldsymbol{-8}, \boldsymbol{0})$

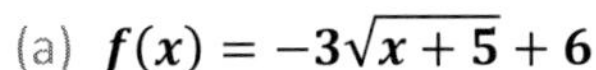

Sketching the Graph of a Radical Function – and Analyzing Characteristics

For each of the functions given below, use transformations to construct a mapping rule and **sketch the graph**. Then, use algebraic processes (as shown on the previous page) to **determine** the indicated **graph characteristics**.

(a) $f(x) = -3\sqrt{x+5} + 6$

i Mapping Rule:

ii Domain:

iii Range:

iv x-intercept:

v y-intercept:

As an exact value, and to the nearest hundredth

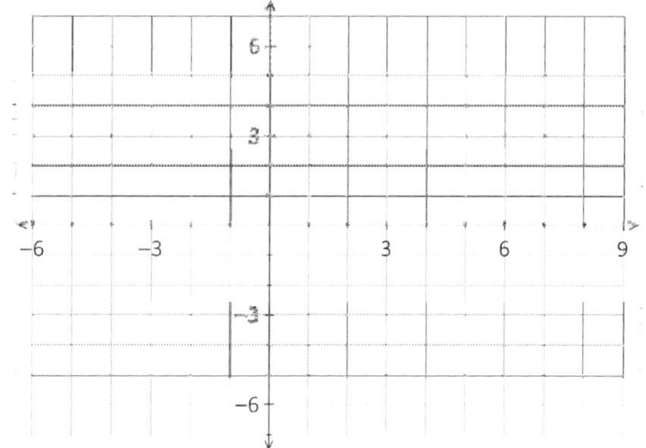

(b) $g(x) = \sqrt{-2(x-8)} - 2$

i Mapping:

ii Domain:

iii Range:

iv x-int:

v y-int:

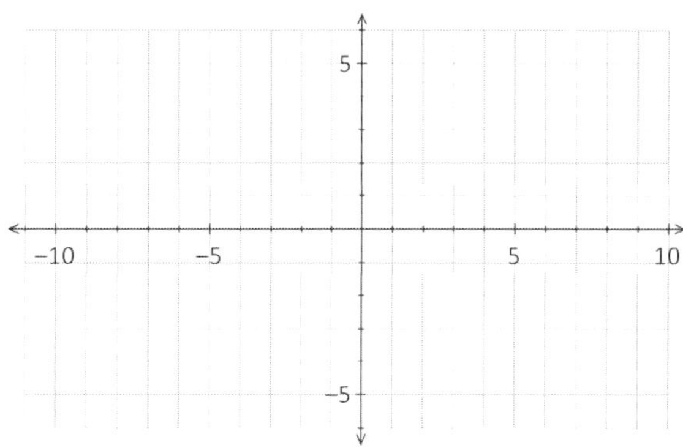

While we **must** be able to **algebraically** determine things like *domain*, *range*, and *intercepts....*

We should also be familiar with how to verify using a graphing calculator!

Let's consider the function $\boldsymbol{y=\sqrt{-2(x-8)}-2}$.

- For the **domain and range**, we need to consider the *start point*.

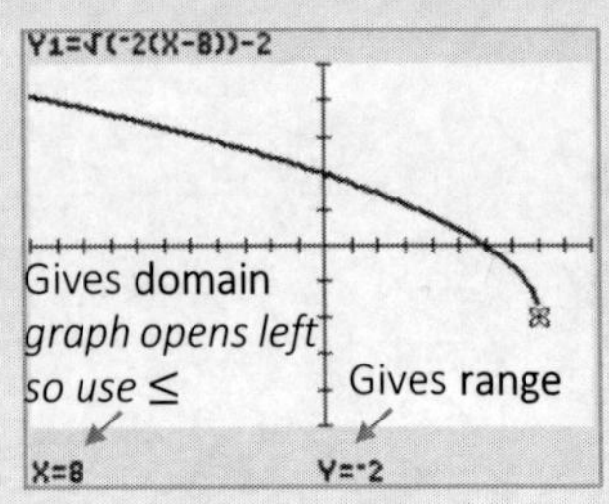

Use [trace] to confirm the start point.

We need to "guess" on the x-coord. of the start point based on the graph, and key in TRACE ... "8" to confirm.

We can now see that the domain is: $\{\boldsymbol{x|x\le 8, x\in\mathbb{R}}\}$

And the range is: $\{\boldsymbol{y|y\ge -2, y\in\mathbb{R}}\}$

- For the **y-intercept**, we again use [trace] and set x to zero.

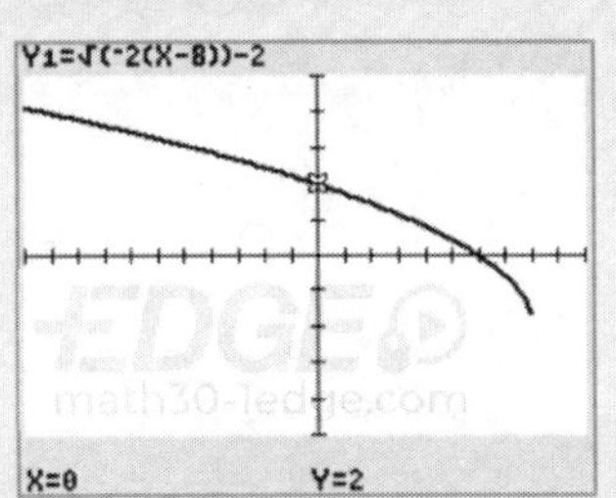

So the $\boldsymbol{y}$-intercept is $\boldsymbol{(0,2)}$

- For the **x-intercept**, we find the zero, which is in the CALC menu. (2nd + TRACE)

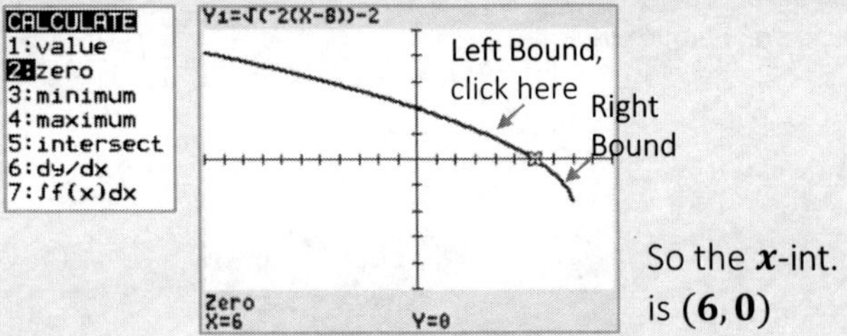

So the $\boldsymbol{x}$-int. is $\boldsymbol{(6,0)}$

Class Example 4.12 *Analyzing Characteristics of a Radical Function*

Use an algebraic process to find the indicated characteristics for the function $y=3\sqrt{-0.5x+9}-12$. Verify graphically.

i Domain:

ii Range:

iii x-int:

iv y-int:

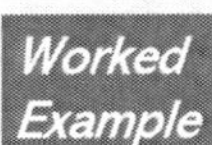

Given the function graphed on the right, determine an equation in the form $y = a\sqrt{x-h}+k$.

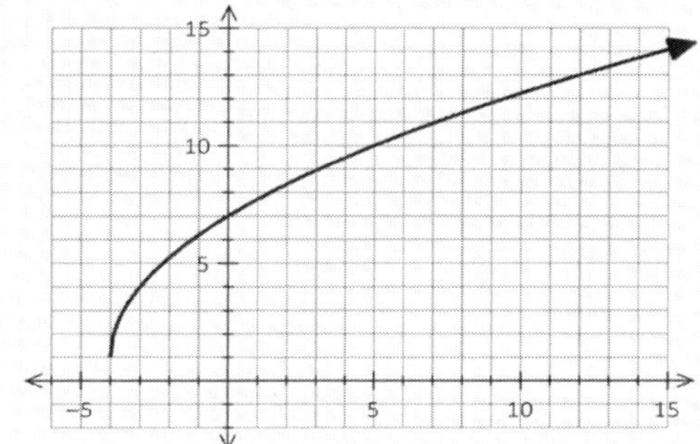

Solution:
- *The values of h, k are given by the* ***start point****.*

  Here, the graph "starts" at $(-4, 1)$

  So our equation is:

  $y = a\sqrt{x+4}+1$

- *Next, substitute in the coordinates of any other point on the graph to solve for "a"*

  Choose a point with identifiable (integer) coordinates, such as $(0, 7)$.

  $(7) = a\sqrt{(0)+4}+1$

  $6 = a\sqrt{4}$ ➧ $6/2 = a$ ➧ $a = 3$

  We've now determined h, k, and a. So our equation is:

  $$\boldsymbol{y = 3\sqrt{x+4}+1}$$

Verify your equation on your graphing calculator.

✓ Be sure to match the window in your calc with what was given.

✓ Verify using TRACE. Here we've confirmed the point $(5, 10)$ is on the graph.

```
Plot1  Plot2  Plot3
\Y1=3√(X+4)+1
\Y2=
```

```
WINDOW
 Xmin=-6
 Xmax=15
 Xscl=1
 Ymin=-2
 Ymax=15
 Yscl=1
 Xres=1
```

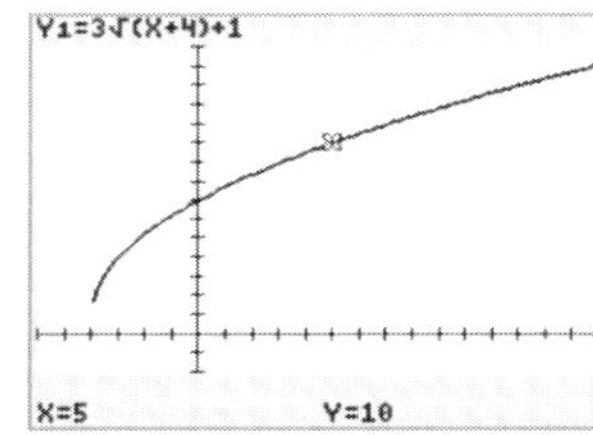

Class Example 4.13 *Analyzing Characteristics of a Radical Function*

Each of the following graphs represents a radical function. All points marked (•) have integer coordinates. Determine an equation for each, in the form stated.

(a) In the form $y = a\sqrt{x-h}+k$

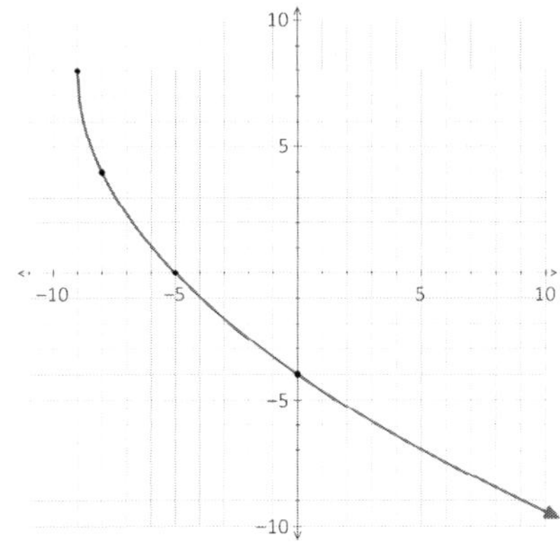

(b) In the form $y = \sqrt{b(x-h)}+k$

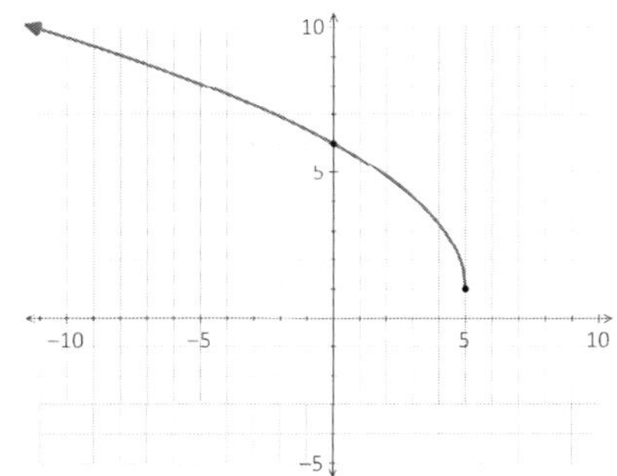

4.1 Practice Questions (Sample, first page only)

SAMPLE of our high-quality, practice questions. Visit **Math30-1edge.com** *for full access.*

1. For each of the functions given below, use transformations to construct a mapping rule for transformation from the basic graph and **sketch the graph**. Then, use algebraic processes to **determine** the indicated **graph characteristics.**

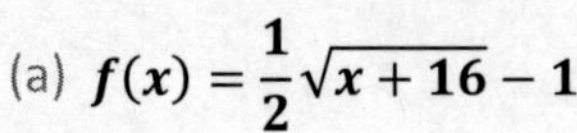

(a) $\boldsymbol{f(x) = \frac{1}{2}\sqrt{x+16} - 1}$

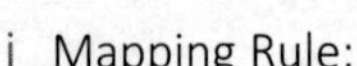

i Mapping Rule:

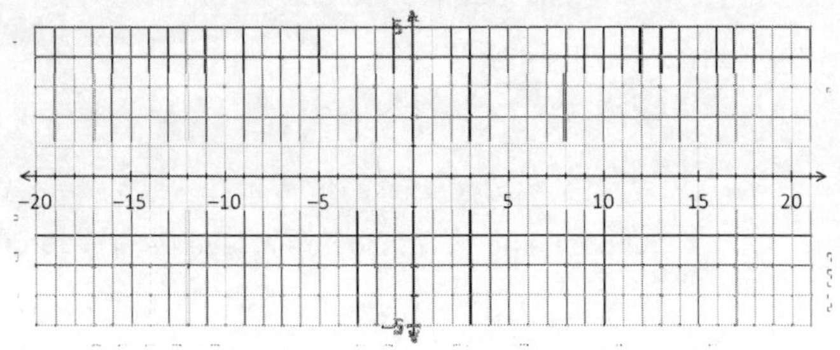

ii Domain:

iii Range:

iv x-intercept:

v y-intercept:

(b) $\boldsymbol{g(x) = \sqrt{-3x+9} - 1}$

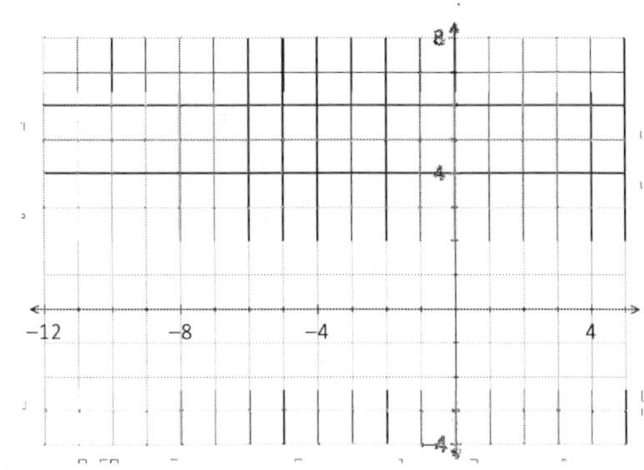

i Mapping Rule:

ii Domain:

iii Range:

iv x-intercept:
As an ***exact value***

v y-intercept:

Answers to practice questions on previous page

1. (a) i $(x, y) \rightarrow (x - 16, \frac{1}{2}y - 1)$

$(0,0) \rightarrow (-16, -1)$

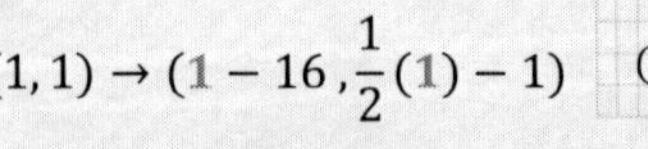

$(1,1) \rightarrow (1 - 16, \frac{1}{2}(1) - 1)$

$\rightarrow (-15, -0.5)$ *And so on...*

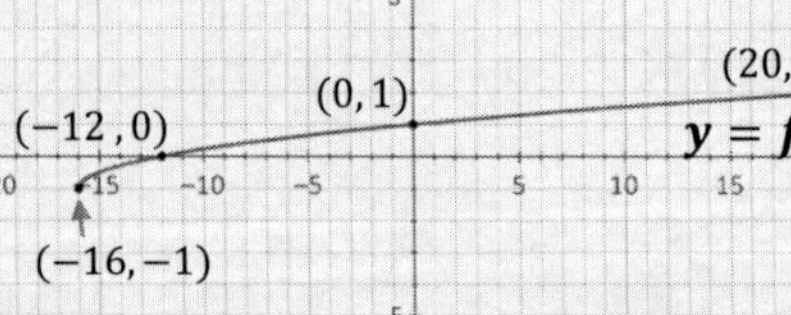

ii $\{x | x \geq -16, x \in \mathbb{R}\}$

iii $\{y | y \geq -1, y \in \mathbb{R}\}$

iv $(-12, 0)$

v $(0, 1)$

(b) i $(x, y) \rightarrow (-\frac{1}{3}x + 3, y - 1)$

ii $\{x | x \leq 3, x \in \mathbb{R}\}$

iii $\{y | y \geq -1, y \in \mathbb{R}\}$

iv $(8/3, 0)$

v $(0, 2)$

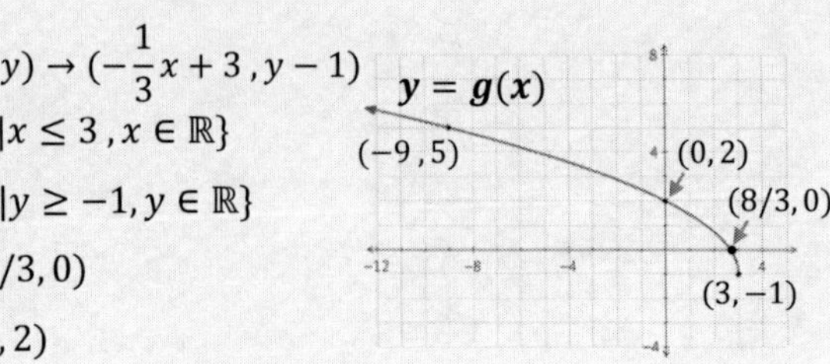

4.2 The Square Root of a Function

Consider the graphs of $f(x) = x - 2$ and $y = g(x)$ below.

In this warm-up you'll be asked to consider the **value** of the functions for some particular x. As a refresher, in this context: *value = y-coordinate*.

For example, the value of $f(x)$ at $x = -1$, that is $f(-1)$, is -3. *On the graph of $y = f(x)$ the y-coordinate when x-coordinate is -1 is -3. We can express this as $f(-1) = -3$.*

1 ➡ Compare the values of the two functions shown, at the following values of x.

$f(18) =$ ____ $g(18) =$ ____

$f(11) =$ ____ $g(11) =$ ____

$f(6) =$ ____ $g(6) =$ ____

$f(3) =$ ____ $g(3) =$ ____

$f(2) =$ ____ $g(2) =$ ____

$f(1) =$ ____ $g(1) =$ ____

$f(0) =$ ____ $g(0) =$ ____

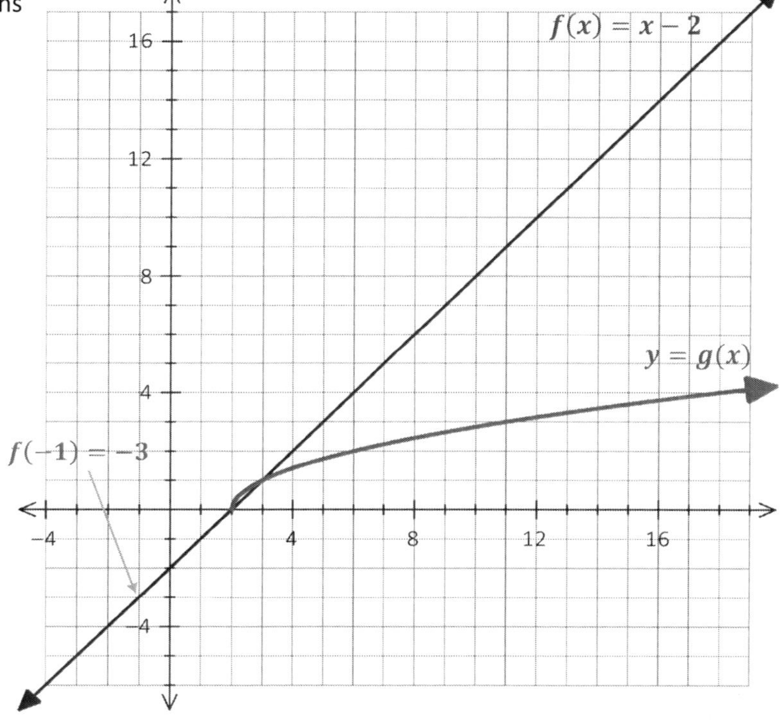

2 ➡ Describe how the values of $g(x)$ compare to those of $f(x)$ for any x.

3 ➡ Complete a mapping rule which describes the transformation of all points on the graph of $f(x)$ to the graph of $g(x)$.

4 ➡ Based on your answer above, provide an equation for $g(x)$, in terms of x. Verify using your graphing calculator.

5 ➡ State the domain and range of each function.

____	____	____	____
$f(x)$ domain	$f(x)$ range	$g(x)$ domain	$g(x)$ range

6 ➡ Explain how the *graph* of $f(x)$ relates to the *domain* of $g(x)$.

7 ➡ Describe the location of the invariant points as the graph of $f(x)$ transforms to the graph of $g(x)$.

In this topic we consider taking the **square root of a function** as a transformation.

In the warmup we saw how a function $f(x) = x - 2$ transformed to $y = \sqrt{f(x)}$, that is $y = \sqrt{x-2}$.

Given the graph of a linear function $y = f(x)$, the graph of $y = \sqrt{f(x)}$ ...

- Can be formed by transforming all points on the graph of $y = f(x)$ by $(x, y) \to (x, \sqrt{y})$
  Take the square root of all y-coordinates
- Has **invariant points** where the y-coordinate on the graph of $f(x)$ is 0 or 1.
  Since those are the only two numbers that don't change when square-rooted! $\sqrt{0} = 0$ and $\sqrt{1} = 1$.
- Is defined wherever $f(x) > 0$.

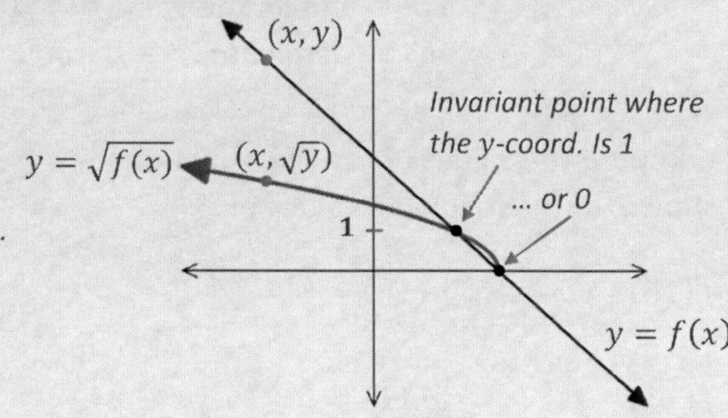

The Three Graph Regions for $y = \sqrt{f(x)}$:

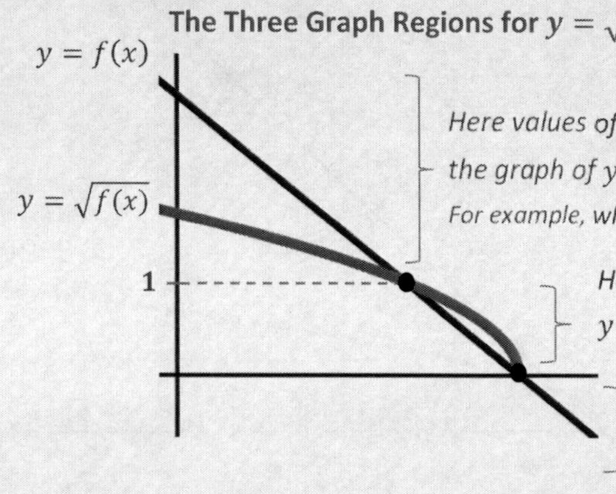

*Here values of $f(x)$ are greater than 1, so values of $y = \sqrt{f(x)}$ are lower - the graph of $y = \sqrt{f(x)}$ is **below** the graph of $y = f(x)$*
For example, when $f(x) = 4$, $\sqrt{f(x)} = 2$

*Here values of $f(x)$ here are between 0 and 1, so values of $y = \sqrt{f(x)}$ are higher – the graph of $y = \sqrt{f(x)}$ is now **above***
For example, when $f(x) = 0.25$, $\sqrt{f(x)} = 0.5$

*Finally the graph of $y = \sqrt{f(x)}$ is **not defined** here, where the graph of $f(x)$ is below the x-axis*
*For example, when $f(x) = -2$, $\sqrt{f(x)} = \sqrt{-2}$ **(undefined)***

Worked Example

Square root of a linear function

Given the function $f(x) = 4 - \frac{1}{2}x$ as shown,

(i) Sketch the graph of $y = \sqrt{f(x)}$,
(ii) state the coordinates of any invariant points, and (iii) give the domain and range.

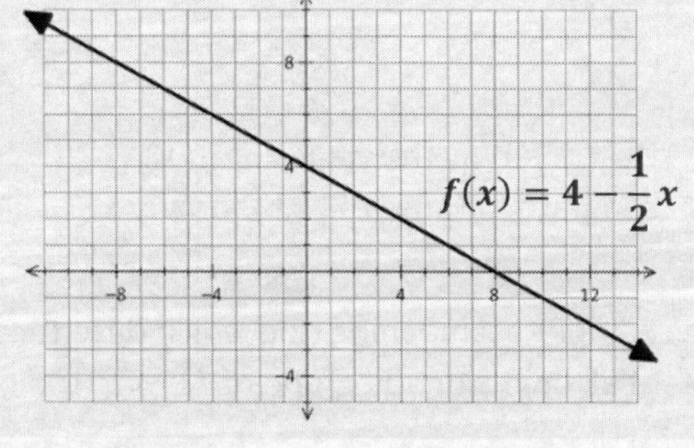

Solution: ❶ Start by plotting (•) all of the invariant points... where the y-coord. on the graph of $f(x)$ is *0 or 1.*

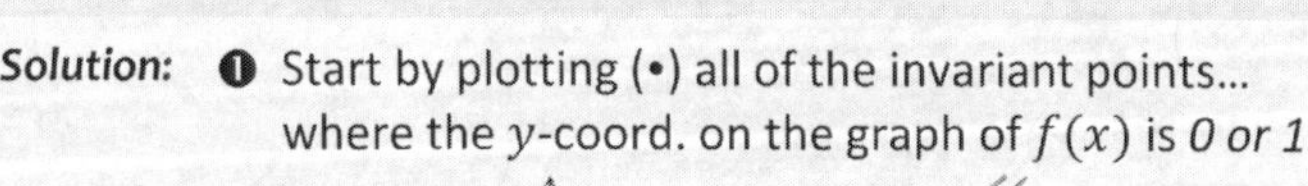

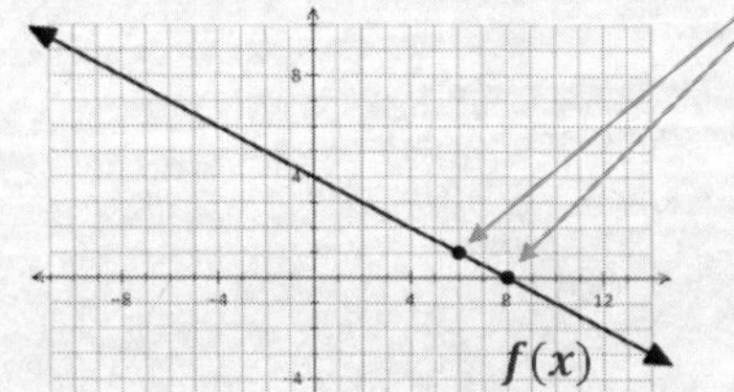

❷ Locate other points on $f(x)$ where the y-coordinate is a perfect square. (4, 9, 16, etc...)

y-coordinate here is 9
And here the y-coord. is 4
$f(x)$

❸ Obtain the points on the graph of $y = \sqrt{f(x)}$ by taking the square root of the y-coordinates of the points identified in the step 2.

(i)

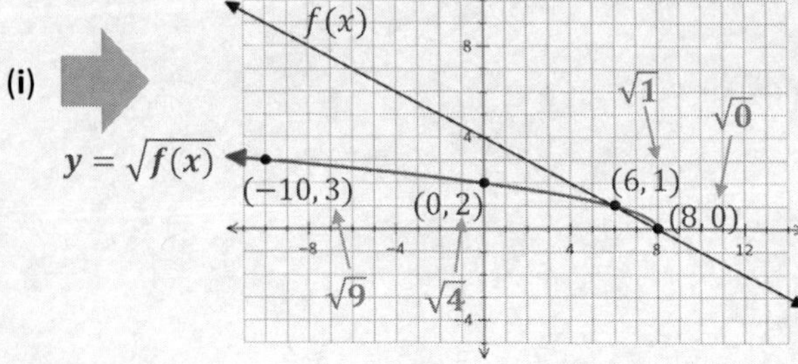

(ii) **Invariant points** are $(8, 0)$ and $(6, 1)$.

(iii) **Domain** is $\{x | x \leq 8, x \in \mathbb{R}\}$ ← where $f(x) \geq 0$

Range is $\{y | y \geq 0, y \in \mathbb{R}\}$

CALC TIP

On the previous page we saw how to construct the graph of a function $y = \sqrt{f(x)}$, given the linear function $f(x) = 4 - \frac{1}{2}x$.

Verify points on the graph of $y = \sqrt{f(x)}$ using TRACE.

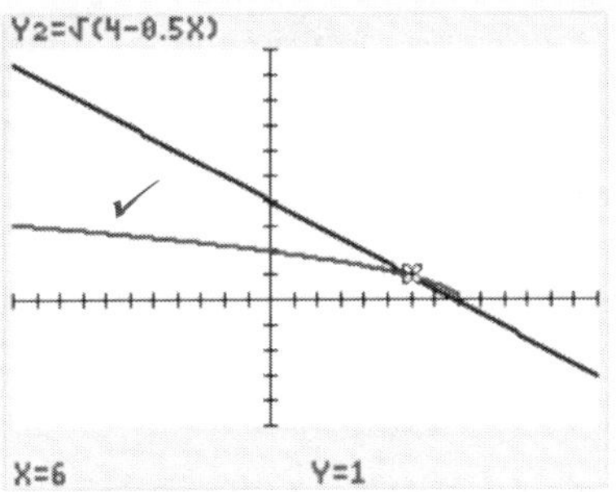

We can verify by graphing $y = \sqrt{4 - (1/2)x}$ in our calculators.

```
Plot1  Plot2  Plot3
■\Y1■4-0.5X
■\Y2■√4-0.5X
```

Graph $y_1 = f(x)$ and $y_2 = \sqrt{f(x)}$

```
WINDOW
 Xmin=-11
 Xmax=14
 Xscl=1
 Ymin=-5
 Ymax=10
 Yscl=1
 Xres=1
```

Match the window to what was given

Class Example 4.21 *Graphing the Square Root of a Linear Function*

Given the graph of the linear function $f(x) = 2x - 6$,

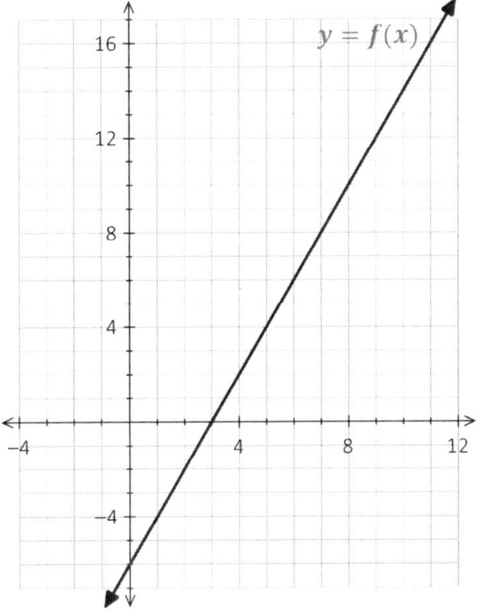

(a) State the domain of the function $y = \sqrt{f(x)}$.

(b) Provide a mapping rule that describes the transformation from $y = f(x)$ to $y = \sqrt{f(x)}$.

(c) State the coordinates of any invariant points.

(d) Sketch the graph of $y = \sqrt{f(x)}$ on the same grid.

(d) State the range of $y = \sqrt{f(x)}$.

Class Example 4.22 *Analyzing Characteristics of $y = \sqrt{f(x)}$*

Given a function $f(x) = 3x + 5$, determine the indicated characteristics of the function $y = \sqrt{f(x)}$.
Use an algebraic approach – *do not graph.*

(a) Determine the domain of $y = \sqrt{f(x)}$.

(b) Determine the coordinates of any invariant points.

Worked Example

Square root of a quadratic function

Given the graph of $f(x) = x^2 - 4$:

(i) Sketch the graph of $y = \sqrt{f(x)}$

(ii) Determine the coordinates of any invariant points

(iii) State the domain and range.

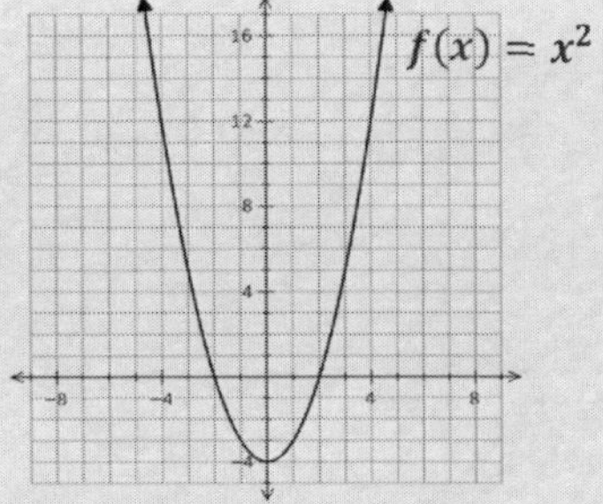

Solution: Obtain the graph of $y = \sqrt{f(x)}$ by taking the square root of all y-coordinates.

❶ Start by plotting (•) all of the invariant points... where the y-coord. on the graph of $f(x)$ is *0 or 1.*

This time there are FOUR invariant points!

Think of where the given graph of $f(x)$ would intersect the imaginary horizontal lines $y = 0$ and $y = 1$.

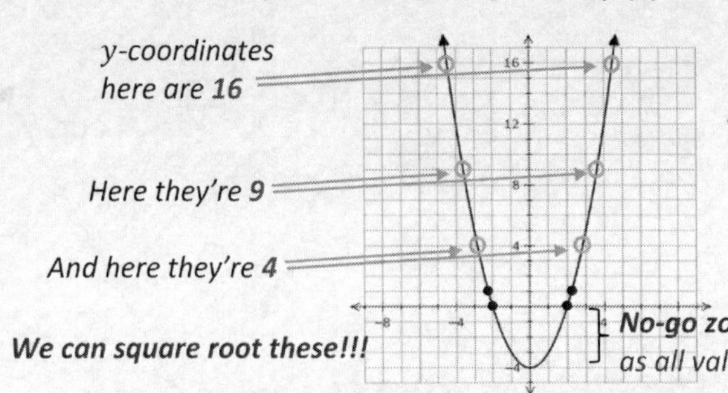

❷ Locate other points on $f(x)$ where the y-coordinate is a perfect square. (4, 9, 16, etc...)

Note that the corresponding x-coords. are irrational.

❸ Obtain the points on the graph of $y = \sqrt{f(x)}$ by taking the square root of the y-coordinates of the points identified in the previous step.

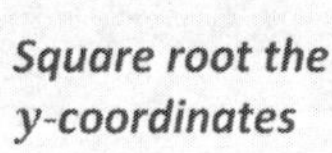

Square root the y-coordinates

Of the corresponding point on $f(x)$... notice the x-coords. align

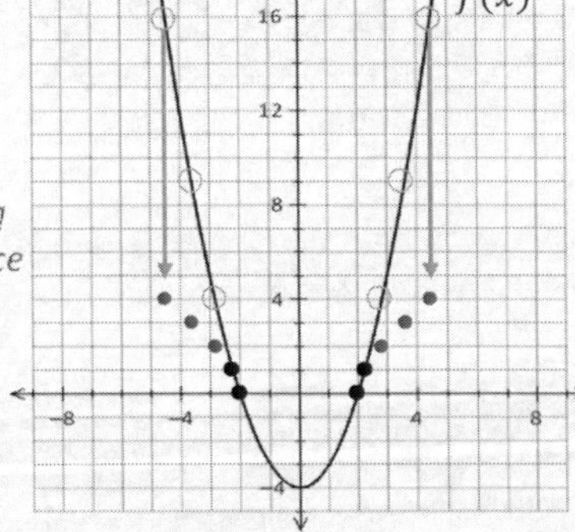

(i)

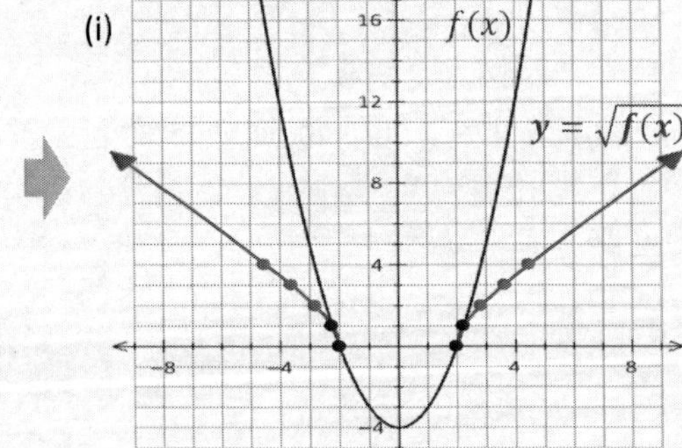

(ii) **Invariant points occur where $f(x)$** (starting function) is equal to 0 or 1.

Since we are transforming all points $(x, y) \to (x, \sqrt{y})$, when the y-coordinate on $f(x)$ is 0 or 1, taking the square root will return the same value as $\sqrt{0} = 0$ and $\sqrt{1} = 1$.

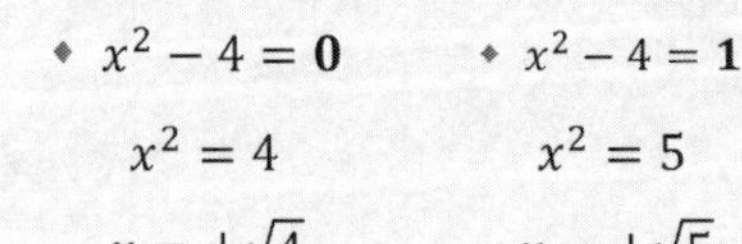

- $x^2 - 4 = \mathbf{0}$
- $x^2 - 4 = \mathbf{1}$

Solve for x

$x^2 = 4 \qquad x^2 = 5$

$x = \pm\sqrt{4} \qquad x = \pm\sqrt{5}$

$x = \pm 2$

Match with y-coordinates $(-2, 0), (2, 0)$ *and* $(-\sqrt{5}, 1), (\sqrt{5}, 1)$

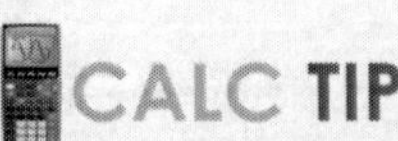

CALC TIP

On your graphing calculator, either:

- Find the intersect of $f(x)$ with 0 and 1
- Or graph $y_1 = f(x)$ and $y_2 = \sqrt{f(x)}$.

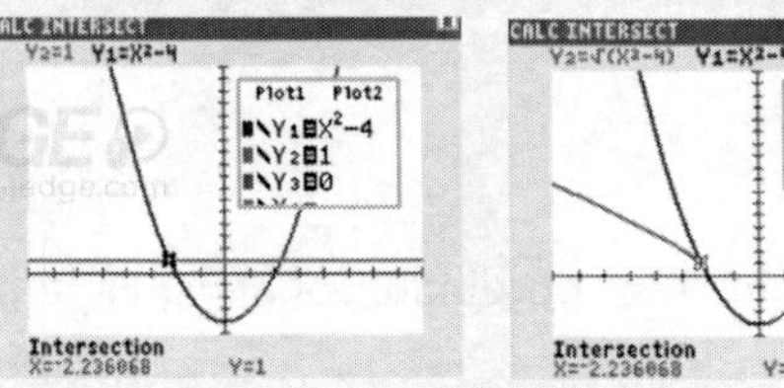

For exact values an algebraic approach is needed

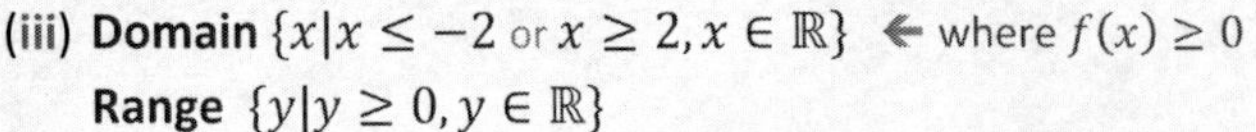

(iii) **Domain** $\{x | x \leq -2 \text{ or } x \geq 2, x \in \mathbb{R}\}$ ← where $f(x) \geq 0$

Range $\{y | y \geq 0, y \in \mathbb{R}\}$

Class Example 4.23 *Graphing the Square Root of a Quadratic Function*

Given the graph of the function $f(x) = \frac{1}{2}(x + 1)^2 - ,2$

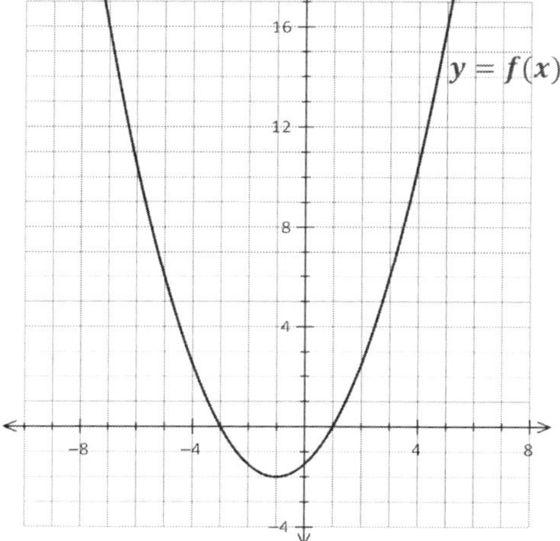

(a) State the domain of the function $y = \sqrt{f(x)}$.

(b) Determine the coordinates of the invariant points in the transformation to $y = \sqrt{f(x)}$.

Provide coordinates both as exacts values and as decimal approximations. (nearest hundredth)

(c) Sketch the graph of $y = \sqrt{f(x)}$ on the same grid.

(d) State the range of $y = \sqrt{f(x)}$.

Class Example 4.24 *Analyzing Characteristics of $y = \sqrt{f(x)}$*

For the function $f(x) = 2(x - 2)^2 - 32$,

(a) Determine the domain of $y = \sqrt{f(x)}$.

(b) Determine the coordinates of the invariant points in the transformation to $y = \sqrt{f(x)}$.

Where applicable provide coordinates as exact values

Let's ***summarize*** the steps involved in sketching the graph of $y = \sqrt{f(x)}$.

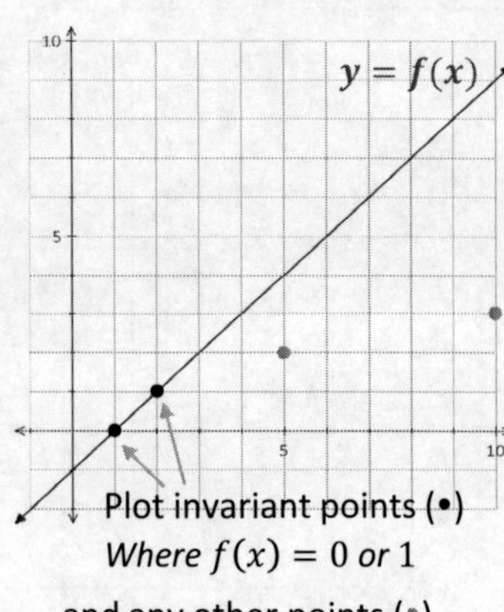

Plot invariant points (•)
Where $f(x) = 0$ *or* 1

...and any other points (•)
Where $f(x) =$ *a perfect square*

Sketch the portion of the graph where $y = \sqrt{f(x)}$ is higher

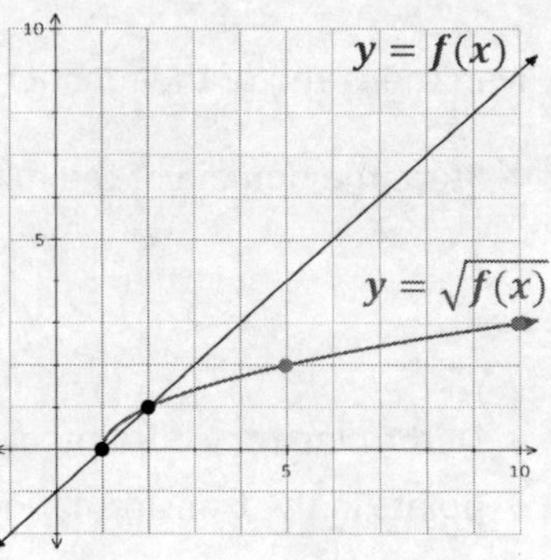

Sketch remainder of the graph

Class Example 4.25 *Graphing the Square Root of a Quadratic Function – Parabola opens downward*

Given the graph of the function $f(x) = 9 - x^2$,

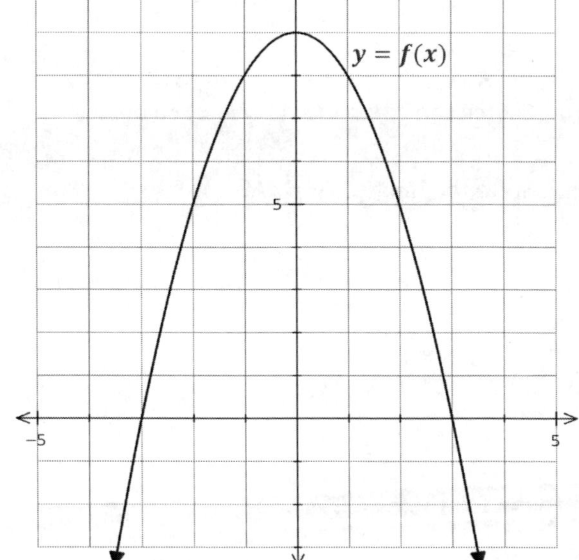

(a) State the domain of the function $y = \sqrt{f(x)}$.

(b) Determine the coordinates of the invariant points in the transformation to $y = \sqrt{f(x)}$.
(Exact values)

(c) Sketch the graph of $y = \sqrt{f(x)}$ on the same grid.

(d) State the range of $y = \sqrt{f(x)}$.

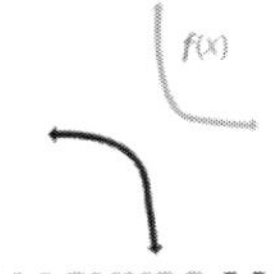

Warm-Up #1 The Graph of $y = \frac{1}{x}$

1 ➡ **Complete** the (partial) **table of values** on the left. *We started for you! Four of the points are already plotted.*

2 ➡ **Sketch** $y = 1/x$ on your graphing calculator, using the window shown.

x: $[-5, 5, 1]$ y: $[-5, 5, 1]$

min max scl

3 ➡ With the help of you calculator, **sketch** the graph by connecting the plotted points in a smooth curve.

4 ➡ *Fill in the blanks:* The graph of $y = \frac{1}{x}$

has a vertical asymptote (V.A.) at _____

and a horizontal asymptote (H.A.) at ______

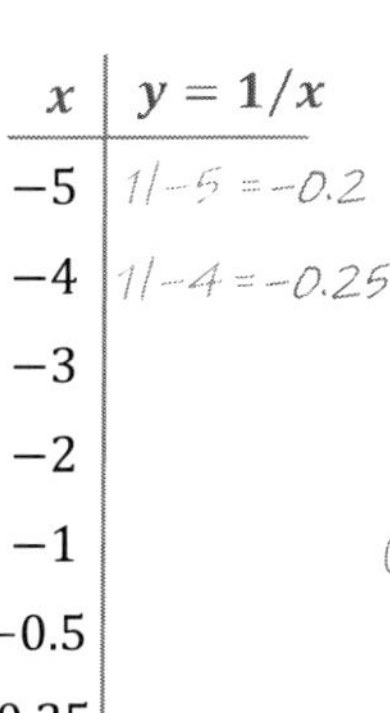

x	$y = 1/x$
-5	$1/-5 = -0.2$
-4	$1/-4 = -0.25$
-3	
-2	
-1	
-0.5	
-0.25	
-0.2	
0	
0.2	$1/0.2 = 5$
0.25	$1/0.25 = 4$
0.5	
1	
2	
3	
4	
5	

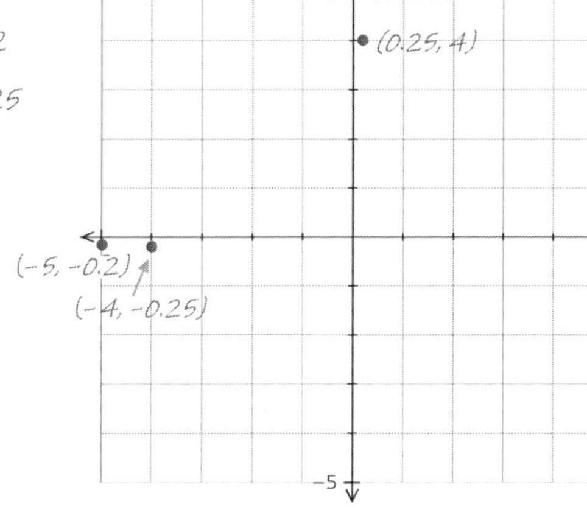

5 ➡ The domain of the function is:

____

And the range is:

____

Warm-Up #2 The Graph of $y = \frac{a}{x}$

1 ➡ Use transformations to sketch the graph of $y = \frac{2}{x}$

The mapping rule is: (x, y) → ______

2 ➡ *Fill in the blanks:* The graph of $y = \frac{2}{x}$

has a vertical asymptote (V.A.) at _____

and a horizontal asymptote (H.A.) at ______

3 ➡ The domain of the function is: ______________

And the range is: ______________

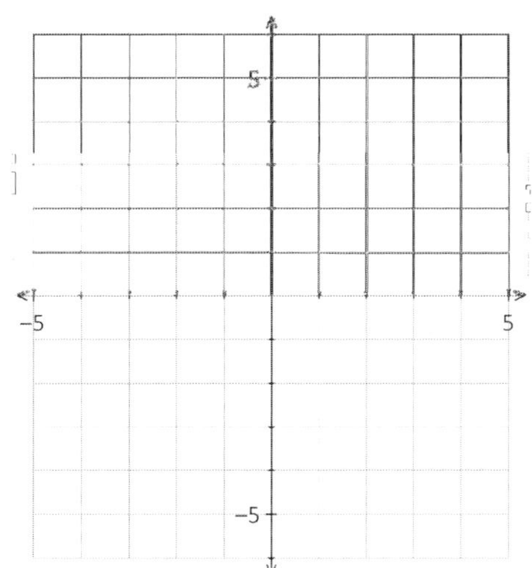

Properties of the Basic Rational Function

The **basic rational function** is $y = \frac{1}{x}$

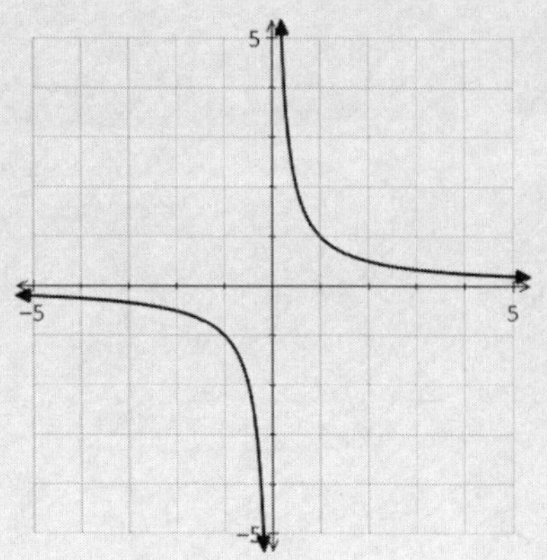

Vertical Asymptote (V.A.) at $x = 0$ (the y-axis)

Horizontal Asymptote (H.A.) at $y = 0$ (the x-axis)

Domain: $\{x|x \neq 0, x \in \mathbb{R}\}$ Range: $\{y|y \neq 0, x \in \mathbb{R}\}$

x-intercept: *none* y-intercept: *none*

These characteristics apply to any function of the form $y = \frac{a}{x}$:

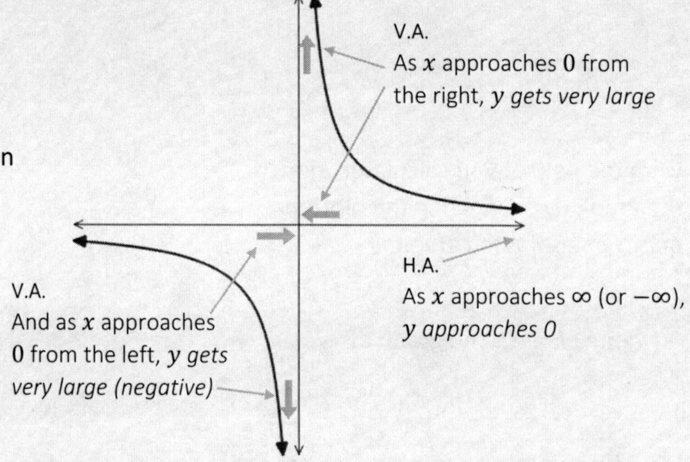

Vertical Asymptotes

- Occur at non-permissible values (NPVs) of x. *(later we'll study an exception!)*
- As values of x approach the NPV (from the left or right), values of y approach either ∞ or $-\infty$.
- Graph can never cross a V.A.

x	$y = \frac{1}{x}$
1	1
0.1	10
0.01	100
0.001	1000
0.000001	1 000 000

← *For example, as we pick smaller and smaller values of x, the y value gets larger and larger!*

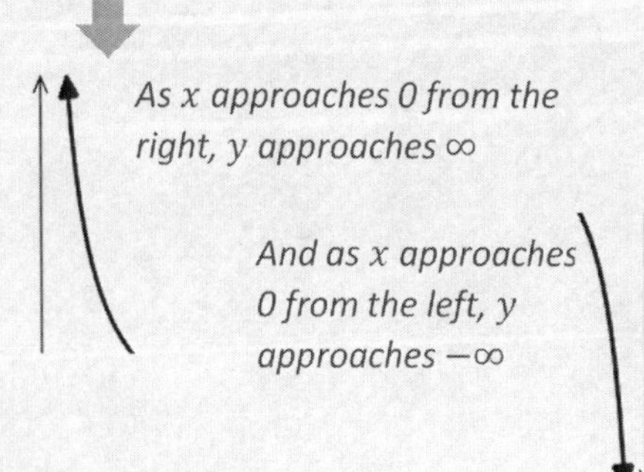

Horizontal Asymptotes

- Occur when y approaches a fixed value when x approaches ∞ or $-\infty$. *(That is, as $|x|$ gets larger and larger)*
- A graph can cross a H.A. *(though often doesn't)*

x	$y = \frac{1}{x}$
1	1
10	0.1
100	0.01
1000	0.001
1 000 000	0.000001

← *For example, as we pick larger and larger values of x, the y values approach 0.*

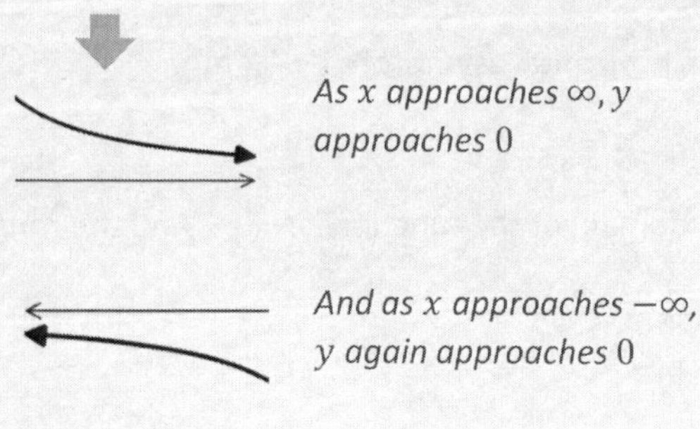

Applying Transformations to the Rational Function

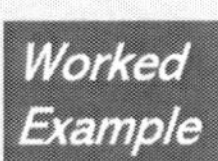

Given the function $f(x) = \dfrac{3}{x+4} - 2$, state the (i) domain, (ii) range, (iii) equation of any asymptotes, and (iv) sketch, labelling the coordinates of any intercepts.

Solution: The basic graph $y = \dfrac{1}{x}$ is transformed by a:

$f(x) = \dfrac{3}{x \boxed{+\,4}} \boxed{-\,2}$ → ....and a vertical. translation 2 units down

❶ Vertical stretch by a factor of 3
(does not affect domain, range, or asymptotes)

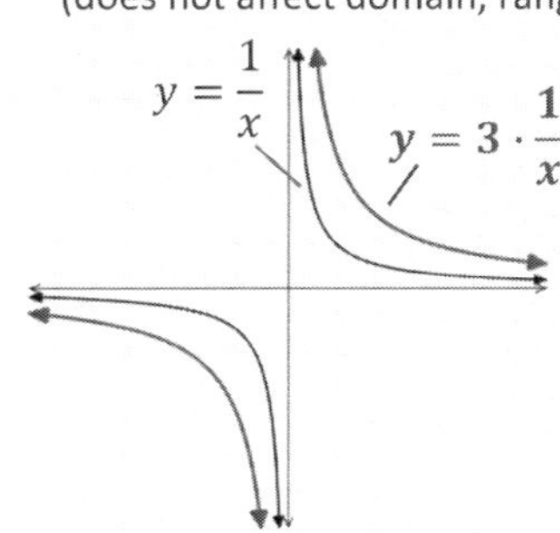

❷ Horizontal translation 4 units left ....

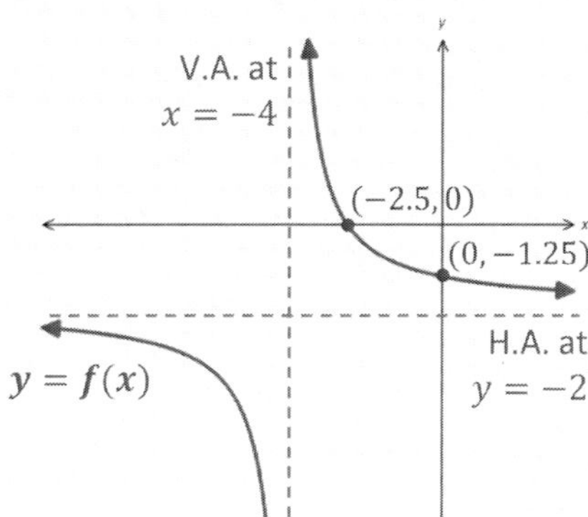

Verify on your graphing calc:

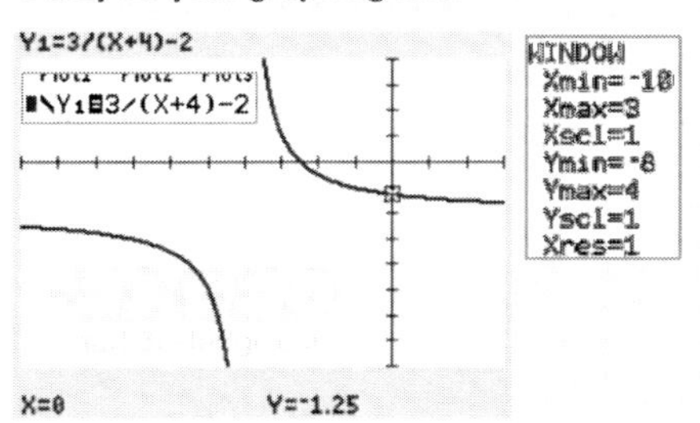

For y-intercept, set $x = 0$:

$$y = \frac{3}{0+4} - 2$$

$$= \frac{3}{4} - 2$$

$$= -\frac{5}{4}$$

➡ $\mathbf{(0, -\frac{5}{4})}$

For x-intercept, set $y = 0$:

$0 = \dfrac{3}{x+4} - 2$ ➡ $2 = \dfrac{3}{x+4}$

➡ $2(x+4) = 3$ ➡ $2x + 8 = 3$

➡ $2x = -5$ ➡ $x = -5/2$

➡ $\mathbf{(-5/2, 0)}$

Class Example 4.31 *Analyzing Characteristics of a Rational Function*

Given the function $f(x) = \dfrac{1}{x-3} + 2$, state the (i) domain, (ii) range, (iii) equation of any asymptotes, and (iv) sketch, labelling the coordinates of any intercepts.

4
−4
4
−4

Rational Functions in the Form $y = \dfrac{a}{x-h} + k$

For a rational function in this form, consider the horizontal and vertical translations to the basic graph $y = \dfrac{1}{x}$

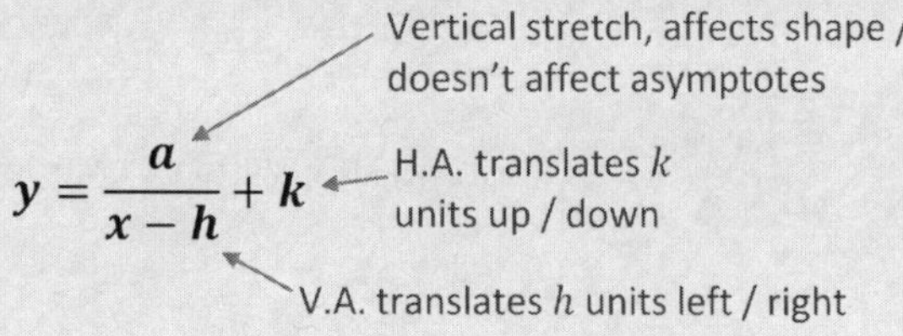

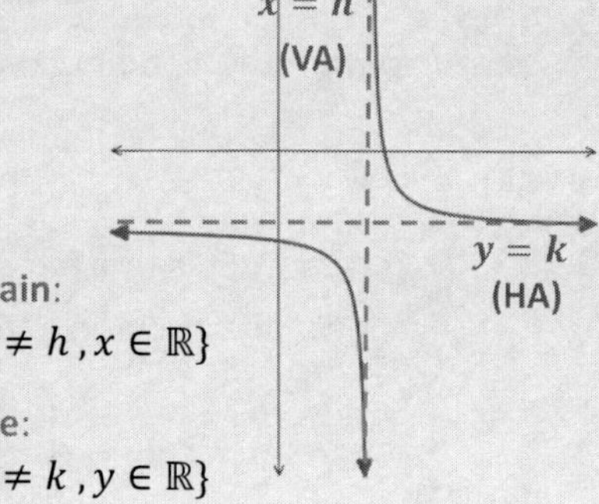

Domain:
$\{x \mid x \neq h, x \in \mathbb{R}\}$

Range:
$\{y \mid y \neq k, y \in \mathbb{R}\}$

Note: Graphs of equations in this form never cross their Horizontal Asymptote.
Though we will encounter some rational functions graphs that DO cross their H.A.!

And of course, no rational function graph can ever cross one of its V.A.s

Class Example 4.32 *Different Forms of a Rational Function Equation*

The graph on the right has integer values for any asymptote or y-intercept.

(a) Determine an equation, in the form $y = \dfrac{a}{x-h} + k$.

(b) Use an algebraic process to convert to the form $y = \dfrac{f(x)}{g(x)}$,
where $f(x)$ and $g(x)$ are linear functions.

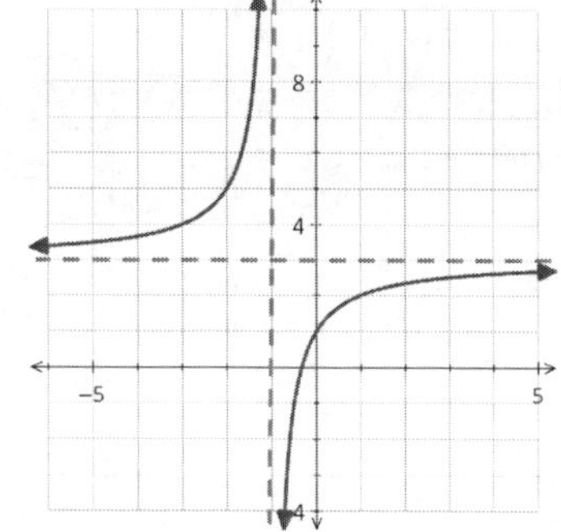

Rational Functions in the Form $y = \dfrac{f(x)}{g(x)}$, where f and g are LINEAR functions.

Recall on a previous example we saw how the characteristics of the graph of the function $f(x) = \dfrac{-2}{x+1} + 3$ relates to the equation.

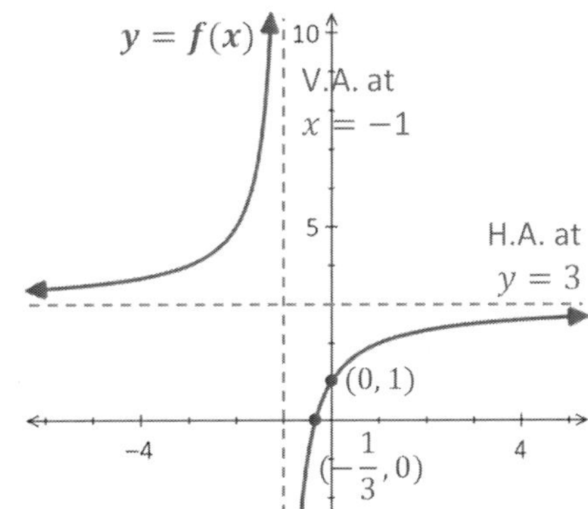

Notice that we can get an ***alternate form*** *of the equation:*

$$f(x) = \frac{-2}{x+1} + \frac{3(x+1)}{x+1}$$ *Combine terms with the common denominator*

$$f(x) = \frac{-2+3x+3}{x+1} \quad \Rightarrow \quad \boldsymbol{f(x) = \frac{3x+1}{x+1}}$$

➔ ***Let's see how the graph characteristics relate this form of the graph….***

$$\boldsymbol{f(x) = \frac{3(x+\frac{1}{3})}{x+1}}$$

The H.A. is now represented by the "a" value, 3*

Factor from the "top" (numerator) gives the x-intercept (x-int. at value of x that makes "top" zero)

Factor from "bottom" (denominator) gives the vertical asymptote (V.A. at value of x that makes "bottom" zero)

When we have the* *<u>same degree</u>* *on the top & bottom, the H.A. is at* $y =$ ***ratio of the leading coefficients*

So above we had: $f(x) = \dfrac{\boxed{3}x+1}{\boxed{1}x+1}$ *H.A. at* $y = 3$

For example, the function $g(x) = \dfrac{4x-1}{3x+5}$

Has a Horizontal Asymptote at $\boldsymbol{y = \frac{4}{3}}$ $\quad g(x) = \dfrac{\boxed{4}x-1}{\boxed{3}x+5} \Rightarrow$

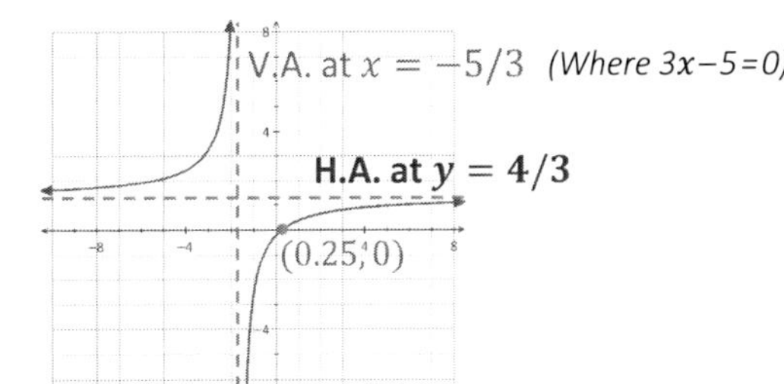

Note that this only works when the numerator and denominator are the same degree. (for now, we'll focus only on examples where the top / bottom are degree 1)

**When there is* *<u>higher degree on the bottom</u>,* *the H.A. is always at* $\boldsymbol{y = 0}$.

For example, $y = \dfrac{-3}{2x+1} \Rightarrow$

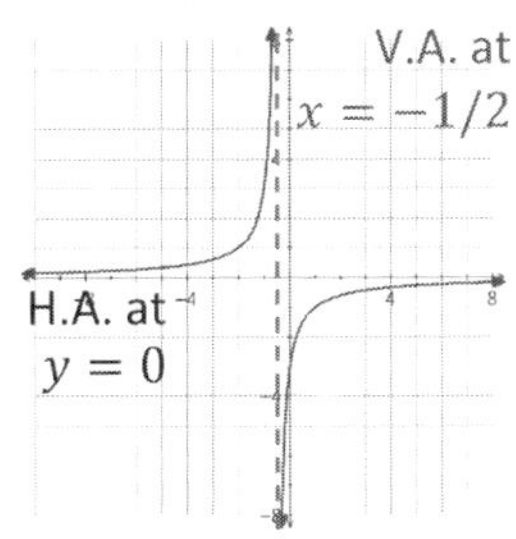

**When there is* *<u>higher degree top</u>,* *the graph will have no horizontal asymptote .*

For example, $y = \dfrac{2x(x-3)}{x-1} \Rightarrow$

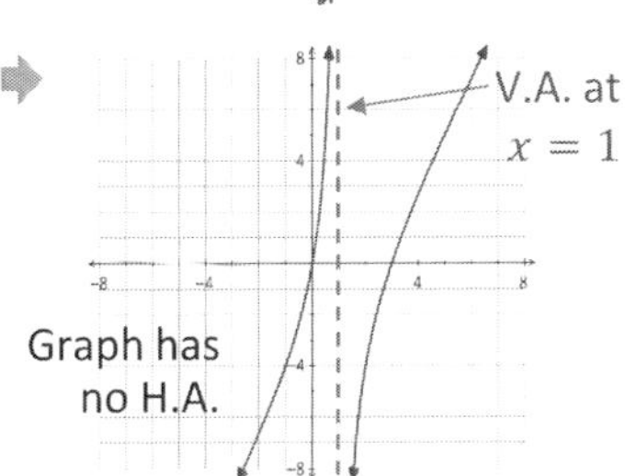

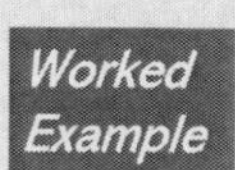

Given the function $f(x) = \dfrac{4x-4}{x+3}$, state the (i) domain, (ii) range, (iii) equation of any asymptotes, and (iv) sketch, labelling the coordinates of any intercepts.

Solution: First prep by factoring the numerator:

Factor from the "top" (numerator) gives the x-intercept

x-int. at x=1

$$y = \frac{4(x-1)}{x+3} \Rightarrow \mathbf{(1, 0)}$$

Factor from the "bottom" (denominator) gives the vertical asymptote

V.A. at x=−3

The H.A. is given by the ratio of the leading coefficients (since same degree top / bottom)

H.A. at y=4

For y-intercept, set $x = 0$:

$$y = \frac{4(0)+4}{0+3}$$

$= 4/3 \Rightarrow \mathbf{(0, \frac{4}{3})}$

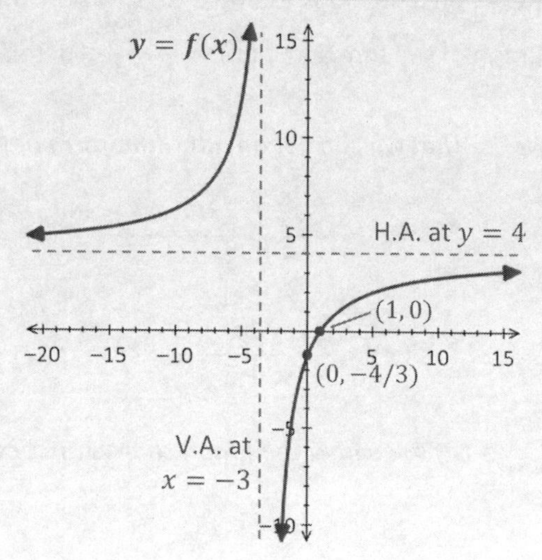

Class Example 4.33 *Characteristics of a Rational Function in the form* $y = \dfrac{f(x)}{g(x)}$

Given the function $g(x) = \dfrac{2x+6}{3x-3}$, determine the (i) domain, (ii) range, (iii) equation of any asymptotes, and (iv) the coordinates of any intercepts. Sketch, labeling all identified characteristics.

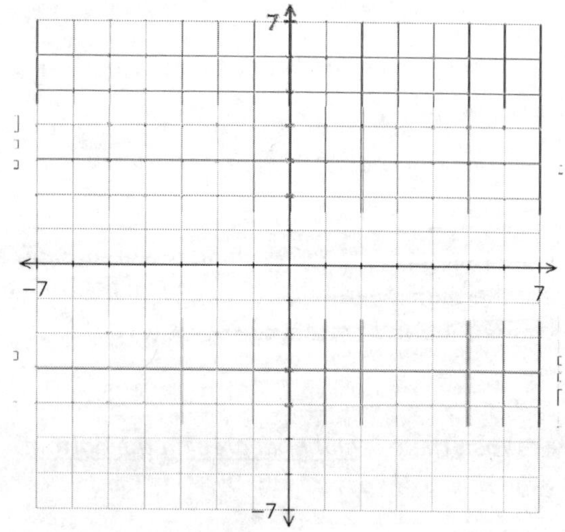

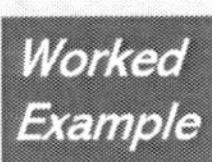

The graph below represents a rational function that can be written in the form $f(x) = \dfrac{g(x)}{h(x)}$, where g and h are linear. Asymptotes and x or y intercept occurs at integer values. Determine an equation of $y = f(x)$.

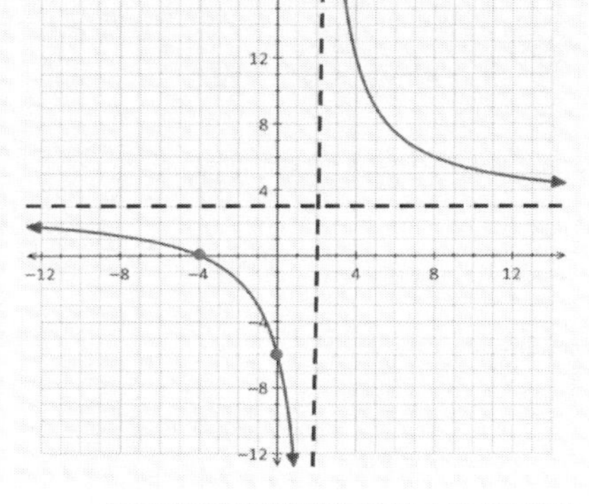

Solution:

$$f(x) = \frac{a(x+4)}{x-2}$$

- x-int gives a factor "on top"
- V.A. gives a factor "on bottom"

- **Always** include vertical stretch ("$\boldsymbol{a}$")
  Use any other point on the graph to solve for it.
  ➔ *Use point* $(\mathbf{0}, \mathbf{-6})$ *to solve for* a.

$$-6 = \frac{a(0+4)}{0-2}$$

$$-6 = \frac{4a}{-2}$$

$$12 = 4a \rightarrow \boldsymbol{a = 3}$$

> So equation is: $f(x) = \dfrac{3(x+4)}{x-2}$

Verify equation graphically:

Match window to that of given graph, remember to put brackets around the numerator / denominator.

```
WINDOW
 Xmin=-12
 Xmax=14
 Xscl=1
 Ymin=-12
 Ymax=16
 Yscl=1
 Xres=1
```

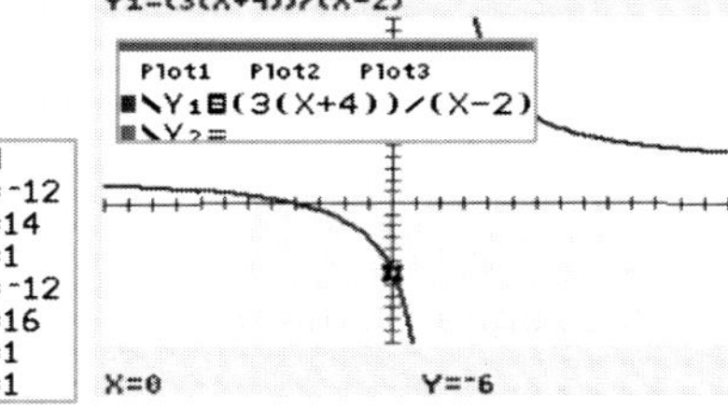

Class Example 4.34 *Determining an Equation of a Rational Function*

The graph below represents a rational function that can be written in the form $f(x) = \dfrac{g(x)}{h(x)}$, where g and h are linear.

(a) Determine an equation for $y = f(x)$.

(b) Determine the equation of the H.A.

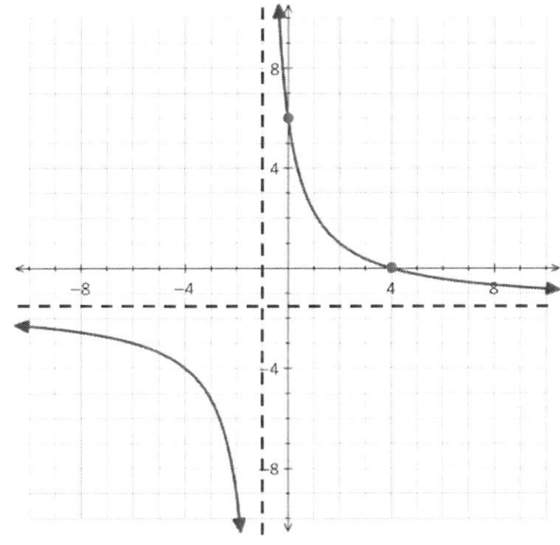

Note: The vertical asymptote and x & y intercepts occurs at integer values.

4.1 Practice Questions (Sample, first page only)

SAMPLE of our high-quality, practice questions. Visit **Math30-1edge.com** for full access.

1. For each of the functions given below, use transformations to construct a mapping rule for transformation from the basic graph and **sketch the graph**. Then, use algebraic processes to **determine** the indicated **graph characteristics.**

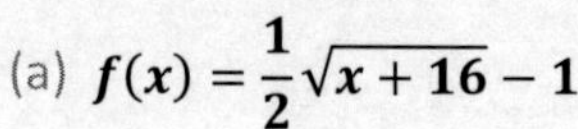

(a) $f(x) = \frac{1}{2}\sqrt{x+16} - 1$

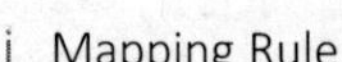

i Mapping Rule:

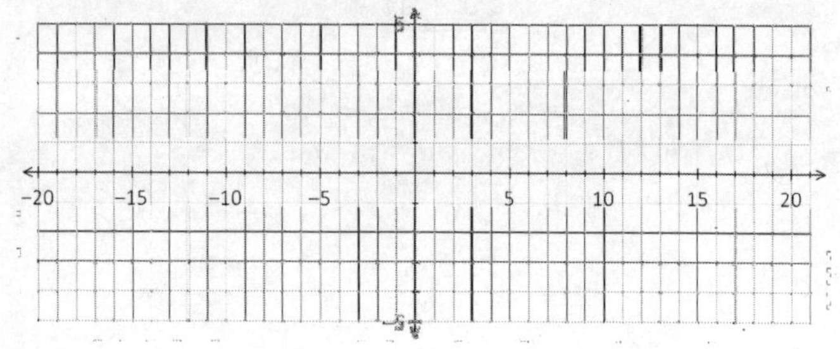

ii Domain:

iii Range:

iv x-intercept:

v y-intercept:

(b) $g(x) = \sqrt{-3x+9} - 1$

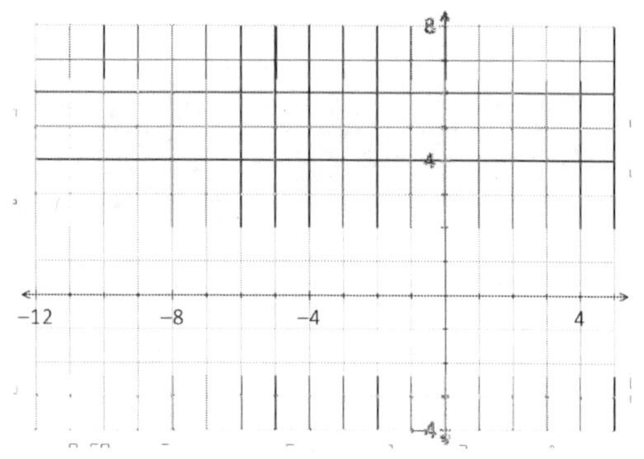

i Mapping Rule:

ii Domain:

iii Range:

iv x-intercept:
As an ***exact value***

v y-intercept:

Answers to practice questions on previous page

1. (a) i $(x, y) \rightarrow (x - 16, \frac{1}{2}y - 1)$

$(0, 0) \rightarrow (-16, -1)$

$(1, 1) \rightarrow (1 - 16, \frac{1}{2}(1) - 1)$

$\rightarrow (-15, -0.5)$ *And so on...*

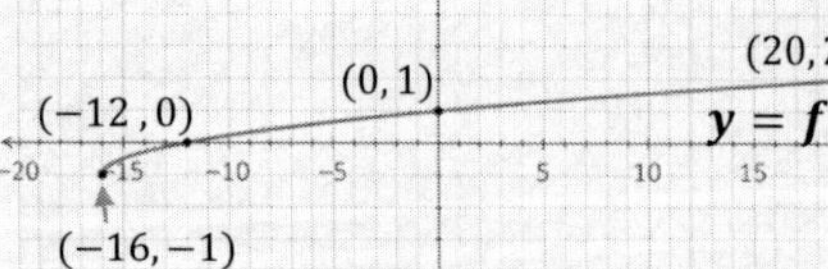

ii $\{x | x \geq -16, x \in \mathbb{R}\}$

iii $\{y | y \geq -1, y \in \mathbb{R}\}$

iv $(-12, 0)$

v $(0, 1)$

(b) i $(x, y) \rightarrow (-\frac{1}{3}x + 3, y - 1)$

ii $\{x | x \leq 3, x \in \mathbb{R}\}$

iii $\{y | y \geq -1, y \in \mathbb{R}\}$

iv $(8/3, 0)$

v $(0, 2)$

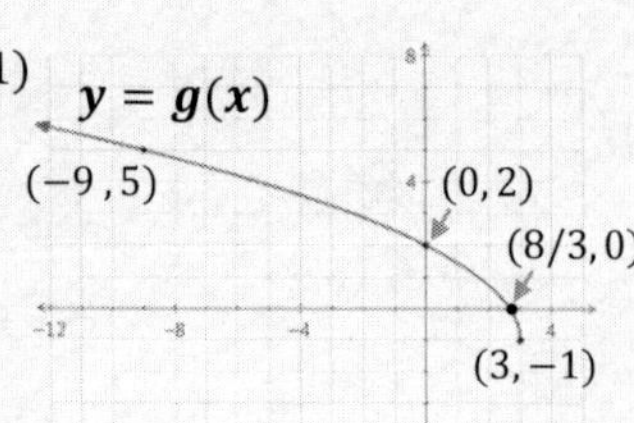

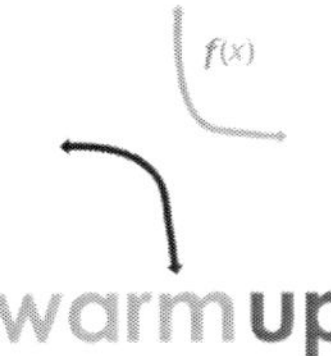

Exploration #1 The Graph of $y = \frac{1}{x^2}$

The graph of $y = \frac{1}{x^2}$ and accompanying table of values are shown.

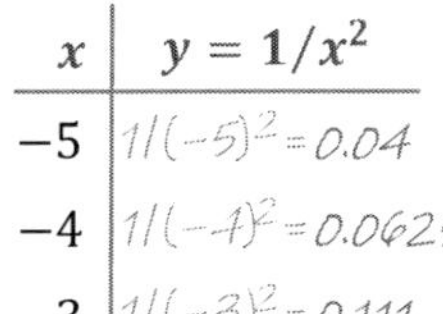

x	$y = 1/x^2$
-5	$1/(-5)^2 = 0.04$
-4	$1/(-4)^2 = 0.0625$
-3	$1/(-3)^2 = 0.111...$
-2	$1/(-2)^2 = 0.25$
-1	$1/(-1)^2 = 1$
-0.5	$1/(-0.5)^2 = 4$
-0.25	$1/(-0.25)^2 = 16$
-0.2	$1/(-0.2)^2 = 25$
0	undefined
0.2	$1/(0.2)^2 = 25$
0.25	$1/(0.25)^2 = 16$
0.5	$1/(0.5)^2 = 4$
1	$1/(1)^2 = 1$
2	$1/(2)^2 = 0.25$
3	$1/(3)^2 = 0.111...$
4	$1/(4)^2 = 0.0625$
5	$1/(5)^2 = 0.04$

1 ➡ **Compare** the graph of $y = \frac{1}{x^2}$ to that of $y = \frac{1}{x}$, shown below.

2 ➡ **Describe** the following characteristics of the graph of $y = \frac{1}{x^2}$:

_____	_____	_____	_____	_____	_____
Domain	*Range*	*V.A.s*	*H.A.*	*x-int*	*y-int*

Exploration #2

Refer to the graphs below to answer to the following questions. For the graph on the left, point P is a local maximum of the first graph, with coordinates $(-0.5, -0.32)$.

$y = \frac{2}{(x+3)(x-2)}$

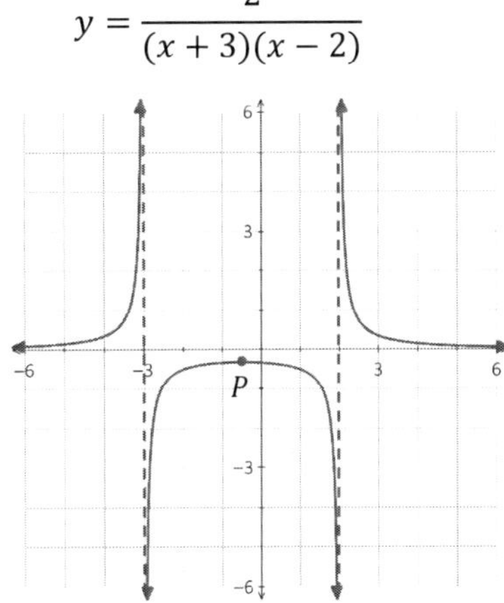

V.A.s:

H.A.:

Domain:

Range:

x-int:

y-int:

Degree of numerator:

Degree of denominator:

Note: Point P is a *local maximum*, with coordinates $(-0.5, -0.32)$.

$y = \frac{x-1}{(x+3)(x-2)}$

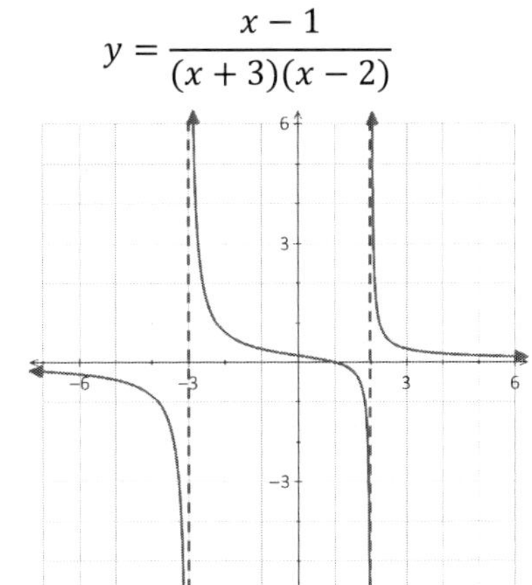

V.A.s:

H.A.:

Domain:

Range:

x-int:

y-int:

Degree of numerator:

Degree of denominator:

1 ➡ **Explain** how each characteristic relates to specific parts of the equation.

x-intercept: *Vertical Asymptote:* *Horizontal Asymptote:*

2 ➡ Based on your analysis for the graph on the right – can a rational function graph cross its horizontal asymptote? .

Rational Functions in the Form $y = \dfrac{f(x)}{g(x)}$, where g is degree two (quadratic)

In our study we'll consider rational functions in the form $y = \dfrac{f(x)}{g(x)}$ where:

- $f(x)$ (the numerator) can be degree 0, 1, or 2
- $g(x)$ (the denominator) can be degree 1 or 2

Consider three examples each with the same (degree two) denominator. As such, they have the same two **vertical asymptotes** at $x = -3$, $x = 2$, and the same **domain**: $\{x | x \neq -3, 2, x \in \mathbb{R}\}$. The coordinates of any local maximum / minimum points, as well as any x-intercepts, are shown.

Note how the numerator affects the graph's horizontal asymptote (H.A.), range, and x-intercepts.

Numerator is degree 0

$$y = \frac{2}{(x+3)(x-2)}$$

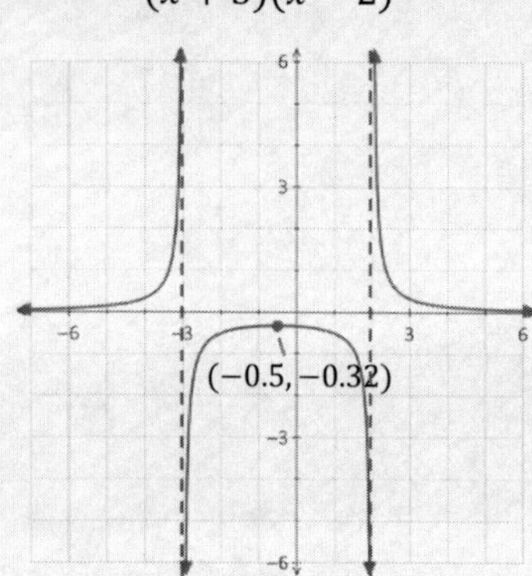

Higher degree in the denominator (at the bottom), so H.A. at $y = 0$

No x-intercept as there is no x-term in the numerator (top)

Range expressed in two parts:

$\{y | y \leq -0.32$ *or* $y > 0, y \in \mathbb{R}\}$

approximate

Numerator is degree 1

$$y = \frac{2x-1}{(x+3)(x-2)}$$

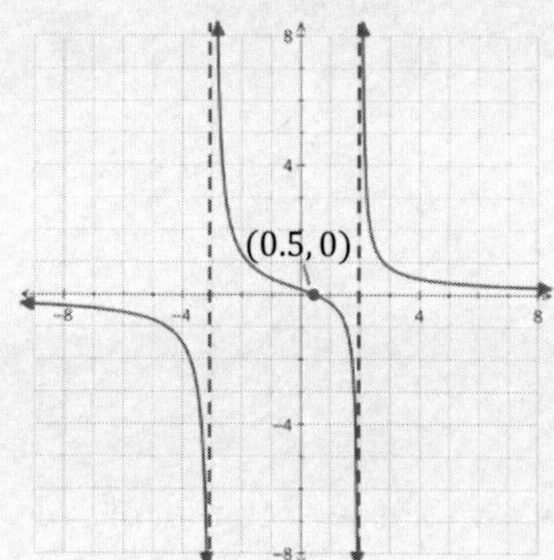

Higher degree in the denominator, so H.A. at $y = 0$

(and graph crosses H.A.!)

x-intercept at $2x - 1 = 0$, which is $(0.5, 0)$

Range is simply $\{y \in \mathbb{R}\}$

Numerator is degree 2

$$y = \frac{-2(x+1)(x-1)}{(x+3)(x-2)}$$

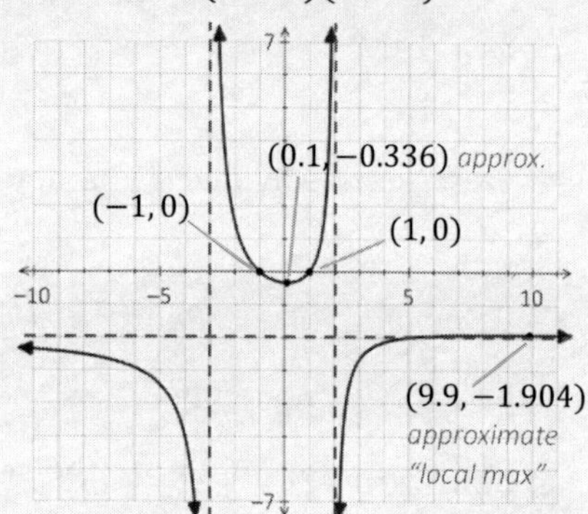

Same degree on the top / bottom, so H.A. at $y = -2$ ← *Ratio of lead coefficients*

(graph crosses H.A.!)

x-ints where $x + 1 = 0$ and $x - 1 = 0$

Range: $\{y | y \leq -1.904$ *or* $y \geq -0.336, y \in \mathbb{R}\}$

Range is found using "MIN" / "MAX" functions on your graphing calculator

Class Example 4.41 *Analyzing Characteristics of a Rational Function*

On the right is the graph of $f(x) = \dfrac{x+2}{x^2 + 3x - 4}$.

Factor to analyze the equation of $y = f(x)$, to determine each of the following. *Sketch / label each characteristic on the graph.* →

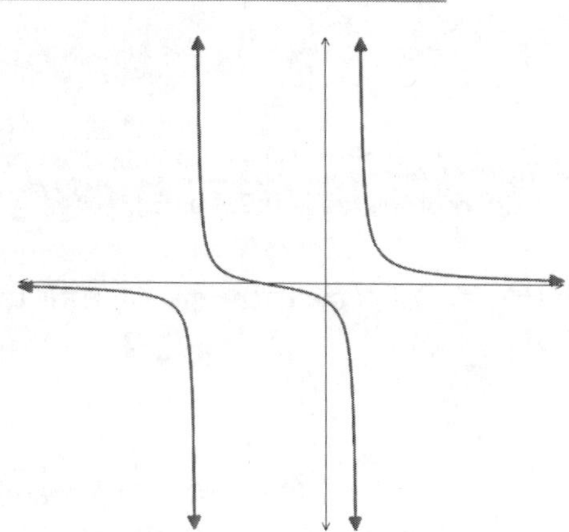

(a) Vertical asymptote(s)

(b) Coordinates of the x and y intercepts.

(c) Horizontal asymptote

(d) Domain and Range

The Three Cases for Horizontal Asymptotes:

Let's summarize how to, at a glance, recognize the horizontal asymptote of a rational function:

❖ **Case 1** – Higher Degree *on the bottom (denominator)*

Example 1 $y = \dfrac{3}{2x+1}$

H.A. at $y = 0$

(along x-axis)

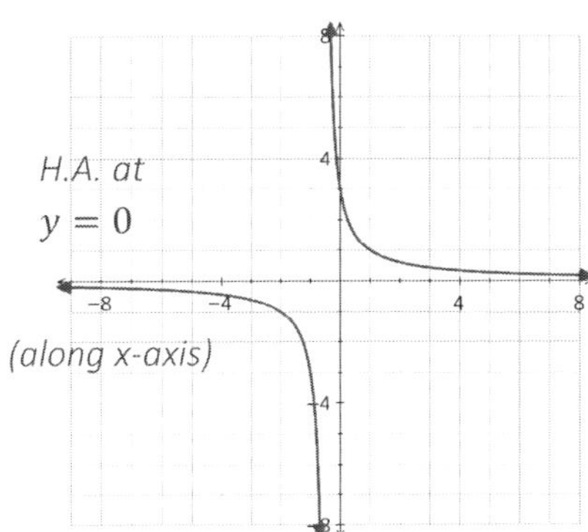

Example 2 $y = \dfrac{x-4}{x^2-x-2}$

H.A. at $y = 0$

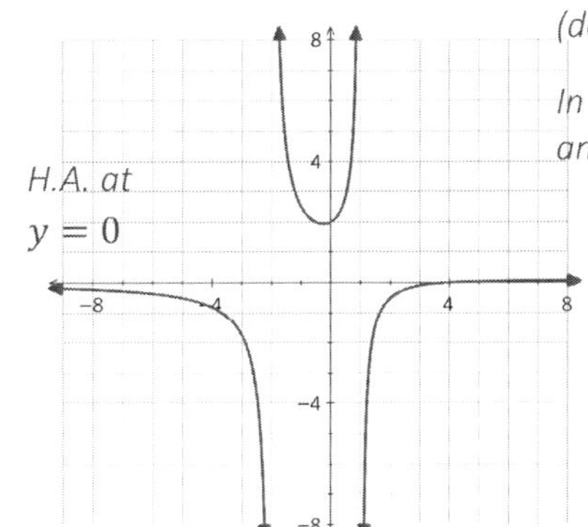

In example 1, the top (the numerator, "3") is degree 0, and the bottom (denominator) is degree 1.

In example 2, the top is degree 1, and the bottom is degree 2.

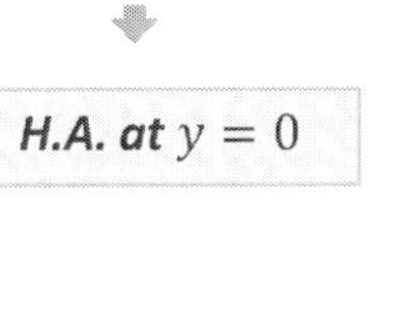

❖ **Case 2** – Same Degree *on the top (numerator) and bottom (denominator)*

Example 1 $y = \dfrac{3x^2+8x-3}{2x^2-7x-4}$

H.A. at $y = 1.5$

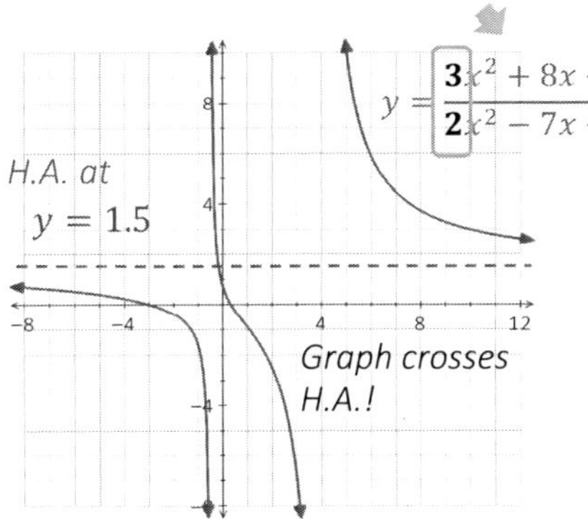

Example 2 $y = \dfrac{-6(x+1)(x-3)}{(2x-3)(x+2)}$

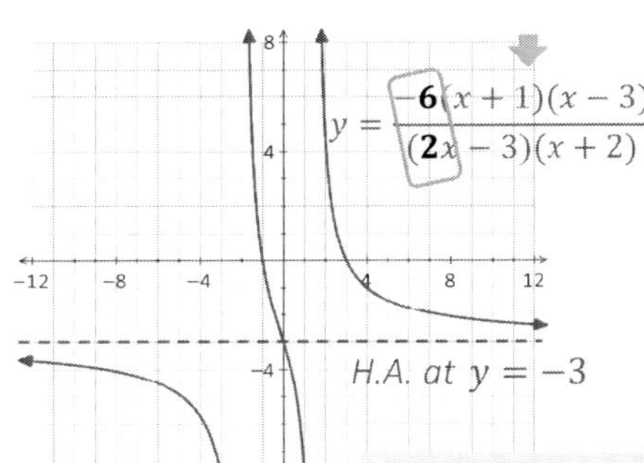

H.A. at y = *ratio of leading coefficients*

❖ **Case 3** – Higher Degree *on the top (numerator)*

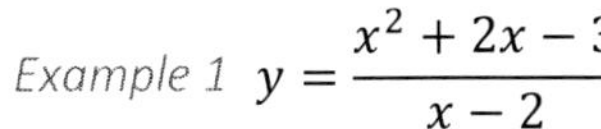

Example 1 $y = \dfrac{x^2+2x-3}{x-2}$

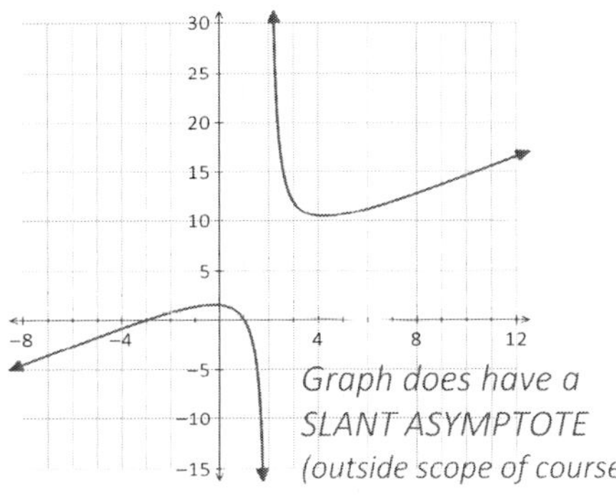

Example 2 $y = \dfrac{(x+2)(x-3)}{x-3}$

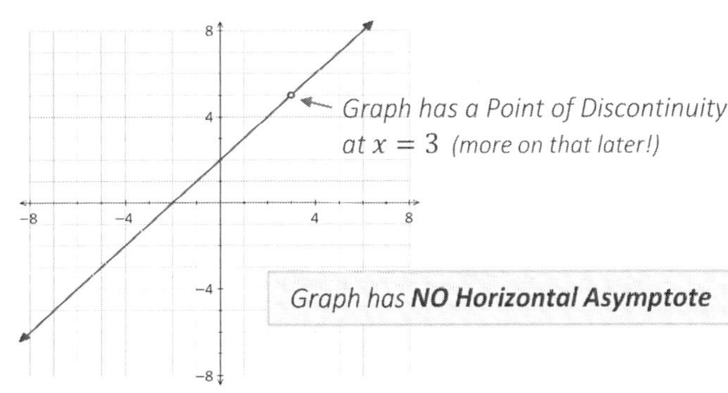

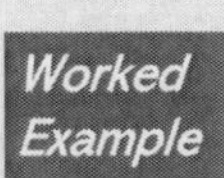

For the function $f(x) = \dfrac{-2x+6}{x^2-x-20}$, use an algebraic process to determine each of the following characteristics: (i) Vertical Asymptotes, (ii) Horizontal Asymptote, (iii) x-intercept, (iv) y-intercept, and (v) Domain and Range. Sketch, labeling all characteristics.

Solution: First prep by factoring:

Factor from the "top" (numerator) gives the x-intercept

x-int. at x−3=0

$$y = \frac{-2(x-3)}{(x+4)(x-5)} \Rightarrow \mathbf{(3,0)}$$

Factors from the "bottom" (denominator) give vertical asymptotes

V.A. at x=−4 and x=5

i $x = -4$ and $x = 5$

ii $y = 0$ *Since higher degree on bottom*

iii $(3,0)$ **iv** $(0,-0.3)$

For y-intercept, set $x = 0$*:*

$$y = \frac{-2(0)+6}{(0)^2-(0)-20}$$

$$= -6/20 \Rightarrow \left(\mathbf{0}, \frac{\mathbf{-3}}{\mathbf{10}}\right)$$

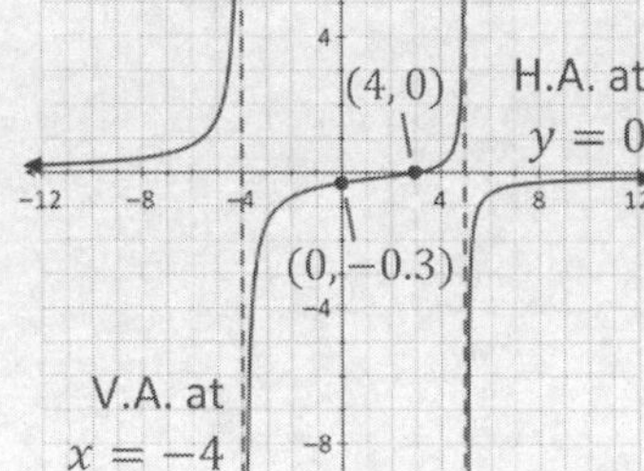

iv Domain is $\{x | x \neq -4, 5, x \in \mathbb{R}\}$

Range is $\{y \in \mathbb{R}\}$ ← *Note that graph crosses H.A. (so no restriction), which we must graph on our Calc to confirm!*

Class Example 4.42 *Analyzing Characteristics of a Rational Function - Quadratic Denominator*

For each of the following functions, use an algebraic process to determine each of the following characteristics: (i) Vertical Asymptotes, (ii) Horizontal Asymptote, (iii) x-intercept, (iv) y-intercept, and (v) Domain and Range. Finally, label all characteristics then use your graphing calculator to help **sketch**.

Need a quick refresher on factoring this type of quadratic? *Flip the page!*

(a) $f(x) = \dfrac{-3x}{2x^2+5x-12}$

i *V.A.(s):*

ii *H.A.*

iii *x-int:*

iv *y-int:*

v *Domain:*

Range:

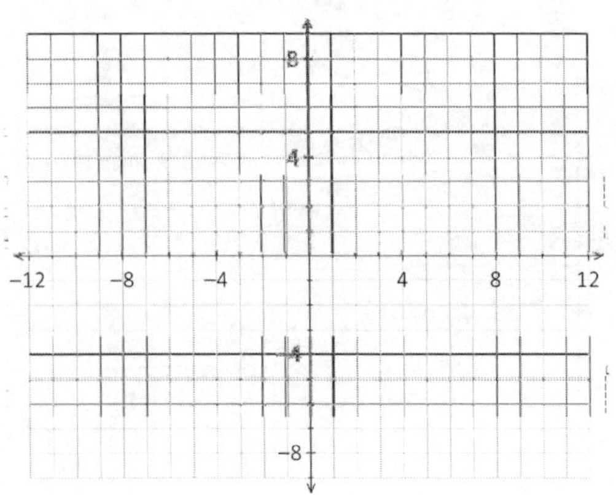

(b) $f(x) = \dfrac{2x^2 - 3x - 2}{3x^2 - 5x - 12}$

i *V.A.(s):*

ii *H.A.*

iii *x-int:*

iv *y-int:*

v *Domain:*

Range:

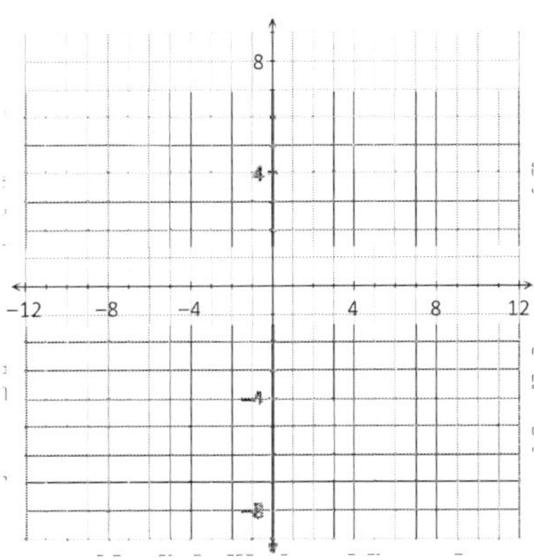

REVIEW – Factoring quadratics in the form $ax^2 + bx + c\,;\, a \neq 1$

In case you've forgotten some of your Math 10C.....

Example 1 – Factor:

$8x^2 + 2x - 3$

$= 8x^2 + 6x - 4x - 3$

$= \boxed{8x^2 + 6x}\boxed{-4x - 3}$

Factor a "2x" from the first two terms — *Factor a "−1" from the last two terms*

$= 2x(4x + 3) - 1(4x + 3)$

$= \mathbf{(4x + 3)(2x - 1)}$

Step 1: Re-write the middle term, "2x", using two terms where the coefficients...

- Multiply to −24 (product of a and c, that is, "8" and "−3")

- Add to 2 (the value for b) ← *Use "6" and "− 4"*

Step 2: Factor by grouping...

Step 3: Note the common binomial factor, 4x+3

You try the following two examples: (answers are on the bottom of the next page)

Example 2 – Factor $6x^2 - x - 12$

Example 3 – Factor $24x^2 - 25x + 6$

The graph of the rational function $y = f(x)$ is shown below. Determine an equation for the function, in both factored and expanded form.

Solution:

Always include a vertical stretch (we'll solve for this later!)

The x-intercept at x=−1 gives a factor in the numerator ("on top"!)

$$y = \frac{a(x+1)}{(x+5)(x-2)}$$

The V.A.s at x=−5 and x=2 give the factors in the denominator ("on bottom"!)

How do we know how many factors go on the "top" / "bottom" ?

- Each x-intercept, gives a factor "on top" *(here there is one)*

- Each V.A. gives a factor "on the bottom" *(here there are two)*

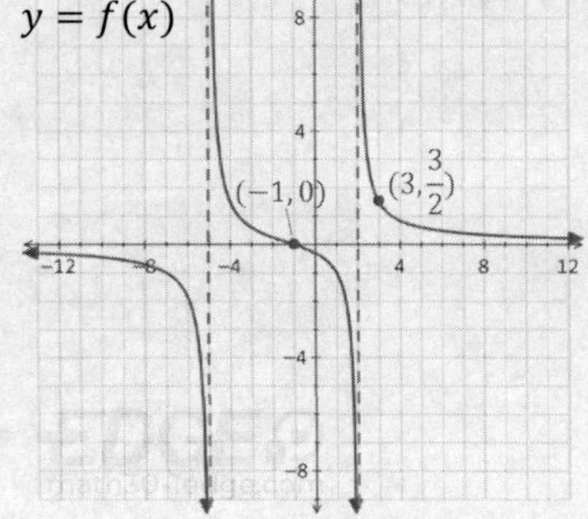

Next, use the point (3, 3/2) to solve for "a":

$$\frac{3}{2} = \frac{a(3+1)}{(3+5)(3-2)}$$

$$2a(4) = 3(8)(1)$$

$$a = \frac{24}{8} \Rightarrow a = 3 \Rightarrow$$

Note: While we can see there is a H.A. at y = 0, we do not consider that in building our equation!

Factored form equation:

$$f(x) = \frac{3(x+1)}{(x+5)(x-2)}$$

And in expanded form:

$$f(x) = \frac{3x+3}{x^2+3x-10}$$

Class Example 4.43 *Determining a Rational Function Equation*

Determine an equation, in both factored and expanded form, given the graphs of each of the following functions.

(a)

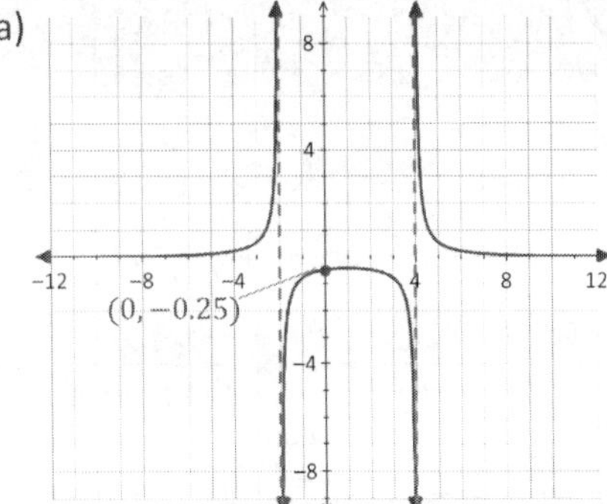

(b)

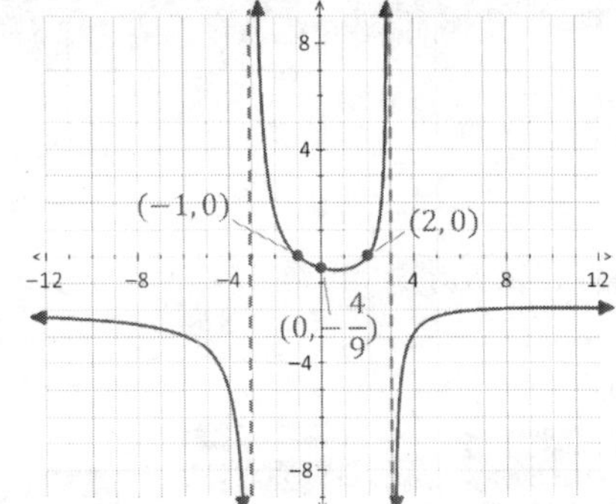

Answers from previous page

Example 2 $6x^2 + 8x - 9x - 12$

$= 2x(3x+4) - 3(3x+4)$

$= \mathbf{(3x+4)(2x-3)}$

Example 3 $24x^2 - 16x - 9x + 6$

$= 8x(3x-2) - 3(3x-2)$

$= \mathbf{(3x-2)(8x-3)}$

A Special Case of the Rational Function – The *Point of Discontinuity*

Exploration #3 Rational Functions where Factors Cancel

The two graphs below represent the functions $f(x)=\dfrac{x^2-x-2}{x-2}$ and $g(x)=\dfrac{x^2-2x-3}{x-2}$.

1 Use your graphing calculator to identify which graph represents the function $f(x)$ and which is $g(x)$. *Label each accordingly*

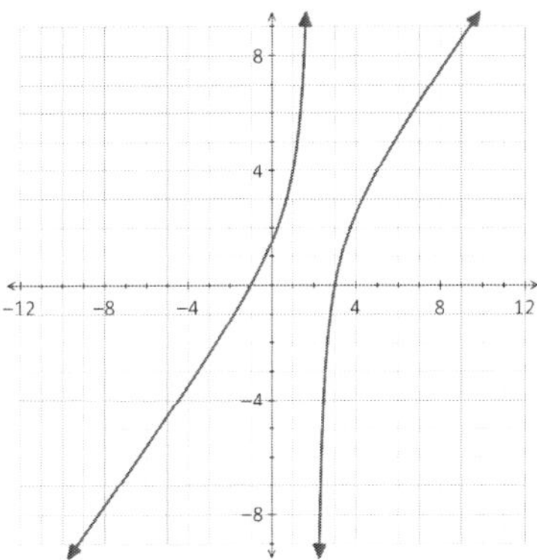

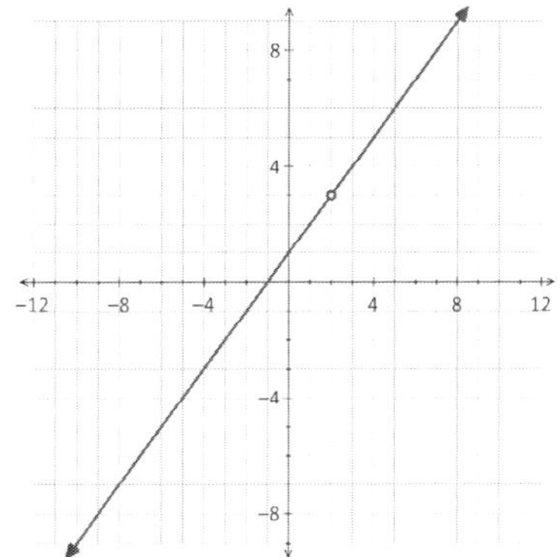

2 State the non-permissible value for each function.

3 For each function - describe the effect on the graph of the non-permissible values.

4 For each function – factor the numerator and simplify where possible. Describe how to determine if the graph of a rational function will have a **point of discontinuity**.

5 Determine where the following function will have a (i) point of discontinuity and (ii) vertical asymptote.

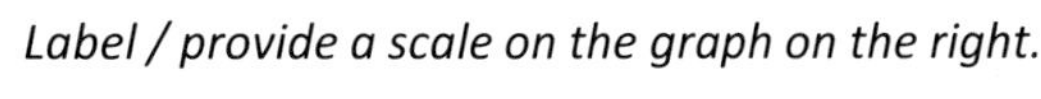

$$y=\frac{x+5}{x^2+3x-10}$$

Label / provide a scale on the graph on the right.

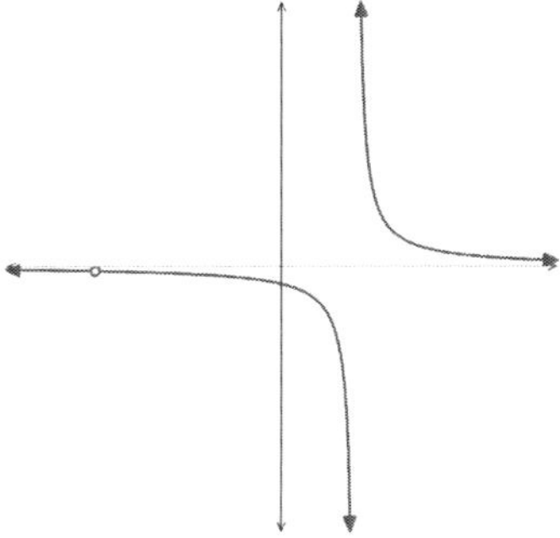

Rational Functions Graphs – The Point of Discontinuity

Rational Functions have **non-permissible values (N.P.V.s)** at any value of x that would make the denominator (bottom) equal to zero.

Non-permissible values have two possible effects on the graph. Consider the function $f(x) = \dfrac{x-2}{x^2+x-6}$

Factoring gives: $f(x) = \dfrac{x-2}{(x+3)(x-2)}$

V.A. at $x = -3$

$y = f(x)$

P.D. at $x = 2$

Coordinates of PD are $(2, 1/3)$

There are two non-permissible values (NPVs); $\boldsymbol{x \neq -3}$ *and* $\boldsymbol{x \neq 2}$.

This NPV comes from a factor in the denominator that doesn't cancel out.

The effect on the graph is a vertical asymptote. (V.A.)

And this NPV comes from a factor in the denominator that DOES cancel.

The effect on the graph is a Point of Discontinuity. (P.D.)

Note that $f(x)$ *simplifies to:* $f(x) = \dfrac{1}{x+1}\,;x \neq -3, 2$ *(We include NPVs)*

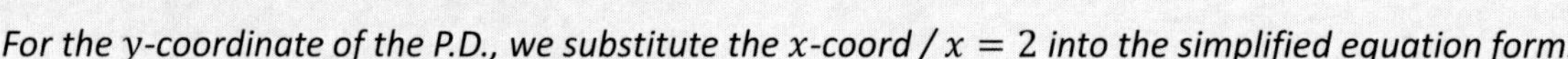

For the y-coordinate of the P.D., we substitute the x-coord / $x = 2$ ***into the simplified equation form.***

$$f(2) = \frac{1}{(2)+1} \Rightarrow = \frac{1}{3}$$

This y-coordinate represents the y value that isn't on the graph!

So the ***range*** *of* $y = f(x)$ *is:* $\{y \mid y \neq 0, y \neq \frac{1}{3}, y \in \mathbb{R}\}$

We kick out the H.A. (graph doesn't cross)

And we kick out the y-coord of the P.D.

A word of caution: We use a lot of shorthand form – such as "*VA*", "*NPV*", and "*PD*"

But a reminder to only use **full terminology** on a test (especially a diploma exam)!

Class Example 4.44 *Analyzing Characteristics of a Rational Function*

For each of the following rational functions, (i) State the non-permissible values, (ii) Describe the effect on the graph of the non-permissible values (V.A. or P.D.), (iii) State the domain, and (iv) State the range. Label all of these characteristics on the grids below, and use your graphing calculator to help sketch each graph.

(a) $f(x) = \dfrac{x^2+7x+10}{x+5}$

i *N.P.V.s*

ii *Effect*

iii *Domain:*

iv *Range:*

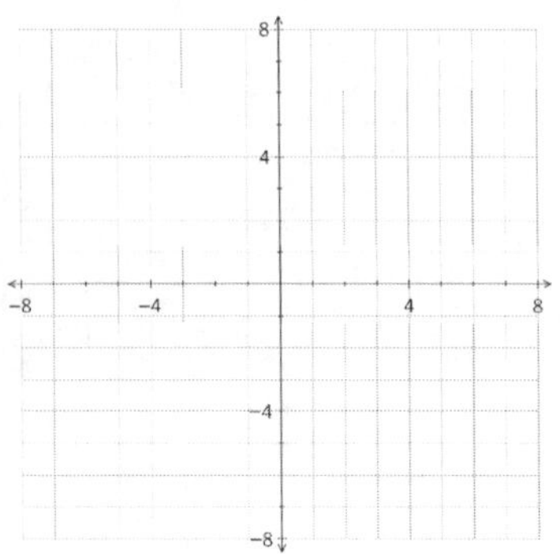

(b) $g(x) = \dfrac{x+2}{2x^2+x-6}$

i *N.P.V.s*

ii *Effect*

iii *Domain:*

iv *Range:*

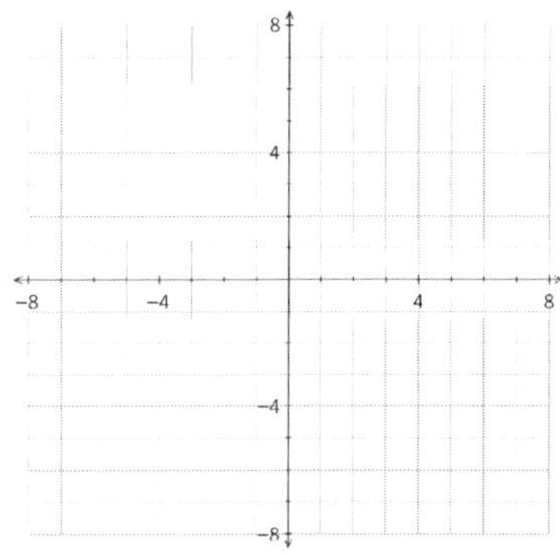

Your graphing calculator isn't always so great at displaying *points of discontinuity*!

Suppose we wished to graph $y = \dfrac{x+2}{2x^2+x-6}$ on our graphing calc. ⟶

Y1=(X+2)/(2X²+X-6)

Plot1 Plot2 Plot3

■\Y1■(X+2)/(2X²+X−6)

X=-2 Y=

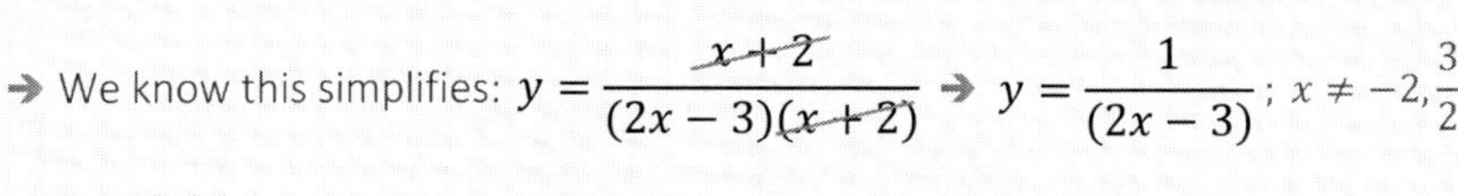

➔ We know this simplifies: $y = \dfrac{\cancel{x+2}}{(2x-3)\cancel{(x+2)}}$ ➔ $y = \dfrac{1}{(2x-3)}$; $x \neq -2, \dfrac{3}{2}$

However we are sure to graph the "pre-simplified" form!

(The only difference is the simplified form function will not have a point of discontinuity at $x = -2$)

Note that we do not see the "hole" in the graph at $x = -2$. However, if we use trace and input "−2", we can see that there is no output / "y" value.

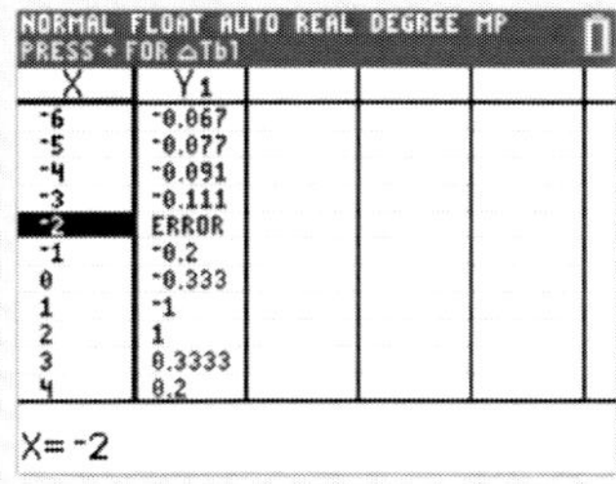

X	Y1
-6	-0.067
-5	-0.077
-4	-0.091
-3	-0.111
-2	ERROR
-1	-0.2
0	-0.333
1	-1
2	1
3	0.3333
4	0.2

X=-2

← We can also confirm that the graph of $y = \dfrac{x+2}{2x^2+x-6}$ has a

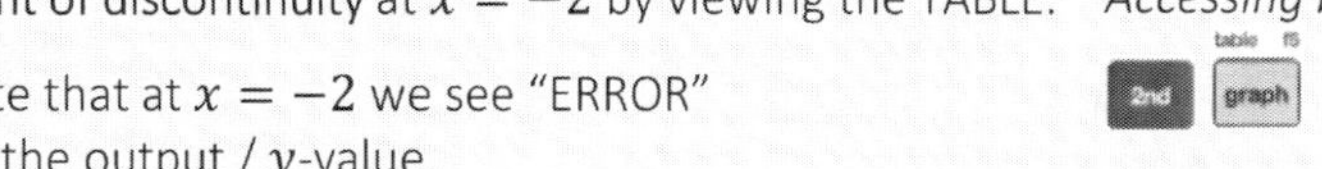

point of discontinuity at $x = -2$ by viewing the TABLE. *Accessing by keying:* 2nd graph

Note that at $x = -2$ we see "ERROR" for the output / y-value.

Summary: On our calculators, the graphs of $y = \dfrac{x+2}{2x^2+x-6}$ and the simplified form $y = \dfrac{1}{(2x-3)}$ look identical!

However graphing the simplified form will miss any P.D.

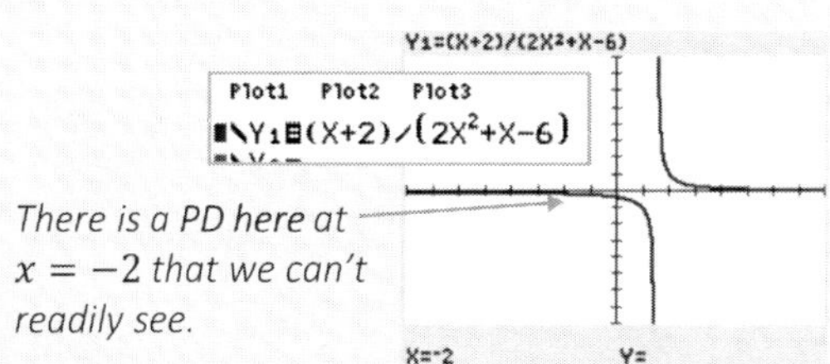

There is a PD here at $x = -2$ that we can't readily see.

By graphing the simplified form, we can obtain the y-coord. of the P.D. (for the RANGE)

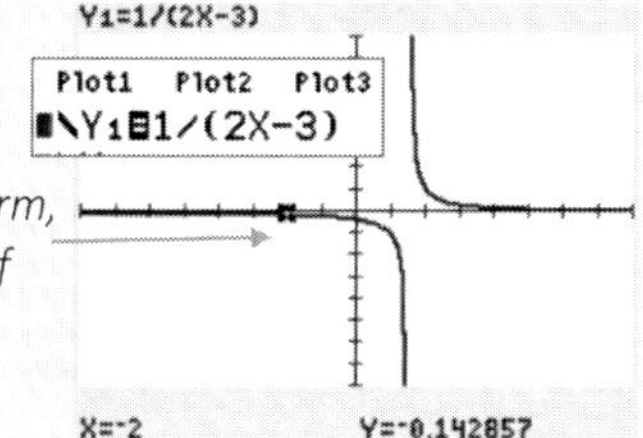

Worked Example The graph of the rational function $y = f(x)$ is shown below. Determine an equation for the function, in both factored and expanded form.

Solution:

Always include a vertical stretch

The P.D. at x=−1 gives a factor on both the top & bottom

$$y = \frac{a(x-1)}{(x+4)(x-1)}$$

The V.A. at x=−4 gives the factor on the bottom only

How do we know how many factors go on the "top" / "bottom" ?

- Each x-intercept, gives a factor "on top" only (none here!)
- Each V.A. gives a factor "on the bottom" only (here there is one)
- Each P.D. gives a factor on both the top and bottom (here there is one)

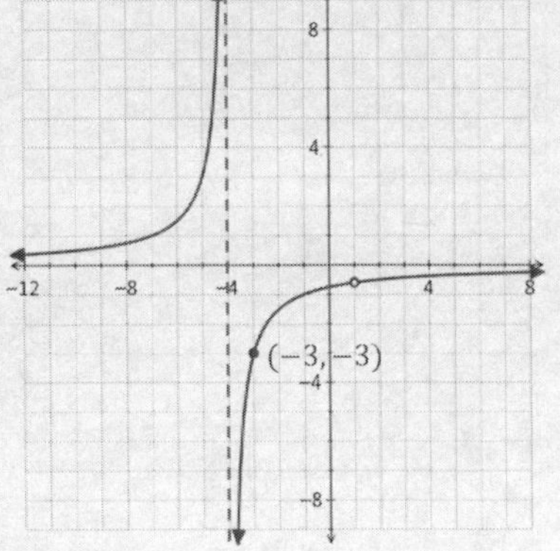

Next, use the point (−3, −3) to solve for "a":

$$-3 = \frac{a(-3-1)}{(-3+4)(-3-1)}$$

$$-4a = (-3)(1)(-4)$$

$$a = \frac{12}{-4}$$

$$a = -3$$

Note: While we can see there is a H.A. at y = 0, we do not consider that in building our equation! (However we do note that there must be "higher degree on the bottom")

Factored form equation:

$$f(x) = \frac{-3(x-1)}{(x-1)(x+4)}$$

And in expanded form:

$$f(x) = \frac{-3x+3}{x^2+3x-4}$$

Class Example 4.45 *Determining the Equation of a Rational Function*

Given the graph of the functions $y = f(x)$ and $y = g(x)$ below,

i - Determine an equation for each function, in both factored and expanded form

ii – Determine the range of each function

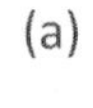

(a)

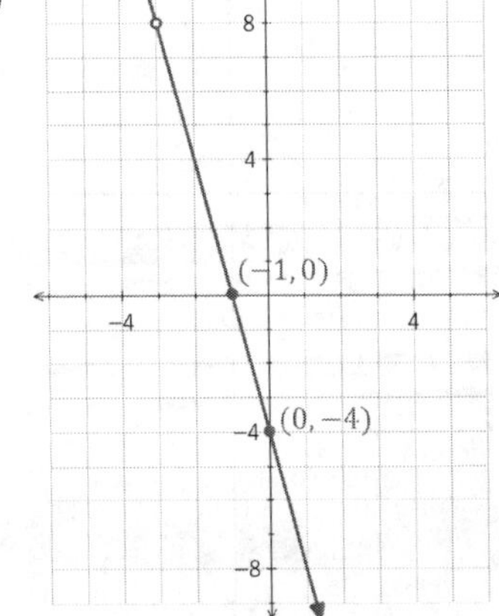

(b)

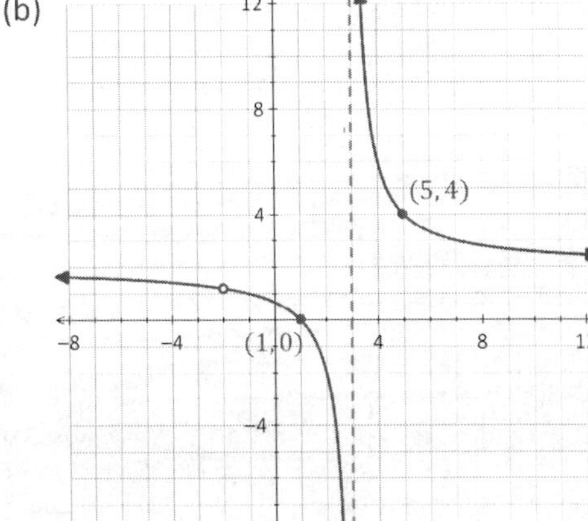

Chapter 5 OPERATIONS on FUNCTIONS

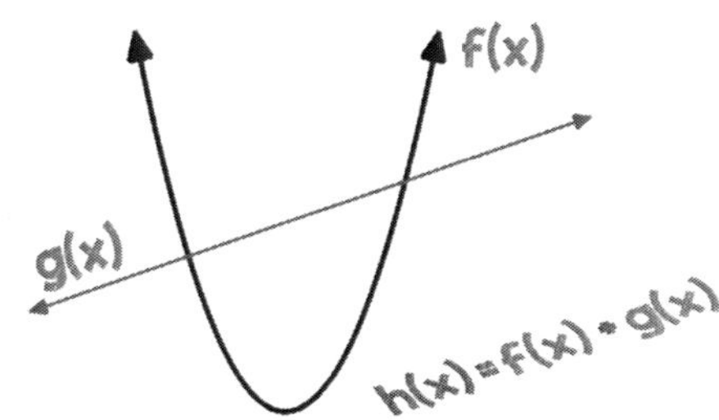

5.1 Adding, Subtracting, Multiplying and Dividing Functions

Functions – in our first four chapters we got in-depth on so many types of functions and their graphs!

Polynomial

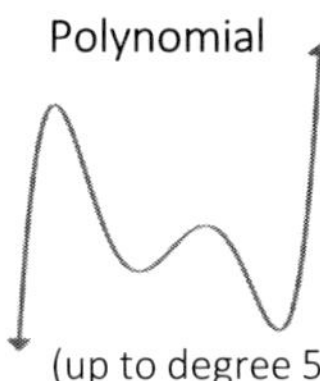

(up to degree 5)

Exponential

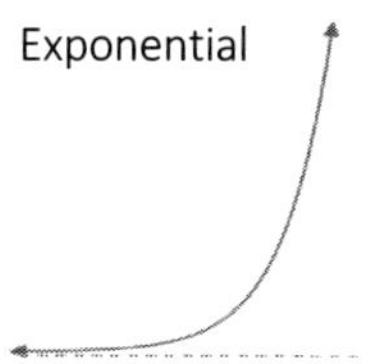

Logarithmic

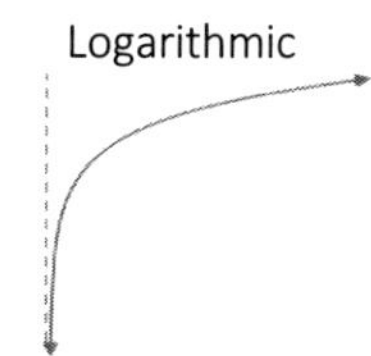

Radical

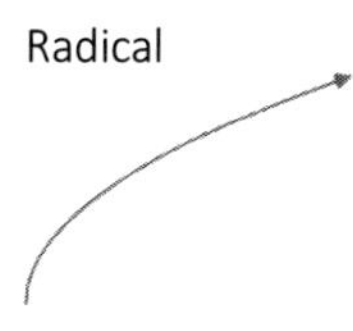

Rational

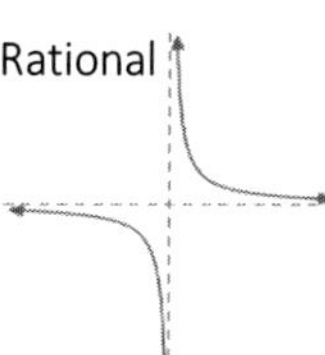

These along with functions we studied in prior courses: (these were reviewed in chapter 1)

Linear

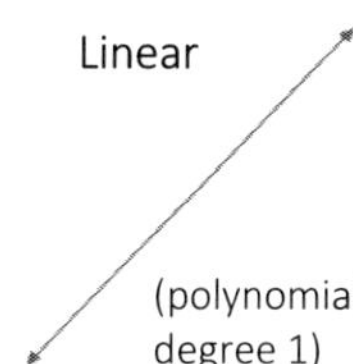

(polynomial degree 1)

Quadratic

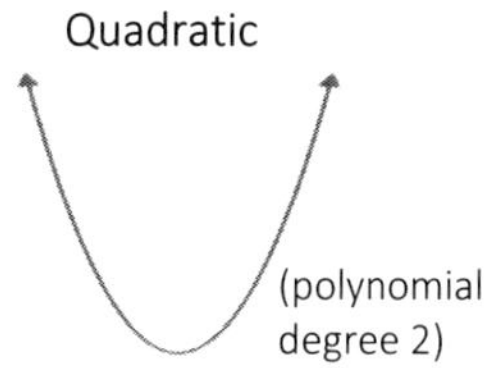

(polynomial degree 2)

Absolute Value

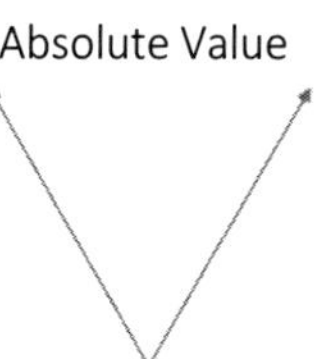

In this chapter we'll apply various types of operations on these functions as well as other functions that could be defined as just a set of ordered pairs, a graph, or a table. It will be a good review of were we've been so far, while tying together some of the core function concepts studied earlier. *Let's get started….*

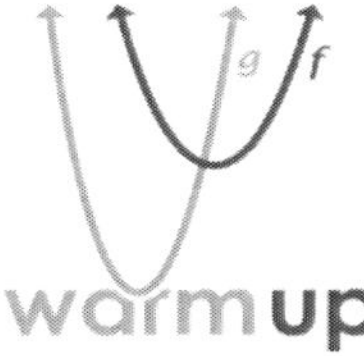

Combining Functions

1 ➡ Use the graph of the three functions on the right to **complete** the table below

x	$f(x)$	$g(x)$	$h(x)$
-4	-9	7	-2
-3			
-2			
-1			
0			
1			
2			
3			
4			
5			

2 ➡ Examine the table to determine the relationship between the values in each column.

3 ➡ State an equation for $h(x)$ in terms of $f(x)$ and $g(x)$.

4 ➡ State an equation for $f(x)$ in terms of $h(x)$ and $g(x)$.

5 ➡ Determine a slope-intercept form equation for $h(x)$.

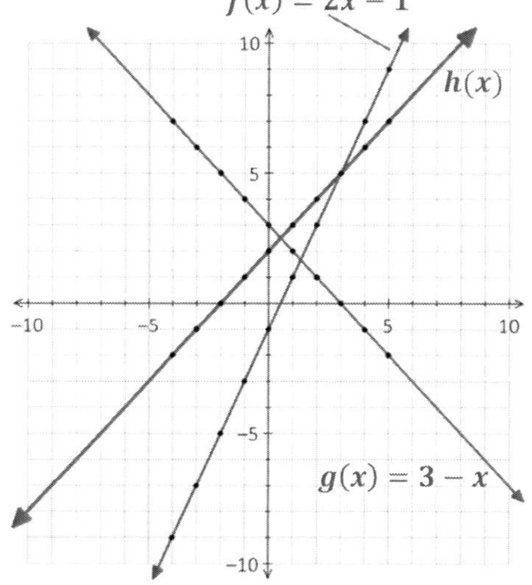

6 ➡ Add the function equations for $f(x)$ and $g(x)$ to show that $h(x)$ is the sum.

Two functions $f(x)$ and $g(x)$ can be combined as follows:

Sum of Functions $y = f(x) + g(x)$ *can also be written* $y = (f + g)(x)$

Difference of Functions $y = f(x) - g(x)$ *can also be written* $y = (f - g)(x)$

Product of Functions $y = f(x) \times g(x)$ *can also be written* $y = (f \cdot g)(x)$

Quotient of Functions $y = \dfrac{f(x)}{g(x)}$ *can also be written* $y = \left(\dfrac{f}{g}\right)(x)$

We can similarly combine functions $f(x)$ and $g(x)$ given their graphs, by adding / subtracting / multiplying / or dividing all corresponding function values. (That is, the y-coordinates)

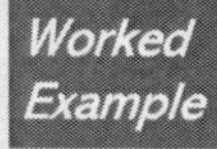

Worked Example The graphs of $f(x) = 2x + 6$ and $g(x) = 3 - x$; $x \leq 3$ are shown on the right.

(a) Sketch the graph of $h(x) = (f - g)(x)$ on the same grid

(b) State the simplified equation for $h(x)$, in terms of x

(c) State the domain of the $h(x)$.

(d) State a simplified equation for $k(x) = (f + g)(x)$ and $m(x) = (g \cdot f)(x)$ *Do not sketch*

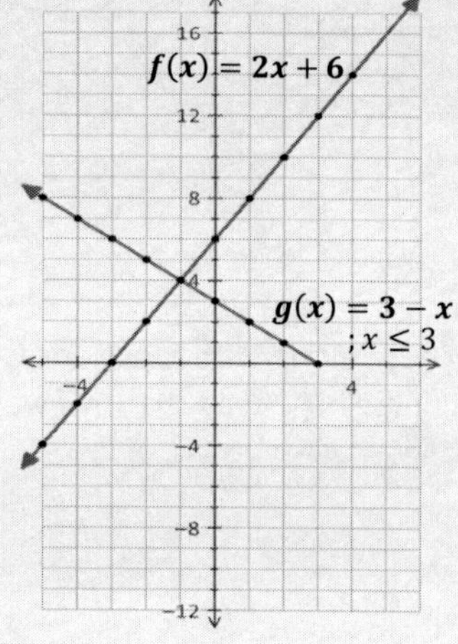

Solution: (a) Start with the "left-most" on each graph.

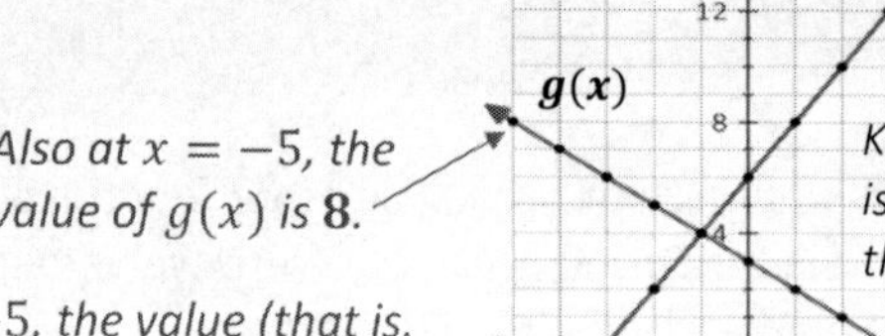

Also at $x = -5$, the value of $g(x)$ is **8**.

→ *When $x = -5$, the value (that is, the y-coordinate) of $f(x)$ is* **−4**.

***So** the value of $h(x)$ at $x = -5$ is $f(-5) - g(-5)$ .... $= \mathbf{-12}$*

$f(-5) = -4$, $g(-5) = 8$

Keep moving right ... $x = 3$ is the last point, beyond that g is undefined.

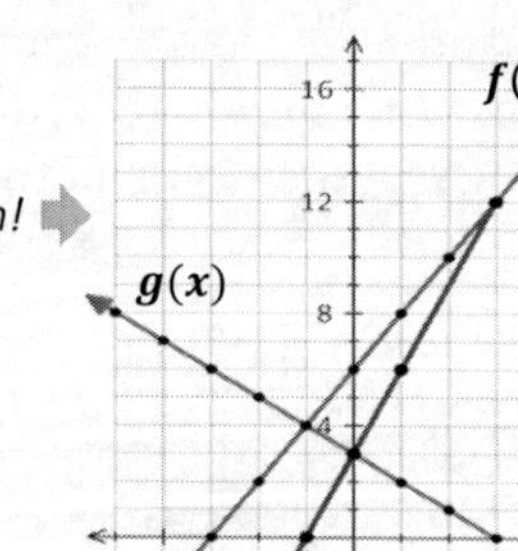

Finished Graph!

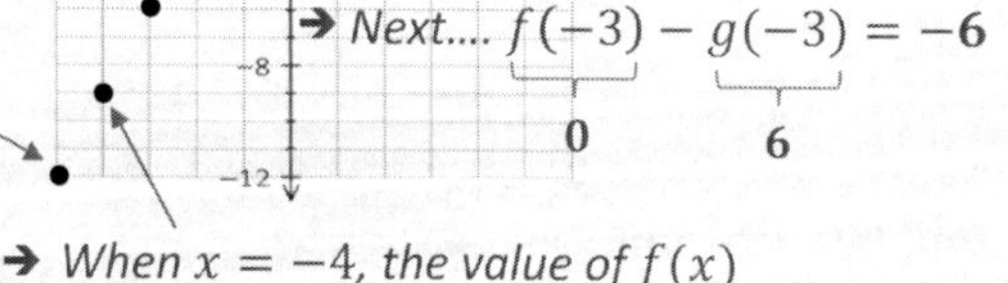

→ *Next.... $f(-3) - g(-3) = -6$*

$f(-3) = 0$, $g(-3) = 6$

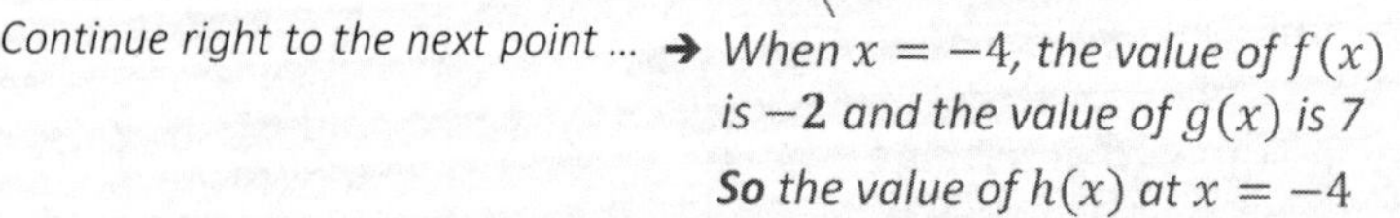

Continue right to the next point ... → *When $x = -4$, the value of $f(x)$ is* **−2** *and the value of $g(x)$ is 7*

***So** the value of $h(x)$ at $x = -4$ is $f(-4) - g(-4)$ .... $= -9$*

$f(-4) = -2$, $g(-4) = 7$

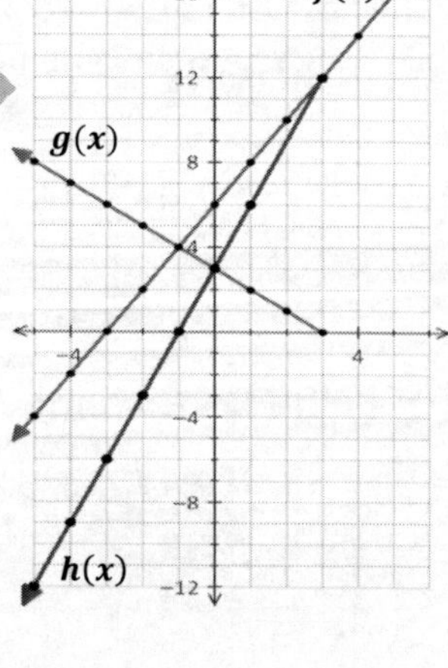

(b) Subtract the functions: $h(x) = \underbrace{(2x + 6)}_{f(x)} - \underbrace{(3 - x)}_{g(x)}$ .... *simplifies to*: $\mathbf{h(x) = 3x + 3}$; $x \leq 3$

Include the same domain restriction as $g(x)$

(c) The domain of $f(x)$ is $\{x \in \mathbb{R}\}$, while the domain of $g(x)$ is $\{x | x \leq 3, x \in \mathbb{R}\}$. So the graph of $h(x)$, which is based on the graphs of both f and g, similarly has a domain restriction. (For any x greater than 3, there are no values of $g(x)$, so our *combined graph stops there*)

(d) $k(x) = (2x + 6) + (3 - x)$

simplifies to:

$\mathbf{k(x) = x + 9}$; $x \leq 3$

$m(x) = (2x + 6)(3 - x)$ → $m(x) = 6x - 2x^2 + 18 - 6x$

simplifies to:

$\mathbf{m(x) = -2x^2 + 18}$; $x \leq 3$ *Remember the domain restriction!*

Class Example 5.11 *Combining Functions given their Graphs*

Given the graphs of the functions $f(x) = x + 1$ and $g(x) = 5 - x$,

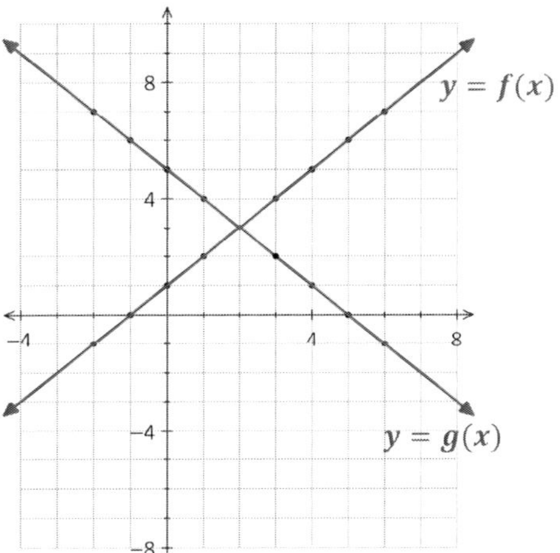

(a) Sketch the graph of $h(x) = (f - g)(x)$, on the same grid.

(b) State the simplified form equation of $h(x)$

(c) State the domain of the three functions f, g, and h.

(d) Determine a simplified equation, in terms of x, for $k(x) = f(x) + 2g(x)$ *do not graph*

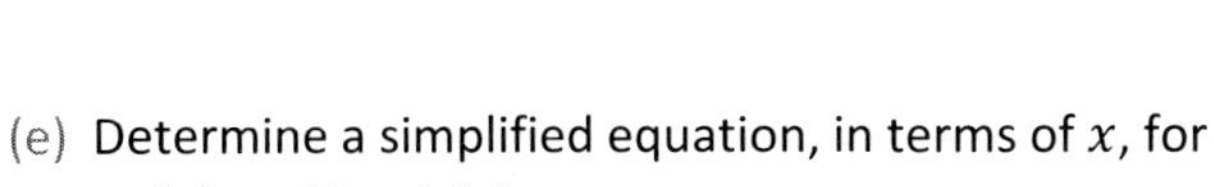

Visit **math30-1edge.com** for solutions to all warm-ups and class examples

(e) Determine a simplified equation, in terms of x, for $m(x) = (f \cdot g)(x)$ *do not graph*

DOMAIN of Combined Functions

The domain of a combined function must contain any restriction pertaining to either original function.

Example 1:

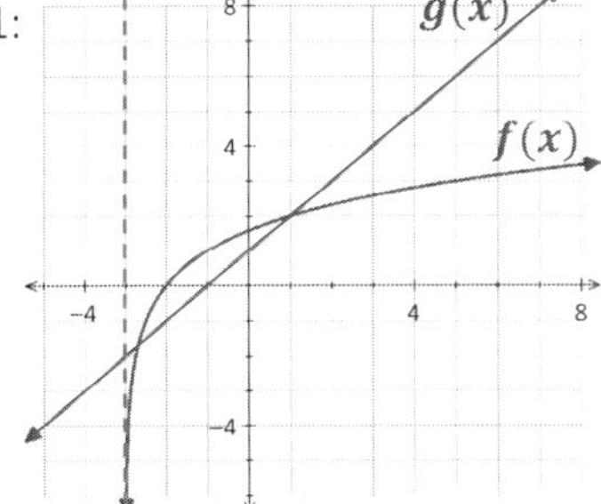

The domain of $f(x)$ is $\{x \mid x > -3, x \in R\}$

The domain of $g(x)$ is $\{x \in R\}$

The domain of any* combined function of f and g is: $\{x \mid x > -3, x \in R\}$

Example 2:

$f(x) = 4x^2(x - 2)$ $\qquad g(x) = \sqrt{5 - x}$

The domain of $f(x)$ is $\{x \in R\}$

The domain of $g(x)$ is $\{x \mid x \leq 5, x \in R\}$

The domain of any* combined function of f and g is: $\{x \mid x \leq 5, x \in R\}$

*For any combined function involving adding, subtracting, or multiplying two functions

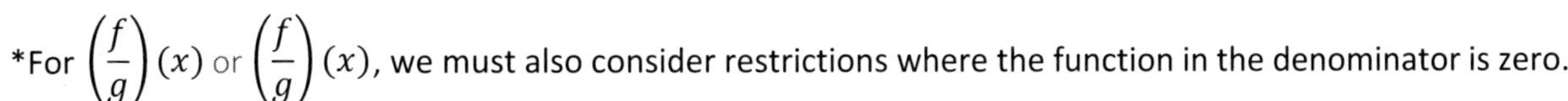

*For $\left(\frac{f}{g}\right)(x)$ or $\left(\frac{f}{g}\right)(x)$, we must also consider restrictions where the function in the denominator is zero.

So for the functions given by the graph in example 1.....

The domain of $\frac{f(x)}{g(x)}$ must also exclude -1, as $g(-1)$ is 0

$\Rightarrow \{x \mid x > -3, x \neq -1, x \in R\}$

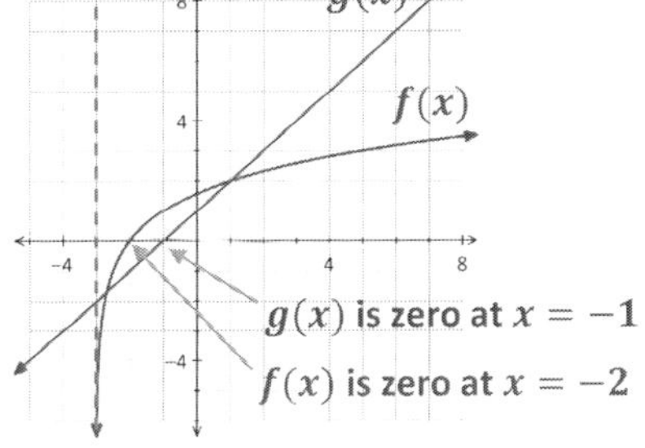

And the domain of $\frac{g(x)}{f(x)}$ must also exclude -2, as $g(-2)$ is 0

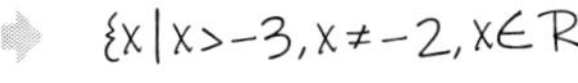

$\Rightarrow \{x \mid x > -3, x \neq -2, x \in R\}$

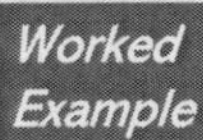

The graphs of $f(x) = x + 3$ and $g(x) = x - 2$ are shown on the right

(a) Sketch the graph of $k(x) = (f \cdot g)(x)$ on the same grid

(b) State the simplified equation for $k(x)$, in terms of x

(c) State the domain of the $k(x)$.

(d) State an equation, and the domain, for $h(x) = \dfrac{f(x)}{g(x)}$
Do not sketch

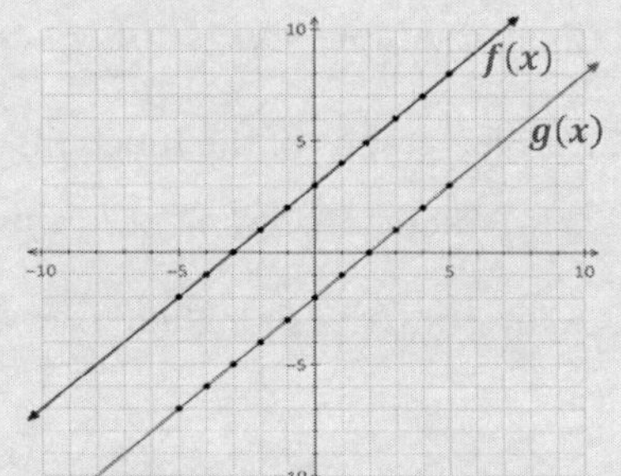

(a) We graph $k(x) = f(x) \times g(x)$ by multiplying all corresponding function values.
(That is, the y-coordinates)

We will stick with x-coordinates between -5 and 5, as beyond that, the function values become too large.

... The y-coord. on the graph of $k(x)$ is obtained by multiplying the corresponding y-cords. on $f(x)$ and $g(x)$.

This process is shown for $x = -4$ and 0.
In your solutions you'd do this for ALL points that fit on the grid! $(-3, -2, -1$, etc)

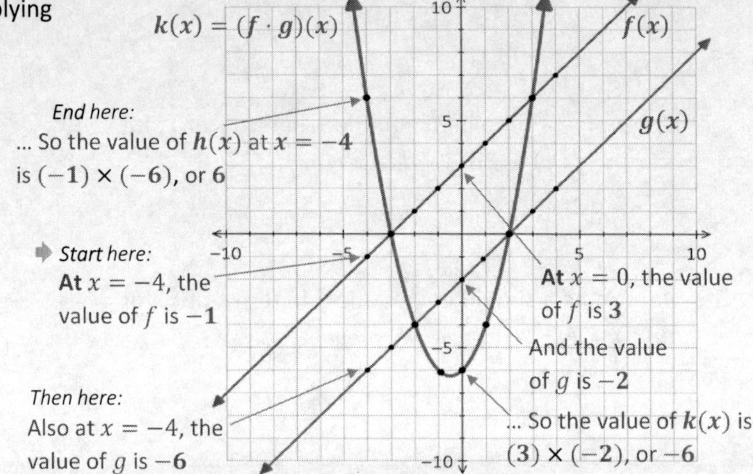

(b) Multiply the functions: $k(x) = \underbrace{(x+3)}_{f(x)} \times \underbrace{(x-2)}_{g(x)}$ .... *simplifies* → $\mathbf{k(x) = x^2 + x - 6}$

(c) The domain of $f(x)$ and $g(x)$ are both $\{x \in \mathbb{R}\}$. As there is no restriction, the domain of $k(x)$ is also $\{x \in \mathbb{R}\}$.

(d) Divide the functions: $\mathbf{k(x)} = \dfrac{x+3}{x-2}$ Domain of $k(x)$ is also $\{x\}x \neq 2,\ x \in \mathbb{R}\}$. ← Restrict value of x that would make the denominator 0

Class Example 5.12 *Combining Functions given their Graphs*

Given $f(x) = \dfrac{2}{x}$, $g(x) = \sqrt{5 - 4x}$ and $h(x) = 3x + 1$, state the domain of each combined function.

(a) $y = (f + h)(x)$

(b) $y = (f \cdot g)(x)$

(c) $y = \left(\dfrac{g}{h}\right)(x)$

(d) $y = \left(\dfrac{h}{g}\right)(x)$

Class Example 5.13 *Combining Functions given their Graphs*

Given the graphs of the functions $y = f(x)$ and $y = g(x)$,

(a) Sketch the graph of $h(x) = (f \cdot g)(x)$, on the same grid.

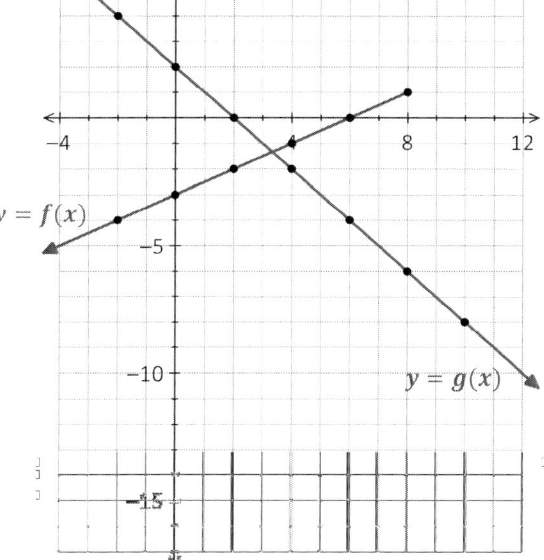

(b) State the slope-intercept form equation for both $f(x)$ and $g(x)$.

(c) State the simplified form equation of $h(x)$

(d) State the domain of the three functions f, g, and h.

(e) State the range of the three functions f, g, and h.

(f) State an equation, for $k(x) = \dfrac{f(x)}{g(x)}$

Do not graph. State any domain restrictions.

When to Include a domain restriction with a function equation

When a restriction on the domain is not implied by the equation itself, we must include it every time we state the equation.

For **example**, the function $f(x)$ above has an equation $f(x) = \frac{1}{2}x - 3$. But if we just left it at that, it would incorrectly imply that the domain was $\{x \in \mathbb{R}|$

So, we must include the restriction immediately afterward: $f(x) = \frac{1}{2}x - 3\,; x \leq 8$ ← *Now it's clear that f(x) is not a line, it cuts off!*

Another example: Consider the function $f(x) = \dfrac{x-3}{x^2-9}$ *Written like this, we are not required to state any domain restriction – we can obtain them by factoring the denominator*

But if we simplify .... $f(x) = \dfrac{\cancel{x-3}}{(x+3)\cancel{(x-3)}}$ ➧ $f(x) = \dfrac{1}{x+3}$

Now we have an issue! Someone looking at this function, not knowing that it came from simplifying (canceling terms), might incorrectly assume that the only restriction is $\boldsymbol{x \neq -3}$.

So, we include restrictions anytime we write an equation where something has been canceled ➧ $f(x) = \dfrac{1}{x+3}\ ;\ x \neq \pm 3$

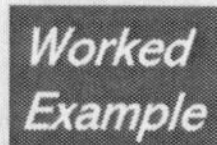

The graphs of $y = f(x)$ and $y = g(x)$ are shown on the right.
$y = g(x)$ is a linear function with a restricted domain.

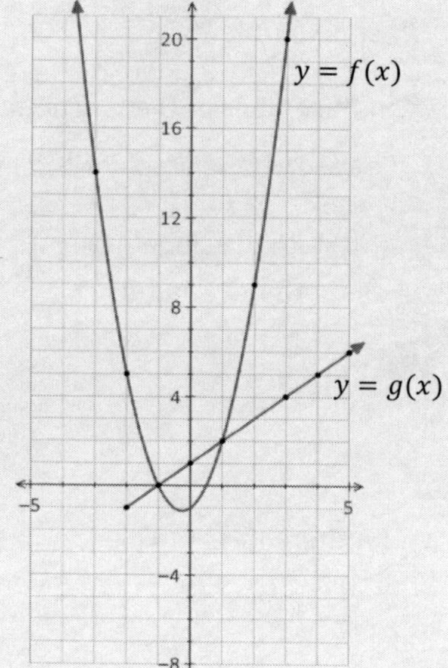

(a) Use the graphs of f and g to sketch $h(x) = \left(\frac{f}{g}\right)(x)$ on the same grid.

(b) Given that the equation $f(x) = 2x^2 + x - 1$, and $g(x) = x + 1$, determine a simplified equation for the combined function $y = h(x)$.
Be sure to include any domain restrictions.

(c) Determine a simplified equation for $k(x) = \left(\frac{g}{f}\right)(x)$
Do no sketch. Include any domain restrictions.

Solution: The graph ...

We first note that g(x) has a restricted domain, the graph starts at −2

So we start there / where both functions are defined ....

(Follow the notes counterclockwise around the graph)

Start here
At $x = -2$ the value of f is **5**

... and at $x = -2$ the value of g is $\mathbf{-1}$

So then...
At $x = -2$ the value of h is $\mathbf{5 \div -1} = \boxed{\mathbf{-5}}$
Plot a point there .... we're on our way!

Next point moving right is at −1
At $x = -1$ the value of f is $\mathbf{0}$ ... and the value of g is $\mathbf{0}$
So the value of h is $\mathbf{0 \div 0}$
(zero divided by zero means Point of discontinuity ... which we will leave for now and determine the y-coord. later)

Keep it going! Next we see that....
At $x = 0$ the value of f is $\mathbf{-1}$ and the value of g is $\mathbf{1}$
So at $x = 0$ the value of h is $\mathbf{-1 \div 1} = \boxed{\mathbf{-1}}$

And then next we see that....
At $x = 1$ the value of f is $\mathbf{2}$ and the value of g is $\mathbf{2}$
So at $x = 0$ the value of h is $\mathbf{2 \div 2} = \boxed{\mathbf{1}}$

Finally...
At $x = 2$ the value of f is $\mathbf{9}$ and the value of g is $\mathbf{3}$
So at $x = 0$ the value of h is $\mathbf{9 \div 3} = \boxed{\mathbf{3}}$

At $x = 3$ the value of f is $\mathbf{20}$ and the value of g is $\mathbf{4}$
So at $x = 0$ the value of h is $\mathbf{20 \div 4} = \boxed{\mathbf{5}}$

(a) **Finished Graph ➔**

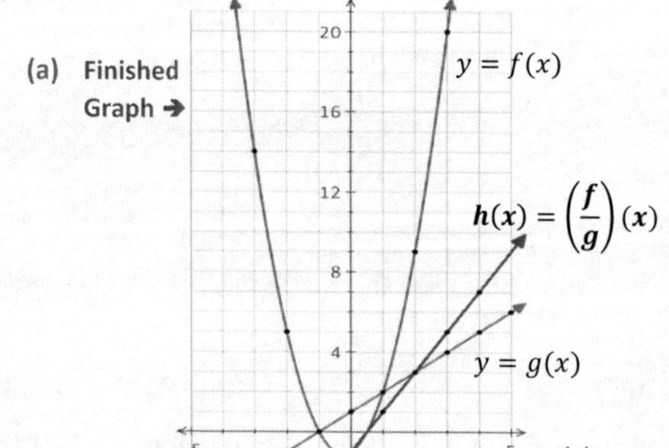

Graph has a point of discontinuity at $x = -1$

(b) $$\mathbf{k(x)} = \frac{\mathbf{2x^2 + x - 1}}{\mathbf{x + 1}}\ ; x \geq -2$$ ← *Must include the domain restriction in the equation*

$$\mathbf{k(x)} = \frac{\mathbf{(2x - 1)}\cancel{\mathbf{(x + 1)}}}{\cancel{\mathbf{x + 1}}}\ ; x \geq -2$$ *Simplify by factoring the numerator:*

$$\boxed{\mathbf{k(x) = 2x - 1};\ x \geq -2, x \neq -1}$$

And now we must also include the new restriction from the canceled factor (can't divide by zero)

(c) $$\mathbf{h(x)} = \frac{\mathbf{x + 1}}{\mathbf{2x^2 + x - 1}}\ ; x \geq -2 \;\Rightarrow\; \mathbf{h(x)} = \frac{\cancel{\mathbf{x + 1}}}{\mathbf{(2x - 1)}\cancel{\mathbf{(x + 1)}}}\ ; x \geq -2$$

$$\Rightarrow \boxed{\mathbf{h(x)} = \frac{\mathbf{1}}{\mathbf{(2x - 1)}}\ ; x \geq -2, x \neq -1, \tfrac{1}{2}}$$

Class Example 5.14 *Combining Functions given their Graphs*

The graphs of $y = f(x)$ and $y = g(x)$ are shown on the right.

- $f(x)$ is a degree two function with a restricted domain.
  Its equation can be written in the form $f(x) = a(x - m)(x - n)$; where $a = 1$ and m and n are zeros.

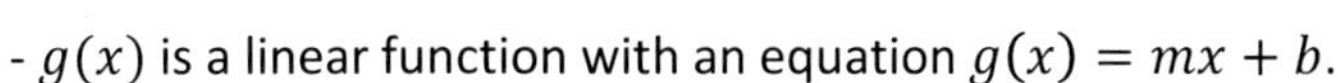

- $g(x)$ is a linear function with an equation $g(x) = mx + b$.

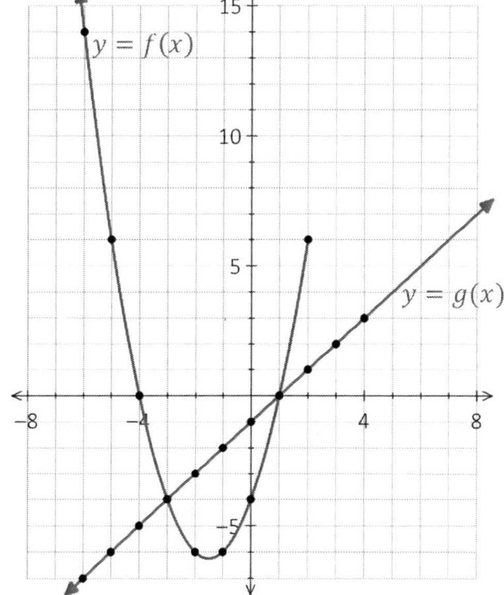

(a) On the same grid, sketch the graph of $h(x) = \left(\frac{f}{g}\right)(x)$

(b) Determine an equation for $f(x)$ and $g(x)$.

(c) Determine a simplified equation for $h(x)$.
*Be sure to include any domain restrictions.

(d) Use the equation to determine the range of $f(x)$, and the graphs to state the range of $g(x)$, and $h(x)$.

(e) Determine an equation for the following combined functions. (Do not graph)
Be sure to include any domain restrictions.

i $y = (f + g)(x)$ ii $y = (f - g)(x)$ iii $y = \left(\frac{g}{f}\right)(x)$

Graphing functions with restricted domains / graphing combined functions on your graphing calculator.

- The previous example included a **function with a restricted domain**. You might ask – can I make a graph like that on my graphing calculator? And the answer is *yes, yes you can*. Here's the steps:

To graph $f(x) = (x+4)(x-1); x \leq 2$

- Put brackets around both the function and the restriction:

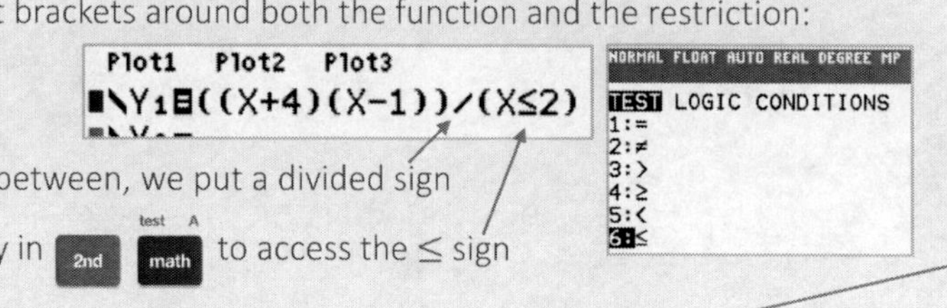

- In between, we put a divided sign

- Key in 2nd math to access the ≤ sign

- GRAPH your resulting, domain restricted function!

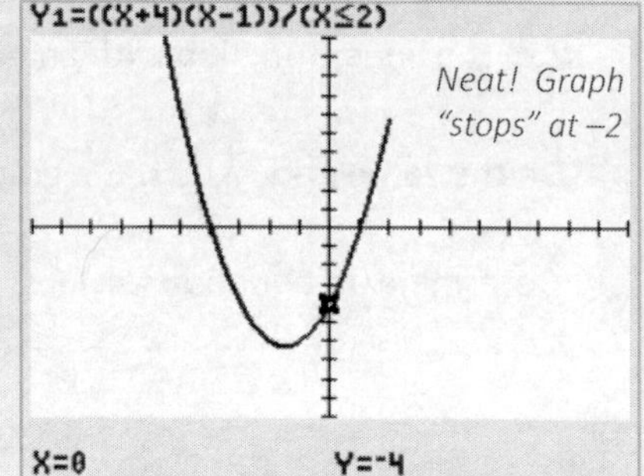

- Next up – **graphing combined functions**. In the previous example, we also had $g(x) = x - 1$, and wished to graph the combined function $h(x) = \left(\frac{f}{g}\right)(x)$.

To graph $f(x)$ (with its restricted domain), $g(x)$, and $h(x)$ all together:

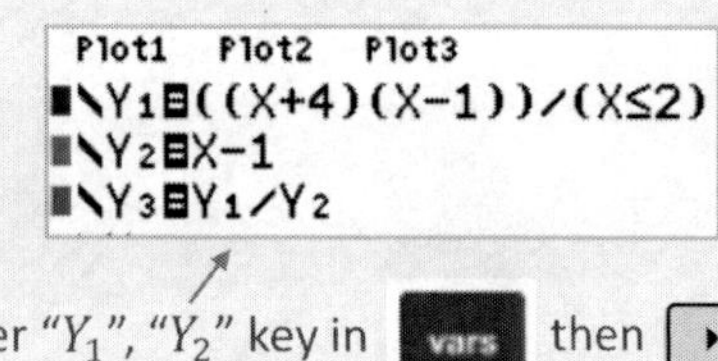

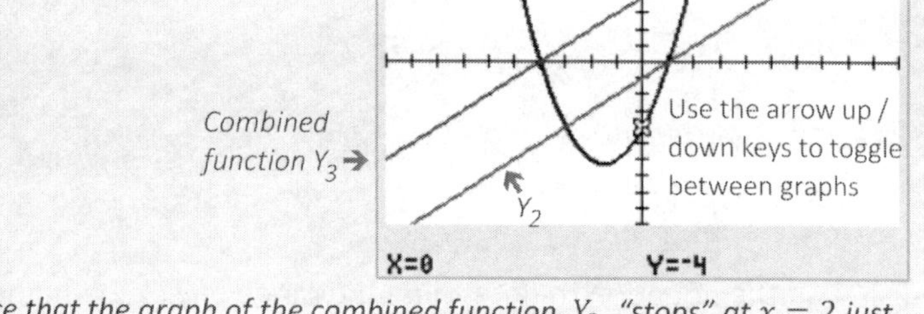

To enter "Y_1", "Y_2" key in vars then ▸

then select #1 **function**

VARS Y-VARS COLOR
1:Function...
2:Parametric...
3:Polar...
4:On/Off...

> That is: VARS ... Y-VARs ... FUNCTION

Here we are dividing Y_1 and Y_2, however we can perform any operation!

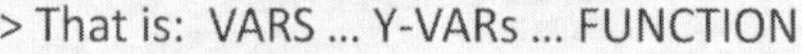

Notice that the graph of the combined function, Y_3, "stops" at $x = 2$ just like the graph of Y_1. (Your calc knows the rules of combining functions!)

You can also compare the function values in TABLE:

NORMAL FLOAT AUTO REAL DEGREE MP
PRESS + FOR ΔTbl

X	Y1	Y2	Y3
-6	14	-7	-2
-5	6	-6	-1
-4	0	-5	0
-3	-4	-4	1
-2	-6	-3	2
-1	-6	-2	3
0	-4	-1	4
1	0	0	ERROR
2	6	1	6
3	ERROR	2	ERROR
4	ERROR	3	ERROR

X= -6

desmos **Exploring Composite Functions Using DESMOS** (free online graphing tool)

Here's how you can graph these same functions using desmos.com

Math, baby!

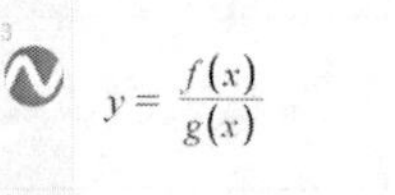

$f(x) = (x+4)(x-1)\{x \leq 2\}$

$g(x) = x - 1$

$y = \dfrac{f(x)}{g(x)}$

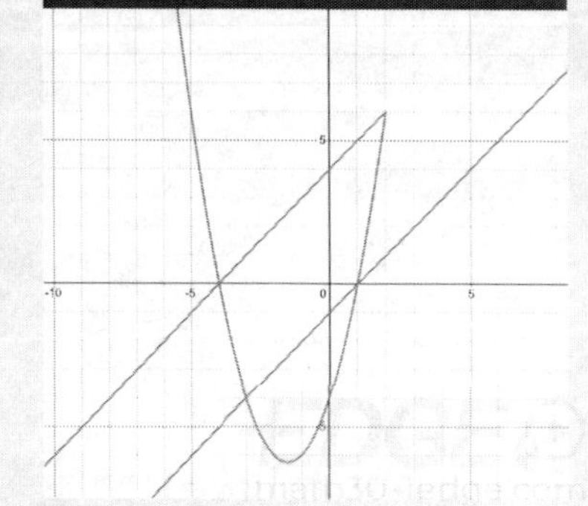

Click on "tools" to adjust the window.

5.2 Composite Functions

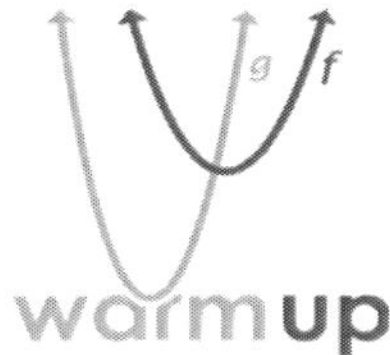

Exploration – Composite Functions

Suppose we define two function functions: $f(x) = x + 2$ and $g(x) = \sqrt{x}$.

1 ➡ Complete the following mapping diagram, such that the output values from f become the input values in g: (To help get things going – one example is done for you, as is #2)

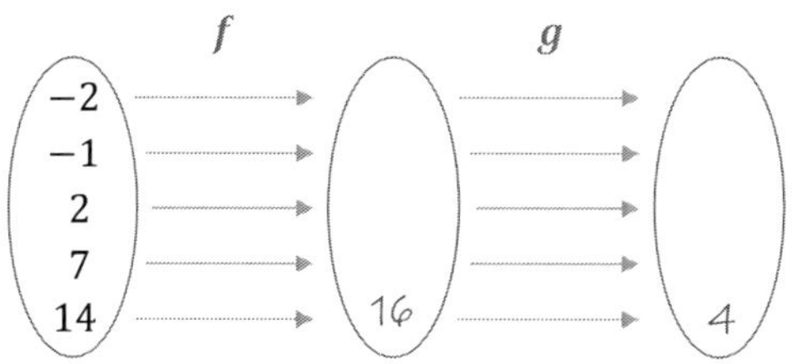

2 ➡ Write a function, $h(x)$, that obtains the result from the first step directly.

$h(x) = \sqrt{x+2}$

Next, let's make those functions: $f(x) = 3x$ and $g(x) = x^2 + 1$

3 ➡ Complete the following mapping diagram, such that the output values from f become the input values in g:

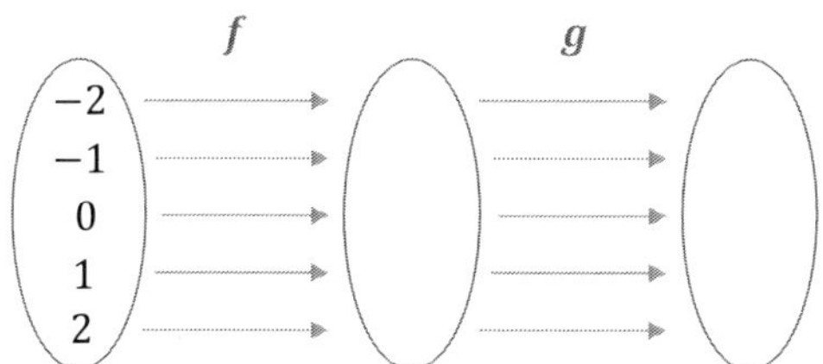

4 ➡ Write a function, $h(x)$, that obtains the result from the first step directly.
Test by substituting some of start points. Is the same result obtained?

Consider this example: Suppose we wish to derive a function that will determine the cost to heat a house on a particular day of the year. The cost to heat the house will depend on the daily average temperature, which in turn depends on the particular day of the year.

We can think of this as involving two functions:

- A **temperature** function, where the input is the day number, and the output the daily average temperature…
- … which then becomes the input in a **cost** function, where the input is average temperature.

Expressing this visually using the function input / output machine model, we have:

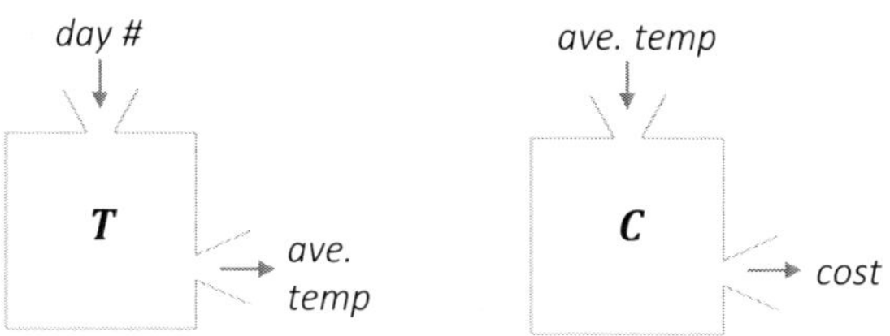

The cost of heating a home, as a function of day number in the year is:
$\boldsymbol{C(T(d))}$

By combining these two relationships into one function, we have performed **function composition**.

A **composite function** $f(g(x))$ is a function that is composed of one function inside another.

Given $\boldsymbol{f(x) = x^2}$ and $\boldsymbol{g(x) = 3x - 1}$ …

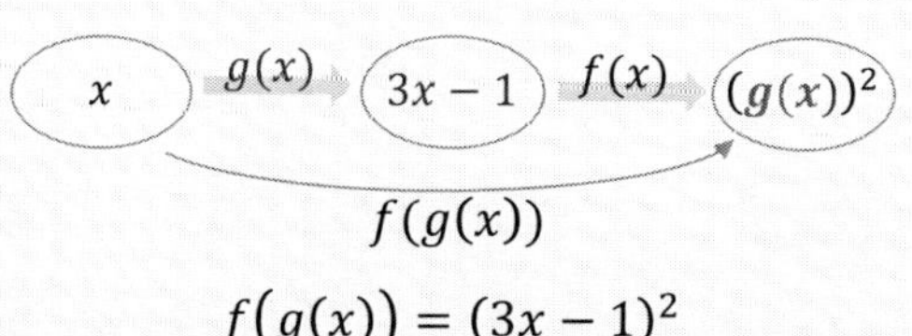

$f(g(x)) = (3x - 1)^2$

- Another way we write $f(g(x))$ is $(f \circ g)(x)$
  $(f \circ g)(x)$ is read: "f of g of x"
- Order matters: $(f \circ g)(x)$ is not the same as $(g \circ f)(x)$
- Substitute the expression for $g(x)$ into $f(x)$ to determine the composite function $(f \circ g)(x)$

Worked Example

Given the functions $f(x) = 2x + 1$, $g(x) = x^2$, and $h(x) = \sqrt{x} - 1$, determine the value of:

(a) $f(g(3))$ (b) $f(f(1))$ (c) $h \circ g \circ f(2)$

Solution:

(a) $f(g(3))$

First evaluate the "inside"

$= 3^2 \Rightarrow = 9$

This becomes the input in *f*:

$= f(9)$

$= 2(9) + 1$

$= \mathbf{19}$

(b) $f(f(1))$

First evaluate the "inside"

$= 2(1) + 1 \Rightarrow = 3$

This becomes the input .... *back into f:* $= f(3)$

$= 2(3) + 1$

$= \mathbf{7}$

(c) $h\left(g(f(2))\right)$

First evaluate the "inside"

$= 2(2) + 1 \Rightarrow = 5$

This becomes the input in *g*:

$= g(5) \Rightarrow = (5)^2 \Rightarrow = 25$

Which becomes the input in *h*:

$= h(25) \Rightarrow = \sqrt{25} - 1 \Rightarrow = \mathbf{4}$

Class Example 5.21 *Composing values*

Given the functions $f(x) = log_2(x)$, $g(x) = \lceil x - 1 \rceil$ and $h(x) = \dfrac{4}{x+2}$, determine the value of:

(a) $(f \circ g)(-7)$ (b) $(g \circ f)(-7)$ (c) $(h \circ h)(0)$ (d) $(g \circ h \circ f)(4)$

Worked Example

Given the functions $f(x) = 4x$, $g(x) = x^2 + 2$, and $h(x) = \sqrt{x+1}$, determine:

(a) $f(g(x))$ (b) $g \circ f(x)$ (c) $h \circ g \circ f(x)$ Also state domain

Solution:

(a) $= f(x^2 + 2)$

Substitute *g(x)* expression for "x" in *f(x)*

$= 4(x^2 + 2)$

$(f \circ g)(x) = \mathbf{4x^2 + 8}$

(b) $= g(f(x))$

$= g(4x)$

Substitute *f(x)* expression for "x" in *g(x)*

$= (4x)^2 + 2$

$(g \circ f)(x) = \mathbf{16x^2 + 2}$

(c) $= h\left(g(f(x))\right) \Rightarrow = h(g(4x))$

Substitute *f(x)* expression for "x" in *g(x)* ...

$= h((4x)^2 + 2)$

... then substitute result into *h(x)*.

$= \sqrt{(4x^2 + 2) + 1}$

$(h \circ g \circ f)(x) = \sqrt{\mathbf{4x^2 + 3}} \quad \{x \in \mathbb{R}\}$

Class Example 5.22 *Composing functions*

Visit math30-1edge.com for solutions to all warm-ups and class examples

Given the functions $f(x) = x^2 - 5$ and $g(x) = 2x - 1$, determine a simplified expression for each:

(a) $(f \circ g)(x)$ (b) $(g \circ f)(x)$ (c) $(g \circ g)(x)$

The **domain** of a composite function $f(g(x))$ is the set of all inputs x, in the domain of $\boldsymbol{g}$, for which $g(x)$ is in the domain of $\boldsymbol{f}$.

➡ We must consider any restrictions on the inside function (prior to any simplification), as well as any restrictions on the final composite function.

For example, consider the functions $\boldsymbol{f(x) = x^2}$ and $\boldsymbol{g(x) = \sqrt{x-4}}$

- An expression for $\boldsymbol{f(g(x))}$ is $(\sqrt{x-4})^2$, which simplifies to ➔ $\boldsymbol{f(g(x)) = x - 4}$

  The inside function has a domain $\{x|x \geq 4, x \in \mathbb{R}\}$

  No added restriction here

  ➡ *Domain is* $\{\boldsymbol{x|x \geq 4, x \in \mathbb{R}}\}$

- And an expression for for $g(f(x))$ is $\boldsymbol{\sqrt{x^2-4}}$

  Here the inside function, x^2*, has a domain* $\{x \in \mathbb{R}\}$

  The composite function has restrictions $x^2 - 4 \geq 0$ which simplifies to a domain of:

  ➡ $\{\boldsymbol{x|x \leq -2}$ *or* $\boldsymbol{x \geq 2, x \in \mathbb{R}}\}$

Note: *Solve* $x^2 - 4 \geq 0$ *graphically*

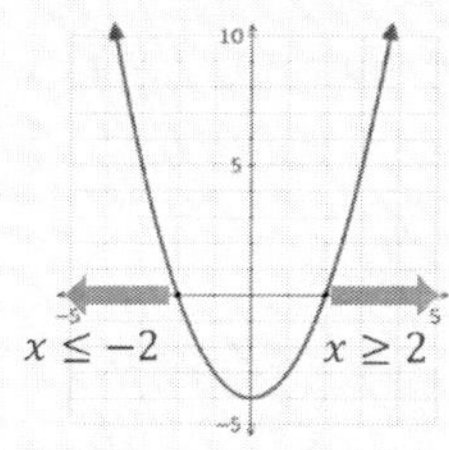

There are two intervals where the graph of $y = x^2 - 4$ *is positive (above x-axis)*

Key Points:

☞ The domain of a composite function $(f \circ g)(x)$ can be different than the domain of either $f(x)$ or $g(x)$

☞ The domain of a composite function cannot exceed the domain of the *inside function*

Class Example 5.23 *Composite functions – working backwards*

(a) If $h(x) = f \circ g(x)$ and $h(x) = (2x-1)^2 + 3(2x-1) - 11$, determine $f(x)$ and $g(x)$.

(b) If $h(x) = f \circ g(x)$ and $h(x) = \sqrt{9 - x^2}$,

i - determine possible expressions for $f(x)$ and $g(x)$

ii – determine the domain of $h(x)$.

Worked Example Given the functions $f(x) = 2x + 3$ and $g(x) = \sqrt{x+1}$, determine each composite function and state the domain: (a) $(f \circ g)(x)$ (b) $(g \circ f)(x)$

Solution: (a) $= f(g(x))$

$= f(\sqrt{x+1})$

Substitute *g(x)* expression for "*x*" in *f(x)*

$= 2\left(\sqrt{x+1}\right) + 3$

➧ $(f \circ g)(x) = \mathbf{2\sqrt{x+1}+3}$ Domain is restricted $x+1 \geq 0$

$\{x|x \geq -1, x \in \mathbb{R}\}$

(b) $= g(f(x))$

$= g(2x+3)$

Substitute *g(x)* expression for "*x*" in *g(x)*

$= \sqrt{(2x+3)+1}$

➧ $(g \circ f)(x) = \sqrt{\mathbf{2x+4}}$ Domain is restricted: $2x+4 \geq 0$

Domain: $\{x|x \geq -2, x \in \mathbb{R}\}$

Class Example 5.24 *Composite functions with restrictions*

Given the functions $f(x) = log_2(x)$, $g(x) = 2x - 3$ and $h(x) = \dfrac{1}{x-1}$, determine each composite function, and state the domain.

(a) $(f \circ g)(x)$ (b) $(g \circ f)(x)$ (c) $g(g(x))$

(d) $h \circ g(x)$ (e) $(g \circ h)(x)$ (f) $f\left(h\left(g(x)\right)\right)$

Class Example 5.25 *Composite of Different forms of Functions*

Three functions are defined as follows. Function ❸ is entirely defined by the values in the table.

❶

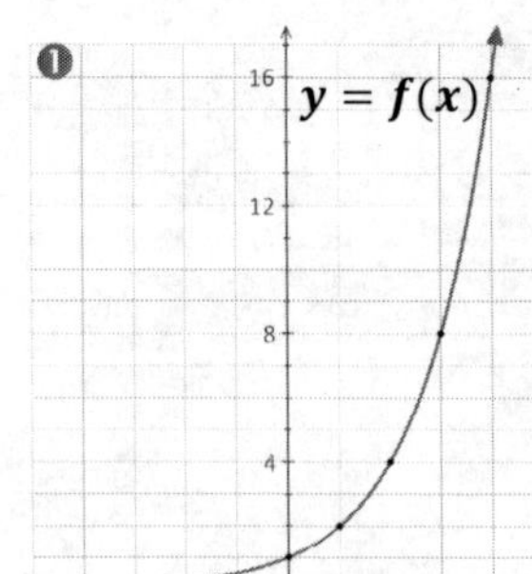

❷ $g(x) = log_3(x+5)$

❸

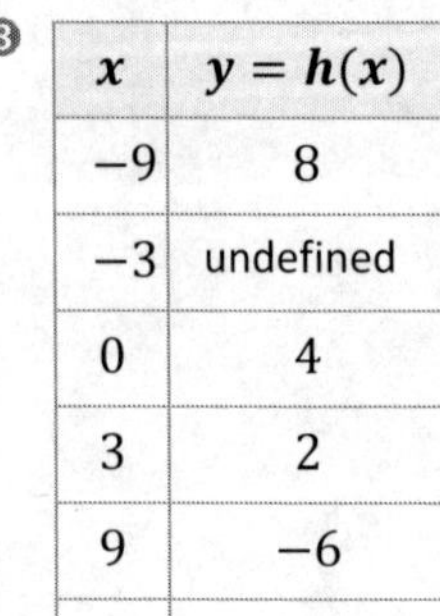

x	$y = h(x)$
−9	8
−3	undefined
0	4
3	2
9	−6
15	−7

(a) Evaluate $(f \circ g \circ h)(0)$.

(b) Determine the domain of $(g \circ h)(x)$.

(c) Determine the domain of $(g \circ f)(x)$.

Exploring Composite Functions on the TI Graphing Calculator

Suppose we had two function, and we wish to see the graph of the composite function.

This time, we'll throw in a domain restriction for each:

$$f(x) = x^2 - 4 \;;\; x \geq -2 \text{ and } g(x) = x - 2;\; x \leq 5$$

To graph the composite function, $y = (f \circ g)(x)$ on your calculator:

❶ Enter $f(x)$ and $g(x)$ in Y_1 and Y_2, respectively.

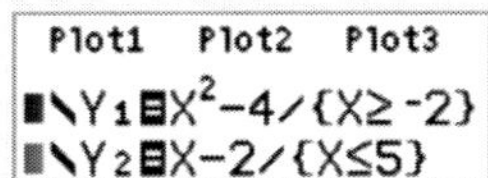

See page 300 for a refresher on restricting the domain of a function in your calc.

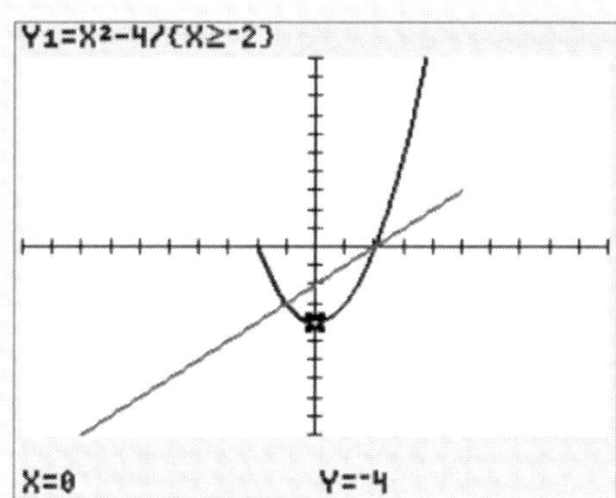

❷ In Y_3, enter the composite function as shown:

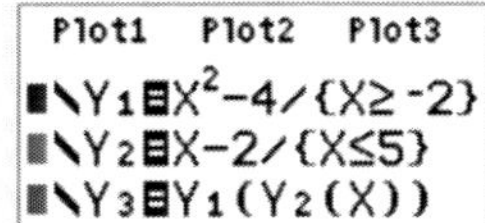

RECALL: To enter Y_1 and Y_2, etc, key in VARS ... Y-VARs ... FUNCTION

OR key in ALPHA ... TRACE →

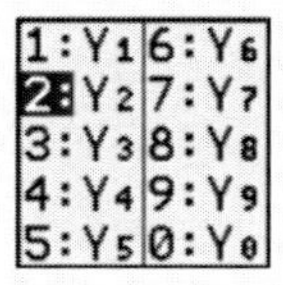

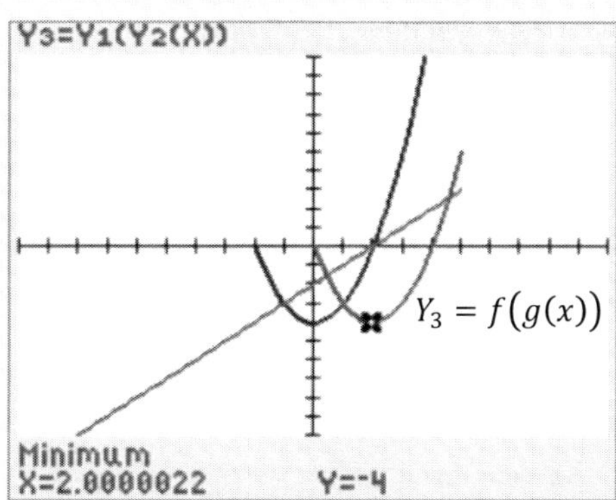

Notice the domain of the composite function is 'automatically' restricted, based on the restrictions on *f*, *g*.

Note the calculator glitch when finding the minimum – decimals are added to the x-coord., which is actually just "2".

To evaluate composed functions at a specific x-value:

Suppose we wish to evaluate $f(g(2))$, using the above functions.

❶ $f(x)$ and $g(x)$ are again inputted in Y_1 and Y_2, respectively.

```
Plot1  Plot2  Plot3
■\Y1■X²-4/{X≥-2}
■\Y2■X-2
```

Note: Only one function can have a domain restriction, otherwise you may get a *data type* error.

❷ Key in **2nd** ... **MODE** to access the home screen.

❸ Key in **ALPHA** ... **TRACE** to access the Y-VAR menu, to input as shown:

```
NORMAL FLOAT AUTO REAL DEGREE MP
Y1(Y2(2))
                              {-4}
```

← Confirmed! The value of $f(g(2))$ is -4

IMPORTANT LEARNING NOTE!

Use this method to check your answers – test questions are often designed to usurp a "calculator only" approach. **Be sure to practice using non-calculator methods.**

desmos Exploring Composite Functions Using DESMOS

Suppose we had two function, and we wish to see the graph of the composite function.

For added fun we'll add a domain restriction for only one of them. *(The first has one built in!)*

$$f(x) = \sqrt{-(x-5)};\ x \geq -4 \text{ and } g(x) = 2x + 1\ ;\ x \leq 2$$

Enter the functions as shown.

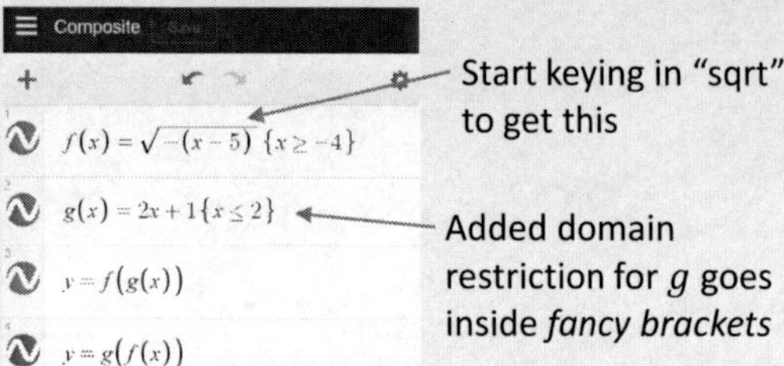

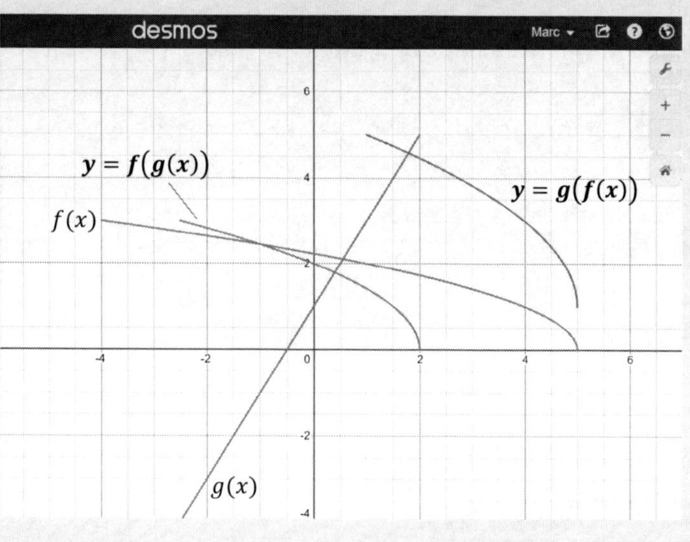

Notice that the domain of the composite functions are automatically restricted, based on the restrictions assigned to f and g

Use desmos (or your graphing calculator) to verify your answers and deepen your understanding – try not to think of it as a way to "do your homework"!

Class Example 5.26 *Combining Functions given their Graphs*

Consider the functions $f(x) = (x+4)^2 - 1; x \geq -4$ and $g(x) = x - 4\ ; x \leq 2$.

(a) Use the graphs to determine the domain for
❶ $y = (f+g)(x)$ and ❷ $y = (f \circ g)(x)$

(b) Verify your answers from (a) using graphing technology

(c) Determine the range of the two functions described in (a)

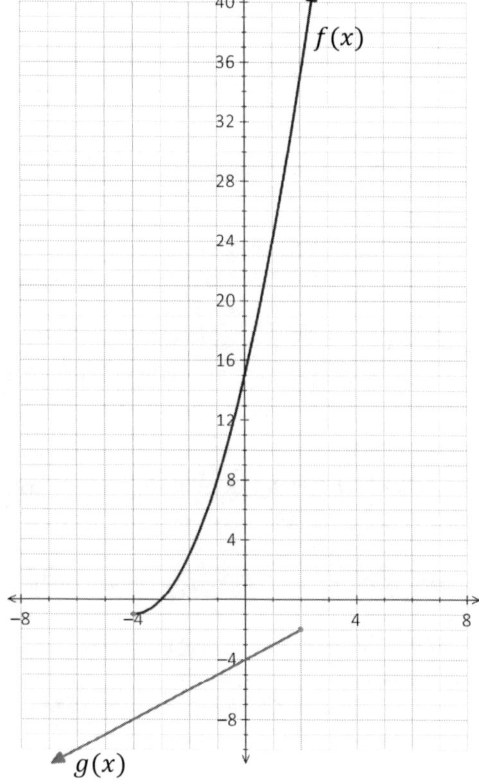

Chapter 6 TRIGONOMETRIC FUNCTIONS

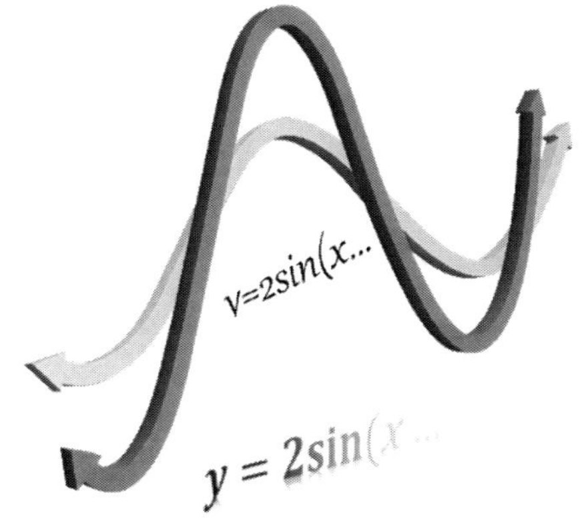

6.1 Radian Measure and Arc Length

Radians – A new Way to Measure Angles

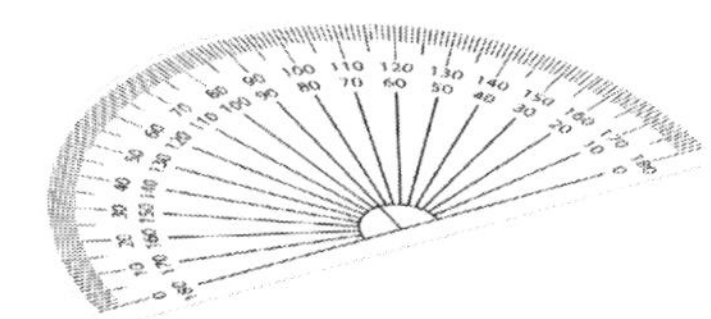

So far in your Trig Career, you've dealt with problems where angles are measured in **degrees**. And it's served you well!

We define one full rotation as $360°$, which likely comes from our ancestors' observation the motion of the sun and stars followed patterns on a 365-day cycle.

For simplicity, they decided to round to 360, which is a good thing, as it's a *highly composite number.*
(360 is divisible by 180, 90, 60, 45, 30, etc)

That said, most of the mathematics and scientific communities use a different angular measure - **radians**.

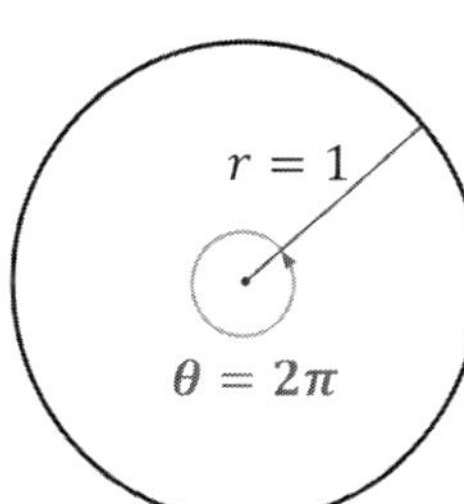

Consider a circle with a radius of one.

The circumference of this circle is:

$$C = \pi \times d \Rightarrow C = 2\pi r \Rightarrow C = 2\pi(1)$$

$$\boldsymbol{C = 2\pi}$$

This 2π value is also the radian measure of the angle θ, representing one full rotation.

Recall that π is the ratio of circumference of a circle to its diameter

$$\pi = \frac{\text{circumference}}{\text{diameter}}$$

π is an irrational number, it's decimal form can only be approximated!

3.14159265358979238....

One rotation is 2π in radians. $\mathbf{360° = 2\pi}$

We need not include "radians" or "rads" as a unit – any angle measure given without a degree symbol is assumed to be in radians!

One radian is the measure of the angle formed by rotating the radius of a circle through an arc equal in length to the radius.

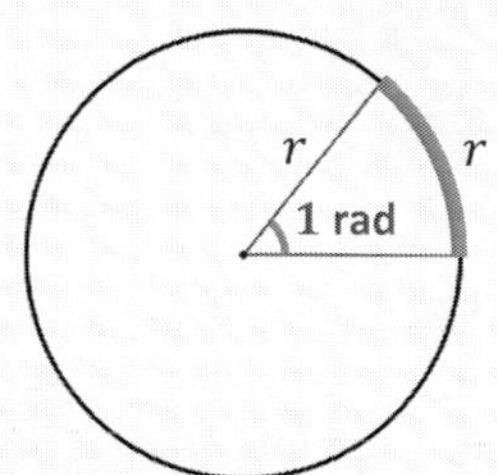

One radian is approximately equal to $\mathbf{57°}$ ($180° \div \pi$)

Since one rotation is $2\pi = 360°$, dividing both sides by 2 gives:

$$\mathbf{180° = \pi} \text{ (rad)}$$

Dividing both sides by π gives:

$$\frac{\mathbf{180°}}{\boldsymbol{\pi}} = \mathbf{1} \text{ (rad)}$$

Dividing both sides by $180°$ gives:

$$\mathbf{1°} = \frac{\boldsymbol{\pi}}{\mathbf{180°}}$$

So, to convert from radians to degrees:

→ ***Multiply the angle by*** $\frac{\mathbf{180°}}{\boldsymbol{\pi}}$

And to convert from degrees to radians:

→ ***Multiply the angle by*** $\frac{\boldsymbol{\pi}}{\mathbf{180°}}$

Worked Examples

Convert each angle **to radians**:

(a) $240°$ (b) $178.4°$ *Round to the nearest hundredth*

Convert each angle **to degrees**:

(c) $-\frac{5\pi}{6}$ (c) 6 rads *Round to nearest whole number*

Solutions: Multiply by $\frac{\pi}{180°}$ to convert to radians

(a) $240° \times \frac{\pi}{180°} = \frac{240°\pi}{180°}$

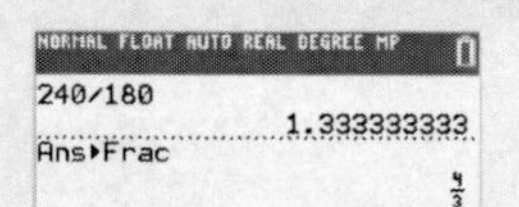

Drop the "π" to reduce 240/180

$= \frac{4\pi}{3}$ *...then add "π" back*

(b) $178.4° \times \frac{\pi}{180°}$

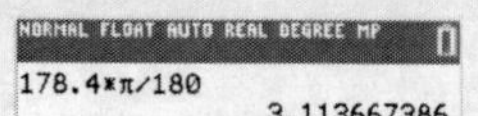

This time, leave the "π" in, since we're answering as a decimal

$\approx$ **3.11 rads**

Note, this is close to "π" (3.14), as 178.4° is closes to 180°

Multiply by $\frac{180°}{\pi}$ to convert to degrees

(a) $-\frac{5\pi}{6} \times \frac{180°}{\pi}$

$= -\frac{5\not\pi}{6} \times \frac{180°}{\not\pi}$

The "π's" cancel out

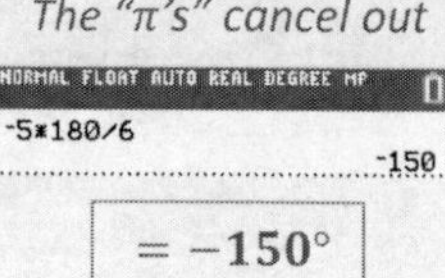

$= -150°$

(b) **6 rads** $\times \frac{180°}{\pi}$

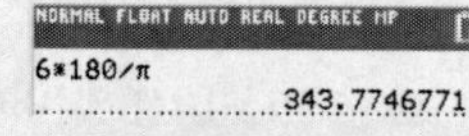

Can't cancel "π's" this time, leave in calculation and round answer

$\approx 344°$

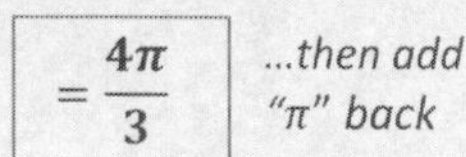

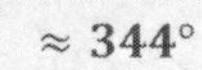

Class Example 6.11 *Converting angle measurements*

Visit math30-1edge.com for solutions to all warm-ups and class examples

Leave answers in terms of a fraction of π for (a) and (b), round (c) to the nearest hundredth, (d) (e) and (f) to the nearest degree.

Convert each of the following angles to radians:

(a) $315°$ (b) $-75°$ (c) $81.8°$

Convert each of the following angles to degrees:

(d) $\frac{2\pi}{3}$ (e) $-\frac{9\pi}{2}$ (f) 2.6 rads

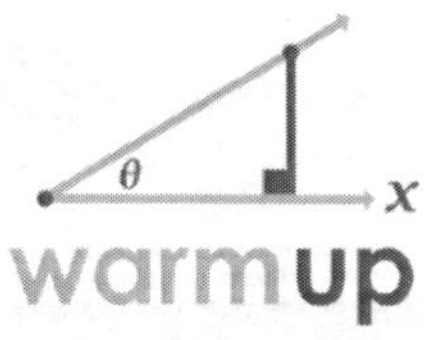

Exploration 1 - Measure of Key Angles in Degrees and Radians

On the previous page we saw how one rotation, or $360°$, is equal to 2π.

1 ➡ The five diagrams below show various key angles θ. Use reasoning to state the measure of each angle shown, in both degrees and radians. *To get you started - the first is done!*

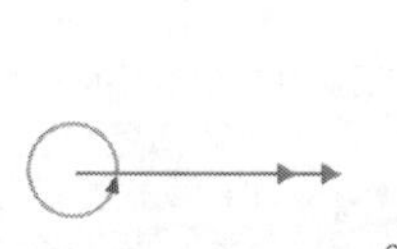

θ (degrees): 360°

θ (radians): 2π

θ (degrees): ____

θ (radians): ____

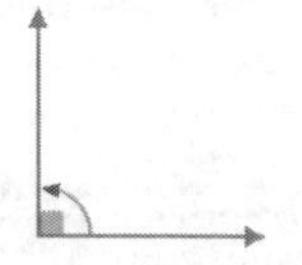

θ (degrees): ____

θ (radians): ____

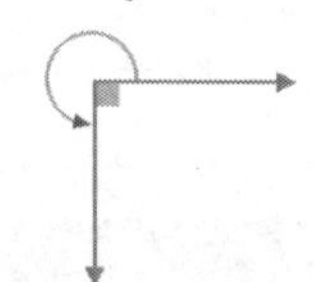

θ (degrees): ____

θ (radians): ____

2 ➡ State the measure θ if the right angle from above is *equally split* as shown here....

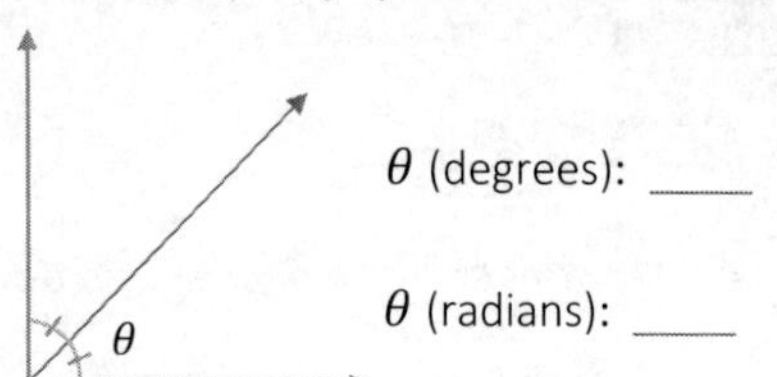

θ (degrees): ____

θ (radians): ____

3 ➡ ... and as shown in each of these:

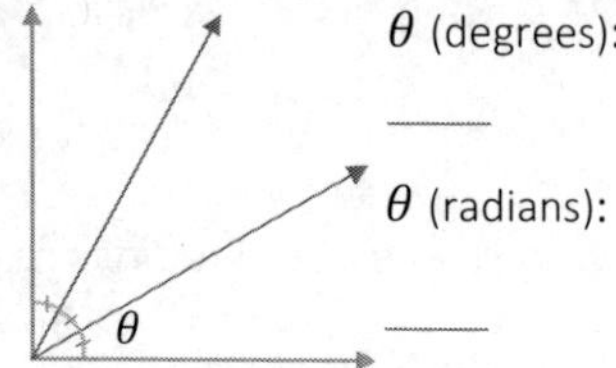

θ (degrees): ____

θ (radians): ____

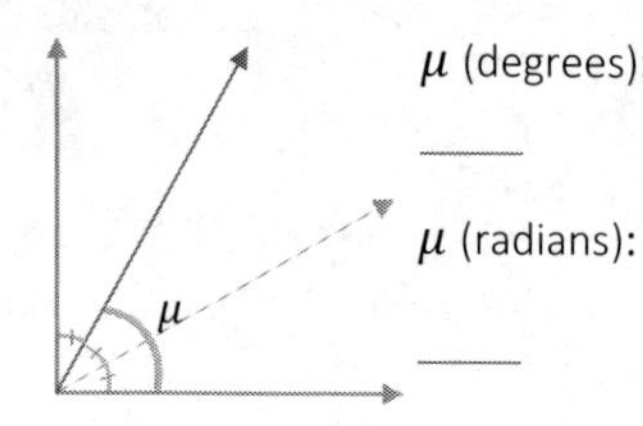

μ (degrees): ____

μ (radians): ____

In our study of trigonometry, we often encounter angles which are multiples of 30° or 45°. As such, it is useful to be familiar with their angular measure in radians.

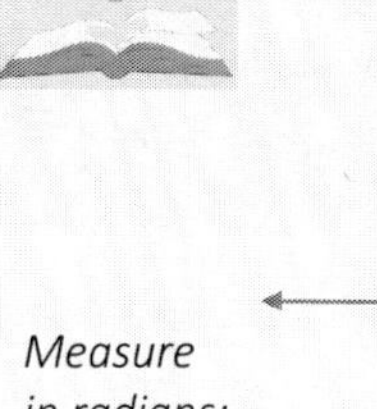

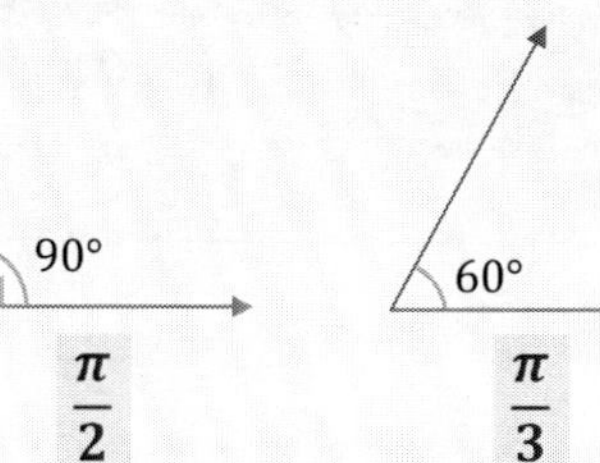

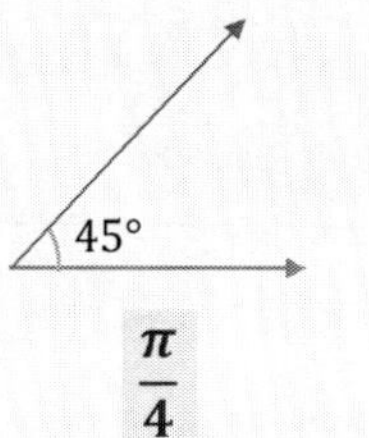

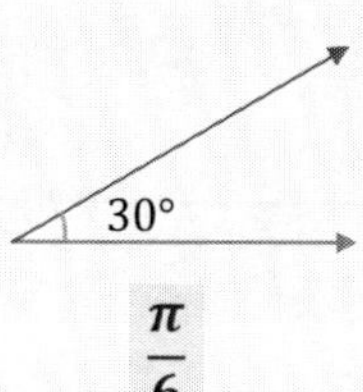

Measure in radians: π | $\frac{\pi}{2}$ | $\frac{\pi}{3}$ | $\frac{\pi}{4}$ | $\frac{\pi}{6}$

90 is half of 180, so in radians, "half of π"

60 is one-third of 180, so in radians, "1/3 of π"

45 is 1/4 of 180, so ... "1/4 of π"

30 is 1/6 of 180, so ... "1/6 of π"

Arc Length

Exploration 2 - Arc Length of a Circle Sector

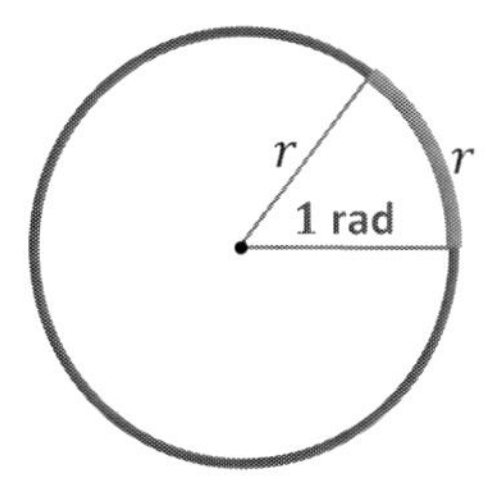

← Recall how we saw that **one radian** is the measure of the angle formed by rotating the radius of a circle through an arc equal in length to the radius.

1 ➡ Determine the measure of one radian, **in degrees**. (Nearest tenth)

Now, what if we **doubled the arc length**, without changing the radius? → How would this affect the sector angle, θ?

2 ➡ Use reasoning to make a prediction of the measure of the sector angle θ. How would it be affected?

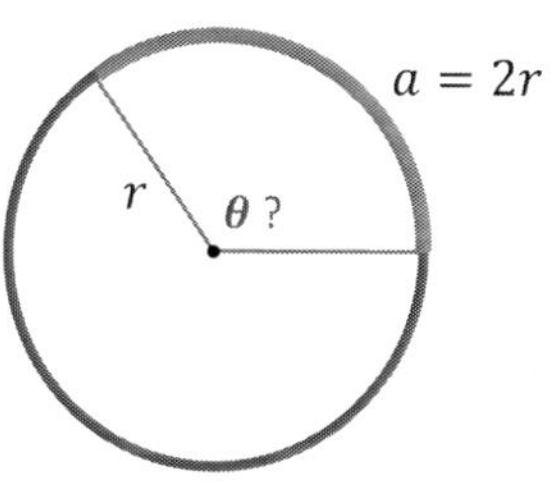

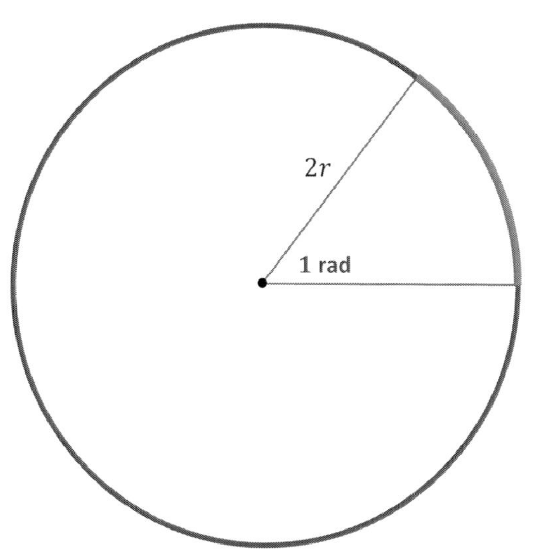

← Finally, say we doubled the angle radius, whilst keeping the sector angle θ the same.

3 ➡ Use reasoning to make a prediction on the length of the sector arc. How would it be affected?

The measure of a sector angle θ is equal to the ratio of the arc length and radius.

This is on your formula sheet: $\boldsymbol{\theta = \frac{a}{r}}$

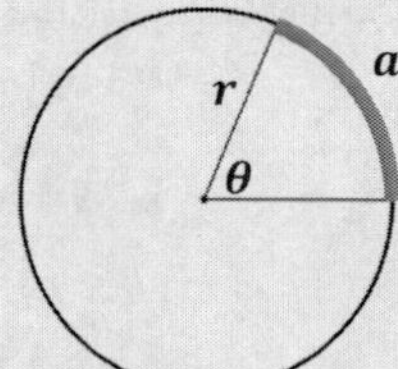

Where

$\boldsymbol{\theta}$ is the measure of the sector angle, *in* ***radians***

$\boldsymbol{a}$ is the length of the arc around the angle

$\boldsymbol{r}$ is the length of the radius

Rearranging gives a formula for **arc length**: $\boldsymbol{a = r\theta}$

Worked Examples Determine each missing value.
For (a), round to the nearest tenth, for (b), round to the nearest ***degree***.

(a)

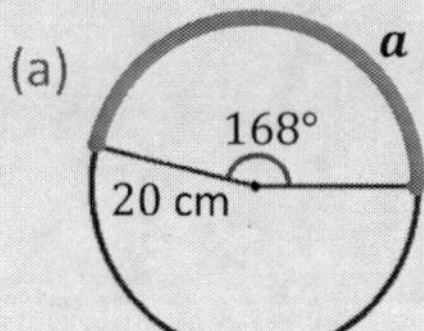

(b)

5.8, 3.6, θ

Solutions:

(a) Use $\theta = \frac{a}{r}$, which rearranges to $\boldsymbol{a = r\theta}$

$$a = (20cm)\left(168° \times \frac{\pi}{180}\right)$$

θ in radians

$$\boldsymbol{a = 58.6} \text{ cm}$$

(b) Use $\theta = \frac{a}{r}$

$$\theta = \frac{3.6}{5.8}$$

$$\approx \mathbf{0.62}$$

this is in ***radians***

$$\boldsymbol{\theta \approx 36°} \leftarrow 0.62 \times \frac{180°}{\pi}$$

to convert to degrees

Class Example 6.12 *Determining values using arc length formula*

Determine each indicated missing value:

(a) Determine the measure of angle θ, correct to the nearest degree.

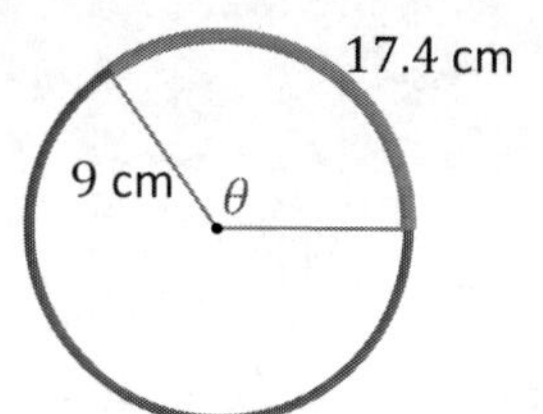

(b) A pendulum swings through an angle of 40°, while forming an arc 8.4 cm in length. Determine the total length of the pendulum, correct to the nearest tenth of a cm.

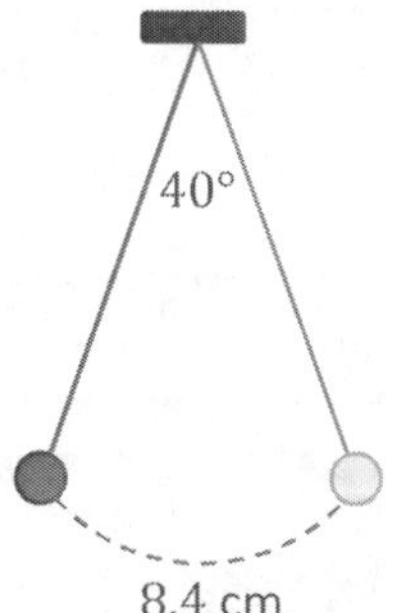

(c) Determine the length of the indicated arc, correct to the nearest tenth of a cm.

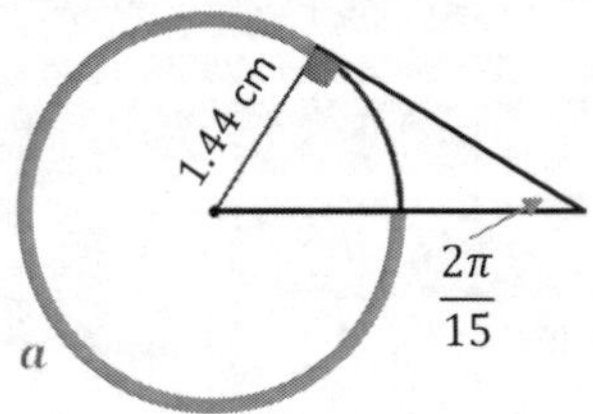

Angles in Standard Position

Exploration 3

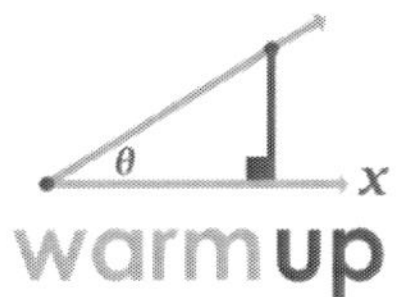

An angle is in **standard position** when its vertex is at the origin and ray forming its **initial arm** is on the positive x-axis.

The diagram on the right shows the θ, measuring $30°$, or $\frac{\pi}{6}$, in standard position. Its **terminal arm** passes through $P(\sqrt{3}, 1)$.

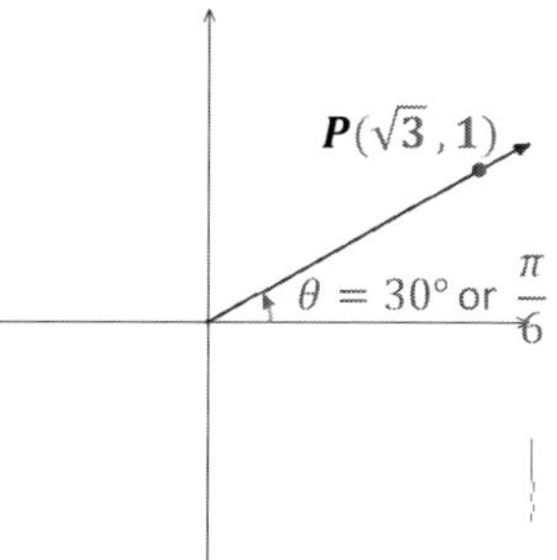

1 ➡ In each diagram below the terminal arm is the same as for θ above. Determine the measure of each angle shown, in both degrees and radians. *Hint: For (c) and (d) the angles are negative.*

(a)

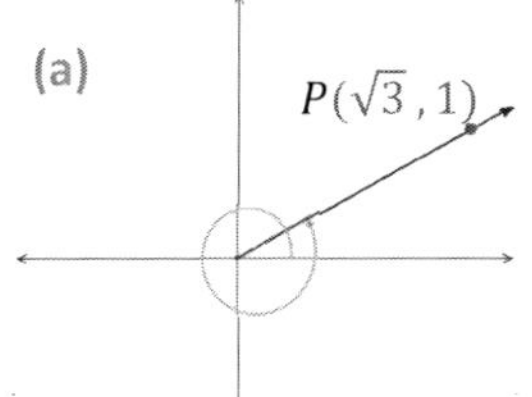

(b)

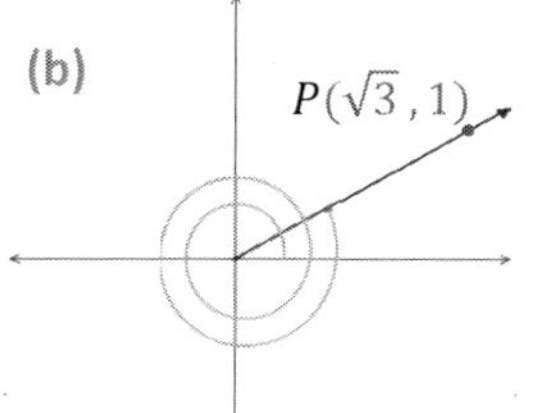

(c)

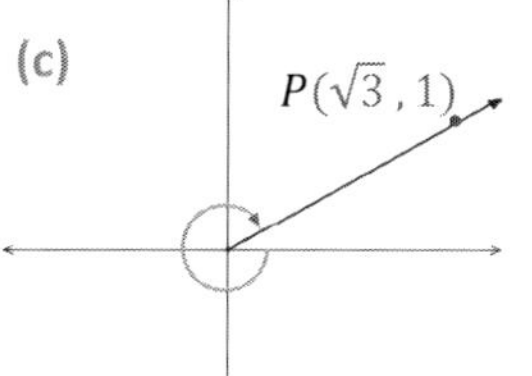

(d)

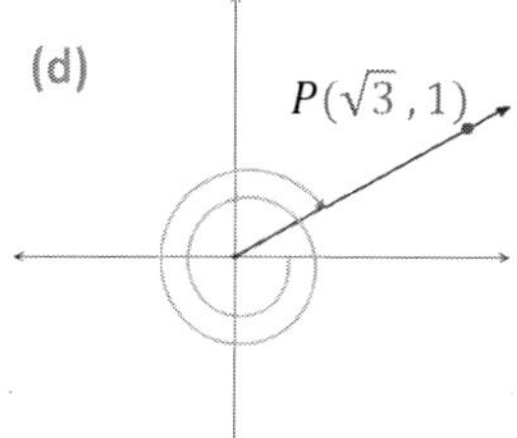

2 ➡ In each shaded region, label all four quadrantal (multiple of $90°$) within one rotation of the coordinate plane.

See diagram on top of next page to confirm your answers!

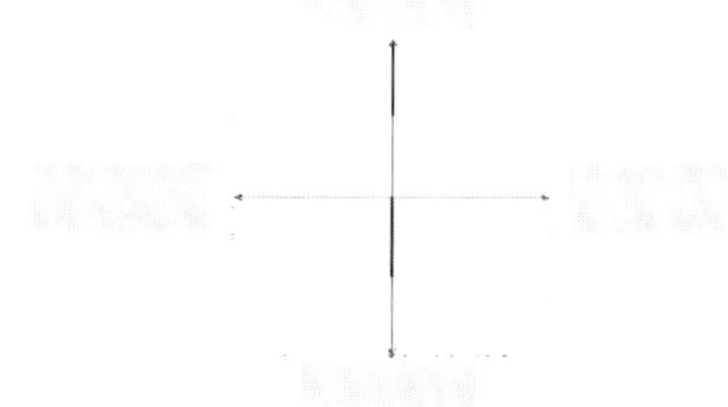

3 ➡ In each diagram below the terminal arm is the same. Determine the measure of each angle shown, in both degrees and radians..

(a)

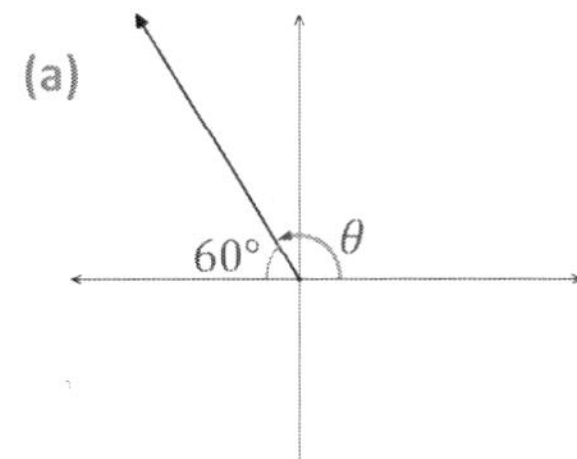

(b)

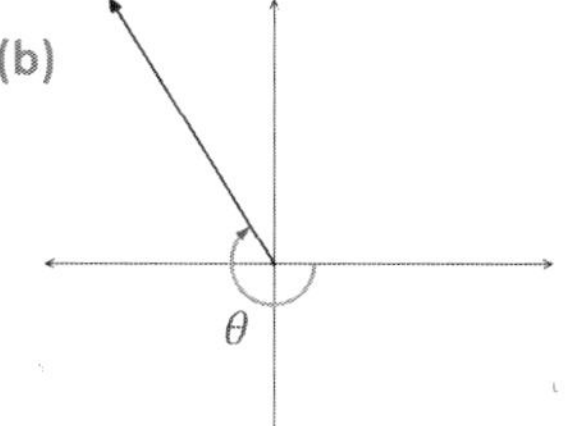

(c)

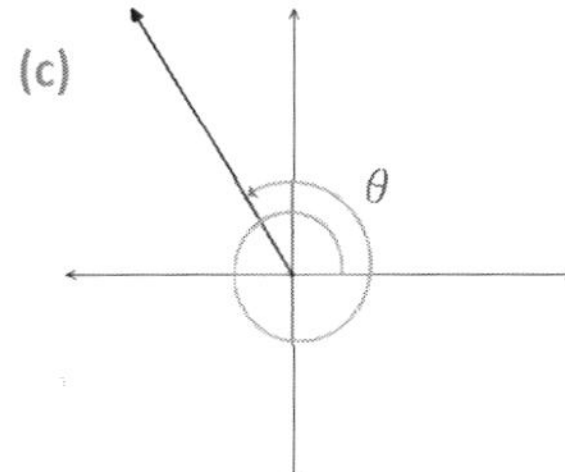

(d)

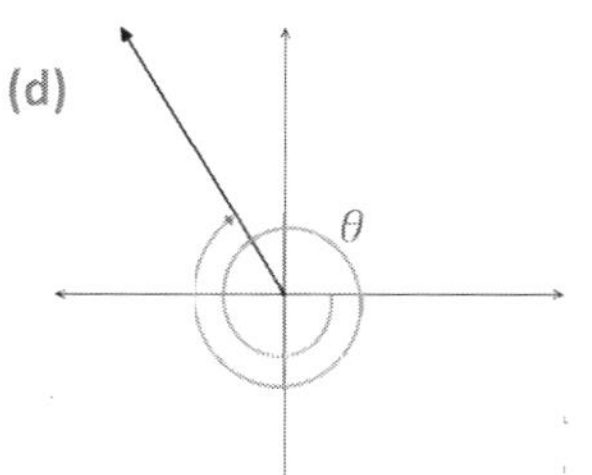

An angle is in **standard position** when its vertex is at the origin and ray forms its **initial arm** is on the positive x-axis.

The angle of rotation is formed by rotating a second ray counter-clockwise, such that this **terminal arm** is in any of the four quadrants.

Angles that have the same terminal arm are called **coterminal angles.**

The smallest positive coterminal angle (within the first positive rotation) is called the **principal angle**.

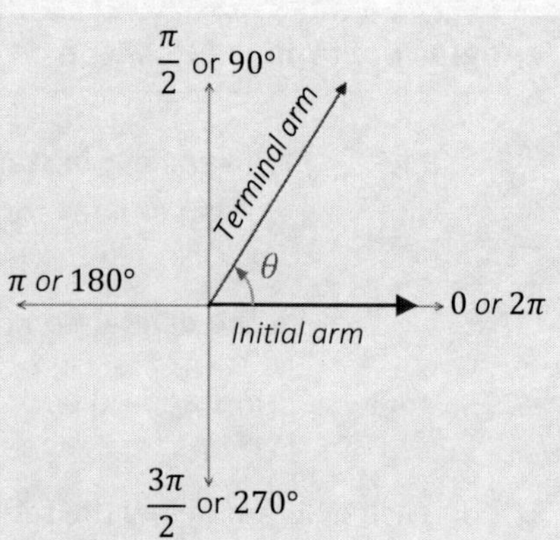

The following three angles are coterminal:

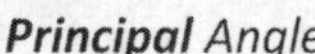

Principal *Angle*

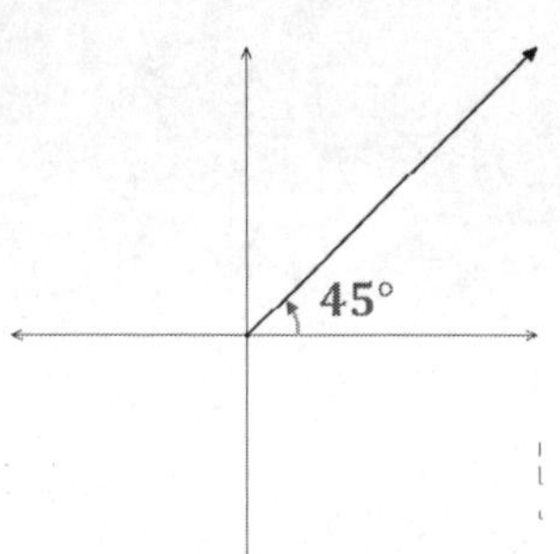

A ***coterminal*** *angle – found by adding one rotation*

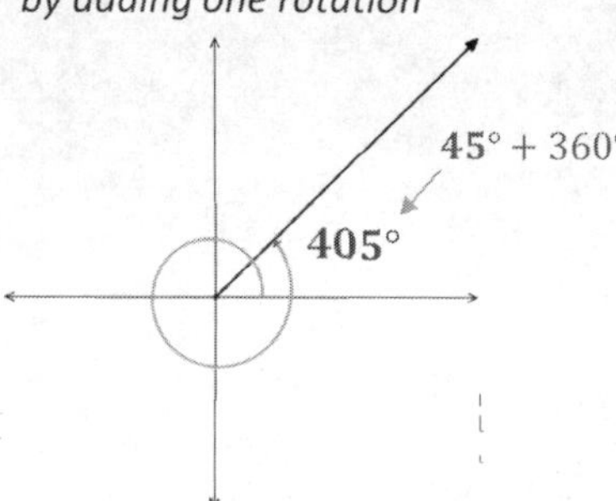

Another ***coterminal*** *angle – this time by subtracting one rotation*

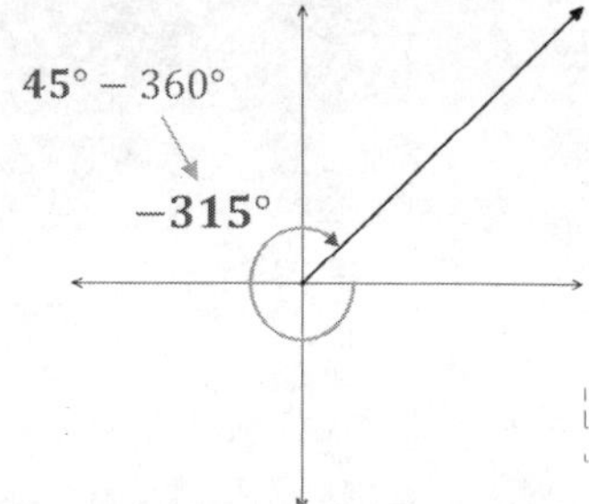

Let's look at that again, this time using RADIANS. (Starting with the principal angle, 45°, which is $\frac{\pi}{4}$):

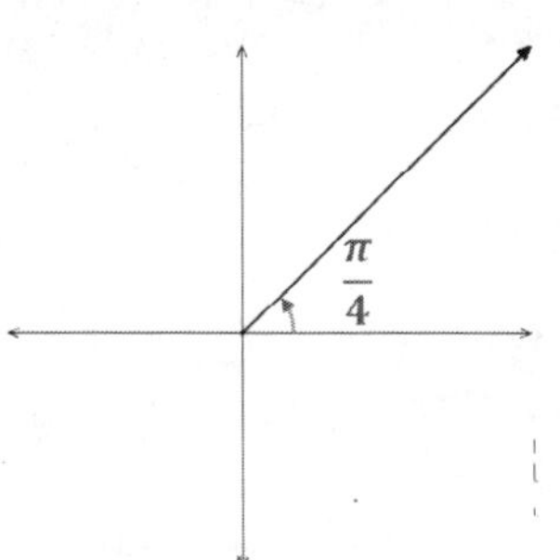

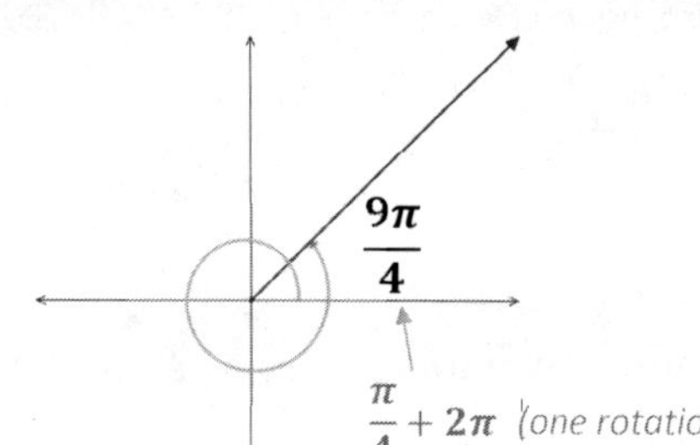

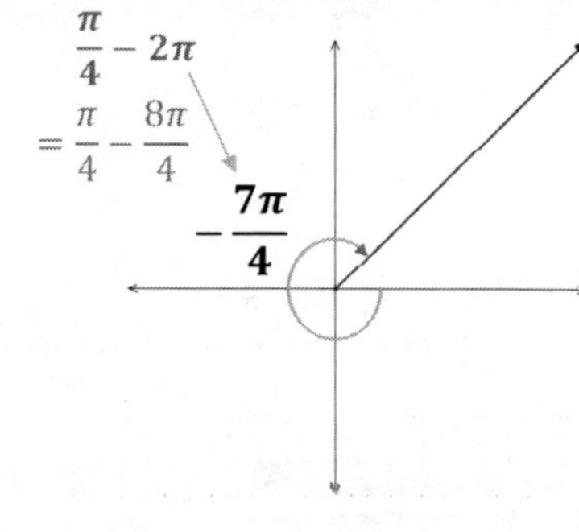

note

When working with radians, students are encouraged to do all work / "think" in radians! (As shown above, in determining the two "closest" coterminal angles to $\frac{\pi}{4}$)

That said, there is always the option of working in degrees / converting! This method is shown below, where we'll again find the closest to coterminal angles to $\frac{\pi}{4}$, in radians.

➡ Finding the closest positive coterminal angle to $\frac{\pi}{4}$, using the CALC / converting to degrees.

```
NORMAL FLOAT AUTO REAL DEGREE MP
45+360
                              405
Ans/180
                             2.25
Ans▸Frac
                              9/4
```

Add the "π" back, this gives $\frac{9\pi}{4}$

➡ And finding the closest negative coterminal angle to $\frac{\pi}{4}$, using the CALC / converting to degrees.

```
NORMAL FLOAT AUTO REAL DEGREE MP
45-360
                             -315
Ans/180
                            -1.75
Ans▸Frac
                             -7/4
```

For an angle in standard position θ, coterminal angles can be found by adding or subtracting any multiple of 360°. (That is, adding / subtracting any number of rotations)

In the diagram on the right, θ_1 is the principal angle, θ_2 is a coterminal angle found by adding 360° once

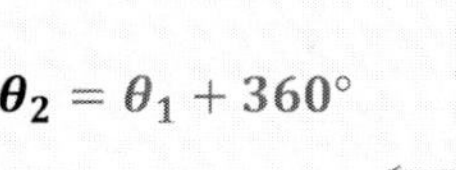

$\boldsymbol{\theta_2} = \theta_1 + 360°$

In the diagram below, θ_3 is a coterminal angle found by adding 360° twice:

$\boldsymbol{\theta_3} = \theta_1 + 360°(\mathbf{2})$

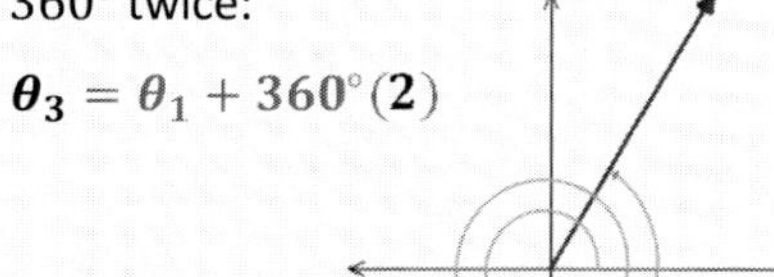

And θ_4 is a coterminal angle found by subtracting 360°

$\boldsymbol{\theta_4} = \theta_1 - 360°(1)$

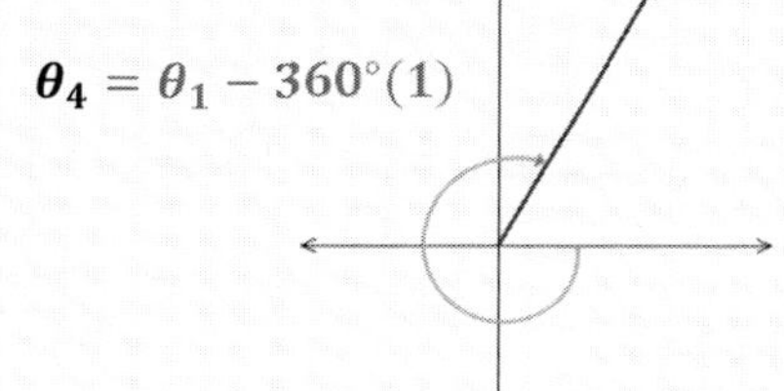

Any angle coterminal to θ_1 can be found by adding any (positive or negative) multiples of 360°.

$\boldsymbol{\theta_n} = \theta_1 + 360°(\boldsymbol{n})$

OR $\boldsymbol{\theta_n} = \theta_1 + 2\pi, \quad n \in I$

Class Example 6.13 *Determining coterminal angles*

Determine the closest three coterminal angles for each – such that two coterminal angles are positive, and one is negative. Sketch each given principal angle. Provide coterminal angles using the same angular measure as given in each question.

(a) $120°$

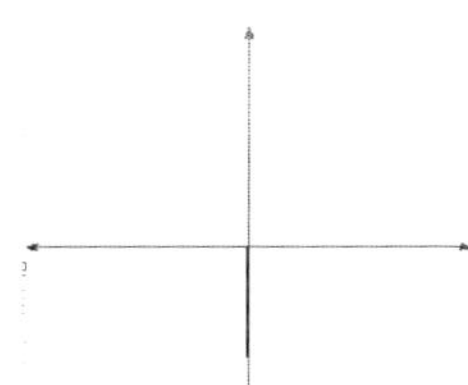

(b) $\frac{5\pi}{4}$

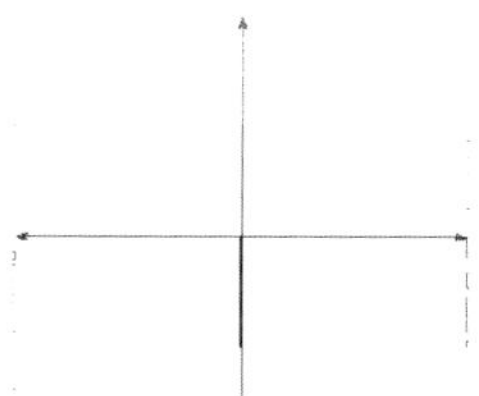

(c) $\frac{\pi}{6}$

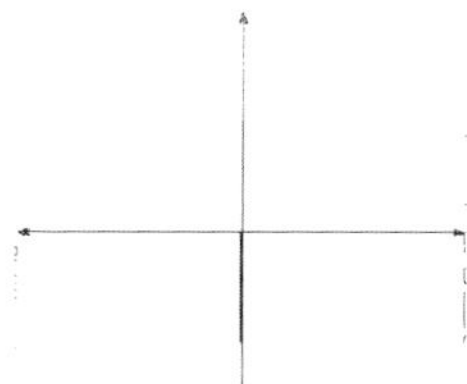

(d) $330°$

Worked Example Given the angle $-\frac{8\pi}{3}$, sketch, and determine the principal angle.

Solution: First find the **principal angle**. We do that by adding / subtracting multiples of one rotation (either $\mathbf{360°}$, or, in this case, $\mathbf{2\pi}$) →

$-\frac{8\pi}{3} + \frac{6\pi}{3}$ ← "2π"

$= -\frac{2\pi}{3}$ *This is still not between 0 and 2π, so we* ***add another*** $\boldsymbol{2\pi}$

$= -\frac{2\pi}{3} + \frac{6\pi}{3}$

$\boxed{= \frac{4\pi}{3}}$ ← ***principal angle***

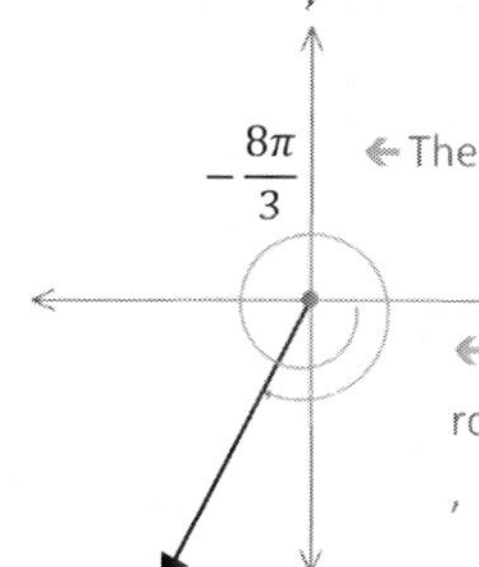

← Then, sketch $-\frac{8\pi}{3}$

← Show one full negative rotation, plus another (negative) $\frac{2\pi}{3}$

$-\frac{\mathbf{8\pi}}{\mathbf{3}} = -\frac{6\pi}{3} - \frac{2\pi}{3}$

"-2π"

See next page for an alternative, calculator method

To find the principal angle, for an angle given in radians such as $-\frac{8\pi}{3}$ ...

Another method is to ***first convert to degrees.***

(Though once again - you are encouraged to "think" in radians!)

Replace π with $\mathbf{180^\circ}$ to express $-\frac{8\pi}{3}$ as $-\mathbf{480^\circ}$ →

Add rotations ($\mathbf{360^\circ}$) until you arrive at an angle **between** $\mathbf{0^\circ}$ **and** $\mathbf{360^\circ}$

Finally, convert back to radians

```
NORMAL FLOAT AUTO REAL DEGREE MP
-8(180)/3
                              -480
Ans+360
                              -120
Ans+360
                               240
Ans/180▸Frac
                               4/3
```

This is $\frac{4\pi}{3}$

Class Example 6.14 *Sketching rotation angles, determining principal angles*

For each of the following, sketch the given rotation angle, and determine the principal angle in the same angular measure unit.

(a) -315° (b) $-\frac{13\pi}{4}$ (c) $\frac{17\pi}{3}$ (d) 960°

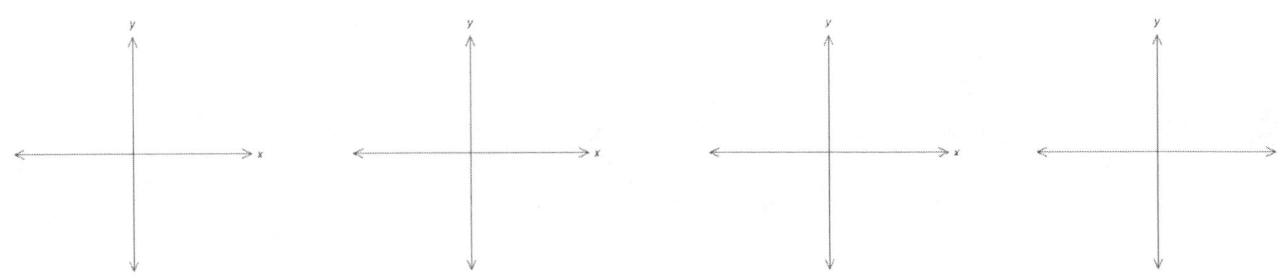

Stating coterminal angles in general form

As we've seen, any angle has an ***infinite number of coterminal angles.*** Each time you make a full rotation from a terminal arm, you arrive back at the same place and can repeat the process.

Angles coterminal with a principal angle θ can be expressed: $\boldsymbol{\theta + 2\pi n} \;;\, n \in I$ or $\boldsymbol{\theta + 360^\circ n} \;;\, n \in I$

Note: $n \in I$ reads: *n is an element of the integers*

Recall that integers are pos or neg whole numbers. This language allows us to represent all possible cases for coterminal angles.

This is another way of saying: *add / subtract any whole number of full rotations!*

Class Example 6.15 *Sketching rotation angles, determining principal angles*

For each of the following angles, state the general form of all coterminal angles. (Start with the principal angle)

(a) -210° (b) $\frac{31\pi}{6}$ (c) $-\frac{7\pi}{4}$

6.2 Trig Ratios of Angles in Standard Position

In previous courses we encountered the three primary trigonometric ratios:

The sine ratio: $\sin\theta = \dfrac{opp}{hyp}$

The cosine ratio: $\cos\theta = \dfrac{adj}{hyp}$

The tangent ratio: $\tan\theta = \dfrac{opp}{adj}$

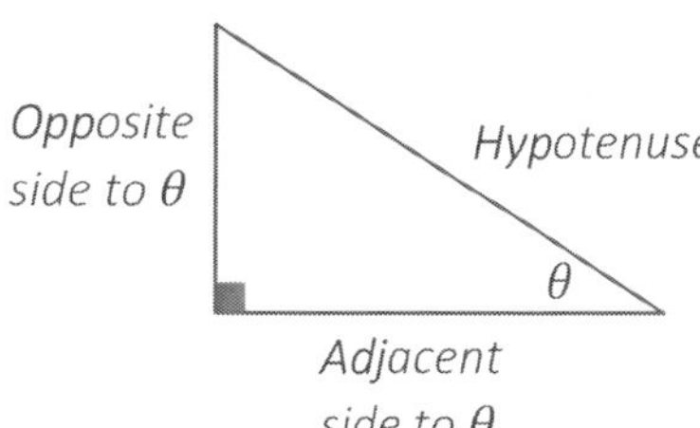

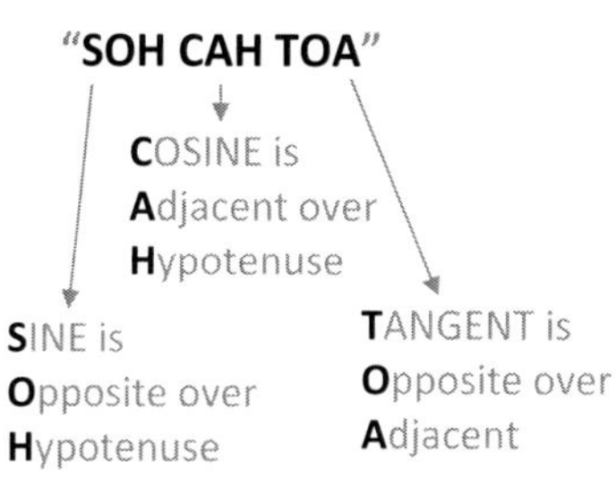

In this chapter we'll study how these primary trigonometric ratios, along with the three reciprocal trigonometric ratios, can be applied to angles in standard position.

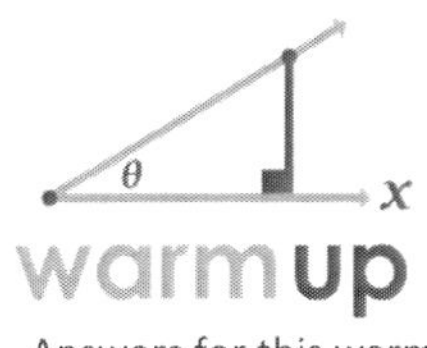

Answers for this warm-up are at the bottom of the next page

Consider and angle in standard position θ, that passes through a point $P_1(3,4)$.

1 ➡ Construct a triangle by drawing a line segment from the point P down to the x-axis. Indicate the right angle with a ■.

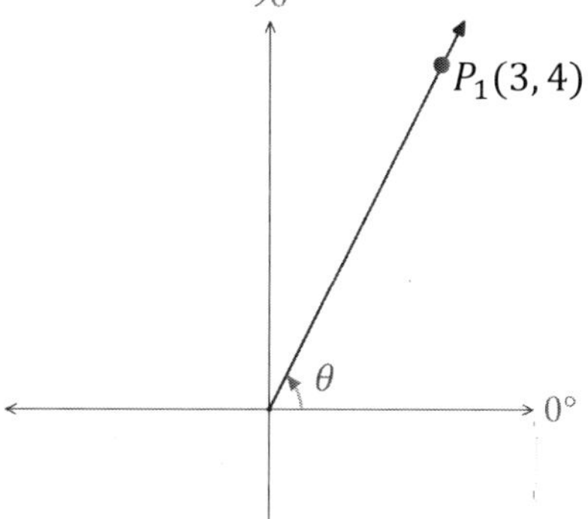

2 ➡ Label the length of each of the sides of the triangle. Use the Pythagorean theorem to determine the length of the hypotenuse.

3 ➡ State the three primary trigonometric ratios of θ; *sine, cosine,* and *tangent.*

For questions 4, 5, and 6, refer to a new angle in standard position β, that passes through $P_2(-3,4)$

4 ➡ Construct a triangle by drawing a line segment, parallel to the y-axis, connecting P_2 to the x-axis. Indicate the right angle with a ■. *Note that β should be **outside** the triangle!*

90°

$P_2(-3,4)$

β

180°

5 ➡ Triangles drawn from an angle in standard position can have *negative side lengths*, as determined by the coordinates / quadrants. Label each side length.
Note that the hypotenuse is always positive.

6 ➡ The angle supplementary to β, located inside the triangle, is the **reference angle**. Label this reference angle α. (that is, "alpha")

7 ➡ State the three primary trigonometric ratios of β; *sine, cosine,* and *tangent.*

8 ➡ Describe how the trig ratio values for β compare to those for θ.

Trig Ratios of Angles in Standard Position

For an angle in standard position θ, that passes through a point $P(x, y)$, the values of the trigonometric ratios of θ can be determined by labeling the sides of the resulting right angle triangle. The hypotenuse can be determined using the Pythagorean Theorem.

$$r = \sqrt{x^2 + y^2}$$

The three **primary** trigonometric ratios are: $\sin\theta = \frac{y}{r}$ $\quad \cos\theta = \frac{x}{r}$ $\quad \tan\theta = \frac{y}{x}$

And the three **reciprocal** trig ratios are: $\csc\theta = \frac{r}{y}$ $\quad \sec\theta = \frac{r}{x}$ $\quad \cot\theta = \frac{x}{y}$

$cosecant = \frac{1}{sin}$ $\quad secant = \frac{1}{cos}$ $\quad cotangent = \frac{1}{cos}$

Each of the three primary trig ratios has a corresponding reciprocal ratio, bringing the total to six trig ratios!

The REFERENCE ANGLE

For an angle in standard position less than 90°, as drawn above, the angle θ is entirely contained within the angle we use to derive the trigonometric ratios. However, that is not the case for angles greater than 90°, as we saw with the second part of the warm-up on the previous page. In such instances we use a **reference angle**, defined as the acute angle made between the terminal arm (the hypotenuse) and the x-axis.

For our purposes we'll call the reference angle $\boldsymbol{\alpha}$ ("alpha"). It's value with respect to θ varies by quadrant:

Quadrant II

Angles between 90° and 180°
(or in radians: between $\frac{\pi}{2}$ or π)

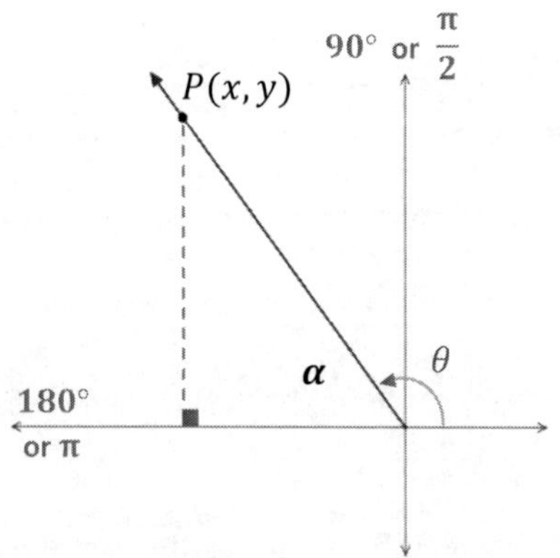

$\boldsymbol{\alpha = 180° - \theta}$

or in radians ... $\pi - \theta$

Quadrant III

Angles between 180° and 270°
(or in radians: between π and $\frac{3\pi}{2}$)

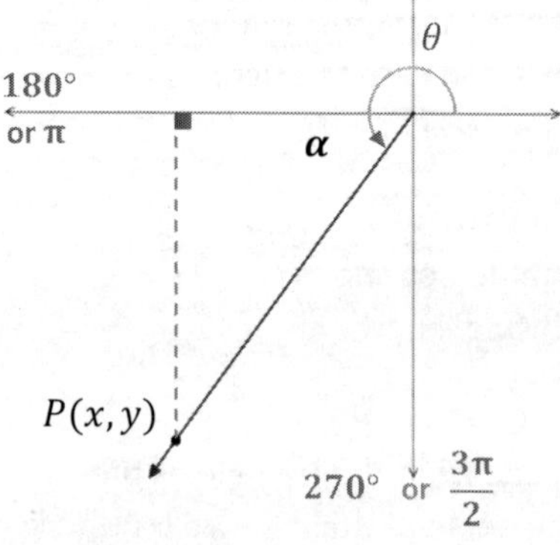

$\boldsymbol{\alpha = \theta - 180°}$

or in radians ... $\theta - \pi$

Quadrant IV

Angles between 270° and 360° 360° or 2π
(or in radians: between $\frac{3\pi}{2}$ and 2π)

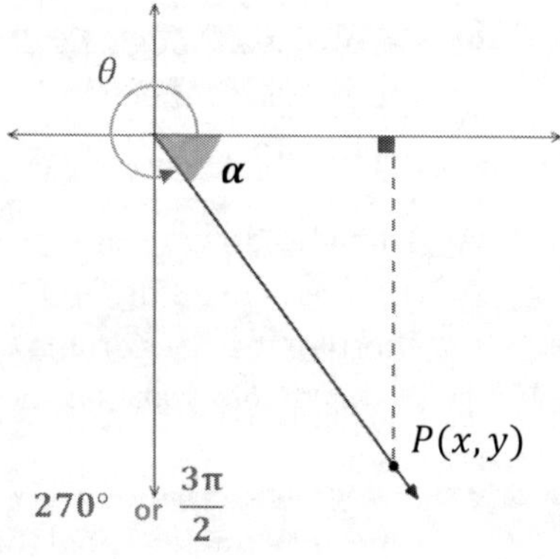

$\boldsymbol{\alpha = 360° - \theta}$

or in radians ... $2\pi - \theta$

In quadrant I, $\boldsymbol{\theta = \alpha}$

There is no need to memorize these. Use reasoning - always find the difference between the angle and the positive or negative $\boldsymbol{x}$-axis!

Worked Example The terminal arm of an angle in standard position θ passes through the point $P(-4, 5)$. **Sketch** to determine the exact value of all six trigonometric ratios of $\angle\theta$.

Solution:

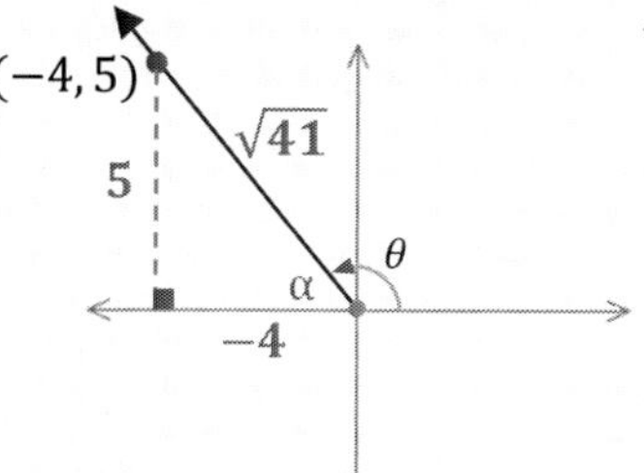

Sketch a diagram, use $r^2 = (-4)^2 + (5)^2$ to determine the length of the hypotenuse. (All sides must be written as exact values – no decimals!)

Since the x-coord. is negative, the length of the adjacent side is negative.

Label $\angle\theta$ from the positive x-axis to the terminal arm.
Note: In this scenario $\angle\theta$ is on the outside of the triangle!

Label $\angle\alpha$, the reference angle, inside the triangle.

Now let's find the value of the three primary trig ratios:

$\sin\theta = \dfrac{5}{\sqrt{41}}$ ← opp / ← hyp $\qquad \cos\theta = \dfrac{-4}{\sqrt{41}}$ ← adj / ← hyp

***Rationalize** the denominator:* $\quad = \dfrac{5}{\sqrt{41}} \times \dfrac{\sqrt{41}}{\sqrt{41}} \qquad = \dfrac{-4}{\sqrt{41}} \times \dfrac{\sqrt{41}}{\sqrt{41}}$

$\boldsymbol{\sin\theta = \dfrac{5\sqrt{41}}{41}} \qquad \boldsymbol{\cos\theta = -\dfrac{4\sqrt{41}}{41}} \qquad \boldsymbol{\tan\theta = -\dfrac{5}{4}}$ ← opp / ← adj

... and each reciprocal trig ratio: $\boldsymbol{\csc\theta = \dfrac{\sqrt{41}}{5}} \qquad \boldsymbol{\sec\theta = -\dfrac{\sqrt{41}}{5}} \qquad \boldsymbol{\cot\theta = -\dfrac{4}{5}}$

$cosecant\theta = \dfrac{1}{\sin\theta} \qquad secant\theta = \dfrac{1}{\cos\theta} \qquad cotangent\theta = \dfrac{1}{\tan\theta}$

Class Example 6.21 *Finding Trigonometric Ratios of Angles in Standard Position*

The terminal arm of an angle in standard position θ passes through the point $P(-5, -12)$. **Sketch** to determine the exact value of all six trigonometric ratios of $\angle\theta$. *Be sure to label both the rotation angle, θ, and the reference angle, α, in your diagram.*

6.2 *Trig Ratios of Angles in Standard Position*

As we've seen, if θ is in quadrant I, all of the trigonometric ratios are positive. However in quadrants II, III, and IV, trig ratios can can also be ***negative***. To determine which, keep your eye on the signs of the x & y coordinates!

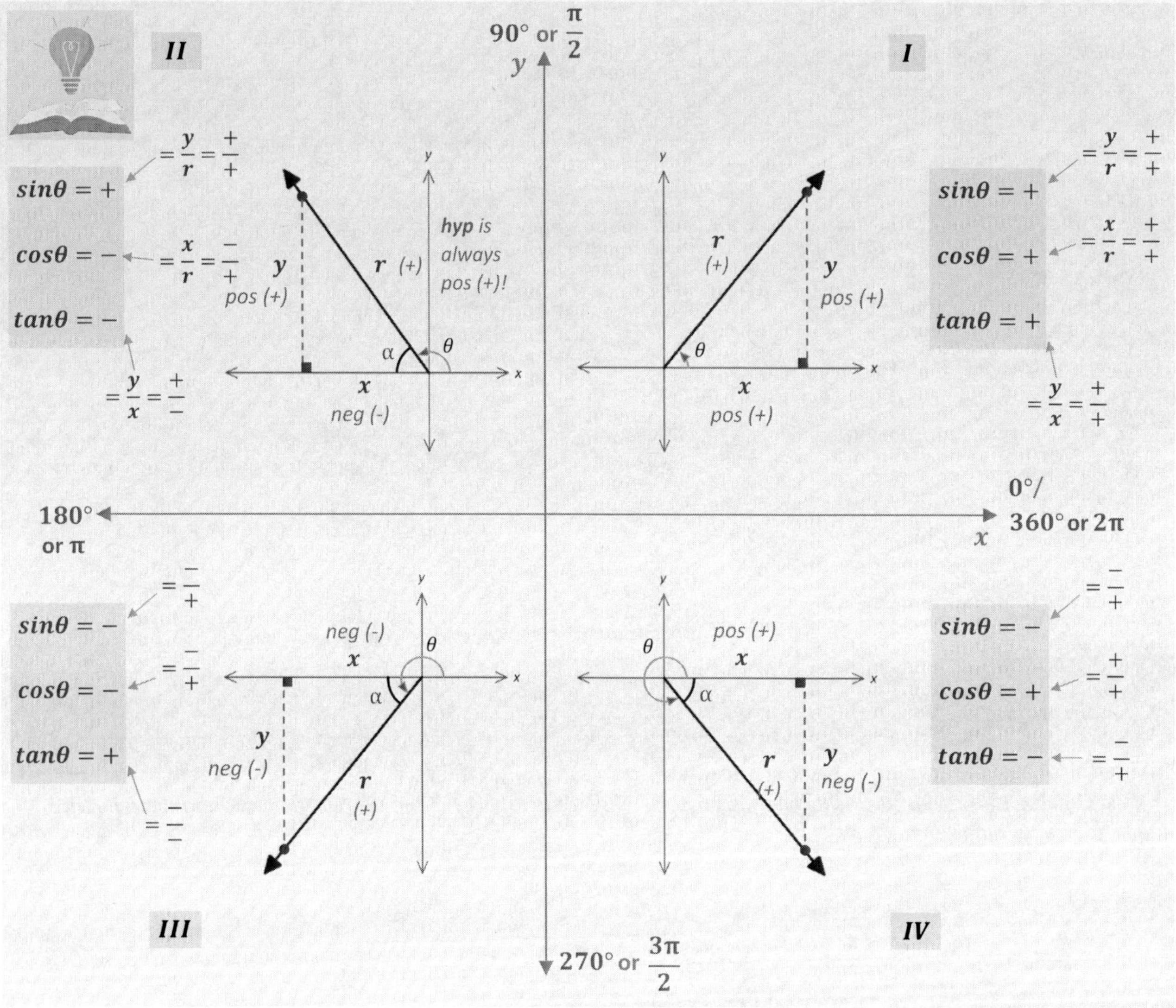

$90°$ or $\frac{\pi}{2}$ y

II

$\sin\theta = +$ $= \frac{y}{r} = \frac{+}{+}$

$\cos\theta = -$ $= \frac{x}{r} = \frac{-}{+}$

$\tan\theta = -$ $= \frac{y}{x} = \frac{+}{-}$

y pos (+), r (+), *hyp is always pos (+)!*, x neg (-), α, θ

I

$\sin\theta = +$ $= \frac{y}{r} = \frac{+}{+}$

$\cos\theta = +$ $= \frac{x}{r} = \frac{+}{+}$

$\tan\theta = +$ $= \frac{y}{x} = \frac{+}{+}$

r (+), y pos (+), x pos (+), θ

$180°$ or π — $0°/360°$ or 2π x

III

$\sin\theta = -$ $= \frac{-}{+}$

$\cos\theta = -$ $= \frac{-}{+}$

$\tan\theta = +$ $= \frac{-}{-}$

x neg (-), y neg (-), r (+), α, θ

IV

$\sin\theta = -$ $= \frac{-}{+}$

$\cos\theta = +$ $= \frac{+}{+}$

$\tan\theta = -$ $= \frac{-}{+}$

x pos (+), y neg (-), r (+), α, θ

$270°$ or $\frac{3\pi}{2}$

The **CAST** rule summarizes the above chart:

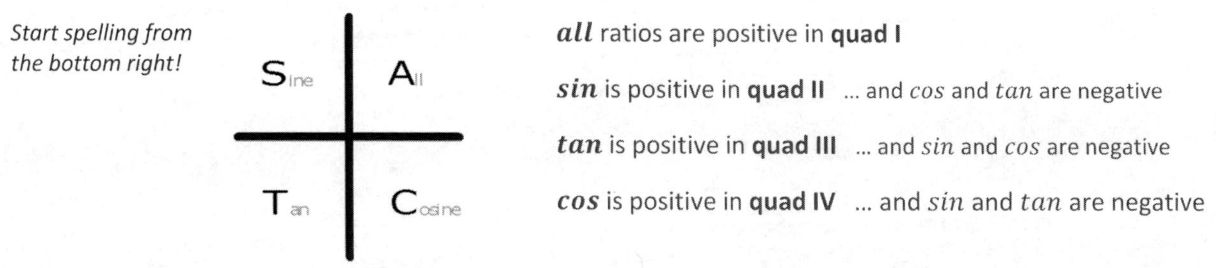

all ratios are positive in **quad I**

sin is positive in **quad II** ... and cos and tan are negative

tan is positive in **quad III** ... and sin and cos are negative

cos is positive in **quad IV** ... and sin and tan are negative

Worked Example If $tan\theta = -5/2$ and $csc\theta < 0$, determine the exact value of $cos\theta$

Sol.: First determine what quadrant $\angle\theta$ terminates in.

We are given that tan is **negative**, and is $tan\theta < 0$ in quads II & IV →

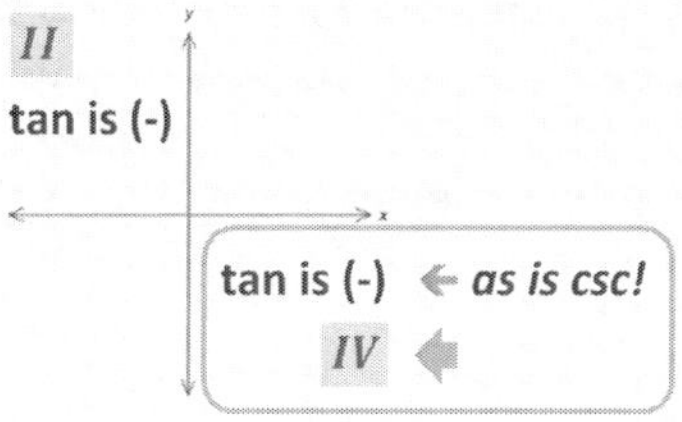

We are also given that csc is negative, which means sin is negative (reciprocal ratio).
Since sin is negative in quads III & IV ... both tan and csc **are negative in quad IV**.

II
tan is (-)
tan is (-) ← *as is csc!*
IV

Now, sketch $\angle\theta$ with a terminal arm in quad IV. Label the sides using $\boldsymbol{tan\,\theta} = \frac{-5}{2}$ ← *opp* / ← *adj*

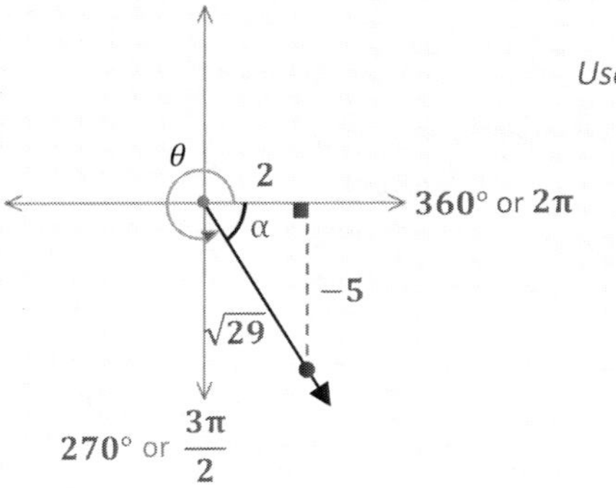

Use $r = \sqrt{(2)^2+(-5)^2}$ *for the hypotenuse.*

θ, 2, α, −5, $\sqrt{29}$, 360° or 2π, 270° or $\frac{3\pi}{2}$

From our diagram we can determine that $cos\theta = \frac{2}{\sqrt{29}}$ ← *adj* / ← *hyp*

Rationalize the denominator: $= \frac{2}{\sqrt{29}} \times \frac{\sqrt{29}}{\sqrt{29}}$

$$\boldsymbol{cos\theta = \frac{2\sqrt{29}}{29}}$$

Class Example 6.22 *Finding Trigonometric Ratios of Angles in Standard Position*

Use reasoning (principal angles and CAST rule) to determine whether each of the following trig ratios is positive or negative. *Approach without using the sine, cos, or tan buttons on your calculator!*

(a) $tan(960°)$

(b) $cos\left(-\frac{19\pi}{6}\right)$

(c) $sec\left(\frac{17\pi}{3}\right)$

Class Example 6.23 *Finding Trigonometric Ratio of Angles in Standard Position*

If $\boldsymbol{cos\,\theta = -\frac{3}{\sqrt{34}}}$ and $\boldsymbol{sin\,\theta > 0}$, draw a diagram to determine the exact value of $\boldsymbol{cot\,\theta}$.

Finding Angles Given their Trigonometric Ratios

We'll next consider problems where we wish to determine the measure of an angle θ in standard position, given either a point on the terminal arm or a trigonometric ratio.

But first, we need to establish the relationship between the value of the trigonometric ratios for θ, and those for α:

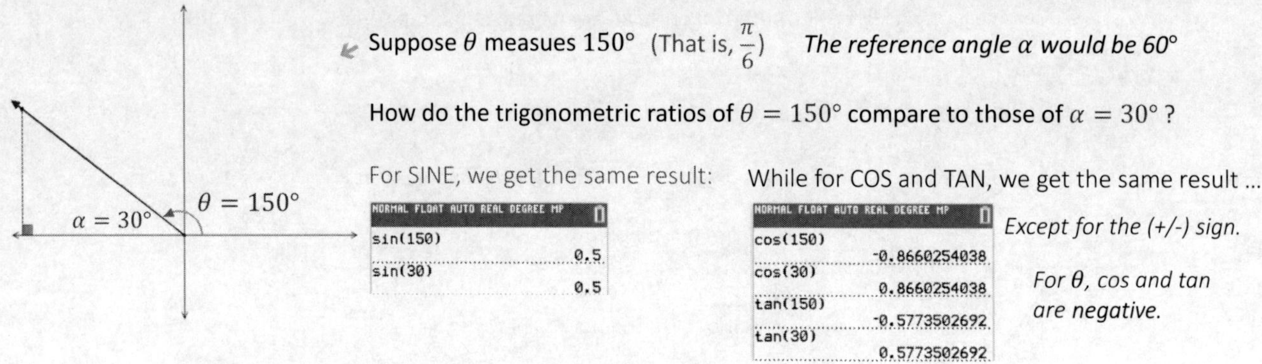

Suppose θ measues $150°$ (That is, $\frac{\pi}{6}$) *The reference angle α would be 60°*

How do the trigonometric ratios of $\theta = 150°$ compare to those of $\alpha = 30°$?

For SINE, we get the same result:

While for COS and TAN, we get the same result ...

Except for the (+/-) sign.

For θ, cos and tan are negative.

In general: For any angle in standard position θ and its reference angle α, the values of the trig ratios are identical, however:

- *For θ the trig ratios can be positive (+) or negative (–), while*
- *For α (being acute) the trig ratios are always positive*

As such, for problems involving determining the angle measure of θ, our method will be to:

Step 1 Determine the reference angle, using INVERSE SINE, INV COS, or INV TAN of the reference angle $\boldsymbol{\alpha}$

Step 2 Use reasoning (based on which quadrant) to determine $\boldsymbol{\theta}$.

Worked Example The terminal arm of an angle in standard position θ passes through the point $P(-4, 5)$. Determine the measure of $\angle\theta$, correct to the nearest degree and hundredth of a radian.

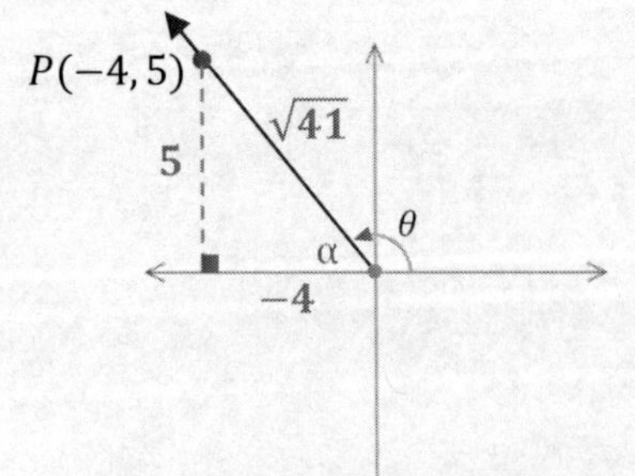

As we saw in our first encounter of this example (page 16), we can construct a triangle using the coordinates of P and the Pythagorean Theorem.

As such, we saw that: $\boldsymbol{sin\theta} = \frac{5}{\sqrt{41}}$ $\quad \boldsymbol{cos\theta} = -\frac{4}{\sqrt{41}}$ $\quad \boldsymbol{tan\theta} = -\frac{5}{4}$

Now to determine $\angle\theta$, we first find the reference angle $\boldsymbol{\alpha}$ *inside the triangle.*

$$\alpha = tan^{-1}\left(\frac{5}{4}\right) \quad \textit{or use} \quad sin^{-1}\left(\frac{5}{\sqrt{41}}\right) \quad \textit{or} \quad cos^{-1}\left(\frac{4}{\sqrt{41}}\right)$$

Note: We are finding $\boldsymbol{\alpha}$, not θ. From the perspective of α (which is acute) – all trig ratios are positive. So we ***drop the negative signs***.

Find $\angle\boldsymbol{\alpha}$ in degree mode:

```
tan⁻¹(5/4)
                 51.34019175
sin⁻¹(5/√41)
                 51.34019175
cos⁻¹(4/√41)
                 51.34019175
```

... and in radian mode:

```
tan⁻¹(5/4)
                 0.8960553846
π-Ans
                 2.245537269
```

$\alpha \approx 51°$ **and we're in quad II, so,**

$\boldsymbol{\theta \approx 129°}$ ← $180° - 51°$

$\alpha \approx 0.896$ **rads**

$\boldsymbol{\theta \approx 2.25}$ ← $\pi - 0.896$

Class Example 6.24 *Finding Trigonometric Ratios Given a Terminal Arm*

The terminal arm of an angle in standard position θ passes through the point $P(-5,-12)$. Complete the diagram from example 6.21 below (label each side, along with both θ and α), to determine the measure of $\angle\theta$.
Answer in both degrees (round to the nearest whole number) and radians. (round to the nearest hundredth)

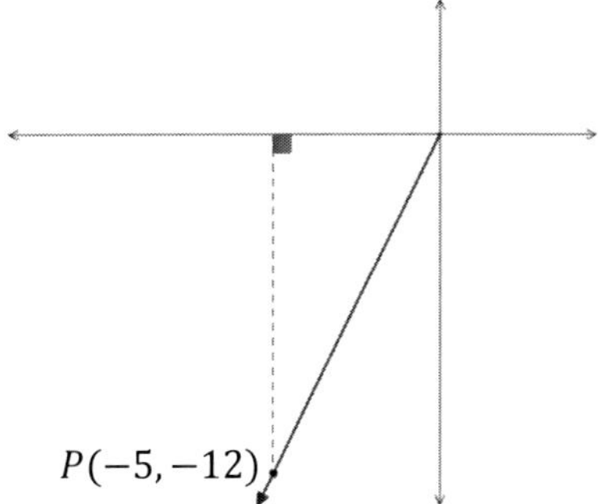

Worked Example If $tan\theta = -5/2$ and $csc\theta < 0$, determine the measure of θ, in degrees (nearest whole number) and radians (nearest hundredth)

A full diagram like we made before is **not** essential, however it is helpful to sketch to see that θ is between $\frac{3\pi}{2}$ and 2π. (See page 357 for reasoning on why θ is in quad IV)

Note that we can use inverse COS or SIN (given the diagram) – as well as TAN!

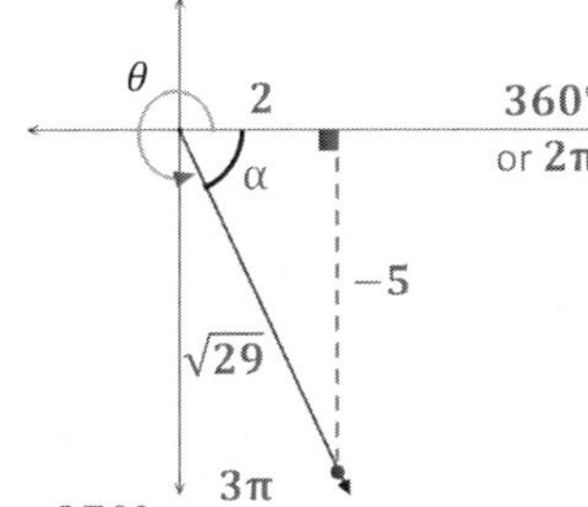

$\alpha = tan^{-1}(5/2)$ ← *remember - drop the negative sign*

$\approx 68°$ calc in **degree mode**

≈ 1.19 calc in **radian mode**

so... $\boldsymbol{\theta \approx 292°}$ or **5.09 rads**

$360° - \mathbf{68°}$ $2\pi - \mathbf{1.19}$

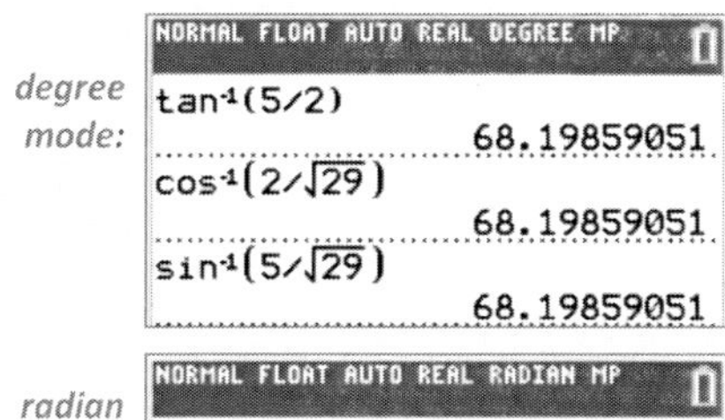

Class Example 6.25 *Finding the Measure of an Angle in Standard Position*

(a) If $\boldsymbol{cos\,\theta = -\frac{3}{\sqrt{34}}}$ and $\boldsymbol{sin\,\theta > 0}$, complete the diagram drawn in example 6.23 to determine the measure of $\angle\theta$ of $\boldsymbol{\theta}$, in degrees (nearest whole number) and radians (nearest hundredth).

(b) If $\boldsymbol{sin\theta = -\frac{3}{\sqrt{58}}}$ and $\boldsymbol{cot\,\theta < 0}$, sketch to determine the measure of $\angle\theta$ (as above, in degrees and radians).

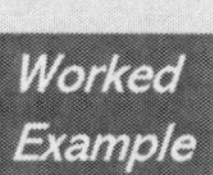

For $\sin\theta = -\frac{1}{2}$; $\{0 \le \theta < 2\pi\}$

(a) Determine the measure all angles θ that satisfy the equation on the given domain

(b) State an expression(s) for all angles that satisfy the equation on $\{\theta \in \mathbb{R}\}$

*State angles as **exact values** in terms of π*

Step 1: Determine the reference angle α, using INVERSE SINE.

$\alpha = sin^{-1}(1/2)$

$= \frac{\pi}{6}$

***remember** - drop any negative sign, for the reference angle all trig ratios are positive*

Step 2: Determine the quadrants where SINE is negative.

S	A
T✓	C ✓

In the *lower quadrants,* III and IV

Step 3: Determine the quadrant III, IV rotation angles for $\alpha = \frac{\pi}{6}$. (That is, 30°)

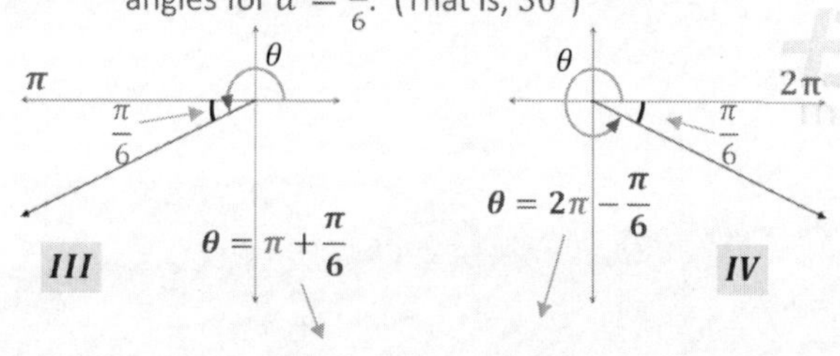

(a) so... $\theta = \frac{7\pi}{6}$ or $\theta = \frac{11\pi}{6}$

(b) $\theta = \frac{7\pi}{6} + 2\pi n$; $n \in I$

or

$\theta = \frac{11\pi}{6} + 2\pi n$; $n \in I$

General solution *– statement covers all coterminal angles for each principal angle.*

To determine the reference angle on your CALC:

We have TWO options, noting that we want our reference angle stated as an exact value:

Option 1: *Radian Mode*

Divide the decimal by "π"

Then convert to FRACTION

Finally, remember to add back the "π"!

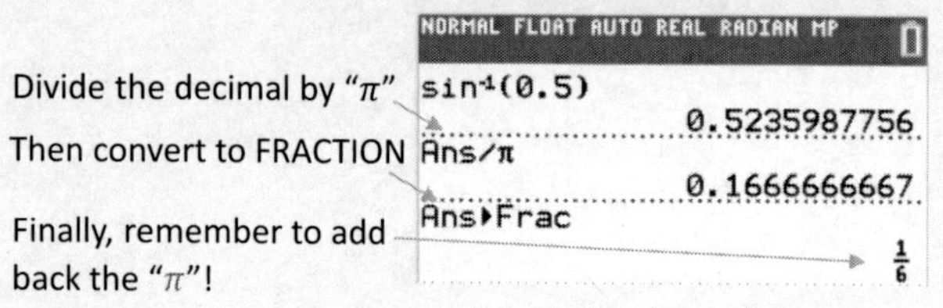

Note: Above method will work only if the angle CAN be expressed "in terms of π"!

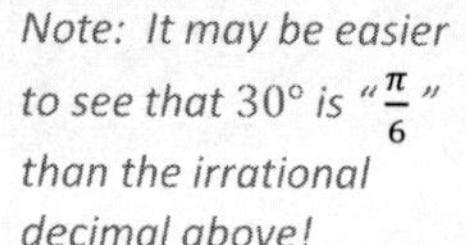

Option 2: *Degree Mode*

Here we get the reference angle as 30°, then convert to radians.

Note: It may be easier to see that 30° is "$\frac{\pi}{6}$" than the irrational decimal above!

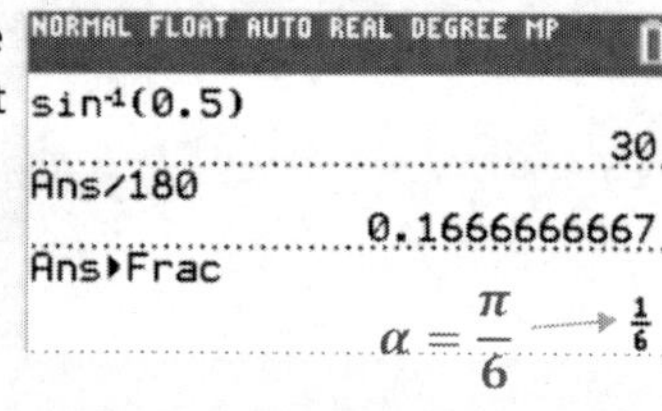

***Note:** Only consider **Degree Mode** for questions where solutions are to given as **exact values**.*

For questions where the radian solutions are to be given "rounded to the nearest hundredth" – go with option 1 / radian mode! (or ... use special triangles!)

Class Example 6.26 *Determining an Angle in Standard Position*

For each of the following equations;

i Determine the measure all angles θ that satisfy the equation on the given domain

ii State an expression(s) for all angles that satisfy the equation on $\{\theta \in \mathbb{R}\}$

(a) $csc\,\theta = -\frac{2\sqrt{3}}{3}$; $\{0 \le \theta < 2\pi\}$

(b) $tan\,\theta = 1$; $\{-\pi \le \theta < \pi\}$

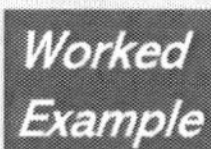

For $\cos\theta = -0.75$; $\{0 \le \theta < 2\pi\}$ *State angles as decimal approximations to the nearest hundredth*
determine the measure all angles θ that satisfy the equation on the given domain

Step 1: Determine the reference angle α, using ➡

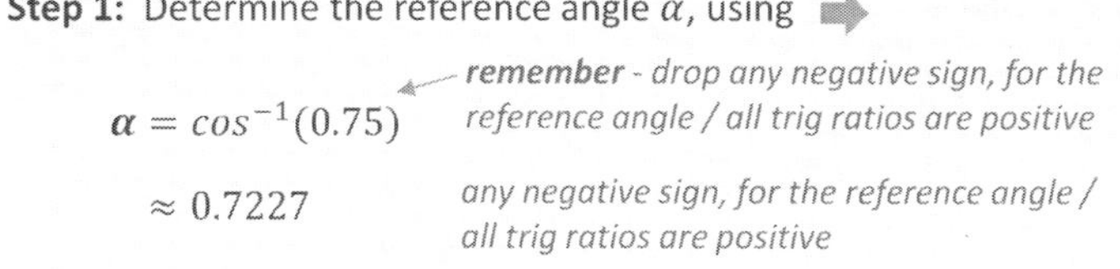

$\alpha = \cos^{-1}(0.75)$ ← ***remember*** *- drop any negative sign, for the reference angle / all trig ratios are positive*

≈ 0.7227 *any negative sign, for the reference angle / all trig ratios are positive*

Radian Mode:

```
NORMAL FLOAT AUTO REAL RADIAN MP
cos⁻¹(0.75)
                    0.7227342478
```

Step 2: Determine the quadrants where COS is negative.

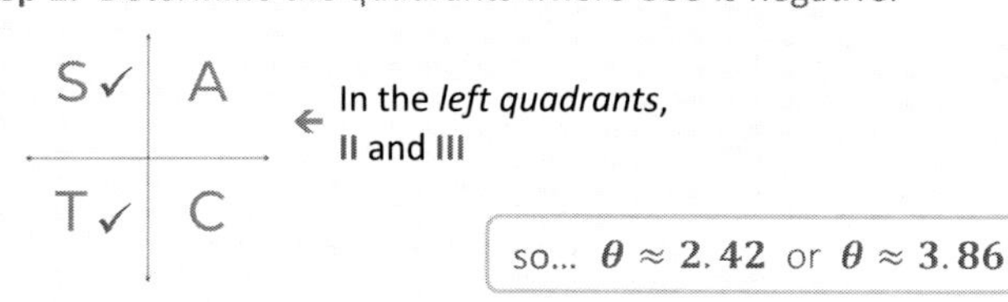

← In the *left quadrants*, II and III

Step 3: Determine the quadrant II, III rotation angles for $\alpha \approx 0.7227$.

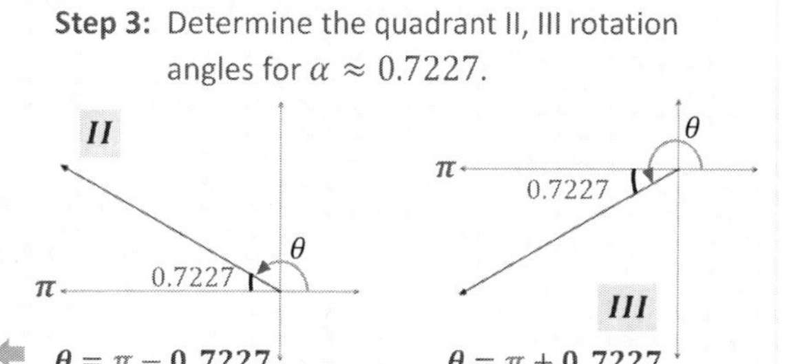

$\theta = \pi - 0.7227$ $\theta = \pi + 0.7227$

so... $\theta \approx 2.42$ or $\theta \approx 3.86$

Class Example 6.27 *Solving a Simple Trigonometric Equation*

Determine the solutions to the following trigonometric equations, on the stated intervals. State all angles as decimal approximations, in the same angular measure as indicated in each interval. *Round to the nearest hundredth.*

(a) $\sec\theta = 3$; $\{0 \le \theta < 2\pi\}$

(b) $\cot\theta = 0.5$; $\{-180 \le \theta < 180°\}$

The Special Triangles

The **special triangles** are used to develop and express the exact values for the trigonometric ratios for 30°, 45°, and 60°. (that is, $\frac{\pi}{6}, \frac{\pi}{4}$, and $\frac{\pi}{3}$)

The special triangles:

Alternate form of the special ◣s (sides are all divided by 2, so that each hypotenuse is 1)

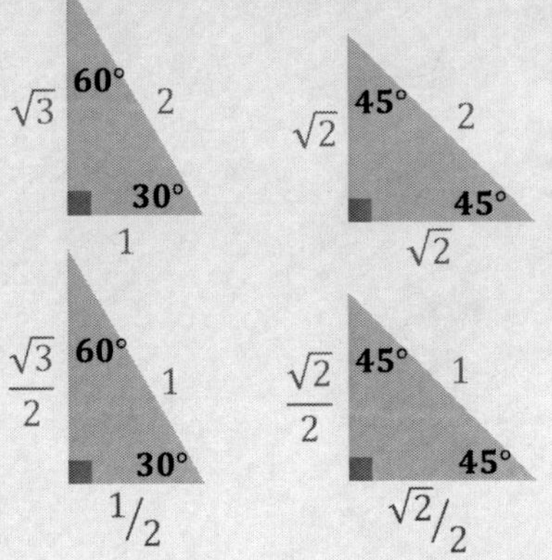

θ	$30°\ /\ \frac{\pi}{6}$	$45°\ /\ \frac{\pi}{4}$	$60°\ /\ \frac{\pi}{3}$
$\sin\theta$	$\frac{1}{2}$	$\frac{\sqrt{2}}{2}$	$\frac{\sqrt{3}}{2}$
$\cos\theta$	$\frac{\sqrt{3}}{2}$	$\frac{\sqrt{2}}{2}$	$\frac{1}{2}$
$\tan\theta$	$\frac{\sqrt{3}}{3}$	1	$\sqrt{3}$

Worked Example Find the exact value of the primary trigonometric ratios for $\frac{5\pi}{6}$.

Solution: SKETCH: $\frac{5\pi}{6}$ is in Quad II, and it's ***reference angle*** is $\pi - \frac{5\pi}{6}$, or $\alpha = \frac{\pi}{6}$.

Alternatively, first convert to 150°, and the reference angle is $\alpha = 30°$

Draw the 30° - 60° - 90° special triangle in quadrant II. ➡

Use the triangle, with a negative base (x-coord is negative in quad II) to find:

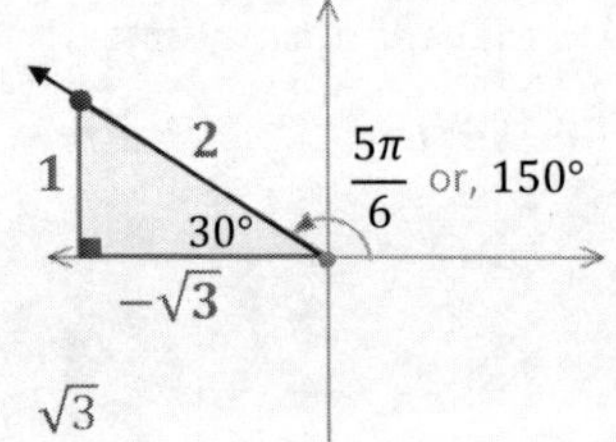

$\sin\frac{5\pi}{6} = \frac{1}{2}$ $\cos\frac{5\pi}{6} = -\frac{\sqrt{3}}{2}$ $\tan 150° = -\frac{1}{\sqrt{3}}$ *Rationalizes to:* $-\frac{\sqrt{3}}{3}$

Class Example 6.28 *Finding Exact Trigonometric Ratios*

Draw the appropriate special triangle to determine the exact value of each trigonometric ratio.

(a) $\sin(120°)$ (b) $\tan\left(\frac{7\pi}{4}\right)$ (c) $\cos\left(\frac{4\pi}{3}\right)$

(d) $\sin(330°)$ (e) $\sec\left(-\frac{5\pi}{4}\right)$ (f) $\cot\left(\frac{19\pi}{6}\right)$

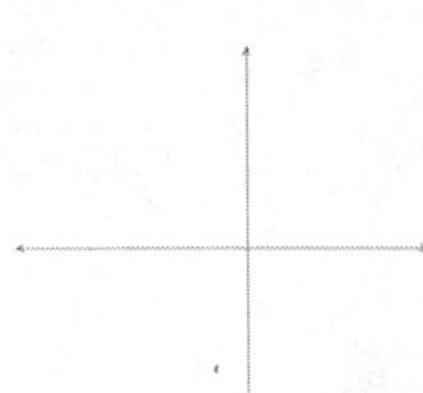

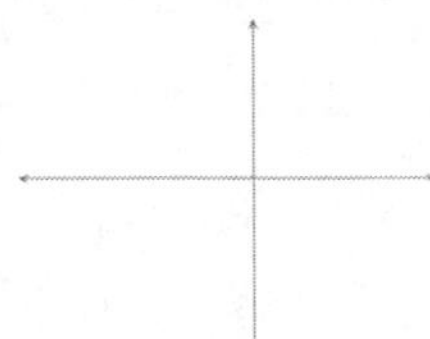

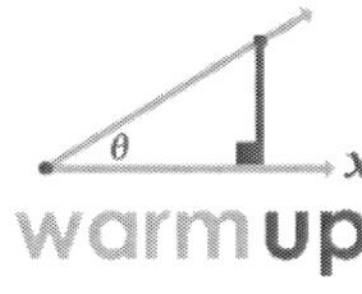

Exploration #1

Both circles in the diagram on the right are centered at the origin. The partially-shown outer circle has a radius of 5, the inner circle a radius of 1.

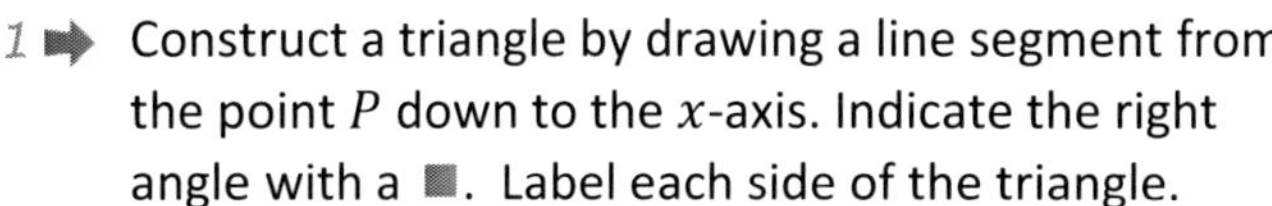

1 ➡ Construct a triangle by drawing a line segment from the point P down to the x-axis. Indicate the right angle with a ■. Label each side of the triangle.

2 ➡ Use the triangle drawn to determine the values of:

$sin\theta =$ $\qquad\qquad$ $cos\theta =$

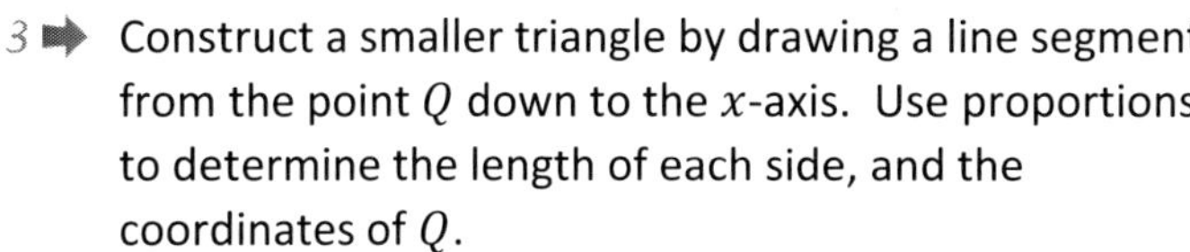

3 ➡ Construct a smaller triangle by drawing a line segment from the point Q down to the x-axis. Use proportions to determine the length of each side, and the coordinates of Q.

4 ➡ Use the smaller triangle to determine the values of: $\quad sin\theta =$ $\qquad\qquad$ $cos\theta =$

5 ➡ How do the values of $cos\theta$ and $sin\theta$ relate to the sides of the triangle when the hypotenuse is 1?

6 ➡ How do the values of $cos\theta$ and $sin\theta$ relate to the coordinates of a point on the smaller circle?

Defining the Unit Circle

So as we can see from the warm-up above, a circle that's centred at the origin with a radius of one has particularly interesting property. Did you notice it?

The coordinates of the points are the cosine and sine values for the corresponding angle in standard position!

A **unit circle** has its centre at the origin, and a radius of 1.

The equation is $\boldsymbol{x^2 + y^2 = 1}$, which comes from the Pythagorean Theorem.

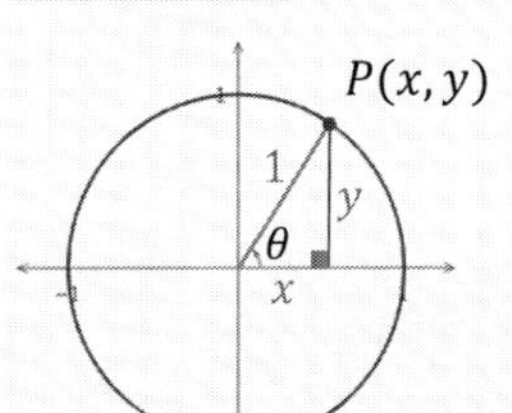

And once again, here's the good part...

← Since the hypotenuse is 1....

$sin\theta = \frac{y}{1}$ ← opp / ← hyp $\qquad$ $cos\theta = \frac{x}{1}$ ← opp / ← hyp

...*The coordinates of any point are* $(x, y) = (\boldsymbol{cos\theta, sin\theta})$

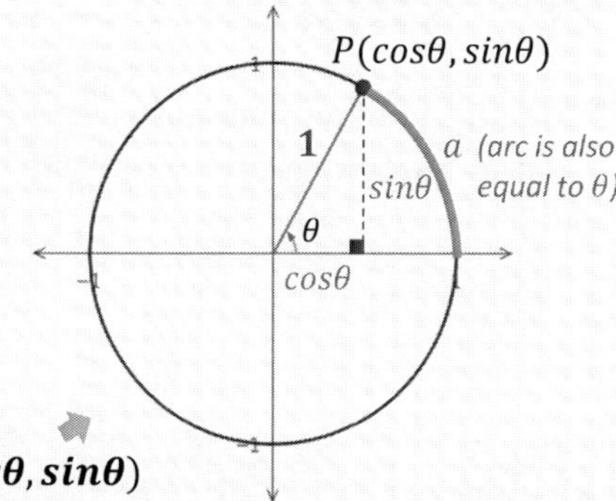

note: As the arc length, a, is equal to $r\theta$, and $r = 1$, the length of any arc on a unit circle is **equal to** the radian measure of the angle in standard position θ

The unit circle is a useful tool for visualizing the values of $\boldsymbol{sin\theta}$ and $\boldsymbol{cos\theta}$. In the diagram above, ↑ try to visualize $\boldsymbol{P}$ as it "rides up" the circle from $\boldsymbol{0}$ to $\boldsymbol{\pi/2}$. Can you see how the x-coord "shrinks" ... from 1 to 0? *That's what's happening to the value of* $cos\theta$*!* And how about the $\boldsymbol{y}$-coord. ($\boldsymbol{sin\theta}$), as θ varies from 0° to 90°?

We'll come back to this concept in the next section.

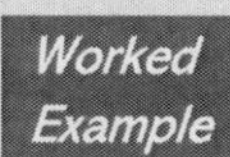

The point $P(0.6, m)$ is at the point of intersection of a circle with a radius of 1 unit, and the terminal arm of an angle θ in standard position. Determine the largest possible measure of $\angle\theta$ on $[0, 2\pi)$, correct to the nearest hundredth of a radian, and the corresponding value of m.

Solution:

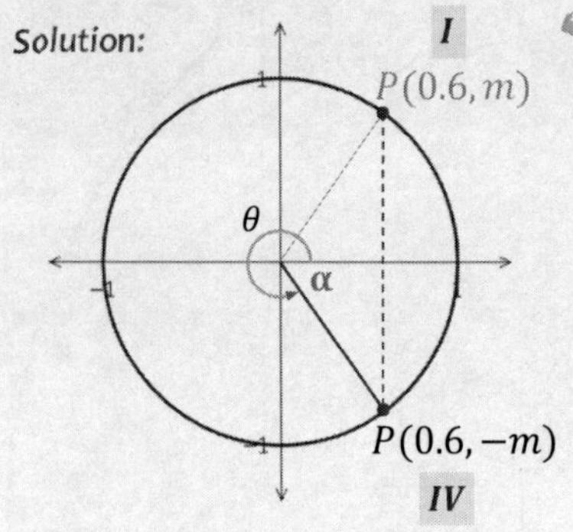

Sketch a diagram to see that there are two possibilities for P, as the x-coordinate can be 0.6 in either quadrant I or IV. However we'll draw $\angle\theta$ in quadrant 4, as we want the *largest possible* measure.

For the reference angle, α, use inverse cos, as we are given that $\boldsymbol{cos\theta = 0.6}$

Remember: *the x-coord* ***is*** *$cos\theta$!*

$$\alpha = cos^{-1}(0.6)$$

$$\alpha \approx 0.927$$

So, $\theta \approx 2\pi - 0.927$

→ $\boxed{\theta \approx 5.36}$

Largest possible θ (Quad IV)

```
NORMAL FLOAT AUTO REAL RADIAN MP
cos⁻¹(0.6)
                    0.927295218
2π-Ans
                    5.355890089
```

Now, for the corresponding value of $\boldsymbol{m}$,

Use: $\boldsymbol{x^2 + y^2 = 1}$

$$(0.6)^2 + m^2 = 1$$

$$m = \pm\sqrt{1 - 0.36}$$

$$\boxed{m = -0.8}$$

Use the neg. option, as θ is in quad ***IV***

Class Example 6.31 *Finding a Trig Ratio using the Unit Circle Equation*

The terminal arm of an angle in standard position θ contains the point $P(x, y)$, where P is on the unit circle . If $sin\theta = \frac{7}{10}$, then use the blank unit circle provided to:

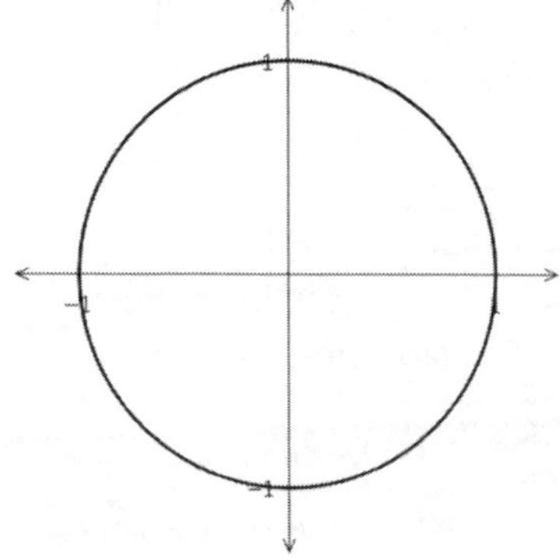

(a) Determine the two possibilities for $cos\theta$, expressed as exact values.

(b) Sketch and determine the two possible measures of θ, correct to the nearest degree, where $0° \leq \theta < 360°$.

Class Example 6.32 *Finding the Coordinates of a Point on the Unit Circle*

The terminal arm of an angle in standard position θ contains the point $P(\frac{1}{\sqrt{2}}, y)$, where P is on the unit circle.

(a) Determine the possible coordinates of P.

(b) Determine the measure of $\angle\theta$, given that $tan\theta < 0$.

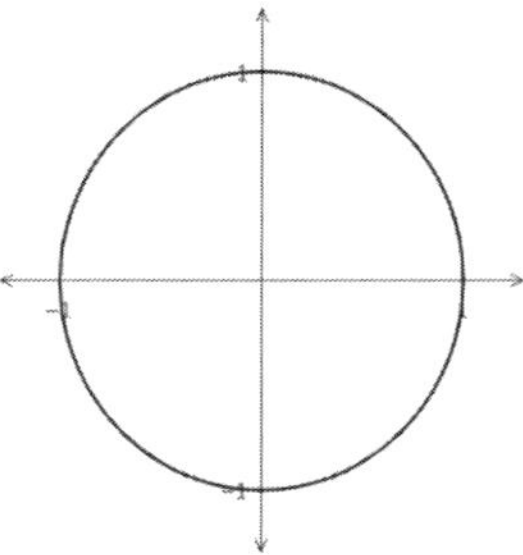

Class Example 6.33 *Applying the Unit Circle - Finding Coordinates*

The terminal arm of an angle in standard position measuring 25° intersects the unit circle at the point $P(x, y)$. The coordinates of P, rounded to the nearest hundredth, are $x = 0.ab$ and $y = 0.cd$.

NR The values of a, b, c and d are, respectively, ____, ____, ____ and ____ .

Class Example 6.34 *Confirming a Point is on the Unit Circle*

For each of the following points P, (i) Determine if the point could be on a unit circle, and if so (ii) Determine the measure of the angle in standard position whose terminal arm passes through P. (correct to the nearest degree)

(a) $P(-\frac{2}{3}, \frac{3}{4})$

(b) $P(0.6, -0.8)$

Relating the Unit Circle and the Special Triangles

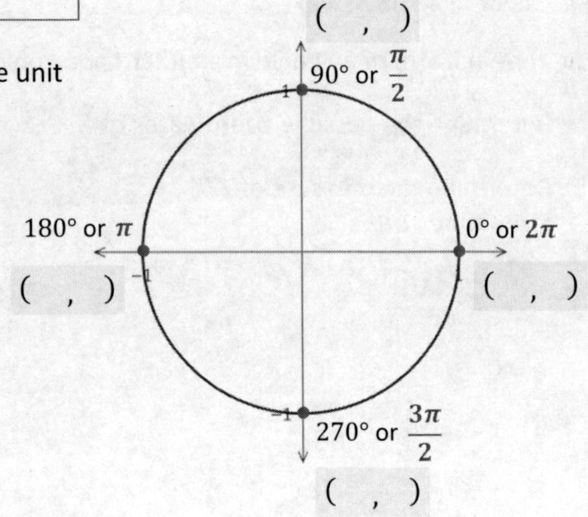

At this point we've seen how the coordinates of any point on the unit circle relate to the values of $sin\,\theta$ and $cos\,\theta$.

The unit circle helps us to ***organize*** and ***visualize*** the various values of $\boldsymbol{cos\theta}$ and $\boldsymbol{sin\theta}$ for θ between 0 and 2π.

➡ Label the coordinates of each point at the indicated quadrantal angles:

Did you see what you just did? You just indicated the values of $\mathbf{cos}$ and $\mathbf{sin}$ for each of those angles.

Remember – on the **unit circle**:

- the $\boldsymbol{x}$-coordinates are the values of $\boldsymbol{cos\theta}$
- and the $\boldsymbol{y}$-coordinates are the values of $\boldsymbol{sin\theta}$

Let's keep it going. Earlier we saw how in trigonometry, we pay extra attention to angels that are multiples of 30° or 45°. (That is, $\frac{\pi}{3}$ or $\frac{\pi}{4}$) We can refer to the **special triangles** for the corresponding values of sin and cos.

The $\mathbf{30° - 60°}$ *Special*

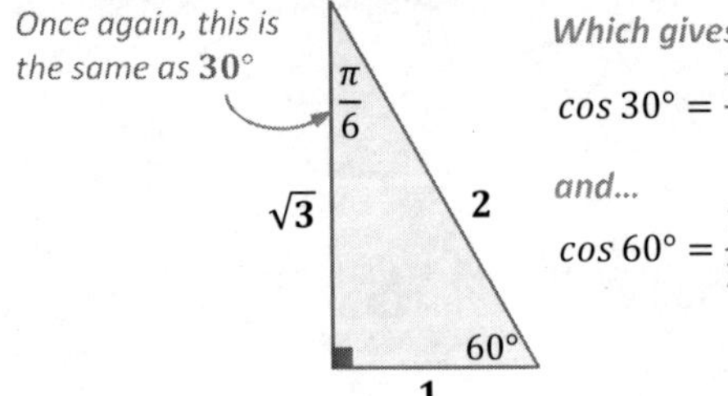

Which gives us:

$cos\,30° = \frac{\sqrt{3}}{2}$ $\quad sin\,30° = \frac{1}{2}$

and...

$cos\,60° = \frac{1}{2}$ $\quad sin\,60° = \frac{\sqrt{3}}{2}$

The $\mathbf{45°}$ *Special*

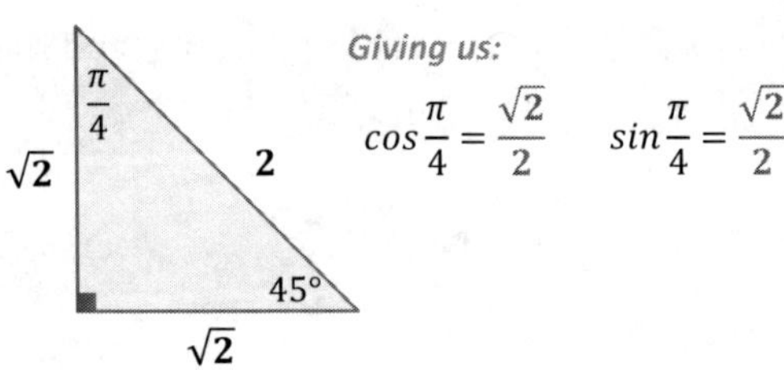

Giving us:

$cos\frac{\pi}{4} = \frac{\sqrt{2}}{2}$ $\quad sin\frac{\pi}{4} = \frac{\sqrt{2}}{2}$

➡ Use the trigonometric ratios defined above to label the coordinates of each unit circle point / at the indicated quadrant I angles:

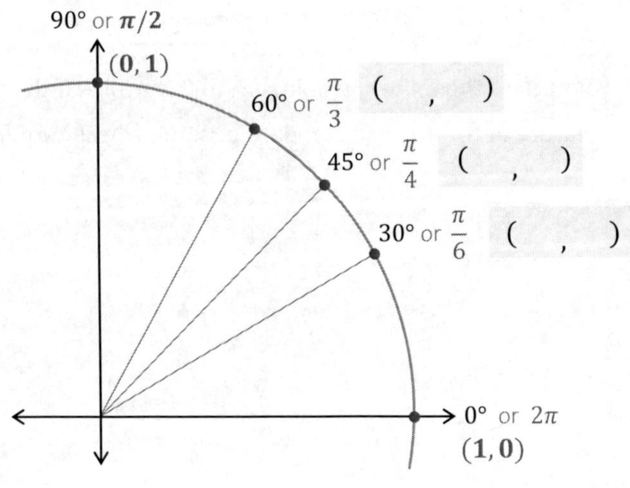

Complete the unit circle below. Start with the quadrantal angles and coordinates from the last page. Then, use the special triangles (visually summarized on the right) and reasoning (quadrants in which coords are pos. / neg.) to label all remaining angles and coords.

Label all angles *inside* the circle (a space is provided for the degree and radian measure) and all coordinates *outside.*

See how to complete you unit circle here!

90° or $\pi/2$

60° or $\frac{\pi}{3}$

45° or $\frac{\pi}{4}$

30° or $\frac{\pi}{6}$

0° or 2π

1

$\frac{\sqrt{3}}{2}$

$\frac{\sqrt{2}}{2}$

$\frac{1}{2}$

$1/2 = 0.5$

$\sqrt{2}/2 \approx 0.707 \ldots$

$\sqrt{3}/2 \approx 0.866 \ldots$

y

x

Now that we've made it – What are we going to *Do With It?*

We can use the unit circle to easily* visualize and determine the exact value of any trigonometric ratio that is a multiple of $30°$, $45°$, or $60°$. *(While the last angle in that list is redundant, it's included to remind us that there are three non-quadrantal angles per quadrant represented on the unit circle!)*

This shall be demonstrated with a series of worked examples....

We can visualize by labeling a "mini-unit circle" (with a special triangle drawn to help visualize):

Worked Example 1 Determine the exact value of $\cos\theta$ and $\sin\theta$ for $\theta = \dfrac{2\pi}{3}$.

Solution: Refer to your unit circle, look up the angle $\dfrac{2\pi}{3}$

☞ $\cos\dfrac{2\pi}{3}$ is the x-coordinate

☞ $\sin\dfrac{2\pi}{3}$ is the y-coordinate

$$\cos\frac{2\pi}{3} = -\frac{\mathbf{1}}{\mathbf{2}} \qquad \sin\frac{2\pi}{3} = \frac{\sqrt{\mathbf{3}}}{\mathbf{2}}$$

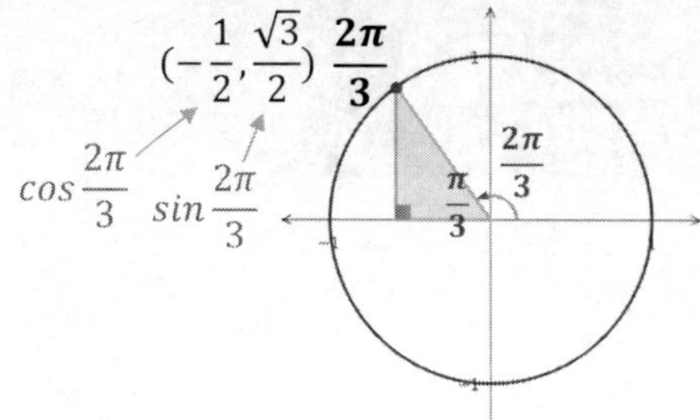

Next, let's look at how the unit circle can be used for the ***tan*** ratio.

Worked Example 2 Determine the exact value of $\tan 330°$.

Solution: Refer to your unit circle, look up the angle $330°$.

☞ For the $\tan$ ratio, use $\dfrac{"y"}{x}$

$$\tan 330° = \frac{-\frac{1}{2}}{\frac{\sqrt{3}}{2}}$$

*"y" / sin 330°, the **opp** side*

*"x" / cos 330°, the **adj** side*

30°
330°
$(\frac{\sqrt{3}}{2}, -\frac{1}{2})$
$\cos 330°$
$\sin 330°$

$$= -\frac{1}{2} \times \frac{2}{\sqrt{3}} \Rightarrow = -\frac{1}{\sqrt{3}} \Rightarrow = -\frac{1}{\sqrt{3}} \times \frac{\sqrt{3}}{\sqrt{3}} \Rightarrow = -\frac{\sqrt{\mathbf{3}}}{\mathbf{3}}$$

Rationalize the denominator

Finally, we'll look determining **reciprocal trigonometric ratios** of *non-principal angles*.

Worked Example 3 Determine the exact value of $\cot\dfrac{11\pi}{2}$.

Solution: First determine the principal angle: $\dfrac{11\pi}{2} - \underbrace{\dfrac{4\pi}{2}}_{2\pi} - \underbrace{\dfrac{4\pi}{2}}_{2\pi} = \dfrac{\mathbf{3\pi}}{\mathbf{2}}$

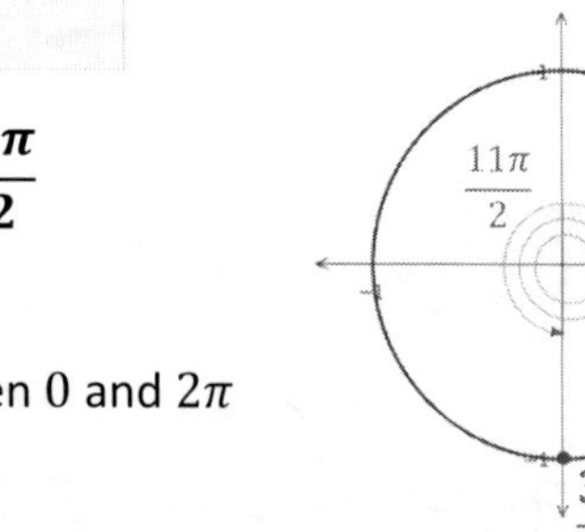

Refer to your unit circle, use the coordinates at the principal (between 0 and 2π and labeled on the unit circle) coterminal angle to $\dfrac{11\pi}{2}$, that is, $\dfrac{3\pi}{2}$.

$\frac{3\pi}{2}$ $(0,-1)$
$\cos\frac{3\pi}{2}$ $\sin\frac{3\pi}{2}$

☞ For the $\cot$ ratio, use $\dfrac{"x"}{y}$, at $\dfrac{3\pi}{2}$.

$$\cot\frac{11\pi}{2} = \frac{0}{-1} \Rightarrow = \mathbf{0}$$

"x" at $\frac{3\pi}{2}$; *"y" at $\frac{3\pi}{2}$*

note

We know that $\cot\theta = \dfrac{1}{\tan\theta}$, and $\tan\dfrac{3\pi}{2}$ is undefined, so we might expect $\cot\dfrac{3\pi}{2}$ to be undefined as well.

But as we see above, $\mathbf{\cot\frac{3\pi}{2}}$ (or $\mathbf{\cot\frac{\pi}{2}}$) **is equal to 0.** *Stick with the identity $\cot\theta = \dfrac{\cos\theta}{\sin\theta}$ when evaluating the **cot** ratio!*

Class Example 6.35 *Finding Exact Values of Trigonometric Ratios using the Unit Circle*

For each of the following,

(i) Where appliable, state in terms of an equivalent **primary** trig ratio of a **principal** angle, and in terms of the unit circle coordinates. *Try to visualize the value of the coordinates (cos and sin values) for each angle!*

(ii) Plot the approximate position on the unit circle provided (label the angle and coordinates), and

(iii) State the exact value of the trigonometric ratio.

(a) $cos120°$

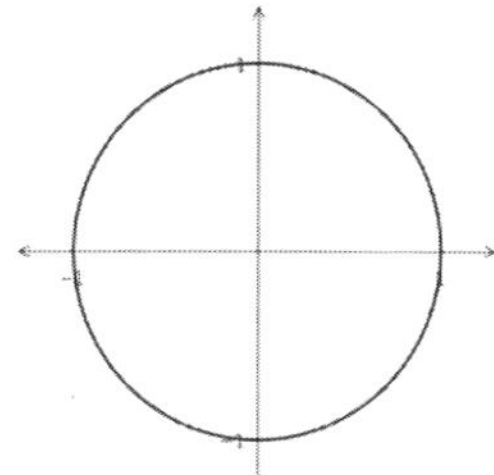

(b) $sin\frac{5\pi}{3}$

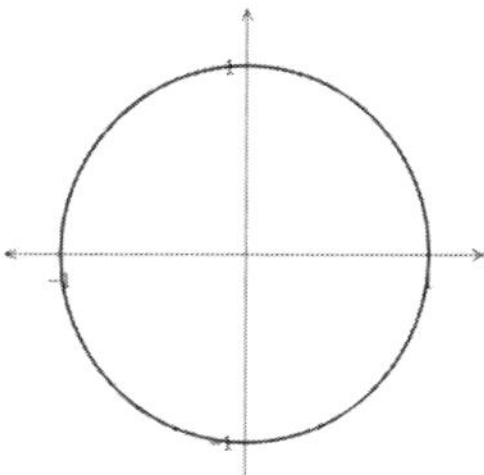

(c) $cos3\pi$

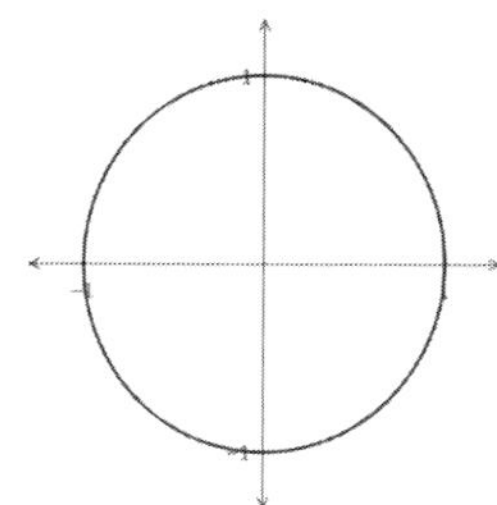

(d) $cos210°$

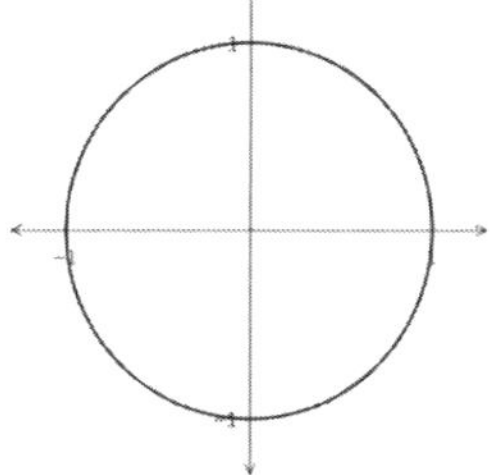

(e) $sin\frac{5\pi}{2}$

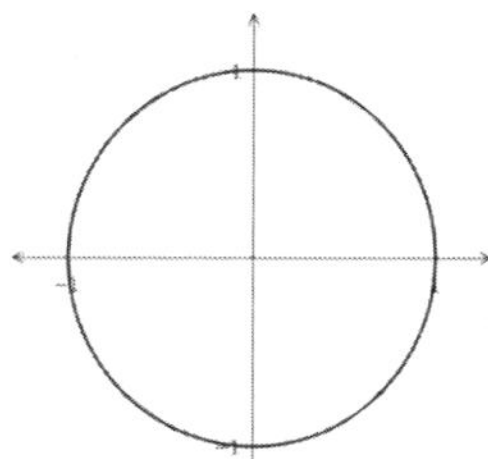

(f) $cos\frac{13\pi}{4}$

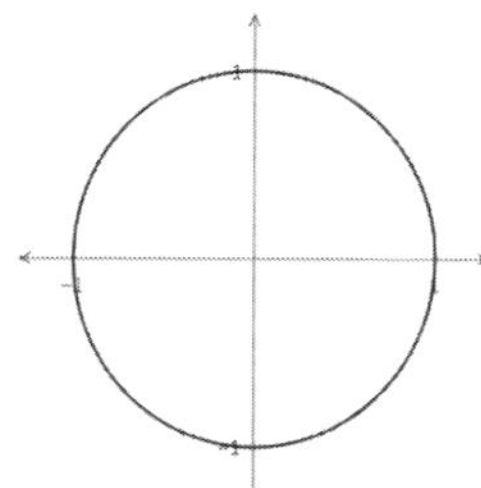

(g) $csc(-\frac{17\pi}{6})$

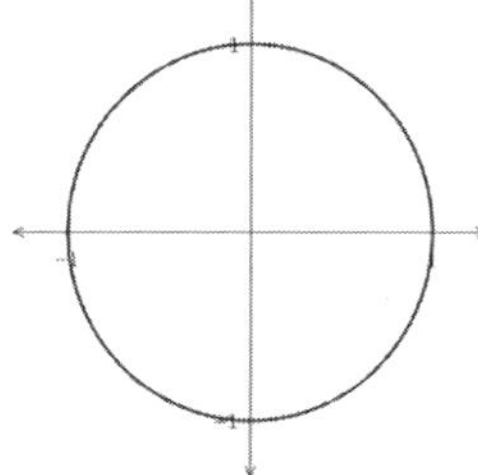

(h) $tan(-\frac{5\pi}{3})$

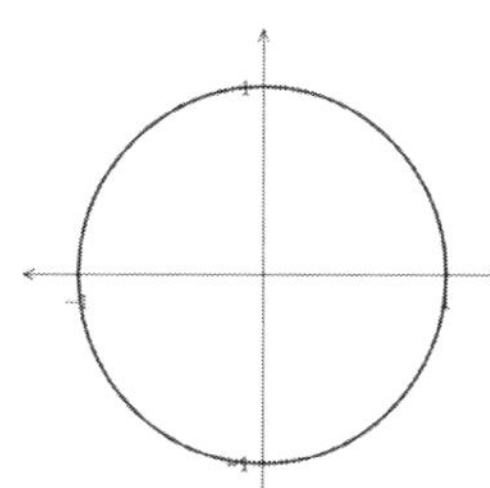

(i) $sec\frac{\pi}{2}$

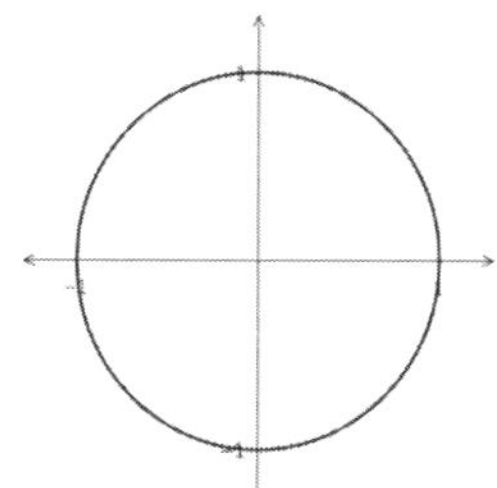

Worked Example

Use the unit circle to (i) solve each equation on the interval $[0, 2\pi)$ and (ii) state a **general solution** for all solutions on $\{\theta \in \mathbb{R}\}$.

(a) $\cos\theta = \frac{\sqrt{3}}{2}$ (b) $\csc\theta = -\sqrt{2}$

Solution: **(a)** Refer to the unit circle; where is the x-coord. (cosine) equal to $\frac{\sqrt{3}}{2}$?

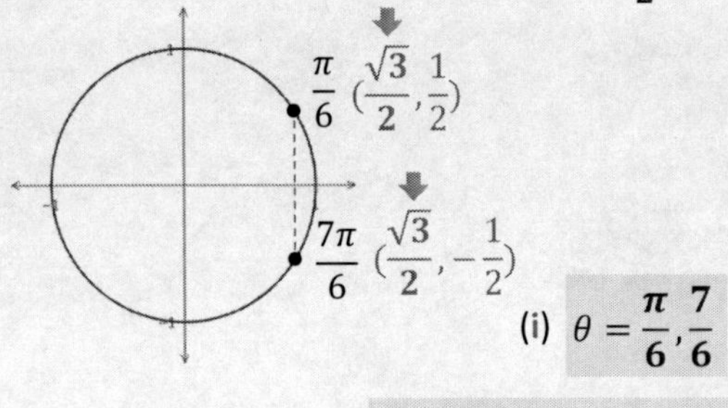

(i) $\theta = \frac{\pi}{6}, \frac{7}{6}$

(ii) $\theta = \frac{\pi}{6} + 2\pi n$; $n \in I$

$\theta = \frac{7\pi}{6} + 2\pi n$; $n \in I$

Read as: *Plus any multiple of "2π"*

(b) First re-write in terms of a primary trig ratio:

Take reciprocal of both sides, rationalize: $\sin\theta = -\frac{1}{\sqrt{2}}$ ➧ $\sin\theta = -\frac{\sqrt{2}}{2}$

Where is the y-coord. (sine) $-\frac{\sqrt{2}}{2}$?

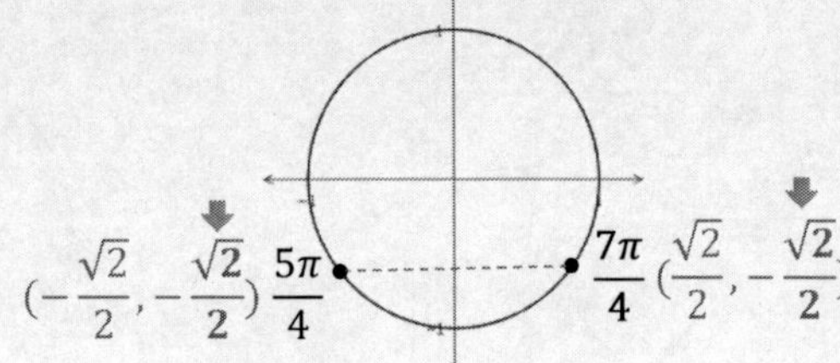

(ii) $\theta = \frac{5\pi}{4} + 2\pi n$, $\theta = \frac{7\pi}{4} + 2\pi n$; $n \in I$

(i) $\theta = \frac{5\pi}{4}, \frac{7\pi}{4}$

Class Example 6.36 *Finding all Trigonometric Ratios using the Unit Circle*

For each of the following equations, use the unit circle to (i) solve each equation on the interval $[0, 2\pi)$ and (ii) state a **general solution** for all solutions on $\{\theta \in \mathbb{R}\}$.

(a) $\cos\theta = -\frac{1}{2}$ (b) $\sin\theta = 0$ (c) $\tan\theta = -1$

6.4 Graphs of Trigonometric Functions

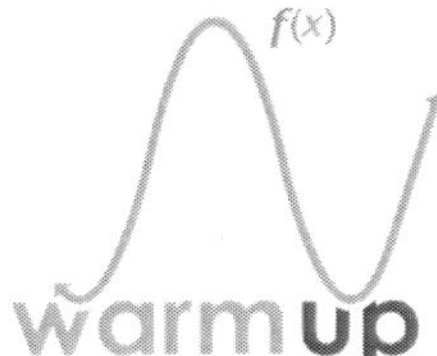

Investigation 1 - The Graph of $y = \sin x$

Consider the values of $\sin\theta$ throughout one rotation on the unit circle. *IE, the y-coordinate*

Through what range of values does it vary?

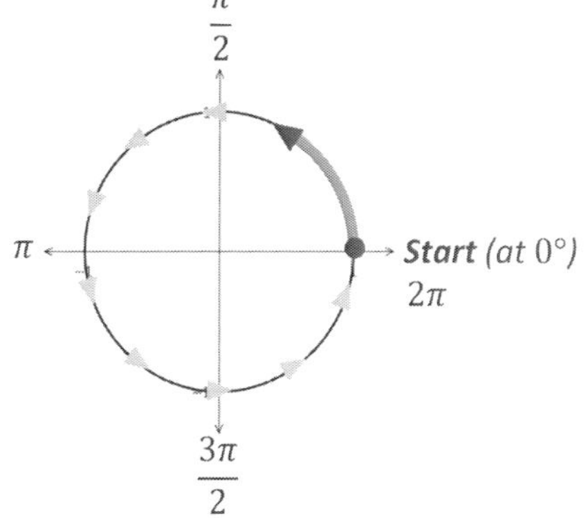

1 ➡ Complete the table of sine values below:

Angles within one unit circle rotation

θ	$\sin\theta$
0	
$\pi/6$	
$\pi/4$	
$\pi/3$	
$\pi/2$	
$3\pi/4$	
π	
$5\pi/4$	
$3\pi/2$	
$7\pi/4$	
2π	

Angles outside $[0, 2\pi]$

θ	$\sin\theta$
$5\pi/2$	
3π	
$7\pi/2$	
4π	
$-\pi/2$	

2 ➡ Complete the table of sine values below:

θ interval	Behavior of $\sin\theta$
From 0 to $\pi/2$	*Goes from ____ to ____*
From $\pi/2$ to π	*Goes from ____ to ____*
From π to $3\pi/2$	*Goes from ____ to ____*
From $3\pi/2$ to 2π	*Goes from ____ to ____*
From 2π to $5\pi/2$ *principal angle 0 to $\pi/2$*	*Goes from ____ to ____*
From $5\pi/2$ to 3π *$\pi/2$ to 3π*	*Goes from ____ to ____*

3 ➡ Use the tables above to plot the graph of $y = \sin x$, on the interval $[-\pi/2, 4\pi]$

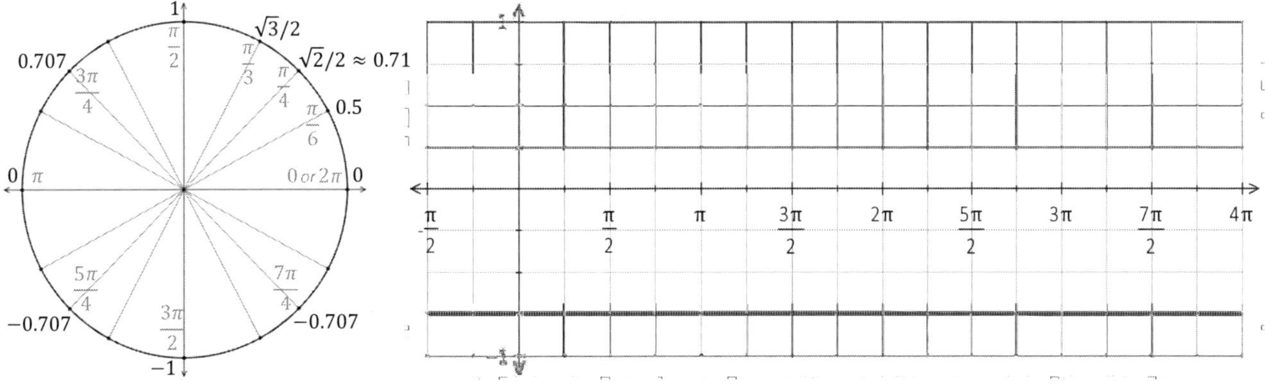

4 ➡ The **PERIOD** of the graph is defined as the horizontal length of one full cycle. State the period.

5 ➡ The **AMPLITUDE** is defined as the vertical distance to the max or the min, from a mid-line through the graph. State the amplitude.

The Graphs of the Primary Trigonometric Functions

The graph of $\mathbf{y = \sin x}$ oscillates as a wave between -1 and 1, repeating forever. As such, it is a **periodic function**, with a period of 2π or $360°$.

Below is the graph of $\mathbf{y = \sin x}$

Note that while this graph is defined for all real values of x, it is shown on the arbitrary window of $[-\pi, 2\pi]$ *(or 0 to 360)*

We can graph in RADIANS:

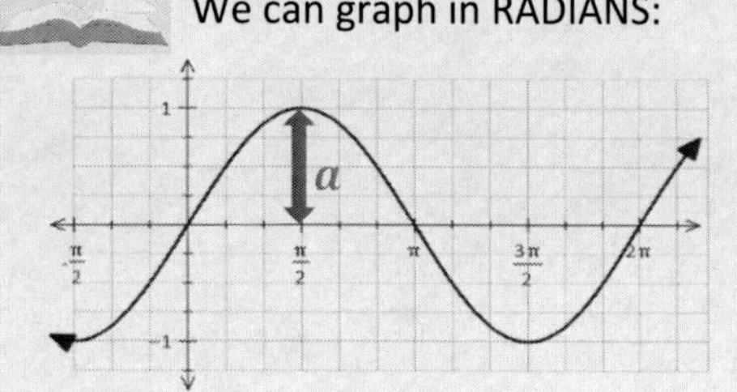

Graph characteristics:

Domain: $\{x \in \mathbb{R}\}$

Range: $[-1, 1]$

Amplitude $\mathbf{1}$

Period: $\mathbf{360°}$ or $\mathbf{2\pi}$

Or in DEGREES:

Same graph! (Different units / horizontal scale)

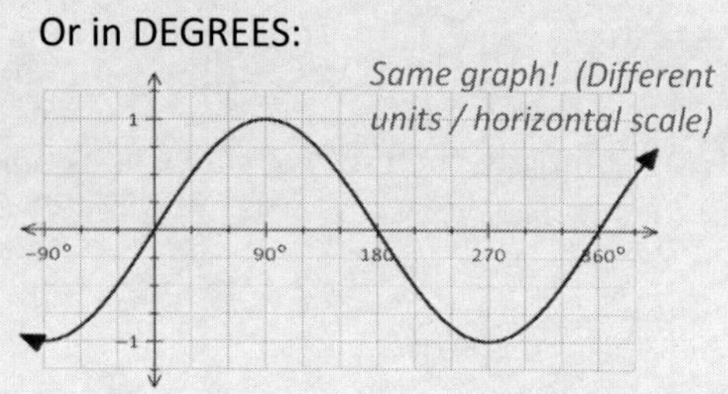

PERIOD: We can confirm the period of $y = \sin x$ (or any other ***sinusoidal function***) from the graph, noting the horizontal distance between *any* two corresponding points. *For example* we can note the distance:

from ***min*** to ***min***

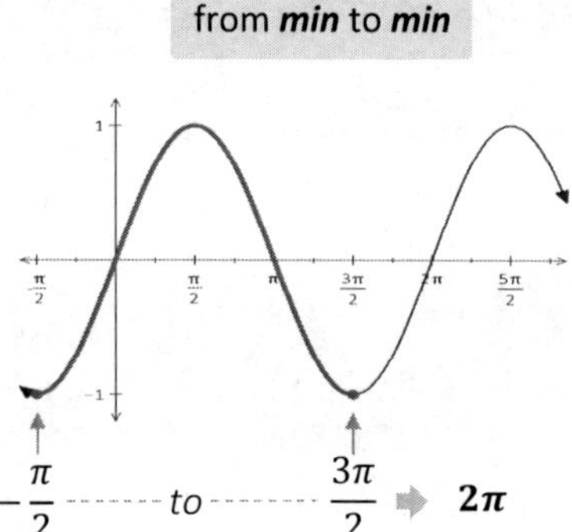

$-\frac{\pi}{2}$ to $\frac{3\pi}{2}$ ➡ $\mathbf{2\pi}$

or ***x-intercept*** to the next*

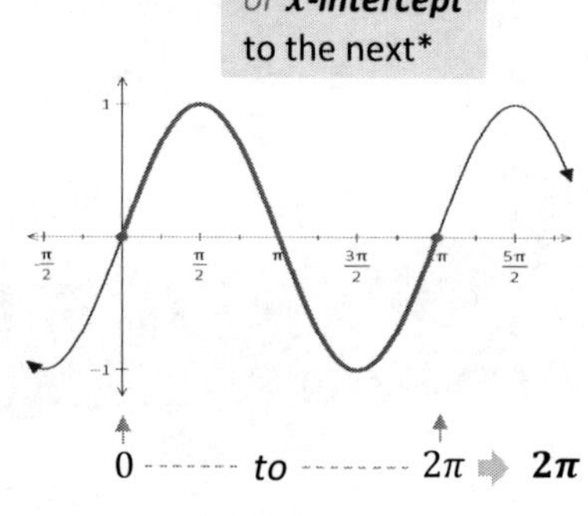

0 to 2π ➡ $\mathbf{2\pi}$

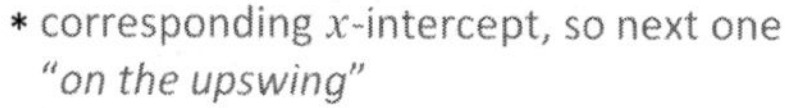

* corresponding x-intercept, so next one *"on the upswing"*

or from ***max*** to ***max***

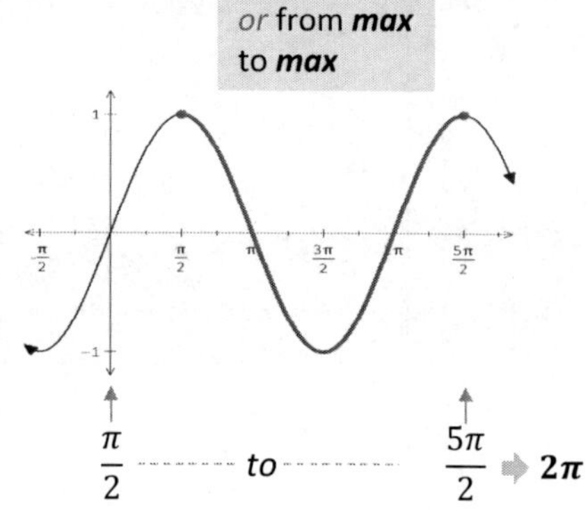

$\frac{\pi}{2}$ to $\frac{5\pi}{2}$ ➡ $\mathbf{2\pi}$

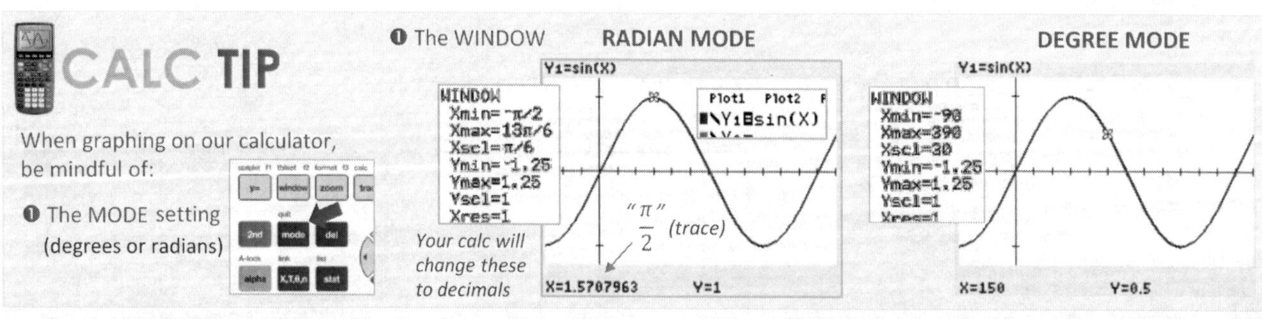

Class Example 6.41 *Vertically stretching the sine function*

Below is the graph of $y = \sin x$ on the interval $[-\frac{\pi}{2}, \frac{5\pi}{2}]$ **(a)** Use transformations to **sketch the graph** of $y = 2\sin x$.

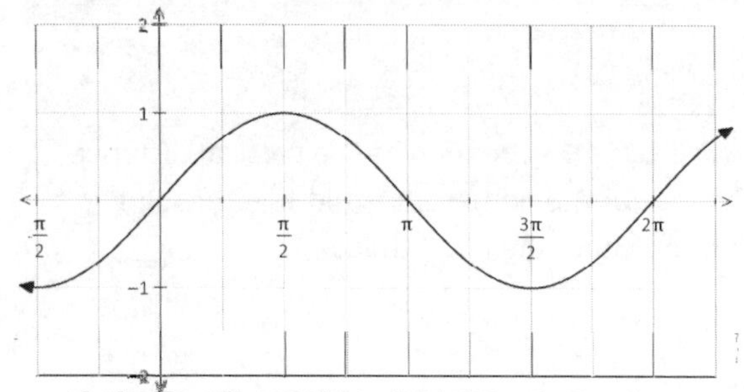

(b) State the following characteristics for the graph of $y = 2\sin x$:

Period	Amplitude	Domain	Range

Class Example 6.42 *Determining a sine function from the range*

A sinusoidal function that can be written in the form $y = a\sin x$ has a range of $[-3, 3]$. Sketch the graph of the function, and state the value of a to state the equation.

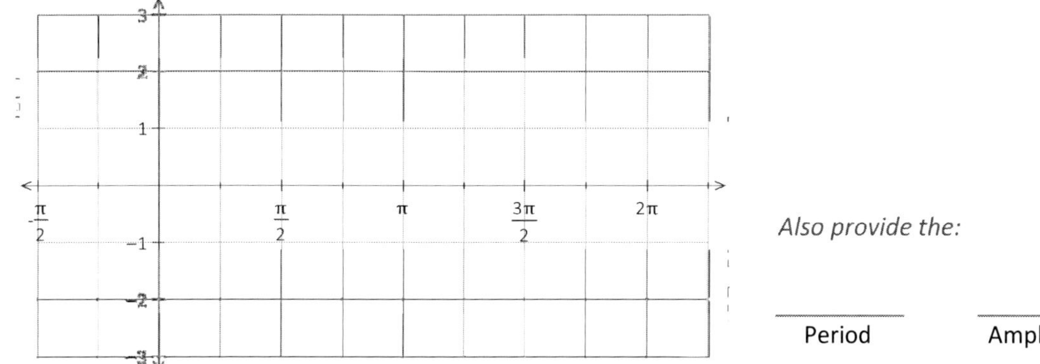

Also provide the:

______ Period ______ Amplitude

Investigation 2 - **The Graph of $y = \cos x$**

Consider the values of $\cos\theta$ throughout one rotation on the unit circle.
Through what range of values does $\cos\theta$ (the x-coordinate) vary?

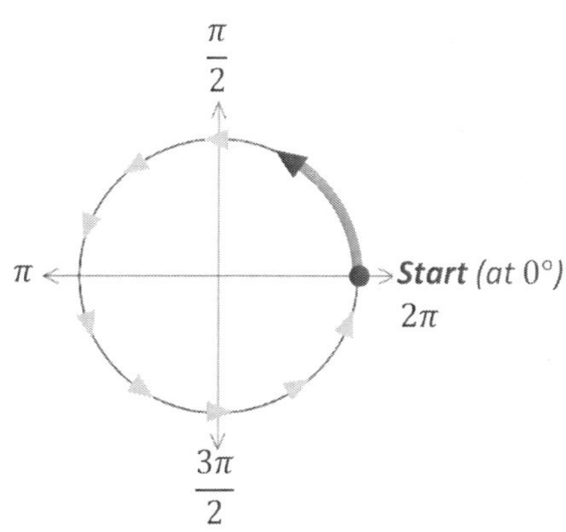

1 ➡ Complete the table of cos values below:

$\boldsymbol{\theta}$ **interval**	**Behavior of $\boldsymbol{\cos\theta}$**
From 0 to $^{\pi}/_{2}$	*Goes from ____ to ____*
From $^{\pi}/_{2}$ to π	*Goes from ____ to ____*
From π to $^{3\pi}/_{2}$	*Goes from ____ to ____*
From $^{3\pi}/_{2}$ to 2π	*Goes from ____ to ____*
From 2π to $^{5\pi}/_{2}$ *0 to π/2*	*Goes from ____ to ____*
From 0 to $-^{\pi}/_{2}$	*Goes from ____ to ____*

2 ➡ Use the table above, along with your graphing calculator, to sketch the graph of $\boldsymbol{y = \cos x}$; on $[-^{\pi}/_{2}, 3\pi]$:

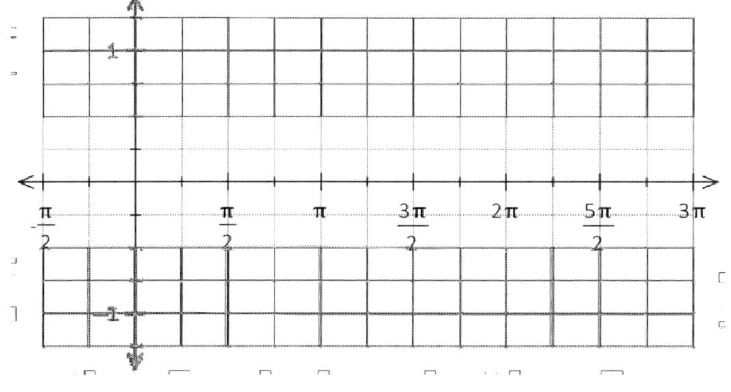

3 ➡ State the indicated characteristics for the graph of $y = \cos x$:

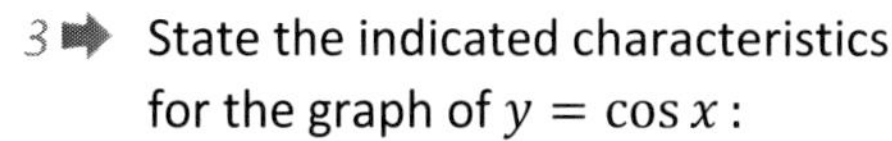

______ Period ______ Amplitude

______ Domain ______ Range

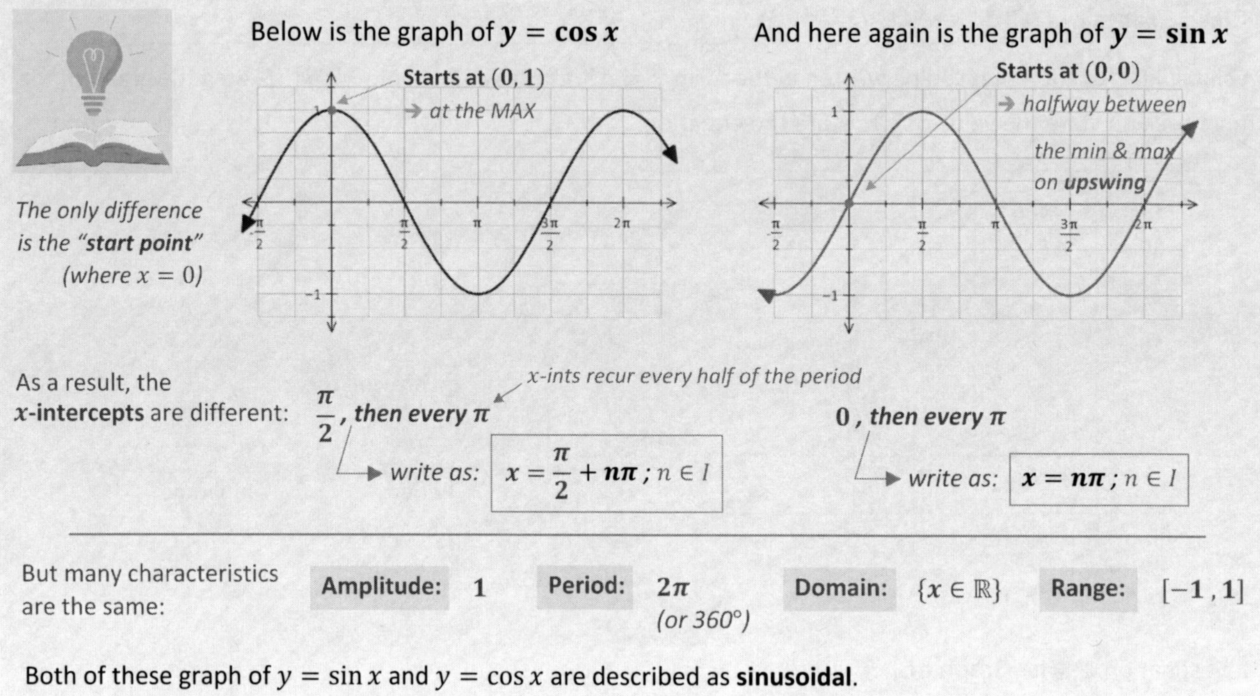

Applying Stretches - The Graphs of $y = a \sin bx$ / $y = a \cos bx$

We now consider the effects of vertically and horizontally **stretching** the graphs of $y = \sin x$ and $y = \cos x$. (As we already brushed on with examples 6.41 and 6.42)

Investigation 3 – **Amplitude:** The Graph of $y = a\sin x$ / $y = a\cos x$

Graphs ❶ and ❷ are obtained by vertically stretching the graph of $y = \sin x$, as shown:

The range of graph ❷ is $[-\frac{1}{2}, \frac{1}{2}]$

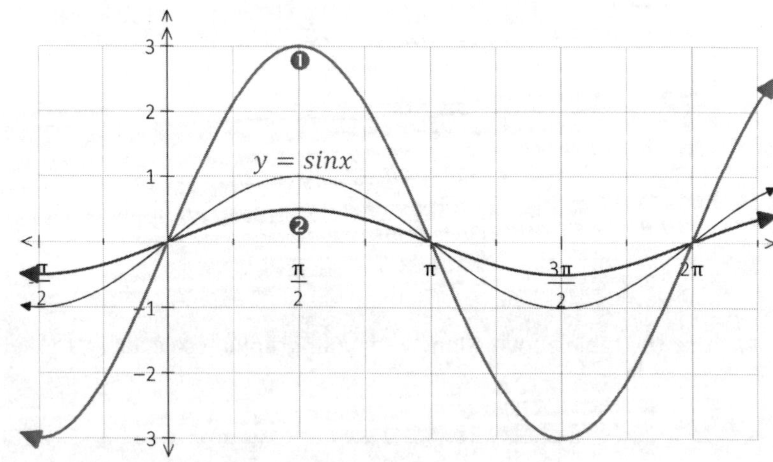

1 ➡ State an equation for the function each graph represents

2 ➡ Explain how the range relates to the equation of the function

The effect **of a** in the equation of
$y = a\sin x$ or $y = a\cos x$

Is a ***vertical stretch***, which affects the **amplitude** ➔

As such, the RANGE is $[-a, a]$

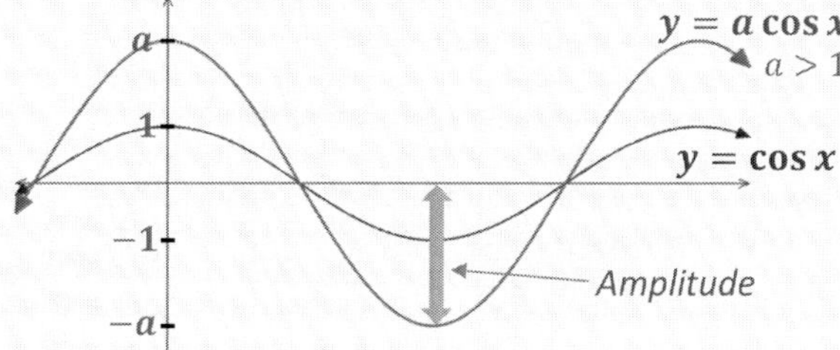

If $a < 0$, there is also a *vertical reflection* ➔

Note that amplitude is always positive, so:

Amplitude = $|a|$

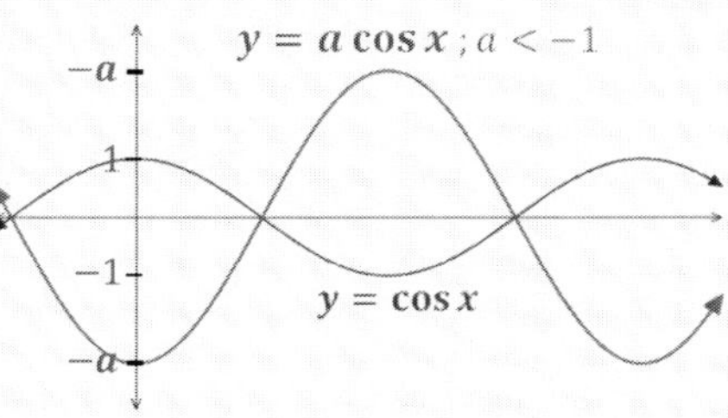

a is the vertical distance from the median line (here the x-axis) and the MAX or MIN.

Or use the formula: $a = \dfrac{max - min}{2}$

Class Example 6.43 *Determining the Amplitude of a Sinusoidal Function*

The following graphs represent functions that can each be expressed in the form $y = a\sin x$ or $y = a\cos x$.
For each; state the amplitude, determine an equation, and state the range.

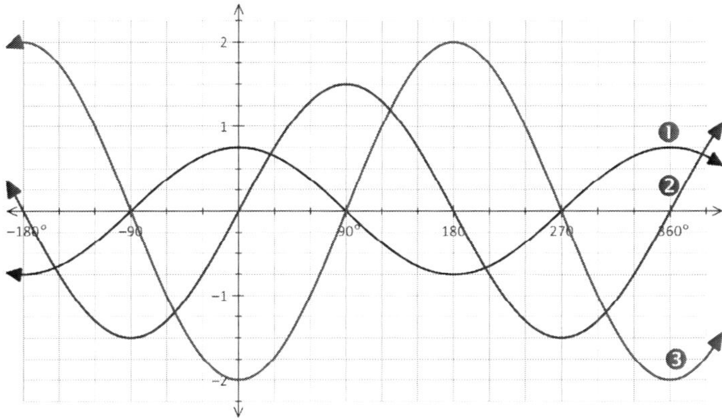

Class Example 6.44 *Determining Graph Characteristics Given a Function Equation*

Determine the domain, amplitude and range for each of the following sinusoidal functions.

(a) $y = 7\sin(x)$

(b) $y = -1.2\cos(x)$

Investigation 4 - **Period: The Graph of $y = \sin bx$ / $y = \cos bx$**

The graph of $y = \sin x$ is shown on the right.

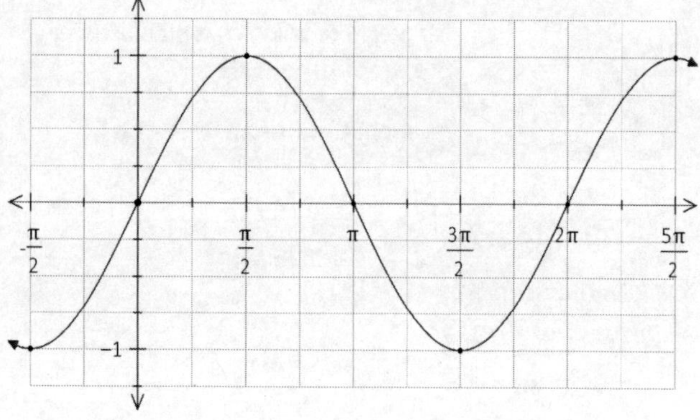

1 ➡ Construct a mapping rule that shows the transformation to $y = \sin 2x$. Transform all points indicated (•) to sketch on the same grid.

2 ➡ State the **period** of the graph of $y = \sin 2x$, in radians.

3 ➡ State the domain and amplitude of the graph of $y = \sin 2x$

4 ➡ State an expression that represents all x-intercepts on the graph of $y = \sin 2x$

The effect **of b** in the equation of $y = \sin bx$ or $y = \cos bx$

Is a ***horizontal stretch***, which affects the **period** ➔

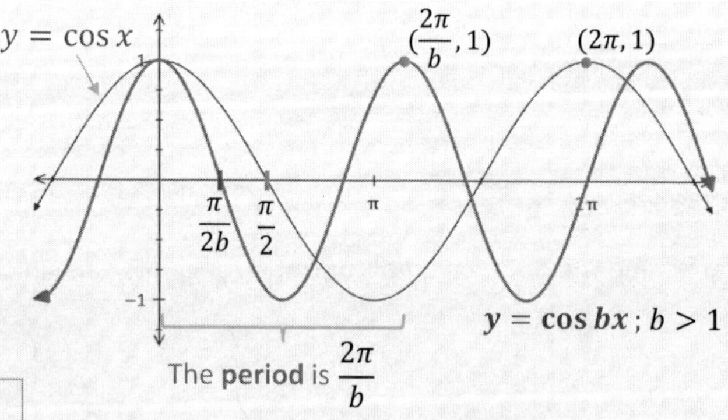

The **period** of the graph is the horizontal stretch factor, $\frac{1}{b}$ multiplied by the *standard period of* 2π

Period $= \frac{1}{b} \times 2\pi$ ➡ **Period** $= \frac{2\pi}{b}$ *or* $\frac{360°}{b}$

The x-intercepts of $y = \cos bx$ are at $\frac{1}{b} \times \frac{\pi}{2}$, then every $\frac{\pi}{b}$ afterward ➔ $x = \frac{\pi}{2b} + \frac{\pi}{b}n\,;\; n \in I$

The first x-int of $y = \cos x$ — *Half of the period of $y = \cos bx$*

Note: *substitute 180° for π when in degrees*

The x-intercepts of $y = \sin bx$ are at $\frac{1}{b} \times 0$, then every $\frac{\pi}{b}$ afterward ➔ $x = \frac{\pi}{b}n\,;\; n \in I$

The first x-int of $y = \sin x$

Worked Example Determine the range, period and x-intercepts for each of the following sinusoidal functions.

(a) $y = 6\sin\left(\frac{1}{5}x\right)$ *Provide period & x-ints in radians*

(b) $y = -2\cos\left(\frac{3}{2}x\right)$ *Provide period & x-ints in degrees*

Sol.: **(a)** The amplitude is 6, so the range is $[-6, 6]$

For **period**, use period $= \frac{2\pi}{b}$ *or...* $\frac{360°}{b}$

Period $= \frac{2\pi}{\frac{1}{5}}$ ➡ $= 2\pi \times \frac{5}{1}$ ➡ **Period $= 10\pi$**

On the basic sine curve, x-ints occur *"every π"*

Here a horizontal stretch, factor of 5 has been applied, so x-ints occur *"**every 5π**"*

So, x-intercepts are at $\boldsymbol{x = 5\pi n}\,;\, n \in I$

(b) The amplitude is 2, so the range is $[-2, 2]$

Period $= \frac{360°}{\frac{3}{2}}$ ➡ $= 360° \times \frac{2}{3}$ ➡ **Period $= 240°$**

On the basic cosine curve, **x-intercepts** occur at $90°$, then "every 180°" (that is, every half-period)

Here a horizontal stretch, factor of $2/3$ has been applied, so x-ints are at $90° \times \frac{2}{3}$, then every $180° \times \frac{2}{3}$

So, x-intercepts are at $\boldsymbol{x = 60° + 120°n}\,;\, n \in I$

Class Example 6.45 *Determining Graph Characteristics*

Determine the period, range, and x-intercepts for each of the following sinusoidal functions.
For (a) and (b), state the period and expression for x-intercepts in radians. **For (c), use degrees.**

(a) $y = 5\sin(4x)$ *radians*

(b) $y = 1.2\cos\left(\frac{2}{3}x\right)$ *radians*

(c) $y = -0.5\sin\left(\frac{4}{3}x\right)$ *degrees*

(a)	(b)	(c)
i *Period:*	i *Period:*	i *Period:*
ii *Range:*	ii *Range:*	ii *Range:*
iii *x-ints:*	iii *x-ints:*	iii *x-ints:*

note The **default** angle measure for trigonometric function graphs is **radians**.
Unless directed otherwise, assume you are to provide the period / domain / intercepts in radians!

Class Example 6.46 *Determining the Period, Range, and x-intercepts of a Sinusoidal Function*

For the sinusoidal function $\boldsymbol{y = -18.5\sin(0.5236x)}$, state (i) the period (round to the nearest whole number), (ii) the range, and (iii) an expression representing the x-intercepts. (round to the nearest tenth)

We'll now consider how to determine the **period** directly **from the graph**, and then use the period to obtain b.

Suppose we'd like to determine the period of a sinusoidal graph, such as this:

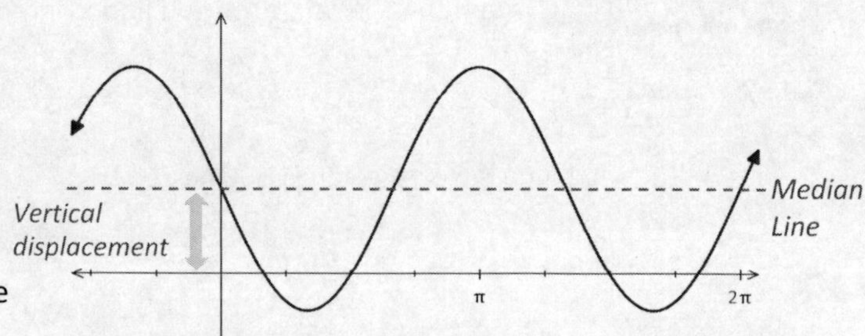

When a vertical translation, or **vertical displacement**, is applied - this distance is represented by the **median line**. (as shown)

We'll further explore vertical displacement (d) in the next section. For now, ***let's focus on the period!***

➔ *And look at the options!* A few of the ways we can **determine the period** is to note the horizontal distance from:

MAX** to **MAX
or MIN to MIN

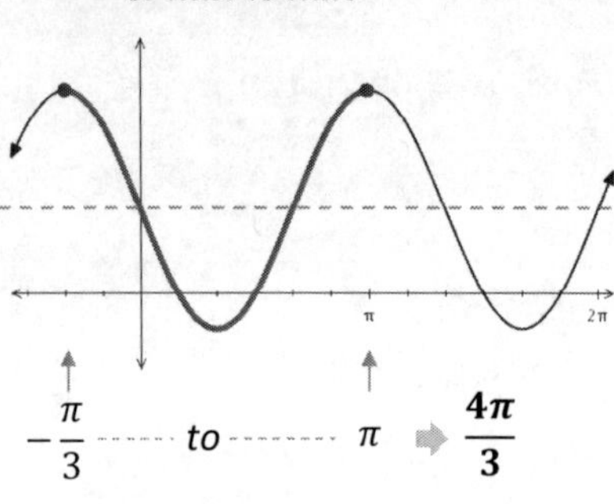

$-\frac{\pi}{3}$ to π ➔ $\frac{4\pi}{3}$

That is, $-\frac{2\pi}{6}$ *to* $\frac{6\pi}{6}$ ➔ $\frac{8\pi}{6}$

x-intercept** to next matching* **x-int.
*(*here both are on the "downswing")*

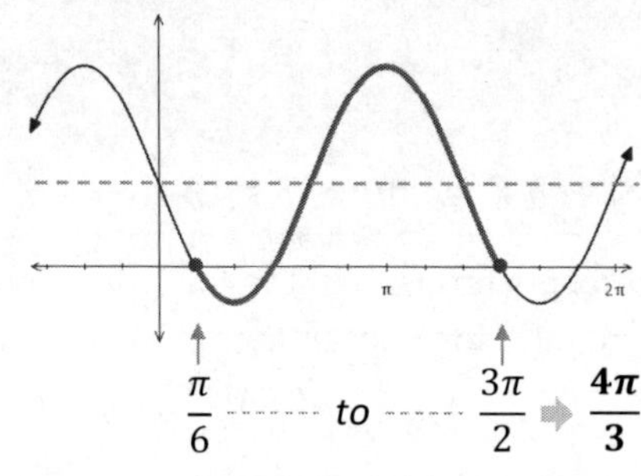

$\frac{\pi}{6}$ to $\frac{3\pi}{2}$ ➔ $\frac{4\pi}{3}$

That is, $\frac{\pi}{6}$ *to* $\frac{9\pi}{6}$ ➔ $\frac{8\pi}{6}$

*Or **any two corresponding points** on the graph!*

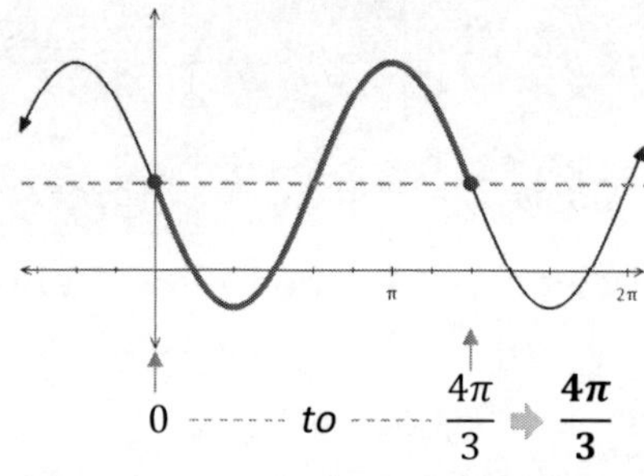

0 to $\frac{4\pi}{3}$ ➔ $\frac{4\pi}{3}$

That is, 0 *to* $\frac{8\pi}{6}$ ➔ $\frac{8\pi}{6}$

So we know that the **period** of the above graph is $\frac{4\pi}{3}$.

To obtain b in the equation
$y = a\sin(bx) + d$ or
$y = a\cos(bx) + d$, use:

$$b = \frac{2\pi}{period} \quad \text{or,} \quad \frac{360°}{period}$$

For the graph above,

$$b = \frac{2\pi}{\frac{4\pi}{3}} \Rightarrow b = 2\pi \times \frac{3}{4\pi} \Rightarrow b = \frac{3}{2}$$

Class Example 6.47 *Determining the Period of a Sinusoidal Function*

Determine the **period** and corresponding "b" value for graph below, to state an equation in the form $y = a\sin bx$ or $y = a\cos bx$.
(From which basic graph is this graph stretched?)

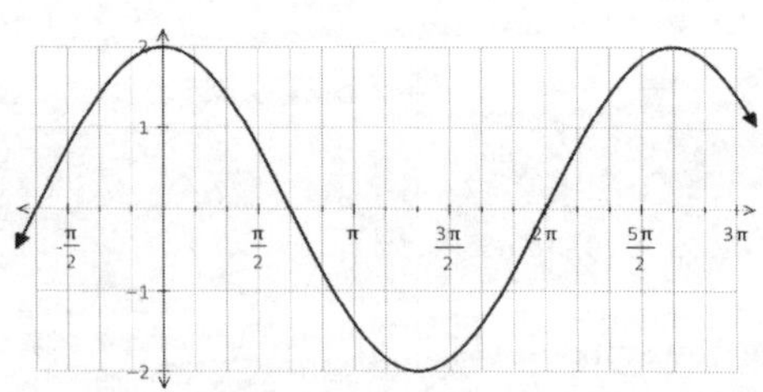

Worked Example Determine the amplitude and period for the following sinusoidal function.

Then, state the a and b values for an equation in the form $y = a\sin bx$ or $y = a\cos bx$.

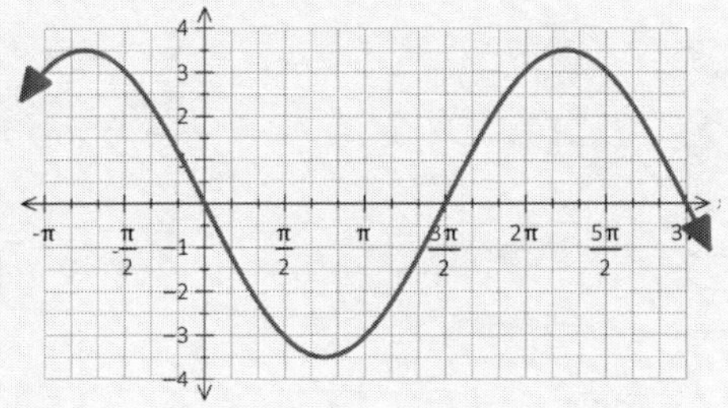

Sol.:

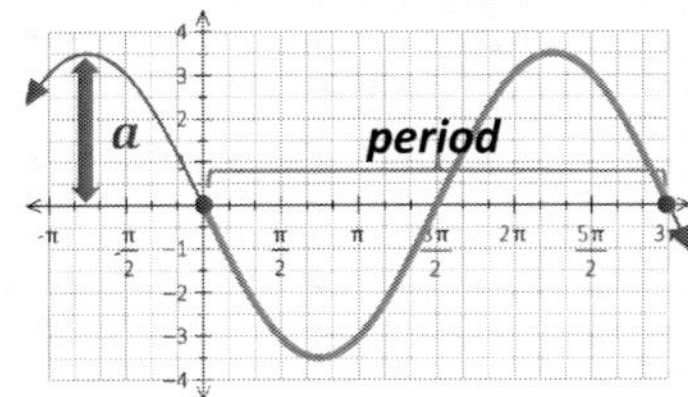

As shown, the amplitude is 3.5. However, this curve is a *vertical reflection* of the basic sine curve. *The point (0, 0) is halfway on the* ***downswing*** *'stead of the upswing.*

So $a = -3.5$.

Also as shown, the period is 3π. Note that here the easiest way to identify the period is to use the *x-intercepts on the downswing.*

So, $b = \dfrac{2\pi}{period} \Rightarrow = \dfrac{2\pi}{3\pi} \Rightarrow b = \dfrac{2}{3}$

Equation:

$y = -3.5\sin\left(\dfrac{2}{3}x\right)$

Class Example 6.48 *Determining the Period and Equation of a Sinusoidal Function*

The following graphs represent functions that can each be expressed in the form $y = \sin bx$ or $y = \cos bx$.

Determine an equation for each, and state the period and an expression for the x-intercepts. (In the angular mesuare given by the graph)

Graph ❶:	*Graph ❷:*
Period:	*Period:*
Equation:	*Equation:*
x-intercepts:	*x-intercepts:*

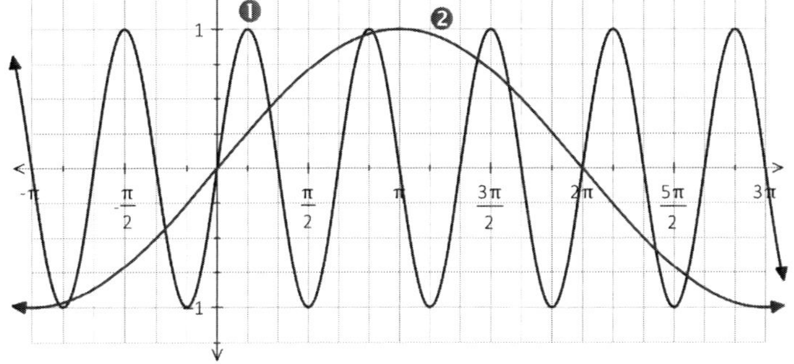

Graph ❸:	*Graph ❹:*
Period:	*Period:*
Equation:	*Equation:*
x-intercepts:	*x-intercepts:*

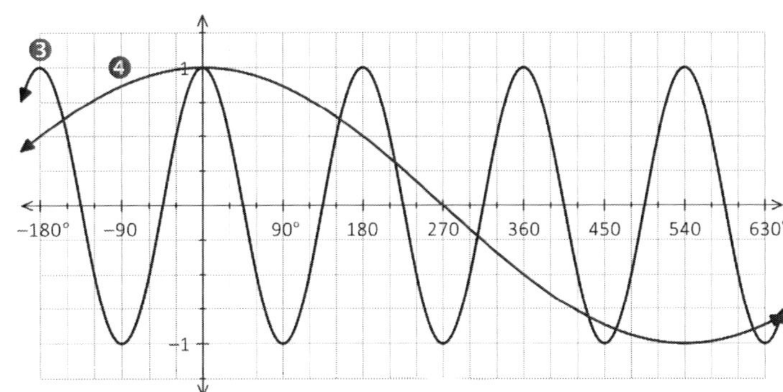

The graph of $y = \tan x$ is shown on the right.

As $\tan x = \dfrac{\sin x}{\cos x}$, the domain is restricted where $\cos x = 0$.

That is, at $\dfrac{\pi}{2}$, *then again every* π

Domain: $\{x \neq \frac{\pi}{2} + n\pi, x \in \mathbb{R}\}$; $n \in I$

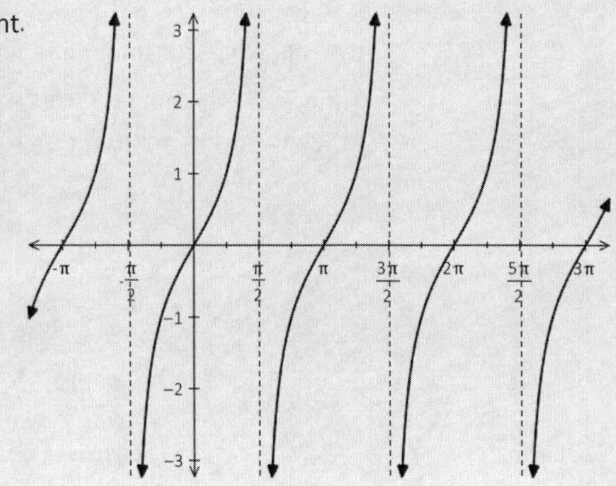

Related to the domain restrictions, the graph has an infinite number of **vertical asymptotes**, which occur at: $x = \frac{\pi}{2} + n\pi$; $n \in I$

The graph has x-intercepts at 0, *then again every* π

x-intercepts: $x = n\pi$; $n \in I$

The **period** of the graph is π *(or 180°)*

As there are no max / mins, there is no amplitude.

Worked Example

Determine the period, domain, range, and expressions for the vertical asymptotes and x-intercepts for $y = 3\tan(2x)$. State the period and x-intercepts in radians.

Sol.: The "standard" tan graph period is π, here there is a horiz. str. factor of 1/2, so **period** is $\frac{\pi}{2}$ ← $\frac{1}{2} \times \pi$

The standard tan graph has restrictions every $\pi/2$, there is a horiz. str. factor of ½, so **domain** is $\{x \neq \frac{\pi}{4} + \frac{\pi}{2}n \,;\, x \in I\}$ ← $\frac{1}{2} \times \frac{\pi}{2}$

The **range** is $\{y \in \mathbb{R}\}$ *(the vert. stretch of "3" has no effect)* **V.A.s** correspond to domain, at $x = \frac{\pi}{4} + \frac{\pi}{2}n; n \in I$

x-intercepts on $y = \tan x$ graph occur every π. Here, apply horiz. stretch of ½, so at $x = \frac{\pi}{2}n$; $n \in I$

Class Example 6.49 *Determining Graph Characteristics of a Tangent Function*

Determine the period, domain, range, and expressions for the vertical asymptotes and x-intercepts for each of the following functions. State the period and expression for x-intercepts in radians.

(a) $y = \tan(3x)$

Period:

Domain:

Range:

V.A.s:

x-intercepts:

(b) $y = 2\tan(\frac{1}{2}x)$

Period:

Domain:

Range:

V.A.s:

x-intercepts:

6.5 Further Transformations of Sinusoidal Functions

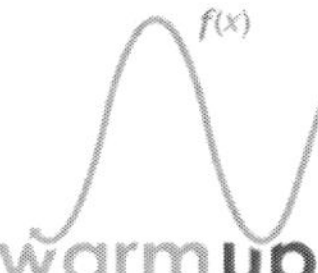

Investigation 1 – **Vertical Displacement**

1 ➡ The graph of $f(x) = \cos x$ is shown below. Use transformations to sketch the graph of $\boldsymbol{g(x) = 2\cos x + 3}$

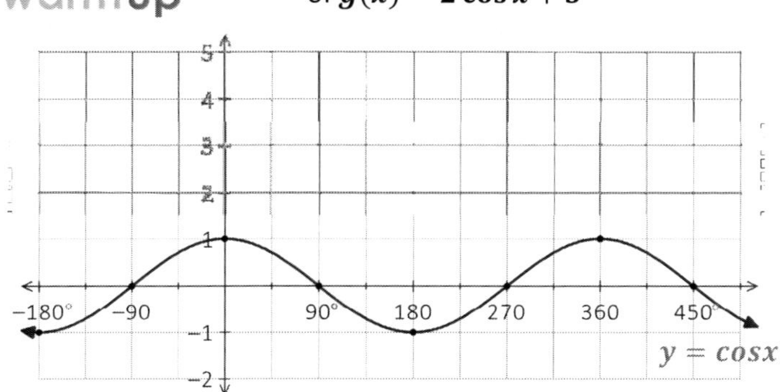

Complete the mapping rule to transform each indicated point (•) on the graph of $y = \sin x$ *Keep it going!*

$(x, y) \rightarrow$

$(-180°, -1) \rightarrow$

2 ➡ On the graph of $f(x) = \cos x$ the x-axis serves as the **median line**, which is a *horizontal line that cuts directly through the middle of a sinusoidal curve*. **Sketch** the dotted median line for your transformed graph $y = g(x)$.

3 ➡ State the range of the resulting function.

4 ➡ Explain how the range of $y = a\cos x + d$ relates to the a and d values.

Vertical Translations – Effect of d in the graph of $y = a\sin(bx) + d$ or $y = a\cos(bx) + d$

So far we've seen how $\boldsymbol{a}$ is a vertical stretch , which affects the **amplitude**, and $\boldsymbol{b}$ represents a horizontal stretch (of $1/b$), which affects the **period**.

Next we'll examine **translations**, which cause the vertical or horizontal shifting of graphs, with no change to the *shape or orientation*.

In the graph of $\boldsymbol{y = a\sin x + d}$, (or $y = a\cos bx + d$)

- $\boldsymbol{d}$ can be visualized as the distance from the **median line** to the x-axis, and
- $\boldsymbol{a}$ the distance from the median line to the max or min point (the **amplitude**).

➔ *The effect of* $\mathbf{d}$ *is a* ***vertical translation****, while* $\boldsymbol{a}$ *is a* ***vertical stretch*** *about the* x*-axis.*

Therefore, the **range** is $[-a + d, a + d]$

As $\boldsymbol{d}$ represents the median value of the function, it can be found using the formula:

$$d = \frac{\text{max + min}}{2}$$

Whereas for $\boldsymbol{a}$ we can use:

$$a = \frac{\text{max - min}}{2}$$

Note: By "max" and "min" we refer to y-coordinates!

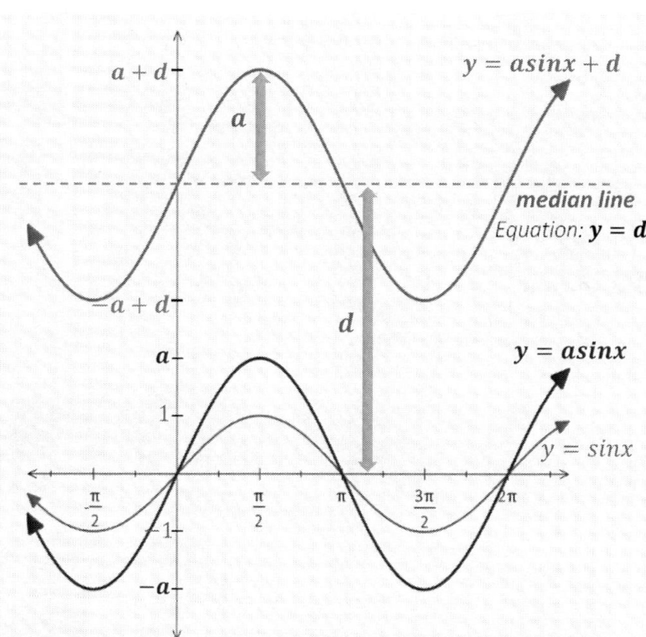

note: Draw your median line so that the distance to the max is the same as the distance to the min!

Worked Example Use transformations to sketch the graph of $y = 4\cos x - 3$.

Solution: Refer to the basic COSINE graph on the right.

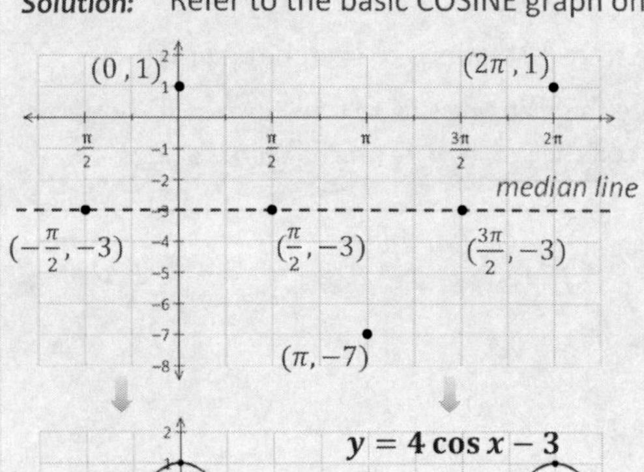

$y = 4\cos x - 3$

Step ❶ $y = 4\cos x \boxed{-3}$

Draw the median line at $y = -3$

Step ❷

Transform pts left to right....

$(x, y) \rightarrow (x, 4y - 3)$

$(-\frac{\pi}{2}, 0) \rightarrow (-\frac{\pi}{2}, -3)$ ← $4(0) - 3$

$(0, 1) \rightarrow (0, 1)$ ← $4(1) - 3$

⋮

Step ❸

Connect points in a smooth curve!

The Basic SINE graph:

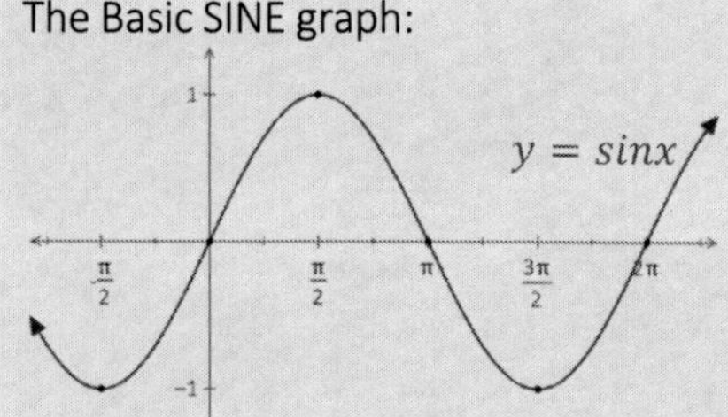

The Basic COS graph:

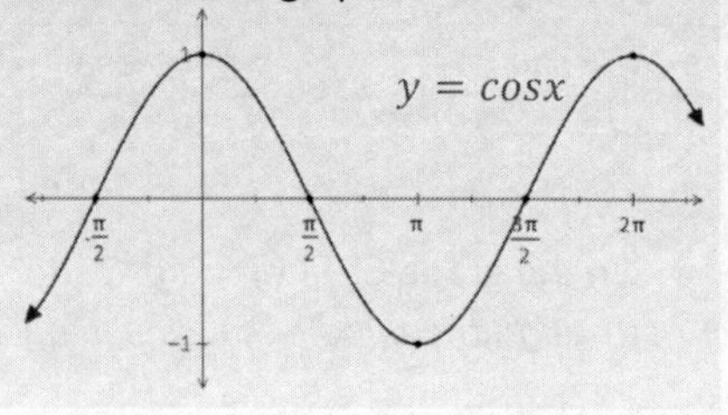

Class Example 6.51 *Determining Graph Characteristics of a Sinusoidal Function*

Use transformations to sketch each of the following graphs and state the indicated characteristics.

(a) $y = -2\cos x + 1$

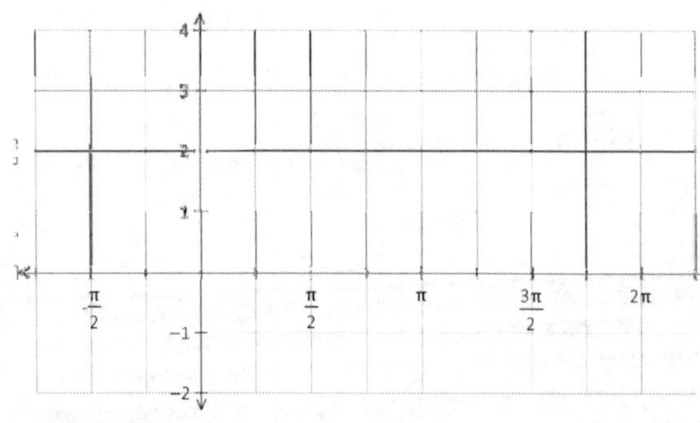

i *Mapping Rule:*

ii *Amplitude*

iii *Period*

iv *Vertical Displacement*

v *Range*

(b) $y = 3\sin(2x) - 1$

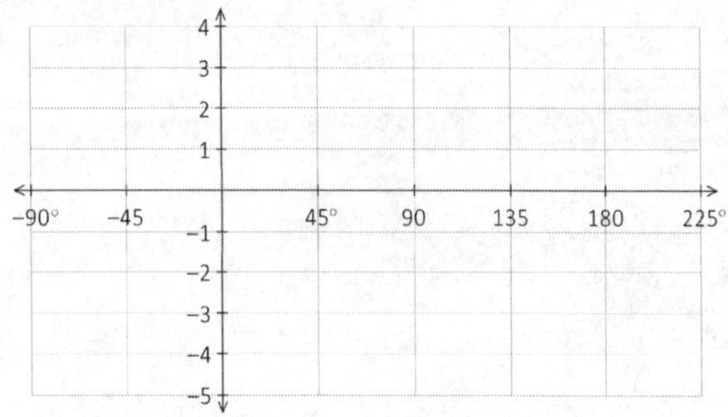

i *Mapping Rule:*

ii *Amplitude*

iii *Period*

iv *Vertical Displacement*

v *Range*

Class Example 6.52 *Determining Graph Characteristics of a Sinusoidal Function*

Without graphing, determine the indicated graph characteristics for each of the following sinusoidal functions:

(a) $y = -2.5\sin(\frac{4}{5}x) + 3$

i *Amplitude:*

ii *Vertical Displacement:*

iii *Range:*

iv *Period:* *in degrees*

(b) $y = 0.21\sin(12x) - 0.47$

i *Amplitude:*

ii *Vertical Displacement:*

iii *Range:*

iv *Period:* *in radians*

Class Example 6.53 *Determining an Equation of a Sinusoidal Function*

Each of the following graphs represent functions that can either be written in the form $\boldsymbol{y = a\sin(bx) + d}$ or $\boldsymbol{y = a\cos(bx) + d}$. Determine one equation for each graph.

(a)

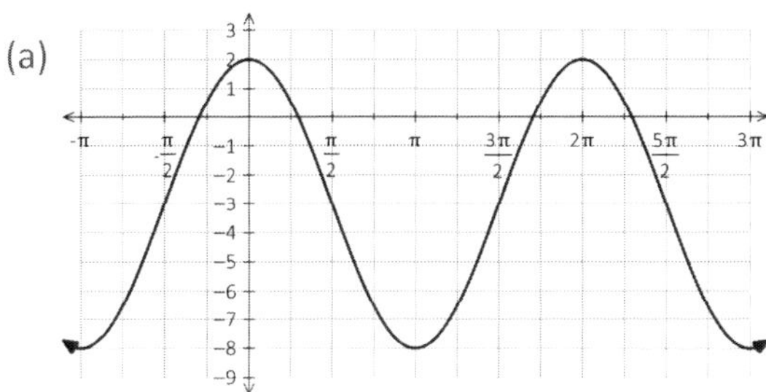

(b)

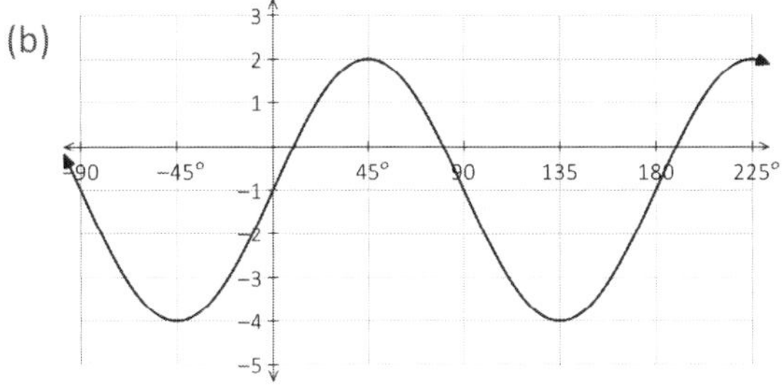

(c)

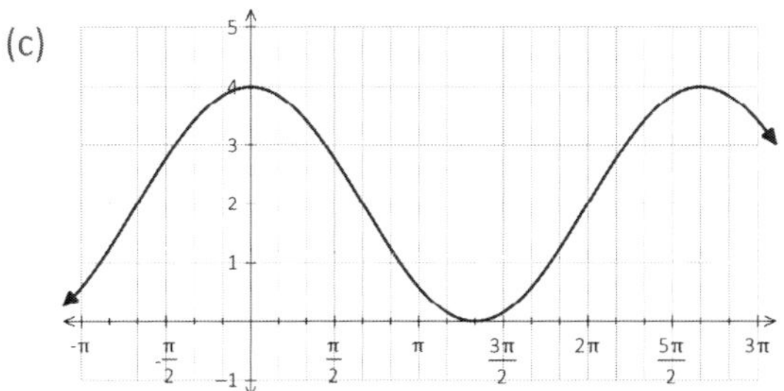

Horizontal Translations – Effect of c on $y = a\sin[b(x-c)]$ or $y = a\cos[b(x-c)]$

Investigation 2 – **Horizontal Phase Shift**

Using the basic graphs provided, sketch each indicated transformed function on the same grid. Verify using your graphing calculator. *Remember to first match the window and set the mode appropriately!*

1 ➡ $y = \cos(x - 60°)$

➡ **Describe** the transformation on the graph of $y = \cos x$

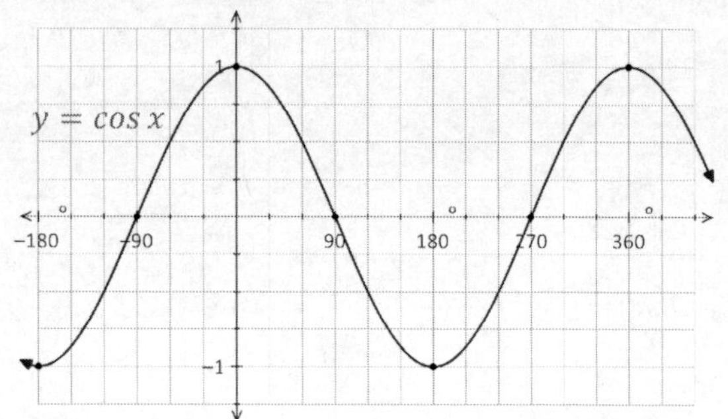

➡ Complete the **mapping rule** from the graph of $y = \cos x$ below, to transform all indicated points (•) on the graph *(keep it going!)*. **Sketch.**

$(x, y) \rightarrow$

$(-180°, -1) \rightarrow$

$(-90°, 0) \rightarrow$

2 ➡ $y = \sin(x + 135°)$

➡ **Describe** the transformation on the graph of $y = \sin x$

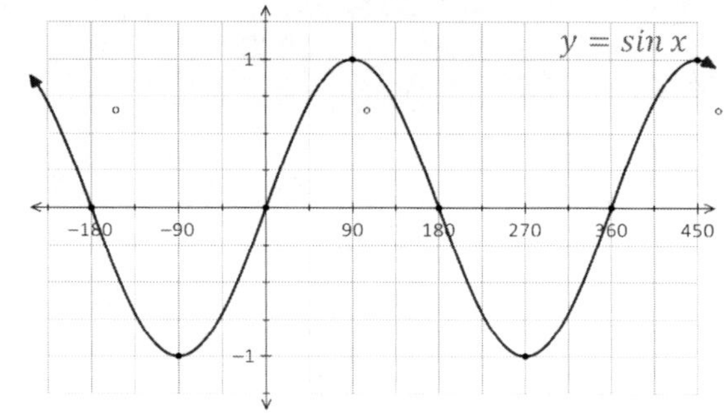

➡ Provide a **mapping rule** from the graph of $y = \sin x$ below, to transform all indicated points (•) on the graph. Sketch.

$(x, y) \rightarrow$

3 ➡ $y = \sin(x - \frac{2\pi}{3})$

➡ **Describe** the transformation.

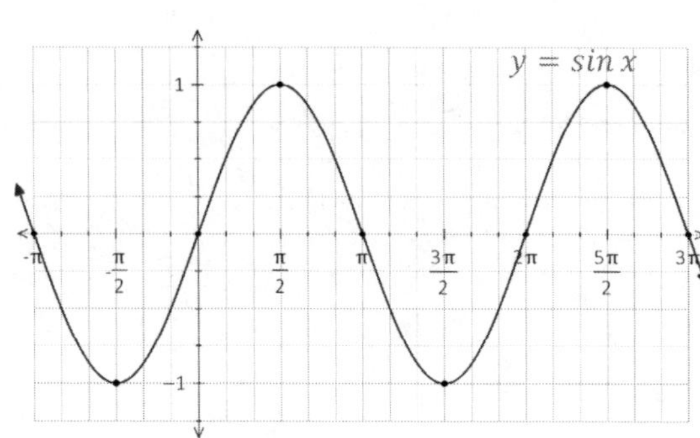

➡ Provide a **mapping rule** from the graph of $y = \sin x$ below, to transform all indicated points (•) on the graph. Sketch.

$(x, y) \rightarrow$

For a function $y = a\sin[b(x-c)]$ or $y = a\cos[b(x-c)]$,

c represents the horizontal translation – in this context known as the **horizontal phase shift**.

The parameters a, b, or d are the same whether a graph is described as a **sine** or **cosine** function. *These parameters don't care who they're transforming!*

☞ However "c" does care! When determining the value of c from a sinusoidal function graph, *we must consider if it's for a sine or cosine function.*

The basic **cosine** graph: $y = \cos x$

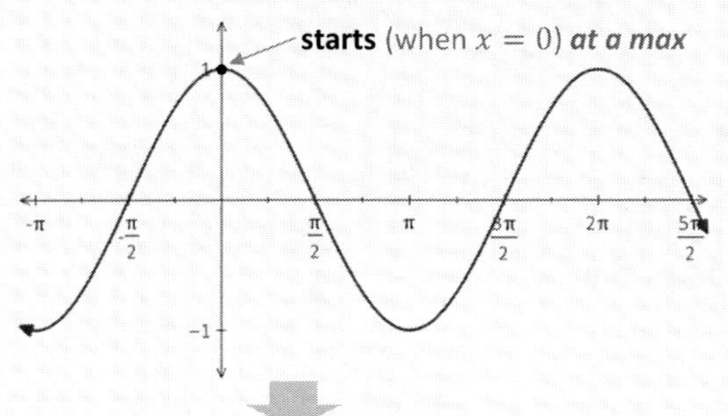

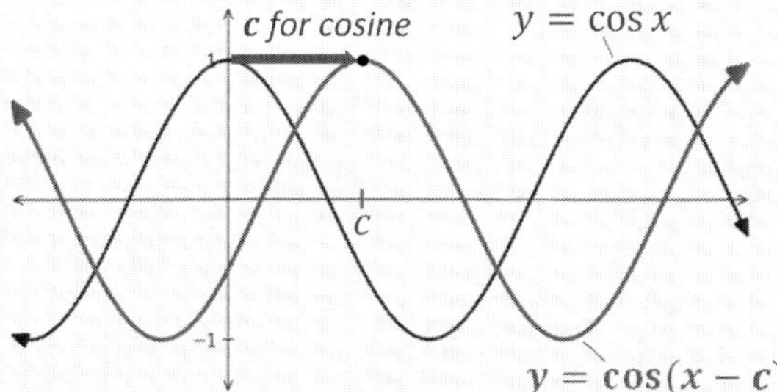

For cosine: c is the horizontal distance from the y-axis to the closest max.

The basic **sine** graph: $y = \sin x$

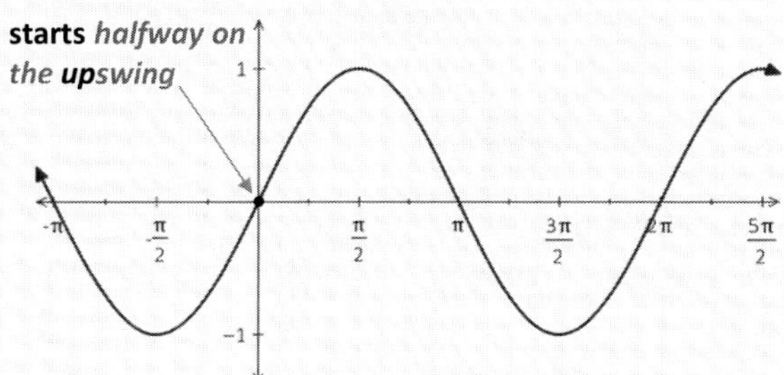

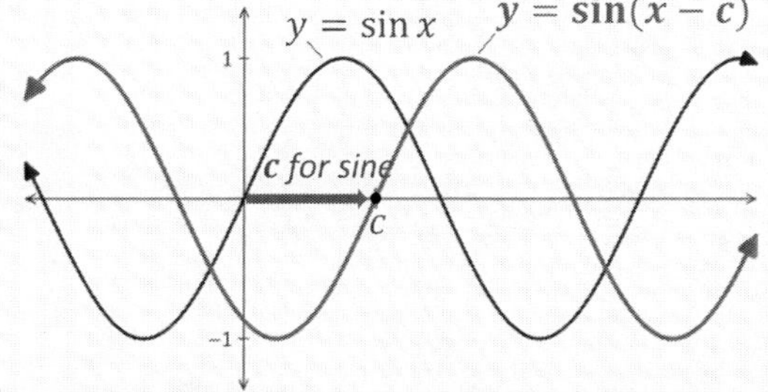

For sine: c is the horizontal distance from the y-axis to the closest point *on the median line - on the upswing.*

We **note that** with the concept of horizontal phase shift – there are no more "sine curves" or "cosine curves".

➔ *Any sinusoidal curve can be written using either a sine or cosine function!*

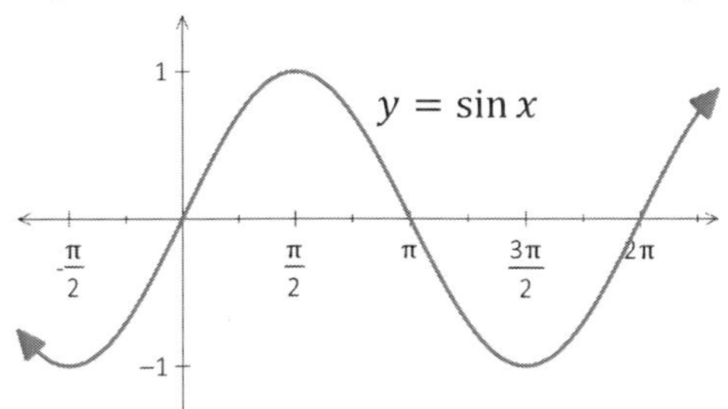

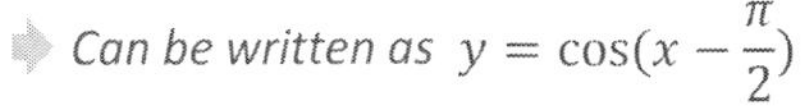
Can be written as $y = \cos(x - \frac{\pi}{2})$

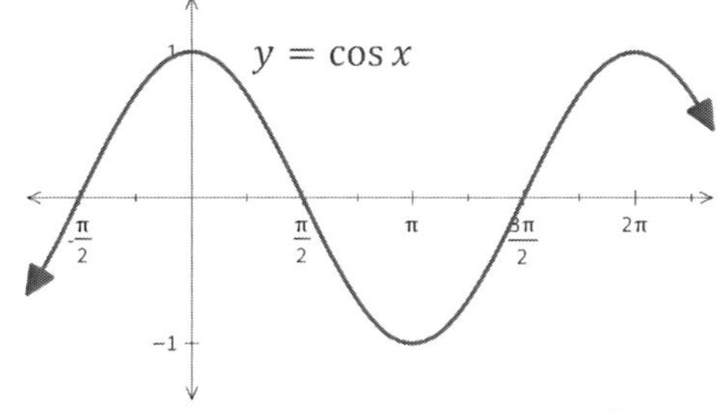

Can be written as $y = \sin(x + \frac{\pi}{2})$

Verify this on your graphing calc:

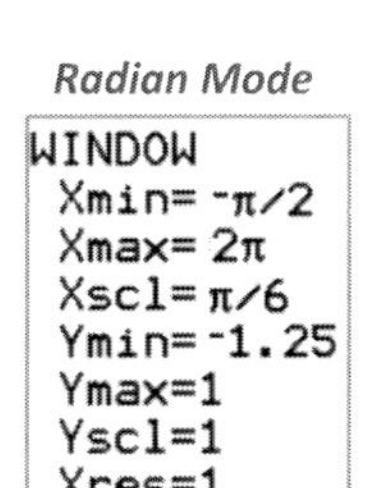

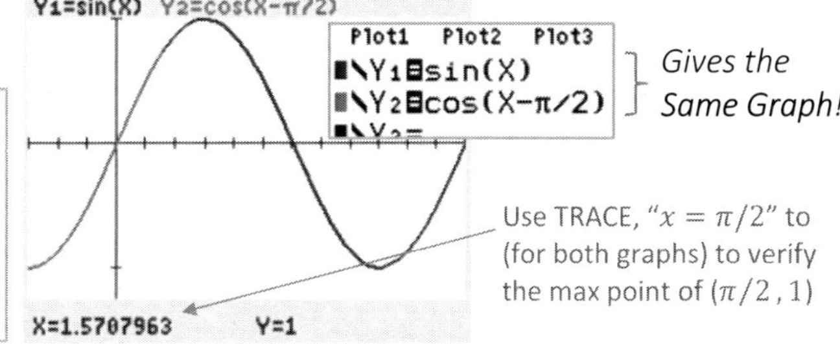

Class Example 6.54 *Determining the Period and Phase Shift of a Sinusoidal Function*

For each of the following sinusoidal functions, describe the (i) period and (ii) horizontal phase shift from the basic graph of $y = \sin x$ or $y = \cos x$:

(a) $y = \cos(4x - 180°)$

(b) $y = \sin(\frac{1}{2}x - \frac{\pi}{6})$

Class Example 6.55 *Using Transformations to Sketch a Sinusoidal Function*

Given each related basic graph, sketch the indicated function on the same grid. Then, provide an alternate sinusoidal equation of *minimum phase shift* for the transformed function, as directed.

(a) i *Sketch* $y = \cos(x + \frac{\pi}{3})$ *below:*

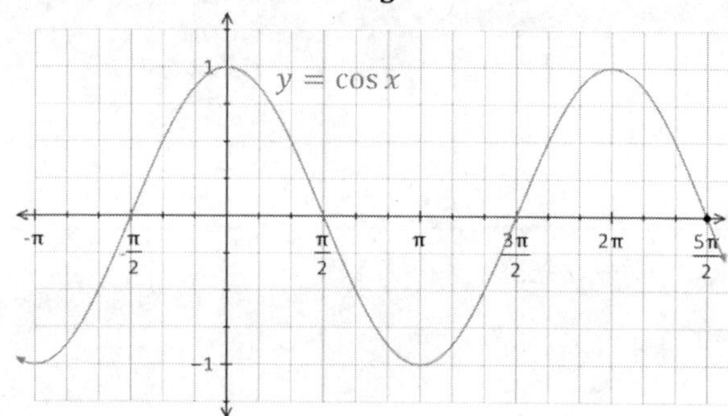

ii *Determine an alternate equation - as a sine function:*

(b) i *Sketch* $y = 2\sin(x - 135°)$ *below:*

ii *Determine an alternate equation - as a cosine function:*

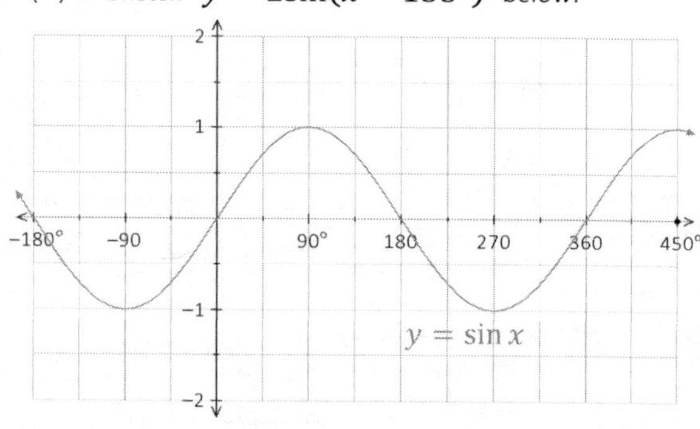

(c) i *Sketch* $y = \sin(x + \frac{\pi}{2}) + 1$ *below:*

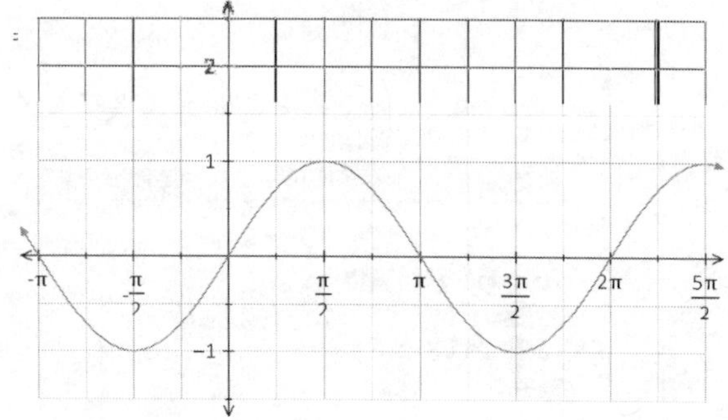

ii *Alternate equation - as a cosine function:*

As we are dealing with *periodic functions*, there are **infinite options** for expressing the "c" value.

For example, consider sinusoidal function $f(x)$ on the right, which is a basic graph with a horizontal phase shift applied.

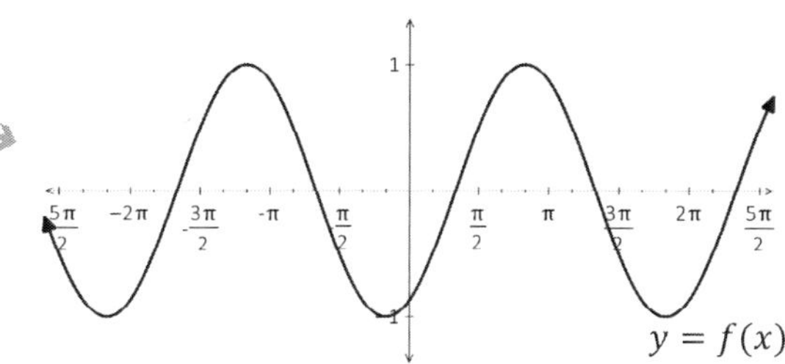

★ A few of the options to express $f(x)$ as a COS function:

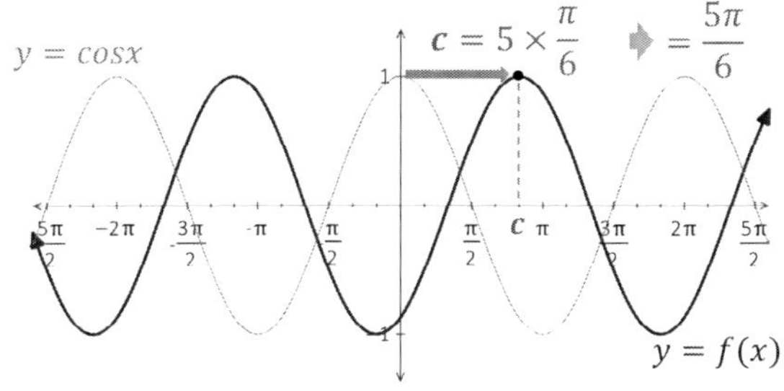

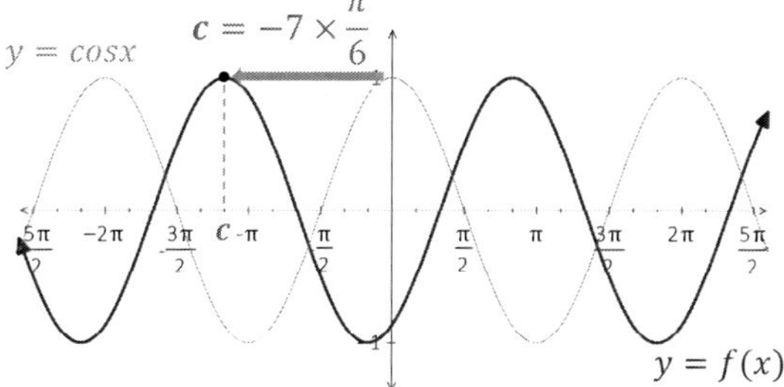

+ **COS** eqn. option 1: (minimum phase shift) $\boldsymbol{f(x) = \cos(x - \frac{5\pi}{6})}$

+ **COS** eqn. option 2: (not minimum phase shift) $\boldsymbol{f(x) = \cos(x + \frac{7\pi}{6})}$

+ COS opt. 3: $f(x) = -\cos(x + \frac{\pi}{6})$

Can you see why? With a vertical reflection, we can use the distance from the min. *Can you spot more options for "c", with or without a reflection?*

★ And here are a few options to express the same graph a SIN function:

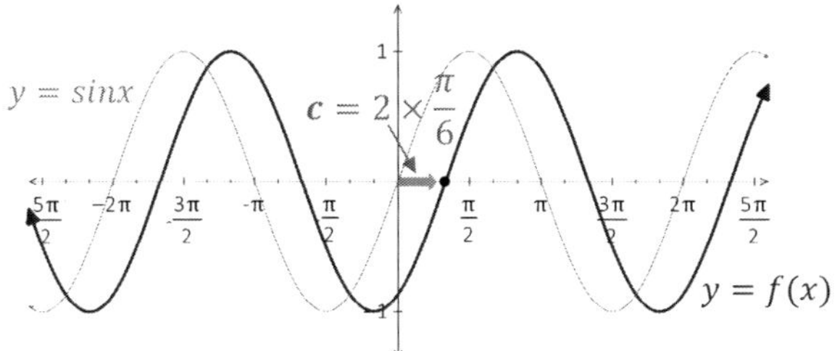

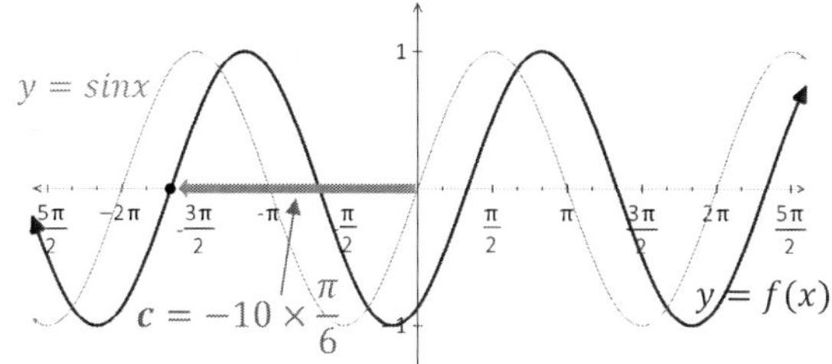

+ **SINE** eqn. option 1: (minimum phase shift) $\boldsymbol{f(x) = \sin(x - \frac{\pi}{3})}$

+ **SINE** eqn. option 2: (not minimum phase shift) $\boldsymbol{f(x) = \sin(x + \frac{5\pi}{3})}$

+ SINE opt. 3: $f(x) = -\sin(x + \frac{2\pi}{3})$ *Can you spot more options for "c", with or without a reflection?*

Worked Example Determine a function in the form of both $y = \sin(x - c)$ **and** $y = \cos(x - c)$ for each of the following graph.

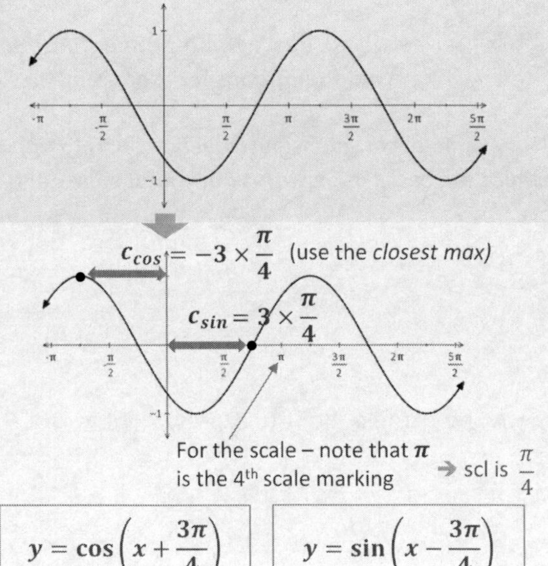

Solution:

For a **cosine** equation, determine the horizontal phase shift using the **max**

For a **sine** equation, use the point **halfway on the upswing**

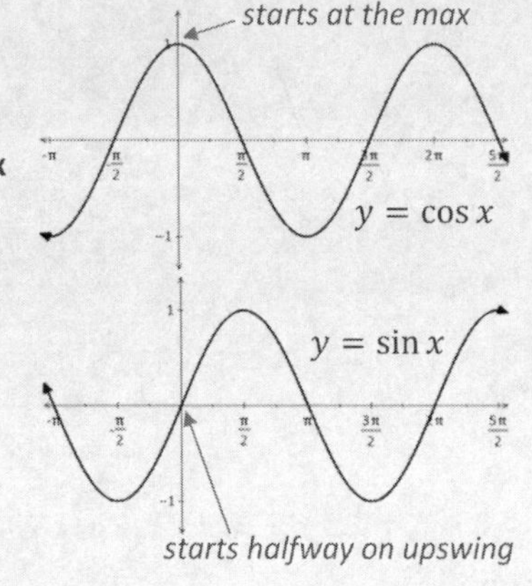

$$y = \cos\left(x + \frac{3\pi}{4}\right) \qquad y = \sin\left(x - \frac{3\pi}{4}\right)$$

Check on your calc! *Both of these equations result in the same graph.*

CALC TIP

```
Plot1 Plot2 Plot3
\Y1=cos(X+3π/4)
\Y2=sin(X-3π/4)
```

← **Graph both** equations together

Match the window to what's given →

```
WINDOW
 Xmin=-π
 Xmax=5π/2
 Xscl=π/4
 Ymin=-1
 Ymax=1
 Yscl=1
 Xres=1
```

To **confirm**, use trace... $x = \pi/4$ We confirm there is a MIN there! →

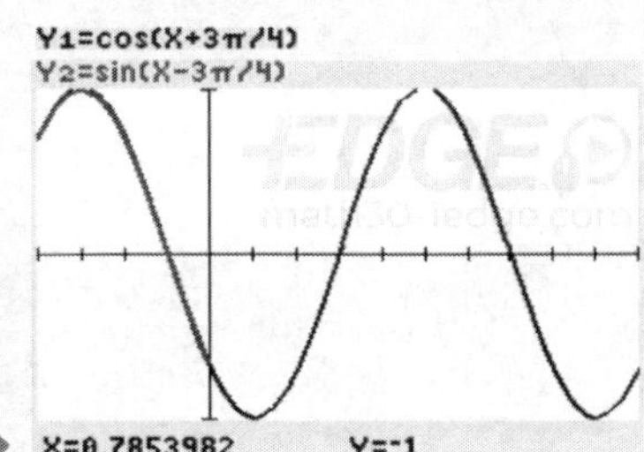

← We should see only **one curve**, one on top of the other

Class Example 6.56 *Determining the Phase Shift / "c" Value of a Sinusoidal Function*

Determine equations for each function, in the form $y = \sin(x - c)$ and $y = \cos(x - c)$, as described below.

(a)

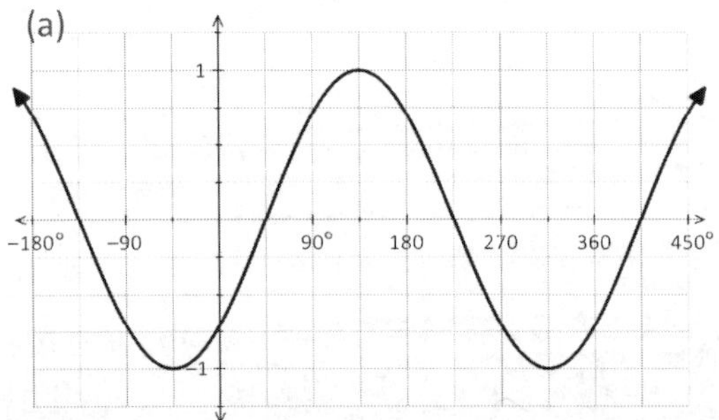

Minimum Phase shift equations, with $a > 0$:

i *COS:*

ii *SIN:*

Alternate COS equation of minimum phase shift with $a < 0$

iii:

(b)

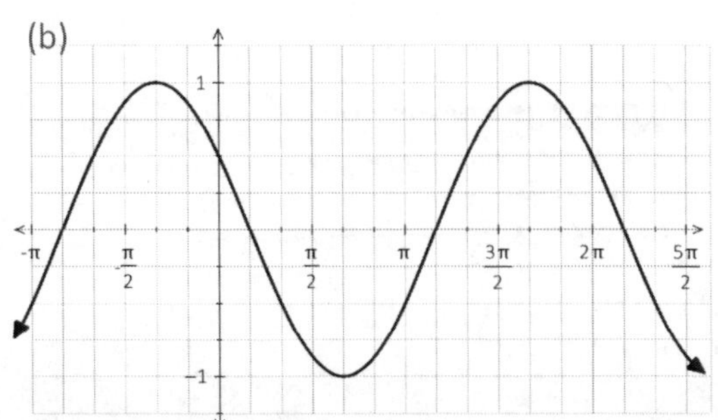

Minimum Phase shift equations, with $a > 0$:

i *COS:*

ii *SIN:*

Alternate SIN equation of minimum phase shift with $a < 0$

iii:

Worked Example Determine a function in the form of both $y = a\sin[b(x-c)] + d$ **and** $y = a\cos[b(x-c)] + d$ for the graph on the right.

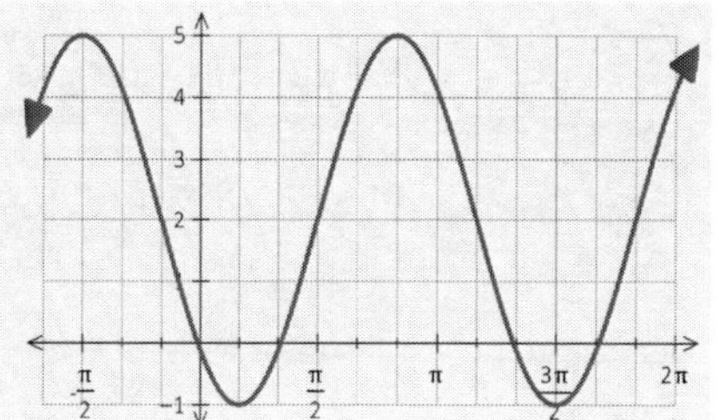

Solution:

Start with the median line, equivalent to the $\boldsymbol{d}$ value.

From there we can see the amplitude is $\mathbf{3}$.

$$d = \frac{max + min}{2} = \frac{5 + (-1)}{2} \Rightarrow \boldsymbol{d = 2}$$

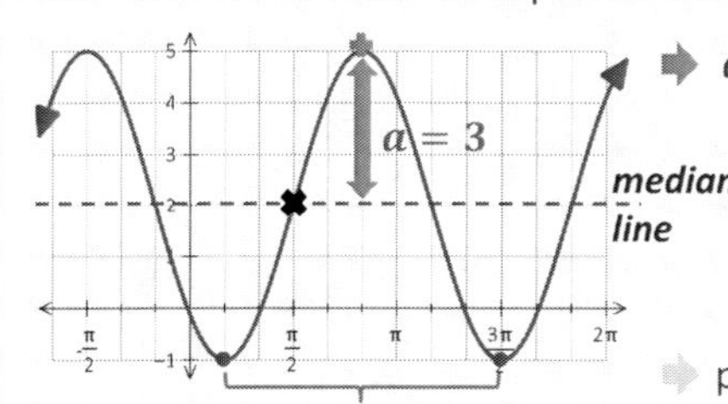

PERIOD is **8** scale markings, each of which is $\frac{\pi}{6}$

$\Rightarrow \boldsymbol{a = 3}$

c_{cos} is represented by ✚ sign $\Rightarrow c_{cos} = \frac{5\pi}{6}$

c_{sin} is represented by ✖ sign $\Rightarrow c_{sin} = \frac{\pi}{2}$

$\Rightarrow$ period $= \frac{8\pi}{6} = \frac{4\pi}{3}$

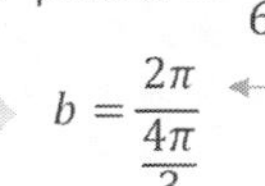

$\Rightarrow b = \frac{2\pi}{\frac{4\pi}{3}}$ ← *use 360° if period is in degrees*

$\Rightarrow \boldsymbol{b = \frac{3}{2}}$

Equations:

$$y = 3\sin\left[\frac{3}{2}\left(x - \frac{\pi}{2}\right)\right] + 2$$

$$y = 3\cos\left[\frac{3}{2}\left(x - \frac{5\pi}{6}\right)\right] + 2$$

Alternative equations with smaller HPS *(and vertical reflection):*

$y = -3\sin\left[\frac{3}{2}\left(x + \frac{\pi}{6}\right)\right] + 2$ *or* $y = -3\cos\left[\frac{3}{2}\left(x - \frac{\pi}{6}\right)\right] + 2$

CALC TIP

Match the window to what's given →

```
WINDOW
 Xmin=-4π/6
 Xmax=2π
 Xscl=π/6
 Ymin=-1
 Ymax=5
 Yscl=1
 Xres=1
```

Graph in RADIAN mode (or, change the window to degrees, and use DEGREE mode)

```
Plot1 Plot2 Plot3
\Y1=3sin(3/2(X-π/2))+2
\Y2=3cos(3/2(X-5π/6))+2
```

↑ **Graph both** equations together

To **confirm,** use trace... $x = 5\pi/6$ *We confirm there is a MAX there!* →

```
Y1=3sin(3/2(X-π/2))+2
X=2.6179939     Y=5
```

← We should see only **one curve**, one on top of the other

Class Example 6.57 *Determining an Equation of a Sinusoidal Function*

Determine a function in the form $y = a\sin[b(x-c)] + d$ **and** $y = \cos[b(x-c)]$ for the graph below.

(a) Use a minimum phase shift for each, where $a > 0$.

(b) For the cosine curve only – use a minimum phase shift, where $a < 0$.

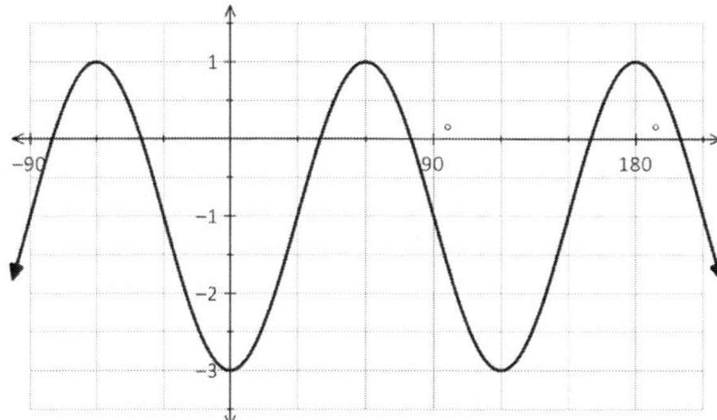

Visual Summary Of Parameters in a Sinusoidal Function

For $y = asin[b(x-c)] + d$ or $y = asin[b(x-c)] + d$

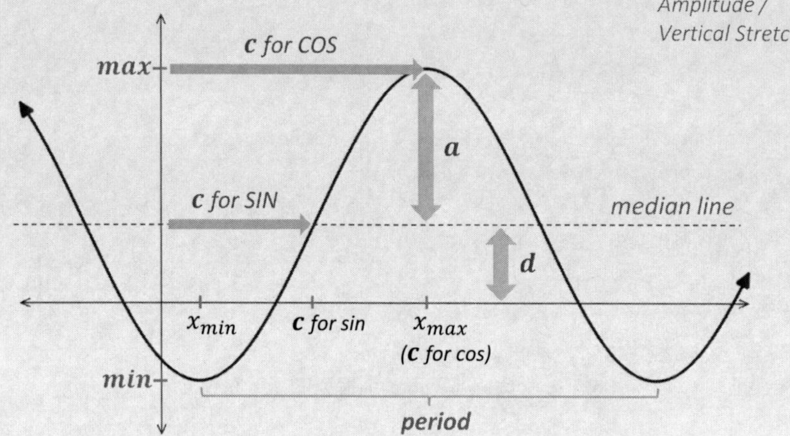

Not on your formula sheet:*

Amplitude / Vertical Stretch: $|a| = \dfrac{max - min}{2}$ *Reflection in x-axis if $a < 0$*

$d = \dfrac{max + min}{2}$ *Vertical displacement*

Horizontal stretch by a factor of $^1/_b$, so period is $^1/_b \times 2\pi$, or:

$$|b| = \frac{2\pi}{period} \text{ or } \frac{360°}{period}$$

Re-arranges to: $period = \dfrac{2\pi}{|b|}$

c for **COS** $= x_{max}$

c for **SIN**** $= \dfrac{x_{min} + x_{max}}{2}$

*Note that none of these formulas are needed to derive the parameters a, b, c, and d – they can all be obtained visually or by using your previous knowledge of transformations. However while not essential these formulas may be helpful!

** For the **c** for sine formula, note on the visual that c is right in the middle of the x-coord of the min and the x coord of the max. (on the upswing of the curve) Also note its relationship to the formula for "d", which uses y-coordinates instead of x.

Class Example 6.58 *Determining an Equation of a Sinusoidal Function – All Parameters*

Determine a function in the form of **both** $y = a\sin[b(x-c)] + d$ and $y = \cos[b(x-c)]$ for each of the following graphs. For the horizontal phase shift, c :

(a) Use the minimum phase shift for both the sine and cosine equation, where $a > 0$.

(b) For the *sine equation only* – use the minimum phase shift, where $a < 0$.

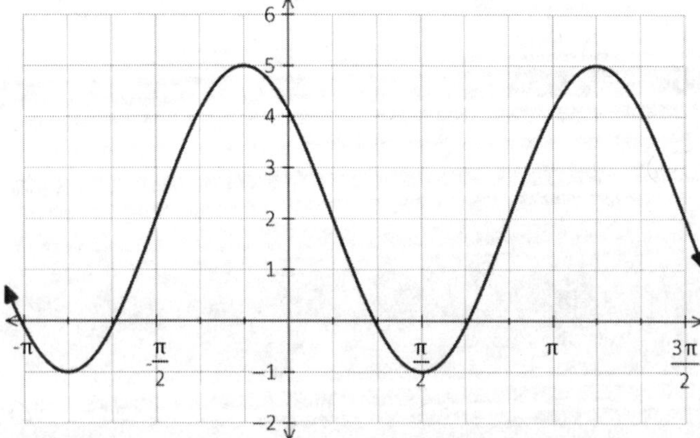

6.6 Modelling Real-World Applications with Sinusoidal Functions

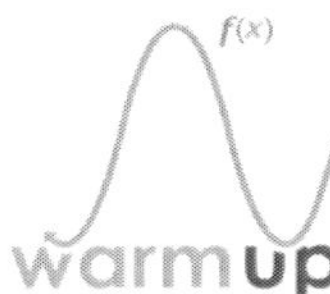

A rock is stuck in a bicycle tire that has a diameter of 40cm. The tire is rotated so that it completes one rotation every 4 seconds. At $t = 0$ the rock is at the lowest position.

A sinusoidal equation that models the height of the rock (in cm) after t seconds can be written in the from $\boldsymbol{h(t) = a\cos[b(x-c)] + d}$ or $\boldsymbol{h(t) = a\sin[b(x-c)] + d}$.

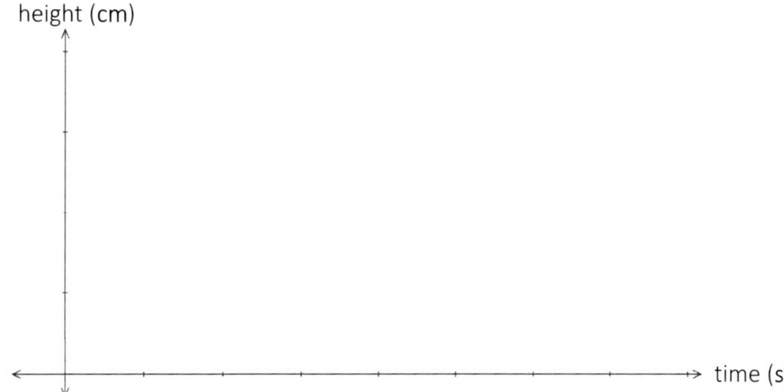

1 ➡ Provide a scale for each axis to sketch a graph showing the height of the rock over two complete rotations.

2 ➡ Determine the following sinusoidal function graph characteristics, as they apply to the path of the rock.

Amplitude: | Vertical Displacement: | Period: | "b" value:

3 ➡ Using a minimum phase shift, state an equation that models the height of the rock, $h(t)$, as a:

i sine equation, with $a > 0$

ii cosine equation, with $a > 0$

iii cosine equation, with $a < 0$

4 ➡ Use one of the equations to determine the height of the nail after 4.5 seconds. (Round to the nearest cm)

Sinusoidal functions can be used to model many real-world scenarios with periodic behavior or wave characteristics. Examples are height involving circular motion (the warm-up above, Ferris wheels), daily temperature or time of sunrise / sunset in a region throughout the year, height of water during tides, etc.

When modelling real-world scenarios we exclusively use **radian mode**.

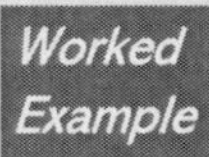

The depth of water, $d(t)$ metres, in a seaport can be approximated by the function $d(t) = 3.5\sin[0.164\pi(t - 1.5)] + 13.4$,where t is in the time in hours.

(a) Sketch a graph of the function over two full cycles.

(b) Use the equation to determine the depth of water after 8 hours. *Round to the nearest tenth*

(c) Use technology to determine the percentage of time the depth is greater than 14 metres

Sol.: (a) Determine a suitable scale for each axis.

For the y-axis (d) consider the range: $[9.9, 16.9]$

$-a + d$ → $-3.5 + 13.4$ $-a + d$ → $-3.5 + 13.4$

➡ So y-scale must go higher than 16.9, we'll use $\mathbf{18}$.

For the x-axis (t) consider the period:

$$\text{period} = \frac{2\pi}{b} = \frac{2\pi}{0.164\pi} \approx 12.2 \text{ hrs}$$

➡ So x-scale must go higher than 2×12.2, we'll use $\mathbf{25}$.

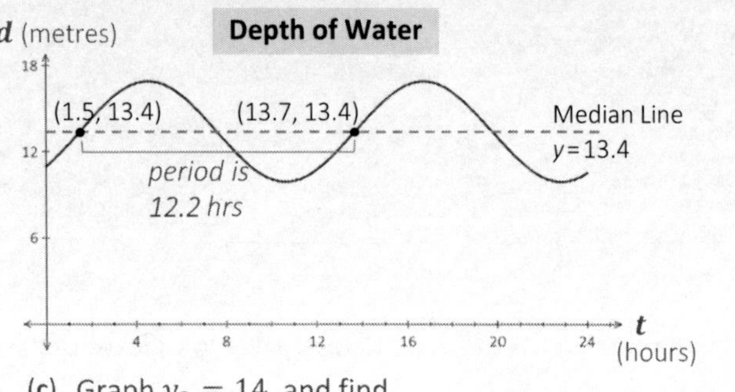

(b) Find $\mathbf{d}$ when $\mathbf{t} = \mathbf{8}$:

$$d(\mathbf{8}) = 3.5\sin[0.164\pi((\mathbf{8}) - 1.5)] + 13.4$$
$$= \mathbf{12.7\ m}$$

Or, sketch the function on your graphing calculator, use TRACE ... $x = 8$:

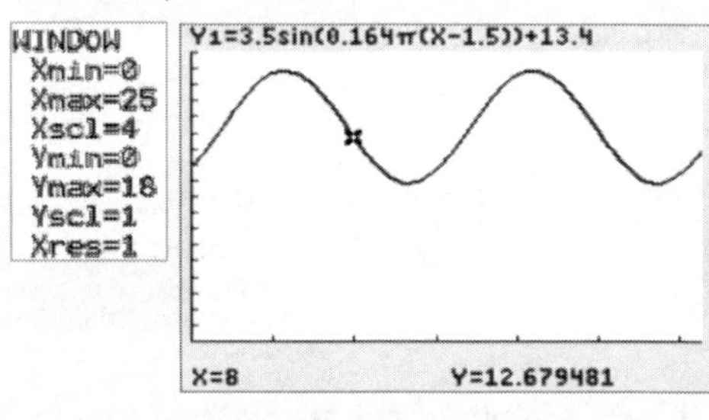

(c) Graph $y_2 = 14$, and find the INTERSECT:

CALC TIP

Y2=14

7.263 hrs

1.834 hrs

Intersection X=1.8343797 Y=14

Plot1 Plot2 Plot3
Y1=3.5sin(0.164π(X-1.5))
Y2=14

NORMAL FLOAT AUTO REAL RADIAN MP

7.263-1.834 5.429 ← # of hours above 14m within one cycle

Ans/12.2 0.445 ← Divided by the period, to get the percentage of time the depth is above 14 m.

➡ **45% of the time**

Class Example 6.61 *Analyzing a Sinusoidal Function – Applications*

In St. John, Newfoundland, the time of sunrise, $t(n)$, on the nth day of the year is given by the formula:

$$t(n) = 1.96\sin\left[\frac{2\pi}{365}(n - 80)\right] + 6.41$$; where n is the number of days after December 31

(a) Use the equation to determine the range and the period of the function.

(b) Determine a suitable scale and sketch the graph over one full cycle.

(c) Predict the time of sunrise on May 9th. (Day 129)

(d) Determine the day # for the first day of the year that the time of sunrise would be 8:00am.

Class Example 6.62 *Modelling a Real-World Situation Using a Sinusoidal Function*

The population of lemmings, $P(t)$, living in a particular area can be modeled by a sinusoidal function in the form $P(t) = a\sin[b(t-c)] + d$ or $P(t) = a\cos[b(t-c)] + d$, where t is the # of months since measurements began.

After 4 months, the lemming population reached the first maximum of 400. After 12 months, the population reached the first minimum of 160.

(a) Label each axis and provide a detailed scale, to sketch a graph that models the lemming population over two full cycles.

(b) Determine each of the following graph characteristics.

i *Amplitude:*

ii *Vertical Displacement:*

iii *Period / "b" value:*

(c) Using a minimum phase shift, state an equation for the lemming population, $P(t)$, as a:

i sine equation, with $a > 0$

ii cosine equation, with $a > 0$

iii sine equation, with $a < 0$

(d) Use one of the equations developed in (c) to predict the lemming population after 5 months. (Round to the nearest whole number)

(e) Use technology to predict the percentage of time during each cycle that the lemming population is below 200. (Round to the whole percentage)

Class Example 6.63 *Modelling a Real-World Situation – Wheel Height*

In the warm-up, we considered the height of a rock stuck in a tire with a diameter of 40cm, rotating once every 4 secs. The height of the rock was modelled by a sinusoidal equation $h(t) = a\sin[b(x - c)] + d$. (Or a cos equation)

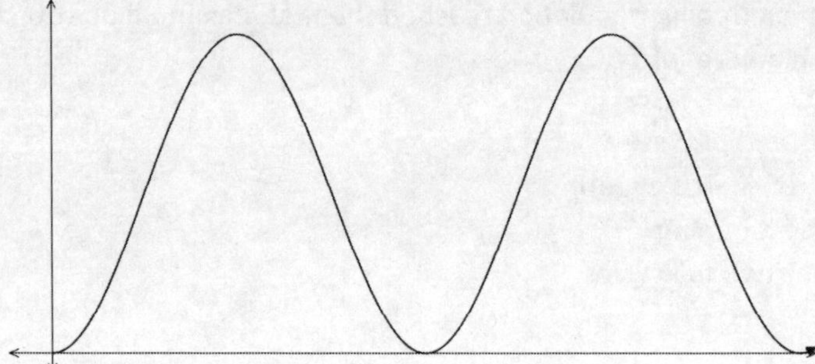

Describe which (a, b, c, or d) of the parameters would change in each of the following scenarios, and how.

(a) The diameter of the tire is smaller, but still rotates once every 4 seconds.

(b) The diameter of tire is unchanged at 40cm, but rotates more slowly.

TRIGONOMETRIC IDENTITIES and EQUATIONS

$\sin(\alpha + \beta) = \sin\alpha\cos\beta + \cos\alpha\sin\beta$

$\cot\theta = \frac{\cos\theta}{\sin\theta}$

$x = \frac{\pi}{6} + n\pi;\ n \in I$

$\sin^2\theta + \cos^2\theta = 1$

$\cos(2\theta) = \sin^2\theta - \cos^2\theta$

7.1 Trigonometric Identities

Welcome to **part II** of Trigonometry! We'll now move into **trigonometric identities**, many basic forms of which are listed on your formula sheet.

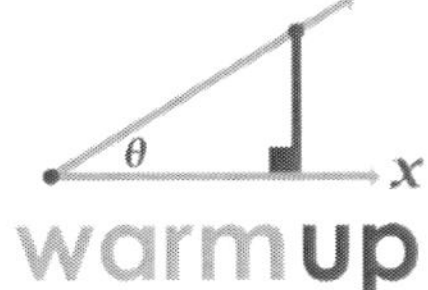

1 ➡ Use your graphing calculator to sketch each of the following trigonometric functions. Match with the correct graph on the right.

❶ $y = \dfrac{\sin x}{\cos x}$

❷ $y = \dfrac{\cos x}{\sin x}$

❸ $y = \sin^2 x + \cos^2 x$

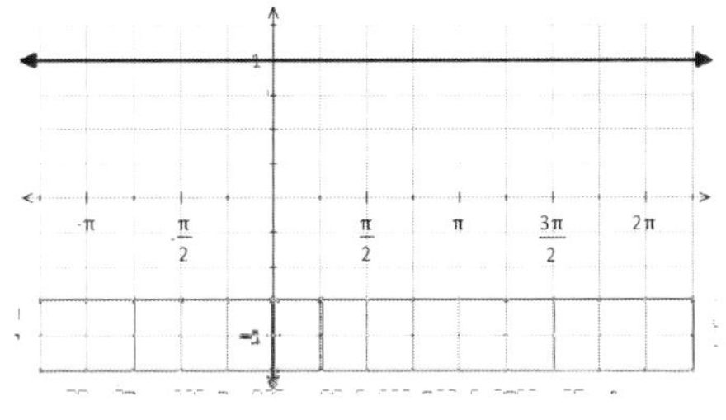

Function # _____

Graph in radian mode,

Match your window settings

```
WINDOW
 Xmin=-5π/4
 Xmax=9π/4
 Xscl=π/4
 Ymin=-4
 Ymax=4
 Yscl=1
 Xres=1
```

Function # _____

2 Refer to your formula sheet to determine a trigonometric identity that applies to each function graphed above.

List each here.

3 ➡ How do the graphs of both sides of an identity relate?

Function # _____

Warm-up is continued on the next page

4 ➡ Use your graphing calculator to sketch the graph of $y = \cos x \tan x$.
Is the curve identical to the graph of $y = \sin x$?

5 ➡ Use TRACE on your calculator to input $x = {}^{\pi}/_{2}$.
What value of y do you get? Explain.

6 ➡ Compare the graphs of ❶ $y = \cos x \tan x$ and ❷ $y_2 = \sin x$ below.
How are they similar? Different?

❶

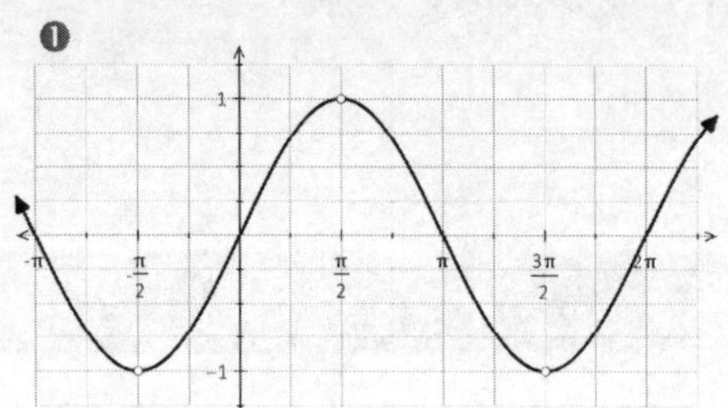

❷

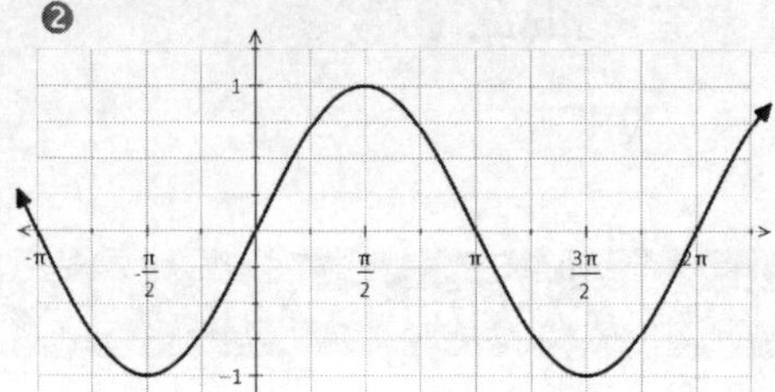

7 ➡ Substitute an equivalent expression for $tanx$
that will allow to you to simplify $cosx\,tanx$.

8 ➡ Substitute an equivalent expression for $cotx$
that will allow to you to simplify $sinx\,cotx$.

An **identity** is an equation that is true for all defined values of the variable in the expressions on both sides. Therefore if two expressions form an identity, they are equivalent.

For example, $(x + 1)^2 = 9$ is an **equation**, which is true when $x = -4$ or $x = 2$

Whereas, $(x + y)^2 = x^2 + 2xy + y^2$ is an **identity** which is true for all real values of x and y.

In chapter 6 we encountered some **trigonometric identities**:

Reciprocal Identities:

$$\csc x = \frac{1}{\sin x} \qquad \sec x = \frac{1}{\cos x} \qquad \cot x = \frac{1}{\tan x}$$

Quotient Identities:

$$\tan x = \frac{\sin x}{\cos x} \qquad \cot x = \frac{\cos x}{\sin x}$$

Trigonometric Identities enable us to write the same expression in different ways. It is often possible to rewrite a complicated expression using a much simpler one. In this section we will:

- **Verify** identities, both **numerically** and **graphically**
- Use algebraic methods to **simplify** trigonometric expressions or **prove** identities

To simplify trigonometric expression, a common first step is to write all functions in terms of ***sin*** and ***cos***.

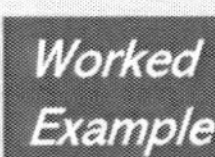

For the equation: $\cos\theta\csc\theta = \cot\theta$

(a) Determine any **non-permissible values,** in radians.

(b) **Numerically verify** that that the equation could be an identify, using $\theta = \frac{\pi}{6}$ and $\theta = 135°$.

(c) **Graphically verify** that that the equation could be an identify, over the domain $-\pi \le \theta < 2\pi$.

(d) **Simplify** the left side of the equation to a single trigonometric expression, equal to the right.

Solutions:

(a) For non-permissible values (NPVs), we must restrict any angles θ that would result in dividing by zero.

$$\cos\theta\csc\theta = \cot\theta$$

$$= \frac{1}{\sin\theta} \qquad = \frac{\cos\theta}{\sin\theta}$$

So, NPV when $\sin\theta = 0$ ➡ $\theta \neq 0, \pi, 2\pi \ldots$

That is, there is a NPV at 0, *then every* π

➡ $\boldsymbol{\theta \neq n\pi\ ;\ n \in I}$

Where on the unit circle is the y-coordinate ("sinθ") equal to 0?

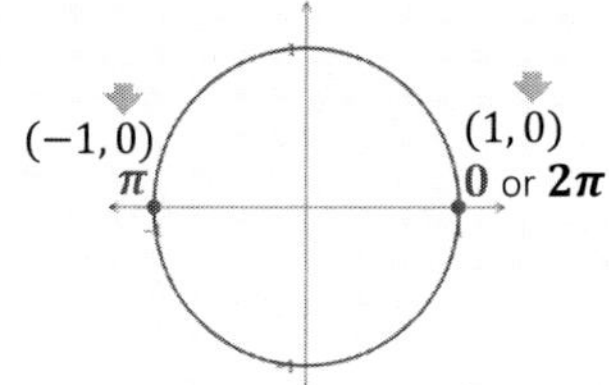

(b) Evaluate both the left side (L.S.) and right side (R.S.) of the equation at the given values θ.

i Substitute $\theta = \frac{\pi}{6}$:

Use the unit circle to evaluate trig ratios for $\theta = {}^{\pi}/_{6}$ ...

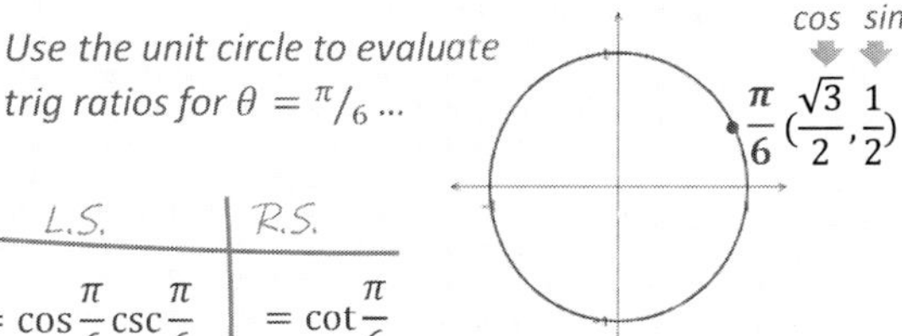

L.S.	R.S.
$= \cos\frac{\pi}{6}\csc\frac{\pi}{6}$	$= \cot\frac{\pi}{6}$
$= \cos\frac{\pi}{6}\left(\frac{1}{\sin\frac{\pi}{6}}\right)$	$= \frac{\cos\frac{\pi}{6}}{\sin\frac{\pi}{6}}$
$= \frac{\sqrt{3}}{2}\cdot\frac{1}{\frac{1}{2}}$	$= \frac{\frac{\sqrt{3}}{2}}{\frac{1}{2}}$
$= \frac{\sqrt{3}}{2}\cdot\frac{2}{1}$	$= \sqrt{3}$ ✓
$= \sqrt{3}$ ✓	

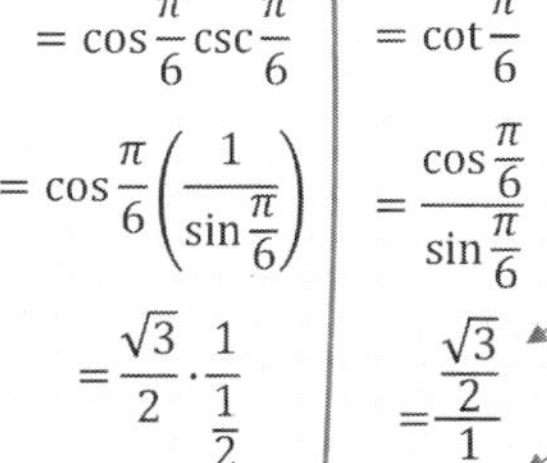

cos is the *x*-coordinate on the unit circle

sin is the *y*-coordinate

Verify on your calc...

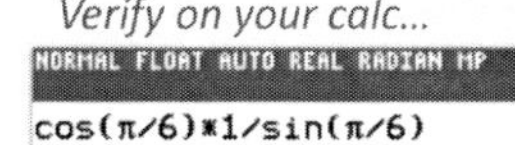

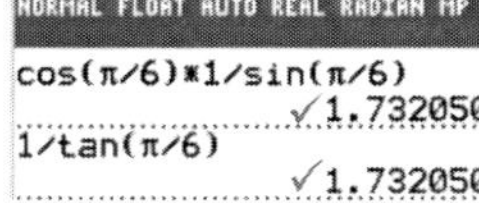

ii Substitute $\theta = 135°$:

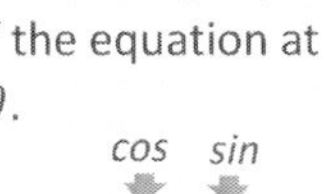

Use the unit circle for $\theta = 135°$ ...

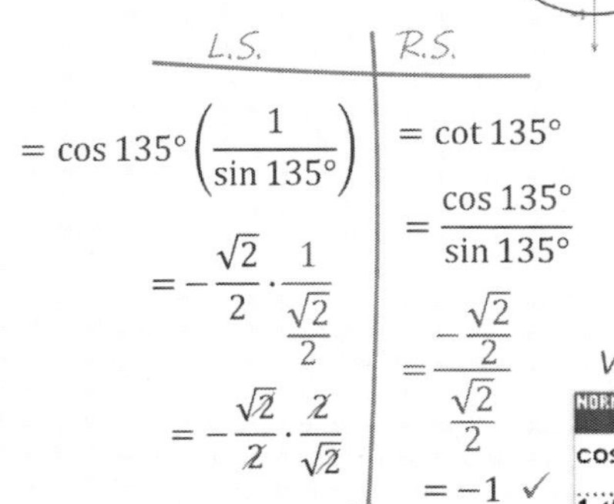

L.S.	R.S.
$= \cos 135°\left(\frac{1}{\sin 135°}\right)$	$= \cot 135°$
$= -\frac{\sqrt{2}}{2}\cdot\frac{1}{\frac{\sqrt{2}}{2}}$	$= \frac{\cos 135°}{\sin 135°}$
$= -\frac{\sqrt{2}}{2}\cdot\frac{2}{\sqrt{2}}$	$= \frac{-\frac{\sqrt{2}}{2}}{\frac{\sqrt{2}}{2}}$
$= -1$ ✓	$= -1$ ✓

Verify on your calc...

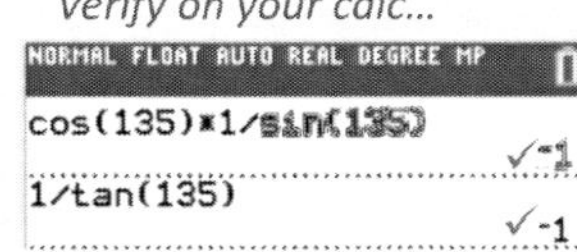

You must be able to verify simplifying exact unit circle values.
(Use your calculator as a second verification)

(c) Graph $y_1 = $ L.S.
$y_2 = $ R.S.

```
Plot1  Plot2  Plot3
■\Y1■cos(X)*1/sin(X)
■\Y2■1/tan(X)
```

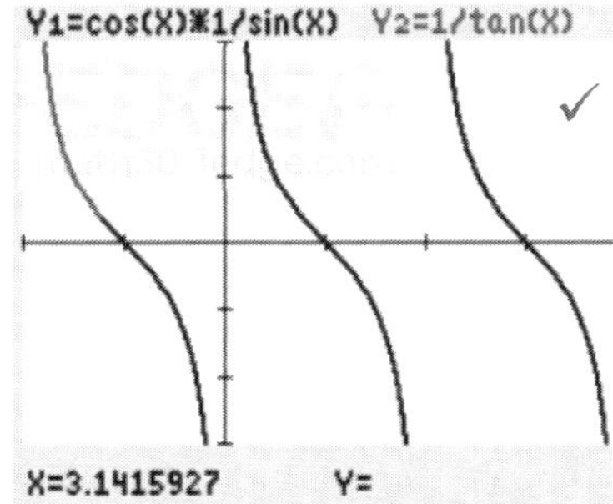

Verify that both the left side and the right side yield the same graph.
(note how the second graph is sketched "over top" of the first)

Note that both graphs have a vertical asymptote (not shown in the graph above) at $x = 0$, then another every π

V.A.s at $x = n\pi\ ; n \in I$

(d) Simplify the ***left side*** by substituting known identities:

$$\text{L.S.} = \cos\theta \cdot \csc\theta$$

$$= \cos\theta \cdot \frac{1}{\sin\theta}$$

$$= \frac{\cos\theta}{\sin\theta}$$

$$= \boldsymbol{\cot\theta}$$ *(same as R.S.!)*

Class Example 7.11 *Verifying Trigonometric Identities*

For the equation: $\sec\theta = \dfrac{\tan\theta}{\sin\theta}$

(a) **Determine any non-permissible values,** in degrees.

(b) **Numerically verify** that that the equation could be an identify, using $\theta = 30°$ and $\theta = \dfrac{2\pi}{3}$.

(c) **Graphically verify** that that the equation could be an identify, over the domain $-180° \leq \theta < 360°$.

(d) **Simplify** the expression on the right side of the equation to a single trigonometric expression.

Class Example 7.12 *Simplifying Trigonometric Expressions*

For the expression $\dfrac{\cot\theta\sec\theta}{\csc\theta}$

(a) Determine any **non-permissible values,** in radians.

(b) Simplify the expression.

Investigation 2 – **The Pythagorean Identity**

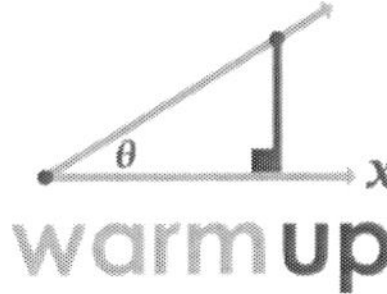

The point P on the right lies on a unit circle.

1 ➡ Use your knowledge of the principles of the unit circle to label the coordinates of P, in terms of the angle in standard position, θ.

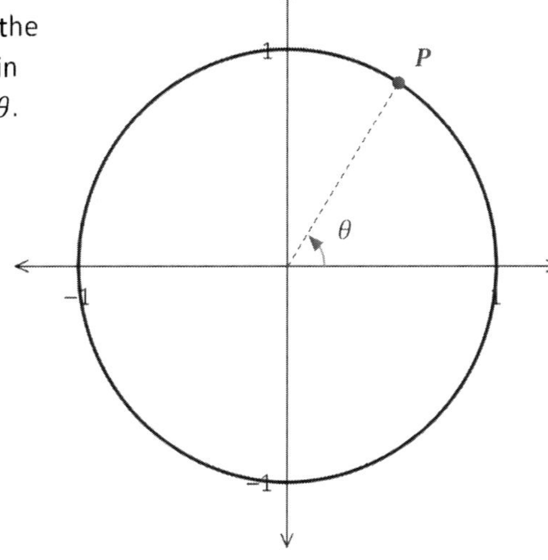

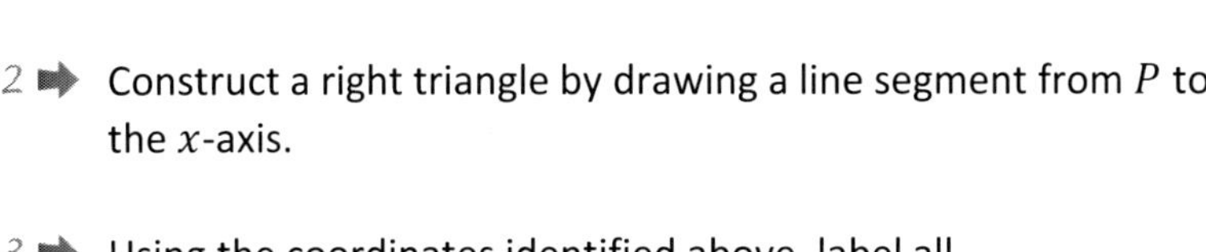

2 ➡ Construct a right triangle by drawing a line segment from P to the x-axis.

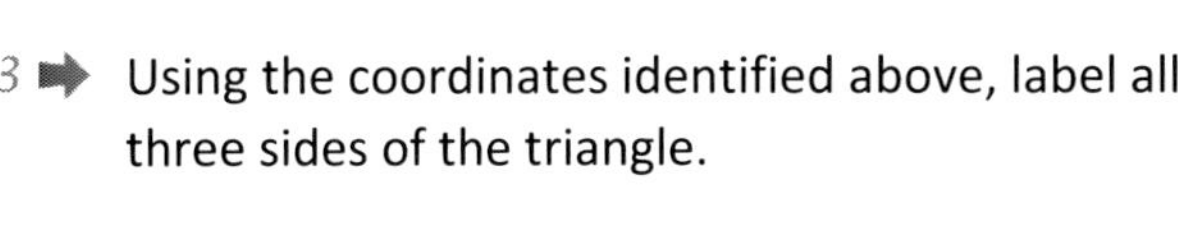

3 ➡ Using the coordinates identified above, label all three sides of the triangle.

4 ➡ Equate the three triangular sides using the Pythagorean Theorem.

The Pythagorean Identity can be visualized by considering any point P on a unit circle.

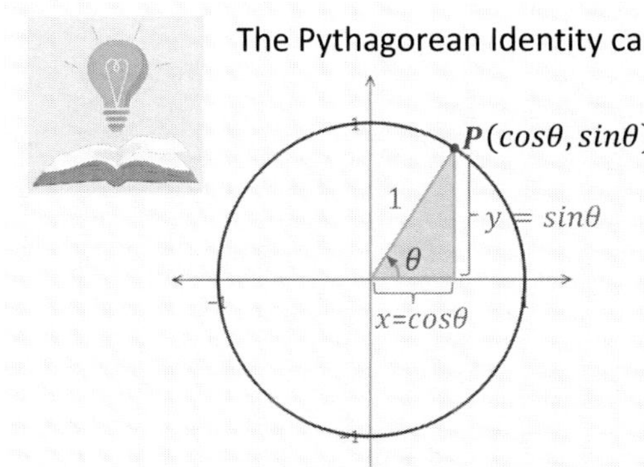

Recall that any point P on the unit circle has coordinates $(\cos\theta, \sin\theta)$, where θ is an angle in standard position whose terminal arm passes through P.

The right triangle formed inside the circle has a hypotenuse of 1, and legs of $\cos\theta$ and $\sin\theta$.

So by Pythagorean Theorem: $(\cos\theta)^2 + (\sin\theta)^2 = (1)^2$

Simplifies to: $\boldsymbol{\cos^2\theta + \sin^2\theta = 1}$

Now whenever we encounter "$\boldsymbol{\cos^2\theta + \sin^2\theta}$" in a trigonometric expression, we can substitute "**1**"

Similarly, we can we can substitute "$\boldsymbol{1 - \cos^2\theta}$" with "$\boldsymbol{\sin^2\theta}$"

And should we encounter "$\boldsymbol{1 - \sin^2\theta}$", we can substitute "$\boldsymbol{\cos^2\theta}$"

Class Example 7.13 *Simplifying Trigonometric Expressions*

Your formula sheet contains the identity $1 + \tan^2\theta = \sec^2\theta$

(a) Verify the identity using $\theta = 60°$

(b) Show that this is related to the Pythagorean Identity by dividing both sides by $sec^2\theta$ and simplifying.

Class Example 7.14 *Simplifying Trigonometric Expressions*

Your formula sheet contains the identity $1 + cot^2\theta = csc^2\theta$

(a) Verify the identity graphically.

(b) Divide both sides of the Pythagorean Identity, $cos^2\theta + sin^2\theta = 1$, by $sin^2\theta$, to show it is equivalent to $1 + cot^2\theta = csc^2\theta$.

There are three forms of the Pythagorean Identity. *Each is on your formula sheet*

$\boldsymbol{cos^2\theta + sin^2\theta = 1}$ $\boldsymbol{1 + tan^2\theta = sec^2\theta}$ $\boldsymbol{1 + cot^2\theta = csc^2\theta}$

We can also re-arrange each of these to get further alternate forms.

$cos^2\theta = 1 - sin^2\theta$ $tan^2\theta = sec^2\theta - 1$ $cot^2\theta = csc^2\theta - 1$

$sin^2\theta = 1 - cos^2\theta$

Class Example 7.15 *Simplifying Trigonometric Expressions*

Simplify each of the following to a single trigonometric function.

(a) $\cos\theta + \tan\theta\sin\theta$

(b) $\cos\theta\,(\sec\theta - \cos\theta)$

7.2 Sum, Difference, and Double-Angle Identities

Investigation 1 – The Sum Formulas

We start with three core intro questions: Given two angles, $\angle\alpha$ and $\angle\beta$, does: $\sin(\alpha+\beta)=\sin\beta+\sin\beta$?

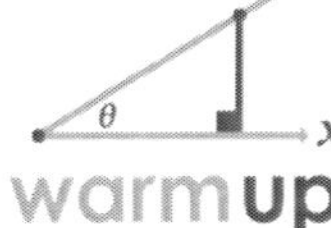

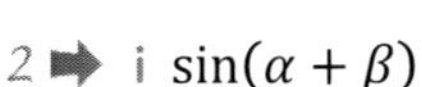

These are good questions! Or they are at the least questions. Let's investigate....

or does $\cos(\alpha+\beta)=\cos\alpha+\cos\beta$?

how about $\tan(\alpha+\beta)=\tan\alpha+\tan\beta$?

1 ➡ Label the coordinates of the 3 points marked on the unit circle. →

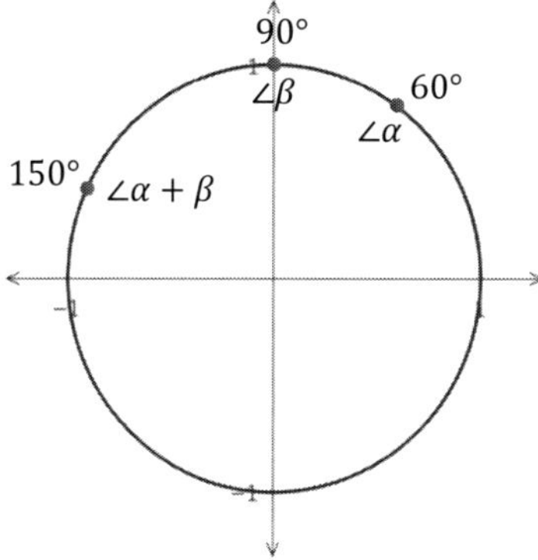

Given $\angle\alpha=60°$ and $\angle\beta=90°$, evaluate each of the following:

2 ➡ i $\sin(\alpha+\beta)$ ii $\sin\alpha+\sin\beta$

3 ➡ i $\cos(\alpha+\beta)$ ii $\cos\alpha+\cos\beta$

4 ➡ i $\tan(\alpha+\beta)$ ii $\tan\alpha+\tan\beta$

5 ➡ Based on your results, what conclusion can we make regarding the core intro questions above?

The trigonometric ratios of angle sums can be found using the sum identities, first developed by ancient mathematician and astronomer Claudius Ptolemy during the 2nd century AD.

$$\sin(\alpha+\beta)=\sin\alpha\cos\beta+\cos\alpha\sin\beta$$

$$\cos(\alpha+\beta)=\cos\alpha\cos\beta-\sin\alpha\sin\beta$$

$$\tan(\alpha+\beta)=\frac{\tan\alpha+\tan\beta}{1-\tan\alpha\tan\beta}$$

We can use these formulas without understanding how they are derived. However, we'll refer to these two core identities to develop further formulas in this chapter.

For the curious, do a web search for "proof of the sum and difference identities". And enjoy!

When dealing with angle measures, you may have noticed we use a lot of Greek letters!

π and θ (theta) are both from the Greek alphabet, as are the two new variables we'll use in this chapter:

α "alpha"
β "beta"

Class Example 7.21 *Simplifying Trigonometric Expressions*

Use the angles $\angle\alpha = 60°$ and $\angle\beta = 180°$ to numerically verify each of the sum identities introduced on the previous page.

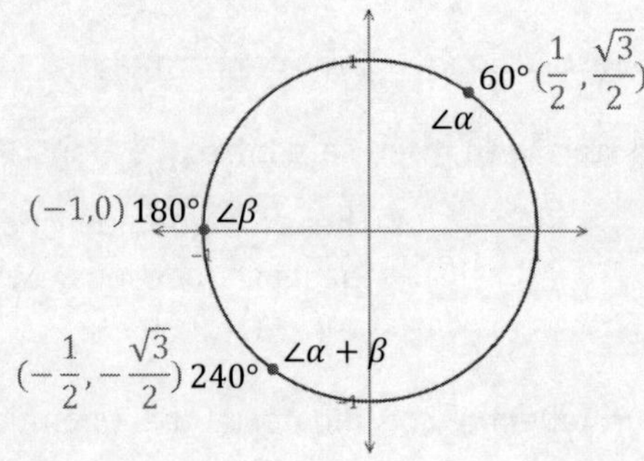

(a) $\sin(\alpha+\beta) = \sin\alpha\cos\beta + \cos\alpha\sin\beta$

(b) $\cos(\alpha+\beta) = \cos\alpha\cos\beta - \sin\alpha\sin\beta$

(c) $\tan(\alpha+\beta) = \dfrac{\tan\alpha + \tan\beta}{1 - \tan\alpha\tan\beta}$

Investigation 2 – **Identities for Negative Angles**

We'll next investigate identities for $\sin(-\theta)$ and $\cos(-\theta)$, so that we'll be ready to look at difference formulas.

1 ➡ Use your unit circle to complete each of the tables below.

θ	$\sin\theta$
30°	
−30°	
135°	
−135°	
240°	
−240°	

θ	$\cos\theta$
30°	
−30°	
135°	
−135°	
240°	
−240°	

2 ➡ What conjecture can we make about the values of $\sin(-\theta)$ and $\cos(-\theta)$?

Enrichment: A function is defined as **odd** if $f(x) = -f(x)$, and the graph is symmetrical about the origin. A function is **even** if $f(-x) = f(x)$, in which case the graph is symmetrical about the y-axis.

3 ➡ Identify the graphs below as the graph of $y = \sin x$ or $y = \cos x$. Label each as odd or even.

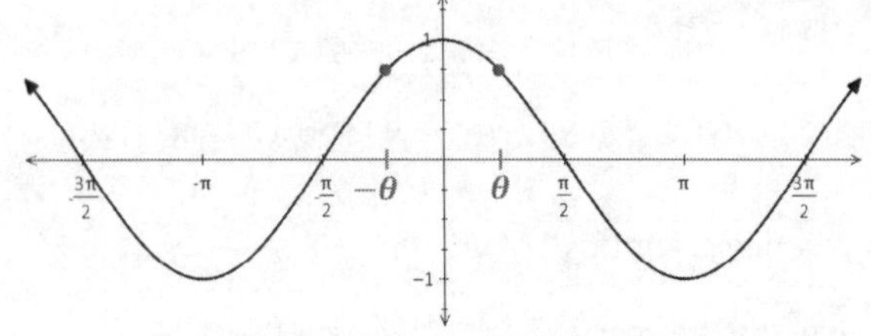

4 ➡ Describe how the negative angle identities can be verified using the graphs of $y = \sin x$ or $y = \cos x$.

Next let's consider both the sum and difference identities.

The Sum and Difference Identities:

$$\sin(\alpha+\beta)=\sin\alpha\cos\beta+\cos\alpha\sin\beta$$
$$\sin(\alpha-\beta)=\sin\alpha\cos\beta-\cos\alpha\sin\beta$$
$$\cos(\alpha+\beta)=\cos\alpha\cos\beta-\sin\alpha\sin\beta$$
$$\cos(\alpha-\beta)=\cos\alpha\cos\beta+\sin\alpha\sin\beta$$
$$\tan(\alpha+\beta)=\frac{\tan\alpha+\tan\beta}{1-\tan\alpha\tan\beta}$$
$$\tan(\alpha-\beta)=\frac{\tan\alpha-\tan\beta}{1+\tan\alpha\tan\beta}$$

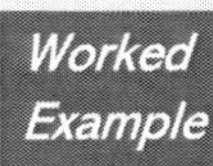

Worked Example

Write each expression as a single trigonometric function:

(a) $\cos 22^\circ\cos 35^\circ-\sin 22^\circ\sin 35^\circ$

(b) $\sin\frac{\pi}{7}\cos\theta-\cos\frac{\pi}{7}\sin\theta$

Solution:

(a) The expression has the pattern for $\cos(\alpha+\beta)$, with $\alpha=22^\circ$ and 35°

$=\cos(22^\circ+35^\circ)$ ➧ $=\mathbf{\cos(57^\circ)}$

(b) The expression has the pattern for $\sin(\alpha-\beta)$, with $\alpha=\frac{\pi}{7}$ and $\beta=\theta$

$=\boldsymbol{\sin\left(\frac{\pi}{7}-\theta\right)}$

Class Example 7.22 *Simplifying Trigonometric Expressions*

(a) Write $\sin\frac{5\pi}{12}\cos\frac{\pi}{12}-\cos\frac{5\pi}{12}\sin\frac{\pi}{12}$ as a single trigonometric function:

(b) Write $\cos\frac{\pi}{5}\cos\frac{2\pi}{7}-\sin\frac{\pi}{5}\sin\frac{2\pi}{7}$ as a single trigonometric function:

Worked Example Determine the exact value of $\cos\frac{5\pi}{12}$.

Solution: Express $\frac{5\pi}{12}$ as the *sum* or *difference* of two unit circle angles. (Converting to degrees could help!)

Converting to degrees could help! $\frac{5\pi}{12}\times\frac{180°}{\pi} \rightarrow = 75°$

$$\cos(75°) \rightarrow = \cos(45° + 30°)$$

$$= \cos 45° \cos 30° - \sin 45° \sin 30°$$

$$= \left(\frac{\sqrt{2}}{2}\right)\left(\frac{\sqrt{3}}{2}\right) - \left(\frac{\sqrt{2}}{2}\right)\left(\frac{1}{2}\right)$$

$$= \frac{\sqrt{6}}{4} - \frac{\sqrt{2}}{4} \rightarrow = \frac{\sqrt{6}-\sqrt{2}}{4}$$

Note that this is but one option – we could have used $120°$ & $45°$, or $210°$ & $135°$, *and so on!*

Apply sum formula with $\alpha = 45°$ and $\beta = 30°$

Result can be verified on your calculator:

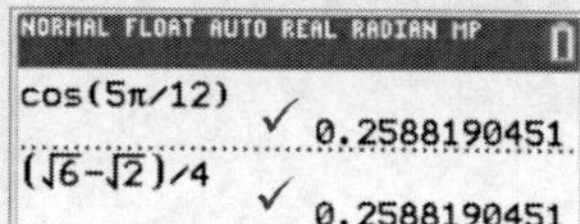

Class Example 7.23 *Simplifying Trigonometric Expressions*

Determine the exact value of each trigonometric expression:

(a) $\cos\frac{11\pi}{12}$

(b) $\tan\frac{\pi}{12}$

Investigation 3 – The Double-Angle Identities

We'll next investigate the double-angle identities, which we can develop using addition formulas.

1 ➡ Apply the sin sum formula with $\alpha = \alpha$ and $\beta = \alpha$, to derive a formula for $\sin(2\alpha)$.

2 ➡ Apply the cos sum formula with $\alpha = \alpha$ and $\beta = \alpha$, to derive a formula for $\cos(2\alpha)$.

3 ➡ Apply the tan sum formula with $\alpha = \alpha$ and $\beta = \alpha$, to derive a formula for $\tan(2\alpha)$.

4 ➡ Use the Pythagorean Identity $sin^2\theta + cos^2\theta = 1$ to derive two alternative forms of the $\cos(2\alpha)$ identity.

i – Formula in terms of cos only

i – Formula in terms of sin only

Summary of identities from this section:

The Double-Angle Identities:

$$\sin(2\alpha) = 2\sin\alpha\cos\alpha$$

$$\cos(2\alpha) = \cos^2\alpha - \sin^2\alpha$$

$$\cos(2\alpha) = 2\cos^2\alpha - 1$$

$$\cos(2\alpha) = 1 - 2\sin^2\alpha$$

$$\tan(2\alpha) = \frac{2\tan\alpha}{1-\tan^2\alpha}$$

The Sum and Difference Identities:

$$\sin(\alpha+\beta) = \sin\alpha\cos\beta + \cos\alpha\sin\beta$$

$$\sin(\alpha-\beta) = \sin\alpha\cos\beta - \cos\alpha\sin\beta$$

$$\cos(\alpha+\beta) = \cos\alpha\cos\beta - \sin\alpha\sin\beta$$

$$\cos(\alpha-\beta) = \cos\alpha\cos\beta + \sin\alpha\sin\beta$$

$$\tan(\alpha+\beta) = \frac{\tan\alpha+\tan\beta}{1-\tan\alpha\tan\beta}$$

$$\tan(\alpha-\beta) = \frac{\tan\alpha-\tan\beta}{1+\tan\alpha\tan\beta}$$

Worked Example Given the diagram of an angle in standard position θ, determine the exact value of:

(a) $\sin 2\theta$ **(b)** $\tan(\theta + \frac{3\pi}{4})$

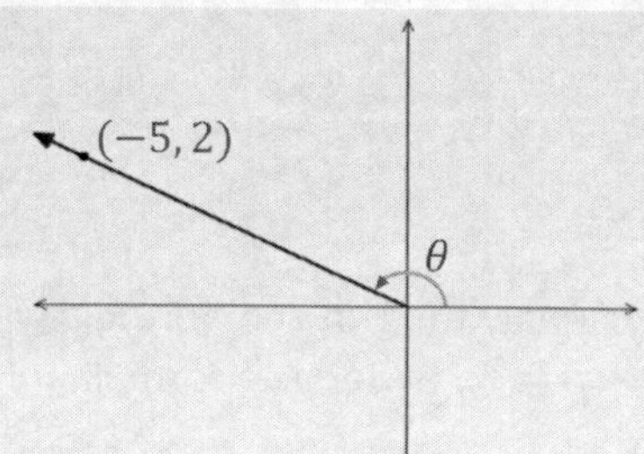

Solution: **(a)** *By double-angle identiy we have:*

$\sin 2\theta = 2\sin\theta\cos\theta$

Substitute:

$$= 2(\frac{2}{\sqrt{29}})(\frac{-5}{\sqrt{29}})$$

$$= -\frac{20}{29}$$

Complete diagram to determine sinθ and cosθ

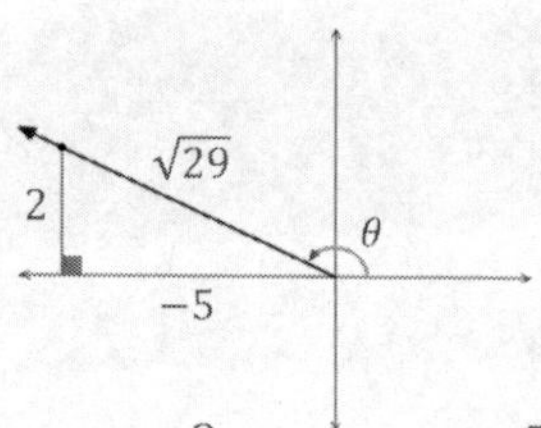

$\sin\theta = \frac{2}{\sqrt{29}}$ $\cos\theta = \frac{-5}{\sqrt{29}}$

(b) *By sum identiy we have:*

$$\tan(\theta + \frac{3\pi}{4}) = \frac{\tan\theta + \tan\frac{3\pi}{4}}{1 - \tan\theta\tan\frac{3\pi}{4}}$$

Use the diagram to determine tanθ, use the unit circle for tan $\frac{3\pi}{4}$

$$= \frac{-\frac{2}{5} + (-1)}{1 - (-\frac{2}{5})(-1)}$$

$$= \frac{-\frac{7}{5}}{\frac{3}{5}} \rightarrow = -\frac{7}{3}$$

Class Example 7.24 *Simplifying Trigonometric Expressions*

Given the diagram of an angle in standard position θ, determine the exact value of:

(a) $\sin(\theta - \frac{5\pi}{6})$ **(b)** $\tan(2\theta)$

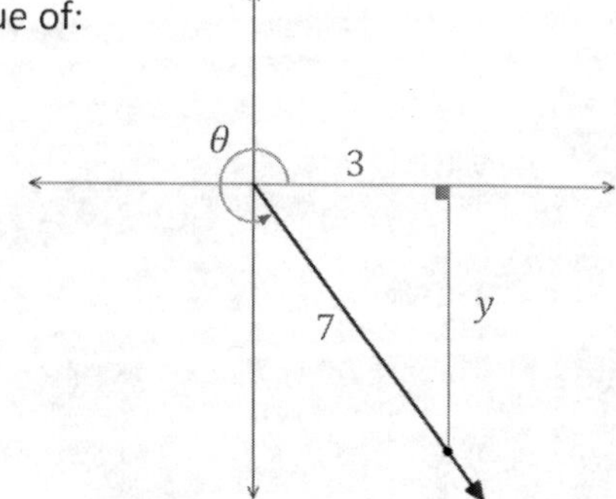

Class Example 7.25 *Simplifying Trigonometric Expressions*

Simplify each of the following to a single trigonometric expression.

(a) $\dfrac{1+\cos 2x}{\sin 2x}$

(b) $\dfrac{-\sin 2x}{\cos x\,(\cos 2x - 1)}$

Class Example 7.26 *Simplifying Trigonometric Expressions*

Use sum and difference formulas to simplify the expression $\sin\left(\frac{\pi}{2}-\theta\right) - \cos\left(\frac{\pi}{2}+\theta\right)$

Worked Example

Given $\boldsymbol{cosA} = \frac{\mathbf{3}}{\mathbf{5}}$ and $\boldsymbol{cosB} = \frac{\mathbf{5}}{\mathbf{13}}$, where $0 \le A \le \frac{\pi}{2}$ and $\frac{3\pi}{2} \le B \le 2\pi$, find the exact value of $\boldsymbol{cos(A+B)}$.

Solution: *Use* $cos(\alpha+\beta) = cos\alpha cos\beta - sin\alpha sin\beta$

Given, $\frac{3}{5}$ Given, $\frac{3}{5}$ *We **need** these two!*

Let's get $\boldsymbol{sinA}$*: ($\angle A$ is in Quad I, since $0 < A < \frac{\pi}{2}$)*

$cosA = \frac{3}{5}$ ← adj, ← hyp

By Pyth. Theorem, the OPP side is "4". $so,\ \boldsymbol{sinA} = \frac{\mathbf{4}}{\mathbf{5}}$

Next get $\boldsymbol{sinB}$*: ($\angle B$ is in Quad IV, since $0 < A < \frac{3\pi}{2}$)*

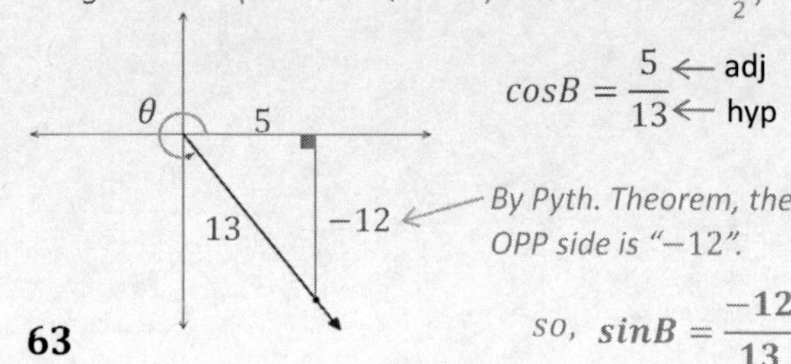

$cosB = \frac{5}{13}$ ← adj, ← hyp

By Pyth. Theorem, the OPP side is "−12". $so,\ \boldsymbol{sinB} = \frac{\mathbf{-12}}{\mathbf{13}}$

So Finally, $cos(\alpha+\beta) = \left(\frac{3}{5}\right)\left(\frac{5}{13}\right) - \left(\frac{4}{5}\right)\left(\frac{-12}{13}\right) = \frac{\mathbf{63}}{\mathbf{65}}$

Class Example 7.27 *Simplifying Trigonometric Expressions*

Given $\cos A = \frac{3}{5}$ and $\sin B = -\frac{7}{25}$, where $\frac{3\pi}{2} \le A \le 2\pi$ and $\pi \le B \le \frac{3\pi}{2}$, determine the exact value of:

(a) $sin(A+B)$

(b) $cos(2A)$

7.3 Proving Trigonometric Identities

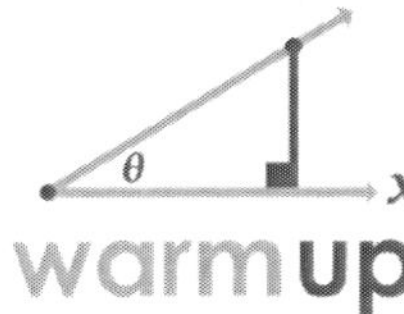

In section 7.1 we saw how a trigonometric **identity** is an equation that is true for all defined values of the variable in the expressions on both sides. In this section we'll differentiate between methods used to **verify** identities, and those used to **prove** them.

Consider the equation: $\cos\theta + \tan\theta\sin\theta = \sec\theta$

1 ➡ Determine the non-permissible values for this identity.

2 ➡ Numerically verify that $\cos\theta + \tan\theta\sin\theta = \sec\theta$ when $\theta = 30°$, and for any other permissible value θ.

3 ➡ Graphically verify that $\cos\theta + \tan\theta\sin\theta = \sec\theta$ could be an identity. Provide a sketch of your graph here, labeling all relevant characteristics.

Do the graphs verify the potential of an identity? Explain your reasoning.

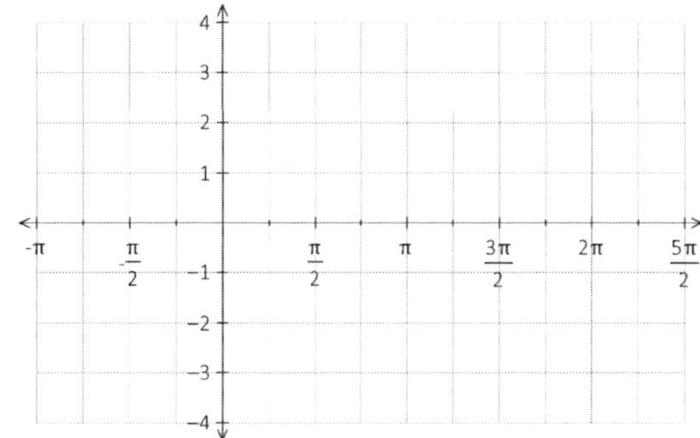

4 ➡ Use the algebraic process described below, on the left side, to **prove** that $\cos\theta + \tan\theta\sin\theta = \sec\theta$ is an identify.

	L.S.	R.S.
	$\cos\theta + \tan\theta\sin\theta$	$\sec\theta$
Re-write so all functions are in terms of sine and cos.	$= \cos\theta + \left(\quad\right)\sin\theta$	
"Prep" the terms – obtain a common denominator	$= \underline{\quad\quad} + \underline{\quad\quad}$	
Combine into a single rational expression	$= \underline{\quad\quad\quad\quad}$	
Substitute the numerator using the Pythagorean Identity.	$= \underline{\quad\quad}$	
Apply the applicable reciprocal identity.		

7.3 *Proving Trigonometric Identities*

Often in mathematics it is useful to write trigonometric equations in alternative forms.

Two trigonometric expressions form an identity if they can be shown as equivalent through algebraic manipulation.

In this section we will further analyze methods to **prove** more complex trigonometric identities. While there is no standard algorithm or steps we can apply each time, there are various strategies we can employ for success.

With practice, these strategies will become routine, and you can go from novice to an identity-proving star!

Strategies For Proving Trigonometric Identities

- First focus on the more complicated side, so that it is made identical to the simpler side.
- Use known identities to make substitutions, including reciprocal, quotient, and double-angle identities. ➔ ***Often it is helpful to express each side in terms of sine and cosine only.***

Be sure to have your formula sheet nearby, to reference core identities that might be useful!

- Quotient or Reciprocal Identities, such as $\tan\theta = \dfrac{\sin\theta}{\cos\theta}$ *or* $\csc\theta = \dfrac{1}{\sin\theta}$

- If squared terms are present, look to use one of the Pythagorean Identities, starting with $\sin^2\theta + \cos^2\theta = 1$

- The Double-Angle Identities for sine and cosine (all three forms!)

- Where applicable, it can be beneficial to **factor expressions**, to simplify and possibly cancel terms.

An example, say we start with: *Here we can common factor...*

$$\cos x + \cos x\tan^2 x \Rightarrow \cos x(1+\tan^2 x) \Rightarrow \frac{\cos x}{\cos^2 x} \Rightarrow = \mathbf{\sec x}$$

Now apply a Pythagorean Identity – this is "$\sec^2 x$*"!*

- Fraction skills are important! Often **common denominators** are needed, and example of which is:

An example, start with: *Re-write "cosx":* *Pythagorean Identity...* *Simplified!*

$$\sec x - \cos x \Rightarrow \frac{1}{\cos x} - \frac{\cos^2 x}{\cos x} \Rightarrow \frac{1-\cos^2 x}{\cos x} \Rightarrow = \frac{\sin^2 x}{\cos x} \Rightarrow = \mathbf{\tan x \sec x}$$

We also often deal with complex fractions. As an example...

$$\frac{\frac{1+\cos x}{\cos x}}{\frac{\sin x}{\cos x}} \Rightarrow = \frac{1+\cos x}{\cancel{\cos x}} \times \frac{\cancel{\cos x}}{\sin x} \Rightarrow = \mathbf{\frac{1+\cos x}{\sin x}}$$

- As you work on one side of the identity, keep the other side in mind as it represents your goal!
- For certain types of identities, none of the above strategies may apply. Should all other methods fail, try **multiplying by the conjugate**. *Example is shown below.*

Start with: *Multiply the numerator / denominator by the conjugate:* *Now notice a Pythagorean Identity in the numerator*

$$\frac{1+\cos x}{\sin x} \Rightarrow \frac{1+\cos x}{\sin x}\frac{\times 1-\cos x}{\times 1-\cos x} \Rightarrow = \frac{1-\cos^2 x}{\sin x\,(1-\cos x)}$$

See example 7.33 (b) for this full identity!

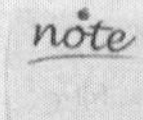

Verifying identities is not the same as solving equations. Techniques used in solving equations, such as adding / subtracting / multiplying / dividing both sides by the same term (moving terms from one side to the other), **should not be used with identities.**

To remind us of this, we typically set up a "T" column for the left side and right side.

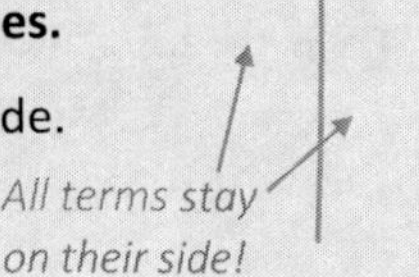

Include a concluding statement at the end of your proof, such as "***LS=RS!***" or **Q.E.D.**

Class Example 7.31 *Simplifying Trigonometric Expressions*

Given each of the following equations;

i Determine any non-permissible values
ii Use an algebraic approach to prove each is an identity

(a) $\cot\theta + 1 = \csc\theta\,(\cos\theta + \sin\theta)$

L.S.	R.S.

(b) $1 - \cos^2 x = \cos x \sin x \tan x$

L.S.	R.S.

(c) $\cot\theta + \tan\theta = \sec\theta\csc\theta$

L.S.	R.S.

When proving an identity, often you will work on the more complex side, using algebraic process to equate it to the simpler side.

However you may also work on **both sides separately**, to get them both equal to some other expression. ➔

L.S.	R.S.
Complex Trig Expression	Also complex
⋮	⋮
Simpler Expression	Same Simpler Expression!

Class Example 7.32 *Simplifying Trigonometric Expressions*

Use an algebraic process to prove that each of the following equations are identities.

(a) $\sin\theta + \cot\theta\cos\theta = \csc\theta$

L.S.	R.S.

(b) $\dfrac{1+\csc\theta}{\cos\theta+\cot\theta} = \sec\theta$

L.S.	R.S.

Class Example 7.32 *Continued*

(c) $\dfrac{\sin 2x}{1-\cos 2x}=\dfrac{1+\cot x}{1+\tan x}$

L.S.	R.S.

(d) $\dfrac{\sin x}{1-\cos x}=\dfrac{1+\cos x}{\sin x}$

L.S.	R.S.

Class Example 7.33 *Factoring to prove an identity*

Use an algebraic process to prove that each of the following equations are identities.

(a) $\dfrac{\sin x}{\csc x + \cot x} = 1 - \cos x$

(b) $\dfrac{\cos x \sin x - \cos x}{\sin^2 x - 1} = \dfrac{1 - \sin x}{\cos x}$

L.S.	R.S.

L.S.	R.S.

(c) For the identity (b) only, determine any non-permissible values.

7.4 Trigonometric Equations

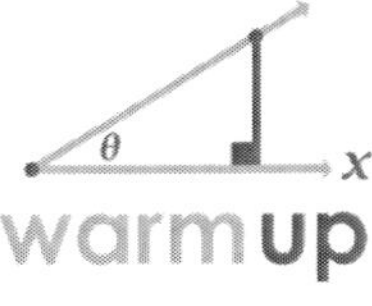

Consider the following trigonometric equations:

i $\sin^2\theta + \cos^2\theta = 1$ ii $2\sin^2\theta - \sin\theta = 0$ iii $\sin 2\theta + \cos\theta = 0$

1 ➡ Explain how equation i is different from equations ii and iii.

2 ➡ Below is the graph of the left side of each of the three equations. Use your graphing calculator to match each equation above with the correct graph. Label each.

equation

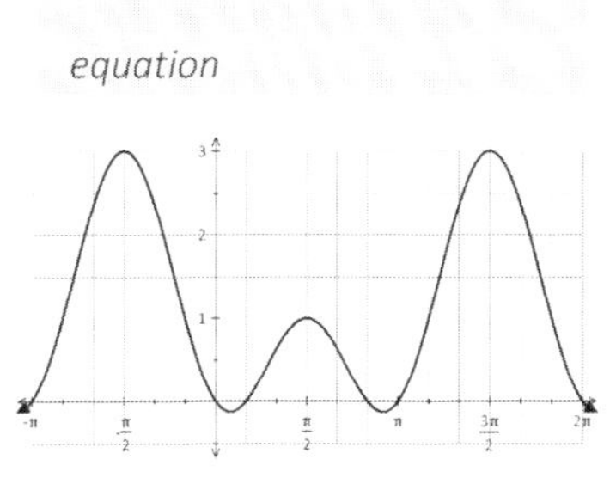

equation

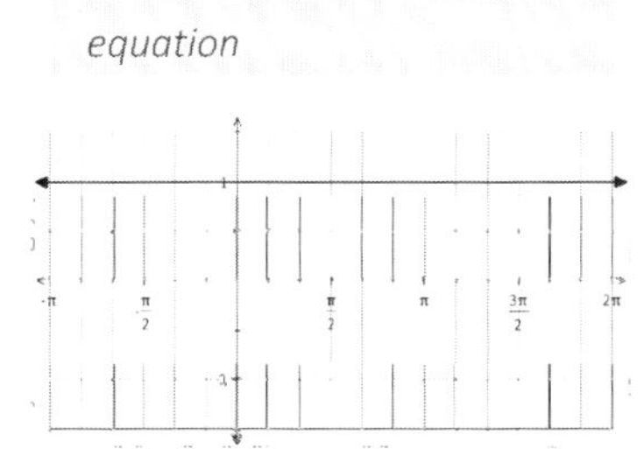

equation

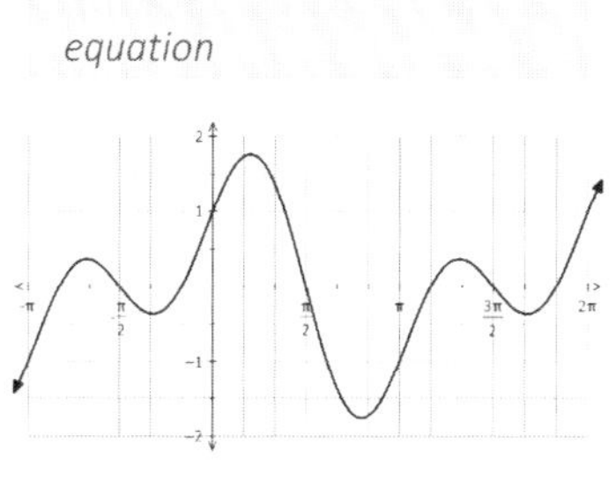

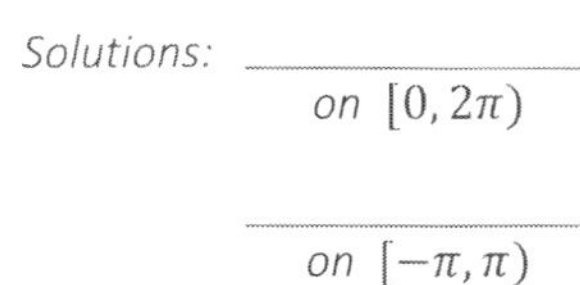

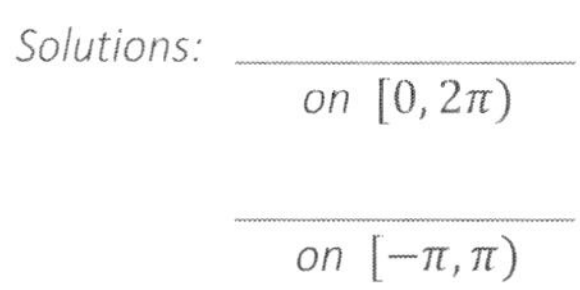

3 ➡ The graphs on the left and right represent equations whose specific solutions can be found by analyzing the graphs. For those two equations, state the solutions on each indicated domain.

Consider the trigonometric equation: iv $2\sin\theta - 1 = 0$; on $0 \le \theta < 2\pi$

4 ➡ Determine the equation solutions by:

- First isolating the trig term
- Referencing the unit circle

Indicate the solutions on the circle ➘

5 ➡ Use an identity to algebraically solve the following equation, on $[0, 2\pi)$:

iii $\sin 2\theta + \cos\theta = 0$

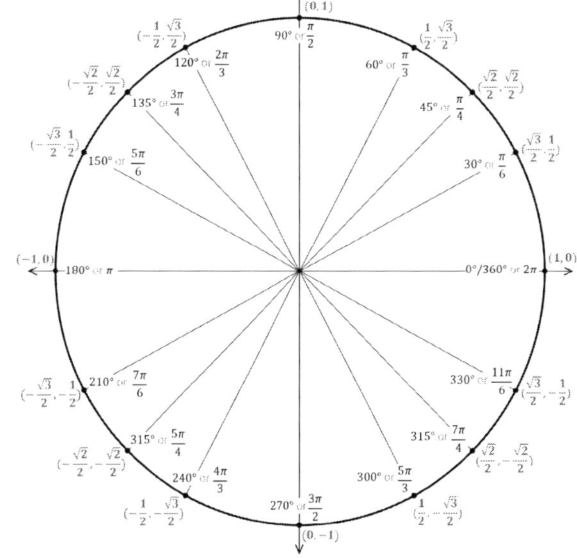

Trigonometric Equations

Earlier in this chapter we studied trigonometric equations that were identities, which were true for any defined variable value.

We'll now consider equations which are only true for certain variable values (solutions); our job will be to determine these solutions. (Sometimes referred to as 'roots' or 'zeros') We can do this either:

- ***Algebraically***, by isolating the trigonometric functions - and using our knowledge of trig functions (example – the unit circle) to solve for the variable.

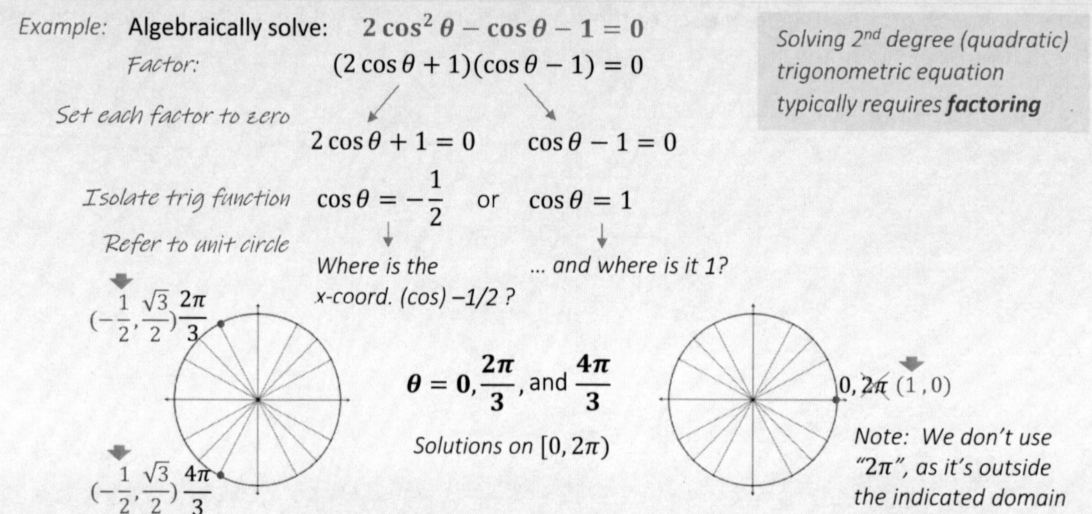

- ***Graphically***, by determining where graphs of $y_1 =$ *Left Side* and $y_1 =$ *Right Side* intersect.

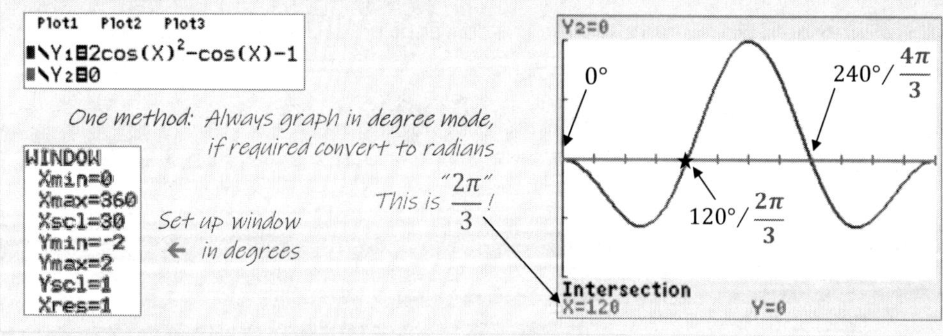

Class Example 7.41 *Solving Trigonometric Equations*

Use an algebraic process to solve $\mathbf{2sin^2\,\theta - sin\,\theta = 0}$; on (a) $0 \le \theta < 2\pi$, and (b) $-2\pi \le \theta < 0$.

General Solutions to a Trigonometric Equation

As trigonometric functions are periodic, trig equations have an infinite number of solutions.

Let's again consider the equation from the previous page:

$$2\cos^2\theta - \cos\theta - 1 = 0$$

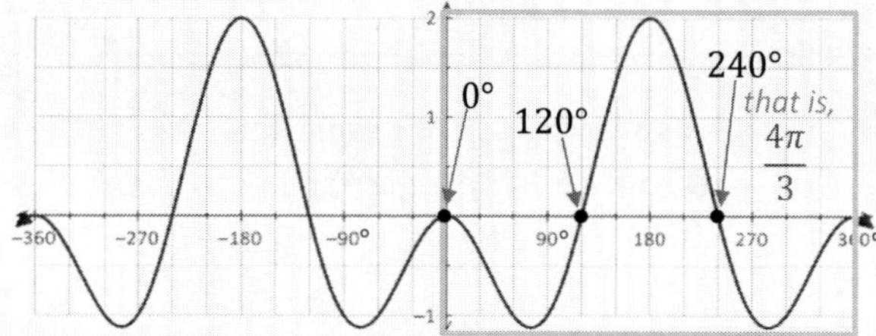

However, if we change the interval to $[-\pi, \pi)$ the solutions become:

$$\theta = -\frac{2\pi}{3}, 0, \text{ and } \frac{2\pi}{3}$$

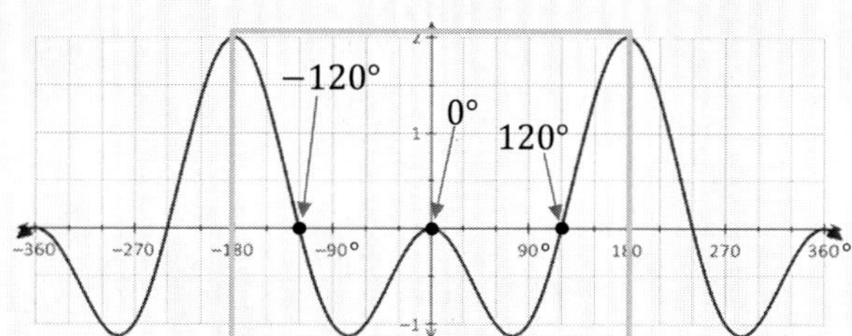

And if we *enlarge* the interval to $[-2\pi, 2\pi)$ the solutions include all of:

$$\theta = -2\pi, -\frac{4\pi}{3}, -\frac{2\pi}{3}, 0, \frac{2\pi}{3}, \text{ and } \frac{4\pi}{3}$$

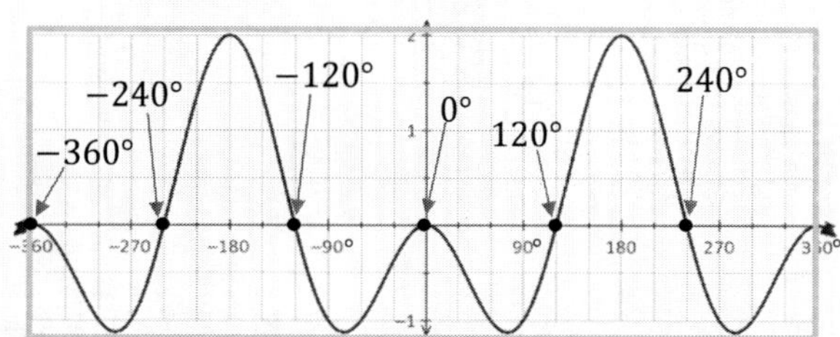

To state a **general solution** to a trigonometric equation, start with the *principal solutions* (between 0 and 2π)...

$$\theta = 0, \frac{2\pi}{3} \text{ and}, \frac{4\pi}{3}$$

... then to each add a statement indicating *"plus any coterminal angle"* (any number of rotations)

$$\theta = 0 + 2\pi n, \frac{2\pi}{3} + 2\pi n \text{ and}, \frac{4\pi}{3} + 2\pi n \Rightarrow \boldsymbol{\theta = 2\pi n, \frac{2\pi}{3} + 2\pi n, \text{ and } \frac{4\pi}{3} + 2\pi n \;; n \in I}$$

Class Example 7.42 *Trigonometric Equations – Stating Solutions*

Given the graph of the periodic trigonometric function $\sin 2\theta + \cos\theta = 0$, shown below,

(a) State the solutions to the equation $\sin 2\theta + \cos\theta = 0$, on the interval $[0, 2\pi)$.

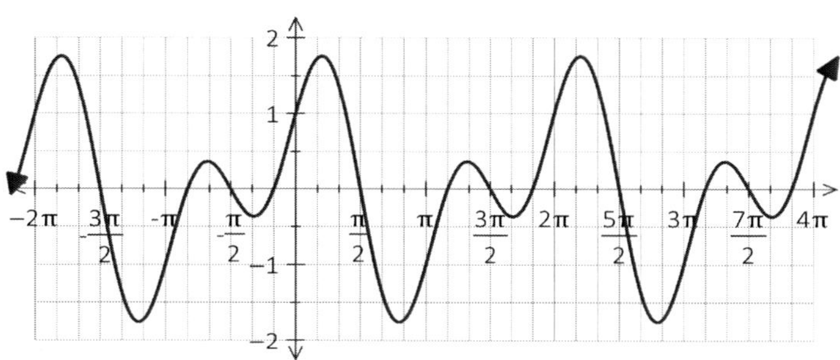

(b) Provide a simplified general solution to $\sin 2\theta + \cos\theta = 0$, in radians.

Class Example 7.43 *Solving Trigonometric Equations – Algebraic & Graphical Solutions*

Consider the trigonometric equation $\mathbf{6\cos^2 x - \cos x - 2 = 0}$;

(a) Algebraically determine the solutions - i on the interval $0 \leq \theta < 360°$, and ii - on the interval $-360° \leq \theta < 0$
Round any approximate solutions to the nearest degree.

(b) State a general solution to the equation

(c) Describe an approach to determine the solutions graphically. Verify your general solution with this method, sketch the graph(s) used in the space provided, labeling each root.

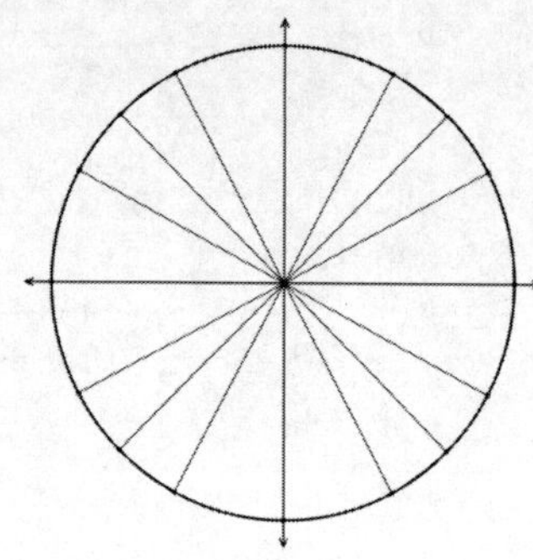

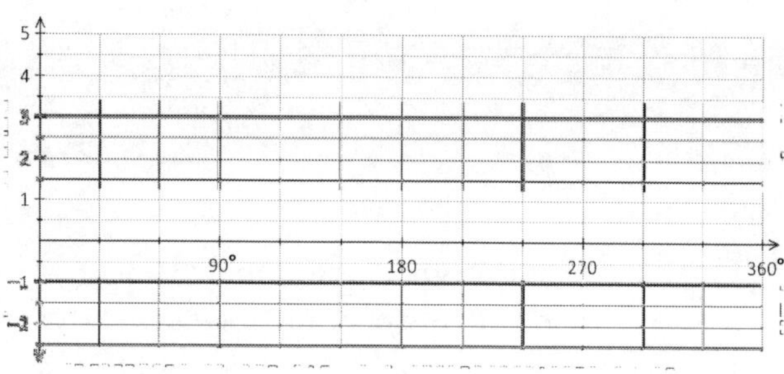

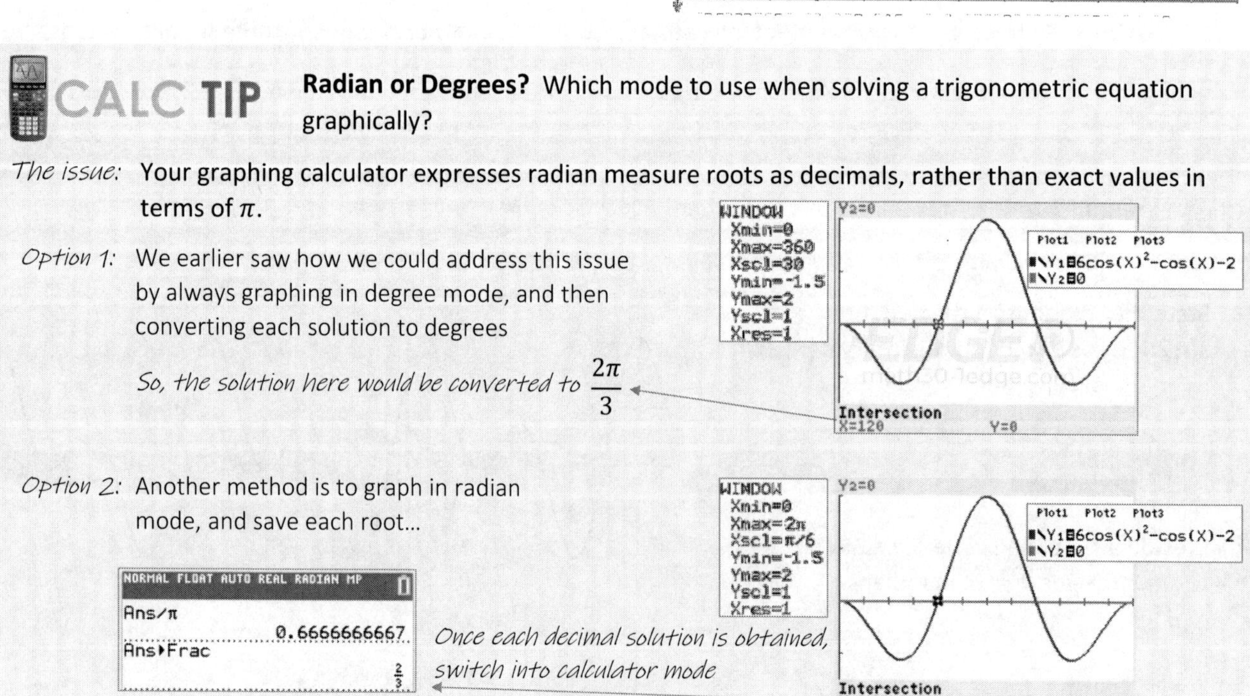

Solving Trigonometric Equations Using Identities

To solve a trigonometric equation algebraically, it is useful to have all terms involving a single trig function.

To achieve this, we'll refer to our known identities to make substitutions as required.

Class Example 7.44 *Solving Trigonometric Equations using Identities*

Consider the trigonometric equation $\mathbf{\sin 2x + \cos x = 0}$;

Algebraically determine the solutions - i on the interval $0 \leq \theta < 2\pi,$ and ii - on the interval $-\pi \leq \theta < \pi$

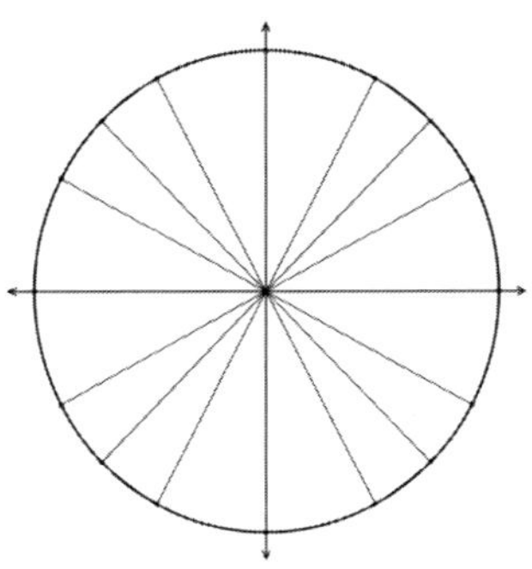

Class Example 7.45 *Solving Trigonometric Equations using Identities*

Consider the trigonometric equation $\mathbf{2\cos^2 x = \sin x + 1}$;

(a) Algebraically determine the solutions - i on the interval $0 \leq \theta < 2\pi,$ and ii - on the interval $-\pi \leq \theta < \pi$

(b) State a general solution to the equation

(c) Verify your solutions graphically

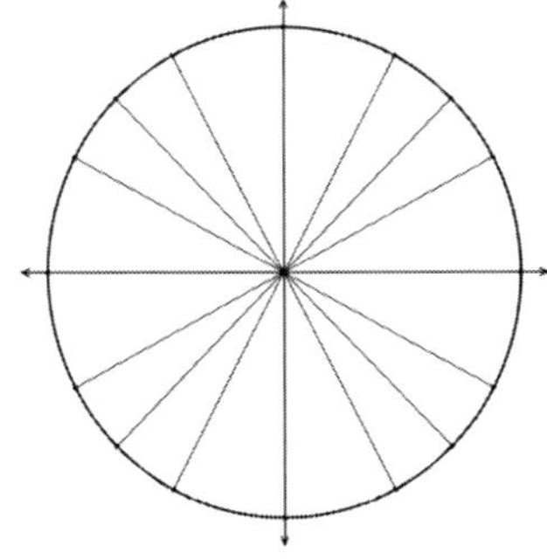

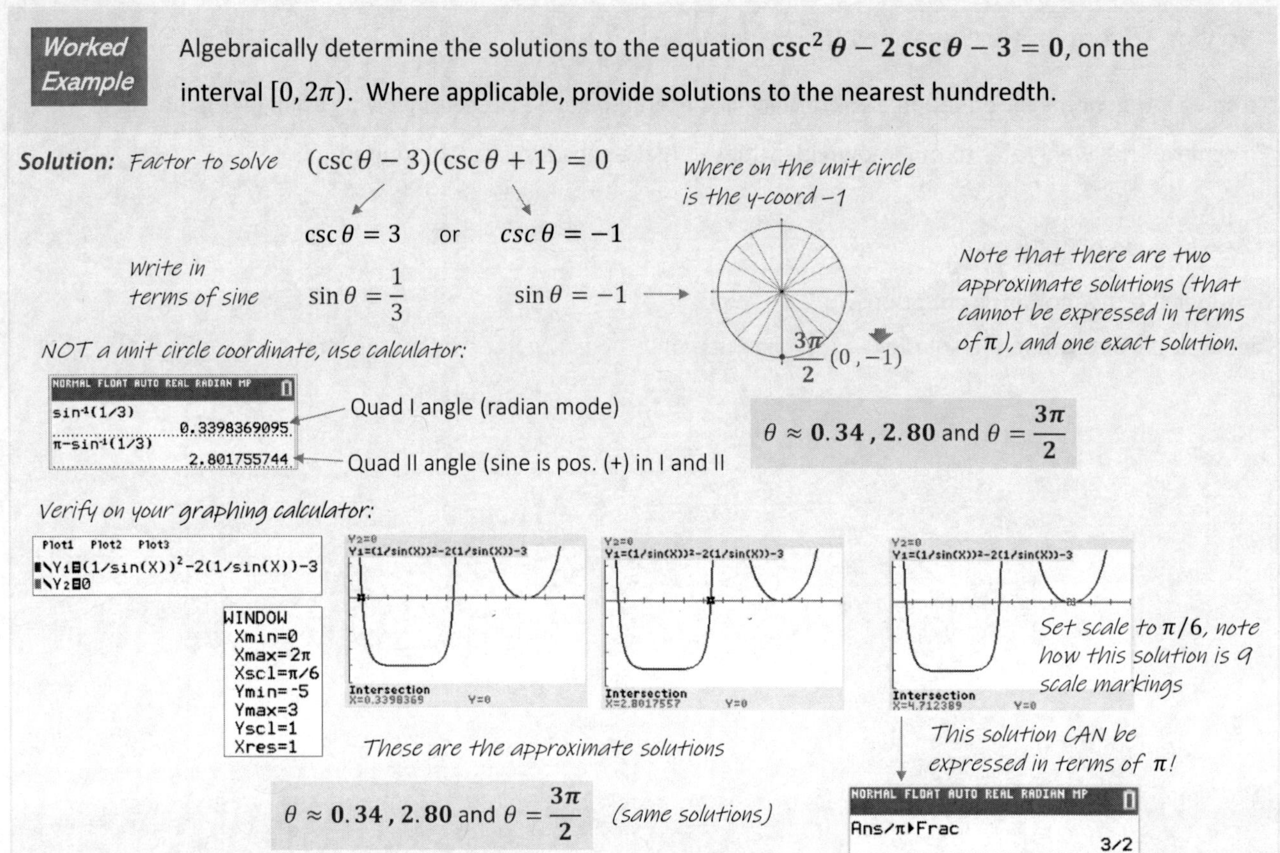

Class Example 7.46 *Solving Trigonometric Equations with Irrational Solutions*

Algebraically determine a general solution, in radians, to the equation $2\sec^2 x - 3 = \tan x$.
Verify your solutions graphically. *State any approximate solutions correct to the nearest hundredth of a radian.*

PERMUTATIONS, COMBINATIONS, AND THE BINOMIAL THEOREM

8.1 The Fundamental Counting Principle and Permutations

Suppose you are at a restaurant that has a breakfast special, where you get scrambled eggs, a choice of three types of protein, and a choice of two types of toast.

Breakfast Special	Enjoy our *Scrambled Eggs*, plus:	**Protein** Choose one of: *Bacon, Sausage,* or *Ham*	**Toast** Choose one of: *Wheat* or *Rye*

1 ➡ Assuming you'll order a complete breakfast, how many different order options do you have? Visually show how you obtained your answer.

2 ➡ The restaurant server informs you that you can also choose between coffee or juice for your beverage. How many different orders consisting of scrambled eggs, a protein, toast, and a beverage are possible?

3 ➡ The restaurant decides to get with the times and offer a plant-based protein option, a tofu patty. How many complete order options are there now (including beverage), with four choices of protein?

The fundamental counting principle (also called the counting rule) is a way to determine the number of outcomes in a problem involving different options at each stage.

If there are a options for the first stage, then b for the second, c for the third (and so on...)

Then the number of possible outcomes is $a \times b \times c \times ...$

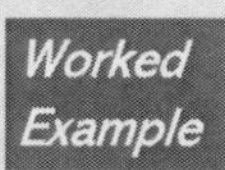

Worked Example

Suppose you planning an outfit for a day at the park. You have four choices for pants, three choices for tops, and will wear either either sandals or sneakers on your feet.

(a) How many possible outfit options do you have?

(b) How many possible outfit options do you have if you must wear sandals?

Solution: (a) **Draw a blank for each stage:**

Label what each represents →

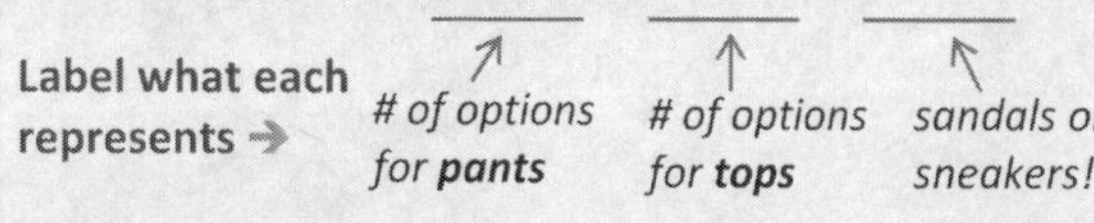

Indicate the number of options for each stage, and *multiply*:

$\underset{pants}{4} \times \underset{tops}{3} \times \underset{shoes}{2} =$ **24 possible outfits**

(b) **Start the same way. But this time, there's only one option for the last blank:**

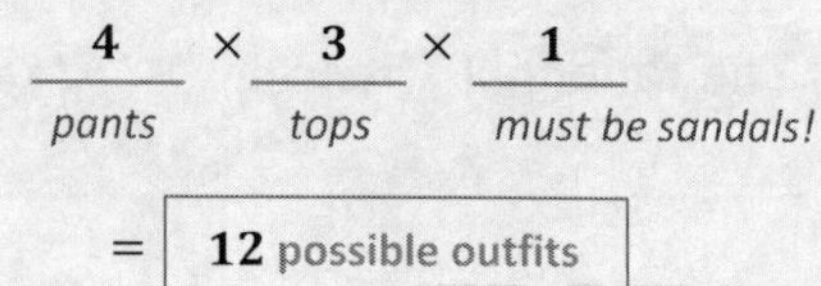

$=$ **12 possible outfits**

Class Example 8.11 *Applying the Fundamental Counting Principle*

Two children, Amelia and Matteo, are each making a birthday card and have some design choices. They can choose from blue, green, or white paper, using pencil crayons or markers for writing, and using chunky or fine glitter.

Amelia will not use green paper, as green is her least favorite colour. While Matteo has decided to use both types of glitter, as he's strongly in a glitter phase.

Determine the number of unique birthday card options for each child.

Fundamental Counting Principle with *Restrictions*

The example above involved a **restriction** – namely that the number of paper options for Amelia was restricted to just two. Problems in this unit feature all sorts of interesting restrictions, to keep you on your toes! (And to make the problems more interesting, and in need of more than just simple multiplication!)

In the following examples, watch for any restrictions, and how they affect the number of choices for any stage.

Worked Example

Website A has a security feature where users must choose a four-character passcode, consisting of two digits followed by two letters.

The passcodes for **website B** are the same as for website A, except that also the first digit must be odd, the first letter must not be O or I, and the second letter cannot be the same as the first letter.

Determine the number of passcodes that can be created for each website.

Solution: **Website A:** **Draw four "blanks"**, for each of the stages of the passcode:

$\underset{1^{st}\ digit}{10} \times \underset{2^{nd}\ digit}{10} \times \underset{1^{st}\ letter}{26} \times \underset{2^{nd}\ letter}{26}$ ← **Multiply** the number of options

note: There are **10 digits** (from 0 to 9) and **26 letters** (from A to Z!)

$=$ **67 600 possible passcodes**

Website B: **Indicate the restriction associated with each option**

$\underset{\text{Must be 1, 3, 5, 7, or 9}}{5} \times \underset{\text{Any digit}}{10} \times \underset{\neq \text{O or I}}{24} \times \underset{\neq \text{first letter}}{25}$

← For the **first letter**, there are 26 letters, minus the two restricted letters, for a total of 24.

For the **2nd letter**, O and I are now okay! But whatever the first letter was is restricted, so there's one less letter to choose from.

$=$ **30 000 possible passcodes**

Class Example 8.12 *The Fundamental Counting Principle with Repeated Options*

A short quiz consists of six multiple choice questions, each with four possible answers A, B, C, or D.

(a) Determine the number of possible answer keys for this quiz.

(b) Suppose the teacher told the class that the answer for question 2 was the same as question 1, which was different than question 3. Determine the number of possible answer keys with these restrictions.

Class Example 8.13 *Counting options for digits*

(a) Determine the total number of three-digit numbers that can be made. *Hint: Consider the number of options you'd have for the first digit – could it be any digit if you're writing a three-digit number?*

(b) Determine the number of three-digit numbers that are lower than 500.

(c) Determine the number of **odd** three-digit numbers that are lower than 500.

Class Example 8.14 *License Plates*

The license plates in a particular jurisdiction originally consisted of any three letters, the first of which must be A, followed by any three digits, the first of which must be either 8 or 9.

Recently this was changed, so that currently each license plate consists of any three non-repeating letters, the first of which must not be O or I, followed by any three digits, the first of which cannot be 1 or 0.

Determine the number of additional license plate options available after the jurisdiction made this change.

Arranging Elements in a Set of Objects

Many problems in this unit involve **arranging** all (or some) of the objects in a set of objects.

Two classic examples involve arranging people, and arranging letters:

How many ways can Adbel, Barack, Cecil, and Dharia be seated in a row of four desks?

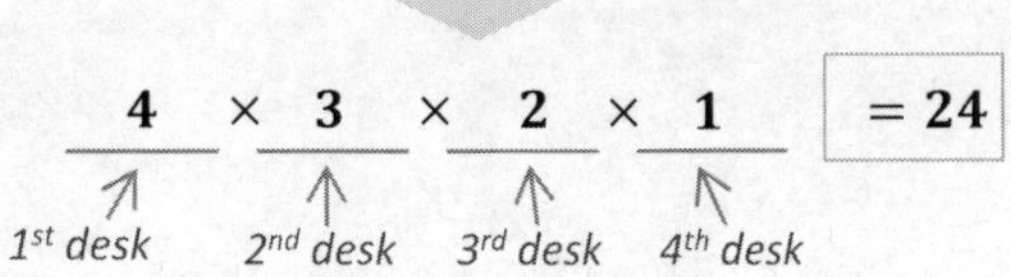

Any of Abdel, Barrack, Cecil, or Dharia can take the first desk – 4 options.

From there, there will be 3 students remaining for the second desk, and so on.

How many unique arrangements can be made using the letters in the word FRUIT?

$5 \times 4 \times 3 \times 2 \times 1 = 120$

1st letter, 2nd letter, 3rd letter, 4th letter, 5th letter

*Option for the first letter are **F**, **R**, **U**, **I**, or **T** – 5 options.*

For each subsequent letter there is one less option, as they are being used up as we proceed.

Notice that with each example above, we start with the number of objects we had to start, then decreased by one as we multiplied down to 1.

The number of ways to arrange $\boldsymbol{n}$ objects is given by: $n \times (n-1) \times (n-2) \times \cdots \times 1$

We refer to this product as $\boldsymbol{n}!$ ← *"n factorial"*

So for the desks example above, the solution is $\mathbf{4}!$. For the FRUIT example, it's $\mathbf{5}!$

You can calculate factorials on your calculator.

Use the math button, and select the PROB menu →

Then select #4

Beats keying in $5 \times 4 \times 3 \times 2 \times 1$ *(exclamation point)*

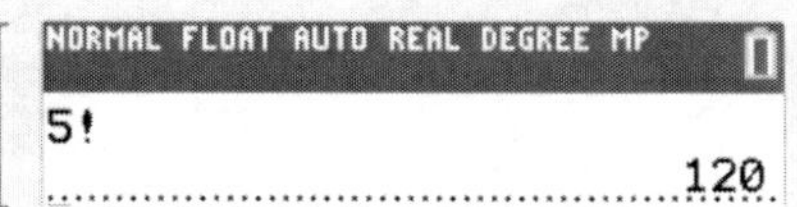

Class Example 8.15 *Arranging a group of people*

Four boys and four girls are to be arranged in a single row for a picture. How many ways can this be done if the boys and girls must alternate? *Hint: Consider two scenarios, start with a boy or start with a girl*

Class Example 8.16 *Arranging Letters*

In how many ways could the letters in the word SENIOR be arranged if:

(a) The first letter must be S

(b) The first and last letters must be vowels *(that is, E, I, or O)*

Simplifying Factorial Notation Expressions

We saw how factorial notation allows us to concisely express long calculations.

For example, $6! = 6 \times 5 \times 4 \times 3 \times 2 \times 1$ ➡ $= \mathbf{720}$

When two factorials are expressed as quotients, they can be simplified without a calculator.

For example

(a) $\dfrac{7!}{5!}$

Start expanding the larger term....

$= \dfrac{7 \times 6 \times \cancel{5!}}{\cancel{5!}}$

Instead of expanding all the way down to 1, express the remaining product as $5!$

... then cancel out with the $5!$ *in the denominator.*

$= 7 \times 6$

$\boxed{= 42}$

(b) $\dfrac{(\boldsymbol{n}+\mathbf{1})!}{\boldsymbol{n}!}$

$= \dfrac{(n+1) \times \cancel{n!}}{\cancel{n!}}$

Expand the numerator, as it's larger.

$\boxed{= \boldsymbol{n}+\mathbf{1}}$

After the $(n+1)$*, this is just* $n!$

*Note that $(\boldsymbol{n}+\mathbf{1})! = (n+1) \times \underbrace{n \times (n-1) \times (n-2) \ldots}_{\text{this is } n! \text{ (cancels out)}}$

Class Example 8.17A *Simplifying Expressions with Factorials*

SET I - Simplify each of the following, to evaluate without a calculator:

(a) $\dfrac{8!}{6!}$

(b) $\dfrac{10!}{7!}$

(c) $\dfrac{9!}{8 \times 6!}$

SET II – Determine the simplest expression for each of the following.

(d) $\dfrac{n!}{(n-2)!}$

(e) $\dfrac{(n+1)!}{(n+3)!}$

Class Example 8.17B *Solving an Equation with Factorials*

Algebraically solve the equation: $\dfrac{(n+1)!}{(n-1)!} = 42$

Defining Permutations

We've already looked at counting the number of ways to **arrange** a set of objects. *(Or, as below, people!)*

*How many ways can **Adbel, Barack, Cecil,** and **Dharia** be seated in a row of four desks?* } There are $4! = \mathbf{24}$ ways to arrange these students

→ Another way of saying this is there are ***24 permutations*** of these students

A **permutation** is an arrangement of all or part of a set of objects, where order matters.
For example: A, B, C and *C, A, B* are two different permutations of the first three letters in the alphabet.

In all there are:
$3! = \mathbf{6}$ permutations

The 6 *Permutations*, or Arrangements, of the letters a, b, *and* c

a, b, c	b, a, c	c, a, b
a, c, b	b, c, a	c, b, a

Permutations of Some (not all!) of the Available Objects

Suppose 7 students are running in a race:

1. **A**bdel
2. **B**arack
3. **C**ecil
4. **D**haria
5. **E**ve
6. **F**rancis
7. **G**arreth

We've already seen how the total number of possible ways the students can finish, 1st through 7th place, is 7!

That is, there are 7! or 5040 permutations of all of the students.

Now, how many ways could we award a 1st, 2nd, and 3rd place medal among the 7 racers?

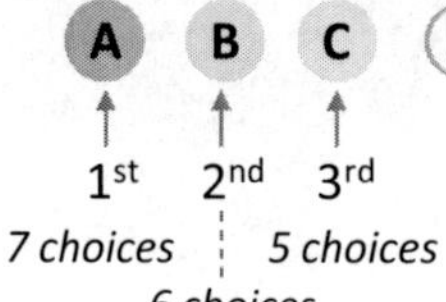

Note how if we start at "1st place" we have 7 choices, which leaves 6 choices for 2nd, and 5 choices for 3rd.

By Fundamental Counting Principle there are: $7 \times 6 \times 5 = \mathbf{210}$ ways to award the medals.

Another Way to Look at This: $7! = \boxed{7 \times 6 \times 5} \times 4 \times 3 \times 2 \times 1$

This is the part we want (the boxed $7 \times 6 \times 5$); *The part we **don't want** is* $4!$

So, the number of permutations of the top 3 students, from a group of 7, is:

Arrangements of all the students → $7!$
"Leftover" students / we don't want → $4!$

$$\frac{7!}{4!} = \frac{7 \times 6 \times 5 \times \cancel{4} \times \cancel{3} \times \cancel{2} \times \cancel{1}}{\cancel{4} \times \cancel{3} \times \cancel{2} \times \cancel{1}} \quad \text{or...} \quad \frac{7!}{(7-3)!}$$

Total number of students, n (the 7)
\# of students being arranged, r (the 3)

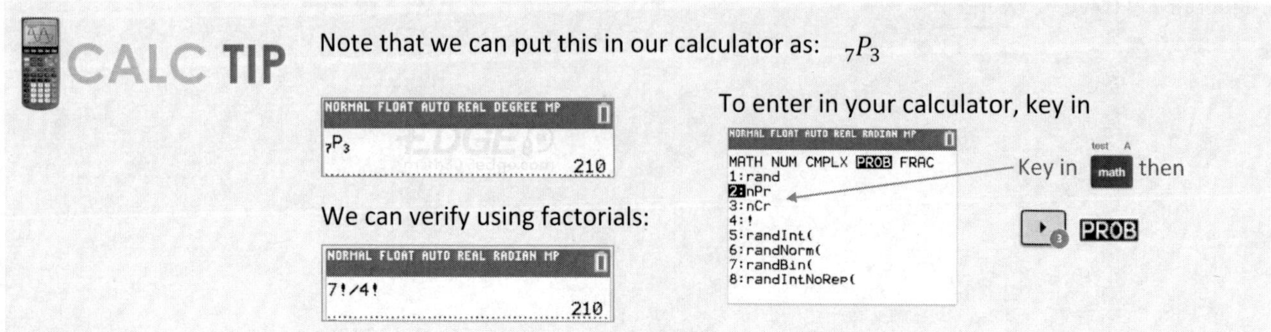

The Permutation Formula

From the racing example on the pervious page, we say that the number of ways r items can be arranged, from a larger group of n items, is:

$$\text{\# of PERMUTATIONS} = \frac{\text{\# of ways to arrange all } n \text{ items}}{\text{\# of ways to arrange items } \textit{not wanted}} \qquad \text{That is: } {}_nP_r = \frac{n!}{(n-r)!}$$

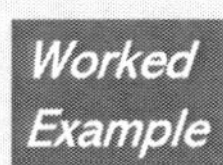

Consider the letters in the word OLYMPICS. Use the permutations formula to determine:

(a) The number of arrangements that can be made using all of the letters

(b) The number of arrangements that can be made using any three of the letters

Solution:

(a) There are 8 letters (n), from which we wish to arrange 8 (r).

$${}_8P_8 = \frac{8!}{(8-8)!}$$

Note that 0! is equal to 1 →

$$= \frac{8!}{0!}$$

$$= 8! \Rightarrow = \boxed{40\,320}$$

(b) There are 8 letters (n), from which we wish to arrange 3 (r).

$${}_8P_3 = \frac{8!}{(8-3)!}$$

# of letters **not** being arranged →

$$= \frac{8!}{5!}$$

$$= 8 \times 7 \times 6 \Rightarrow = \boxed{336}$$

Defining 0!

In the example above (a), from fundamental counting principal we know that the number of arrangements of all 8 letters is 8!

Now, in order for our ${}_nP_r$ formula to work, we must **define 0! to equal 1**.

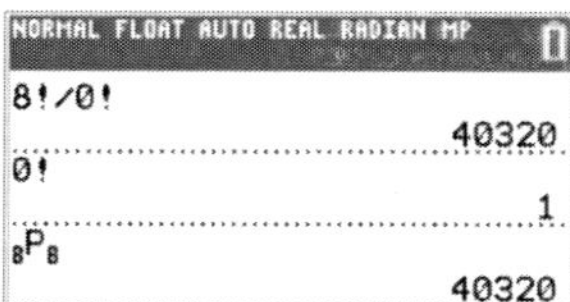

→ For many straightforward permutation's questions like these, we often **stick with the fundamental counting principle.** (That is, we don't often use ${}_nP_r$.)

For the worked example above, our solutions could look like: (a) $\mathbf{8! = 40\,320}$ (b) $\underline{\mathbf{8}} \times \underline{\mathbf{7}} \times \underline{\mathbf{6}} = \mathbf{336}$

Class Example 8.18 *Fundamental Counting Problems as Permutations*

A drama teacher is casting for a play with 3 roles for senior students and 2 roles for junior students. 5 senior students and 6 juniors out. Determine the number of ways the teacher can assign the roles using:

(a) The Fundamental Counting Principle

(b) Permutations

Class Example 8.19 *License Plates, Revisited*

License plates in a particular jurisdiction consist of three different letters followed by two different numbers. If the digit 0 and the letters I and O are not used, determine how many distinct license plates are possible by:

(a) The Fundamental Counting Principle

(b) Permutations

8.1 Practice Questions (Sample, first page only)

SAMPLE of our high-quality, practice questions.
Visit **Math30-1edge.com** for full access.

1. A father is preparing a pasta for his children's meal.
- For the pasta he has a choice of **spaghetti**, **penne**, **rigatoni**, or **orzo**.
- For sauce, he has **marinara** (red) or **alfredo** (white).
- And for meat he has **chicken**, **ground beef**, or **shrimp**.

(a) Assuming he makes one choice each for the type of pasta, sauce, and meat, determine the number of different meal options he can make.

(b) The dad remembers his daughter will not eat anything with white sauce. Determine the number of meal options with this restriction.

2. A child has made an elaborate fort in his room, that has five entrances. How many ways can the child

(a) Enter and exit the fort through any entrance.

(b) Enter the fort and exit through a different entrance.

(c) Enter and exit through the same entrance.

3. Five students enter a classroom with five desks spaced in a single row.

(a) How many ways can the students be arranged in the desks?

(b) How many ways can the students be arranged, if one of the students, Mary, takes the closest desk to the door.

4. Nine athletes line up for a race where the top three will win ribbons.

(a) How many ways can the ribbons be awarded?

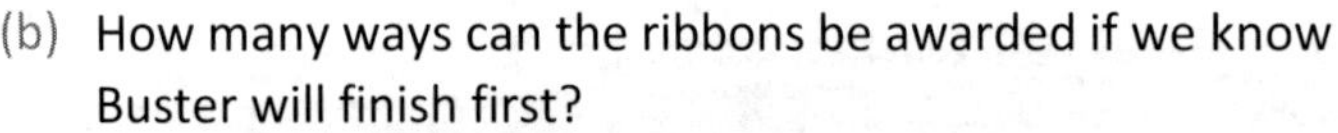

(b) How many ways can the ribbons be awarded if we know Buster will finish first?

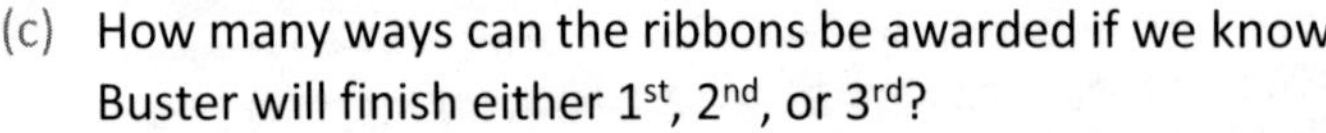

(c) How many ways can the ribbons be awarded if we know Buster will finish either 1st, 2nd, or 3rd?

5. A teacher gives a test where the first two questions are "True / False", followed by 7 multiple choice questions with possible answers A, B, C, or D.

(a) How many different answer keys are there?

(b) The teacher tells his students that true / false questions have the same answer. Further, he mentions that the 2nd multiple choice question has a different answer than the first multiple choice question.

Determine the number of possible answer keys with these restrictions.

Answers to Practice Questions

1. (a) 24 (b) 12 **2.** (a) 25 (b) 20 (c) 5 **3.** (a) 120 (b) 24 **4.** (a) 504 (b) 56 (c) 168
5. (a) 65 536 (b) 24 576

8.2 Problem Solving with Permutations

Warm-Up #1 A family of four; Mom, Dad, and two girls Becky, and Sandra are lining up in a single row for a family photo. M D B S

Dad, always one to throw out challenges, asks his children how many different the family can arrange themselves in a single line.

1 ➡ Determine the number of ways the family members can be arranged in a single line

2 ➡ Determine the number of ways the family members can be arranged if Mom must be on the far left

The photographer suggest they take a few pictures where mom and dad are standing together, with mom on the left and dad on the right. Once again Dad asks how many different ways this can be done.

Sandra, who studies grade 12 math and loves logic problems, asserts that since they are side-by-side in each photograph, Mom and Dad can be counted as one person. M,D B S

3 ➡ Determine the number of ways the family members can be arranged in this fashion, with Mom and Dad together, with mom on the left and dad on the right.

4 ➡ Determine the number of ways the family members can be arranged in this fashion, with Mom and Dad together, on either side of each other.

In this section we'll explore some common types of restrictions in problems involving permutations.

One such restriction is when one (or more) of the objects being arranged *must be in a specific spot*, as with #2 above. In these cases, we can simply **arrange the remaining objects**.

Permutations where Certain Objects are Together

As we saw above, the key to approaching arrangements where some of the objects must grouped together is to count the grouped objects as **one**.

Number of arrangements where some objects *must be grouped together* = # of arrangements, counting grouped objects as one × # ways to arrange the grouped objects among themselves

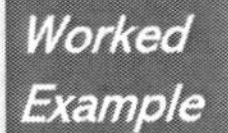

In how many ways can the letters in the word DIPLOMA be arranged if:

(a) The vowels must be together, in alphabetical order (b) The vowels must be together

Solution:

(a) **Count the vowels as one:**

There are 5 letters to arrange

$= 5! \times 1$ ➡ $=$ **120**

Ways to arrange all letters, *counting vowels as one* — Ways to arrange vowels *in alphabetical order*

(b) *This time, vowels can be in any order*

Count the vowels as one:

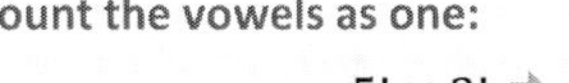

$= 5! \times 3!$ ➡ $=$ **720**

Ways to arrange all letters, *counting vowels as one* — Ways to arrange vowels among themselves

Class Example 8.21 *Arranging members of a Team*

Eight members of a badminton team plus three coaches line up in a single row for a team photo. In how many ways can this be done if:

(a) The coaches must all be together

(b) The coaches must all be together and at the centre of the row

(c) The coaches must be together in the middle of the row, with head coach Steve at the very centre

Worked Example

In how many ways can the letters in the word ORGANIZE be arranged if all four of the vowels must NOT be together.

Solution: **Use an indirect approach:** $= 8! - 5!\,4!$ ➡ $= \boxed{37\,440}$

Arrangements of all the letters, *without restriction* − Arrangements where the vowels ARE together: = Arrangements where the vowels ARE NOT together

R | G | N | Z | OAIE

Class Example 8.22 *Arranging different Books*

The top row of a bookshelf contains 8 books; 4 novels, 2 textbooks, and 2 children's books. In how many ways could these books be arranged if:

(a) The 4 novels, 2 textbooks, and 2 children's books must all be grouped together.

(b) The books can be in any order, except that the textbooks must **not** be together.

Worked Example

In how many ways can the letters in the word OBJECT be arranged, where each arrangement has all the vowels together or all the consonants together.

Solution: **Consider two cases:** $= 5!\,2! + 3!\,4!$ ➡ $= 240 + 144$ ➡ $= \boxed{384}$

Arrangements where all the vowels are together + Arrangements where all the consonants are together

Arrange 5 letters (counting OE as one) { B | J | C | T | OE

O | E | BJCT

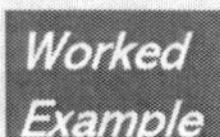

note: For "**or**" problems, consider each case separately and **ADD**

Class Example 8.23 *Arranging Disney Characters*

A child is arranging her unique Disney figures in a single row. She has **four princesses** (Aurora, Tiana, Jasmine, and Moana), **three animals** (Simba, Mushu, and Louis the alligator), and **two villains** (Maleficent and Ursula).

How many ways can these figures be arranged if:

(a) The four princesses must be together, *or* the three animals must be together.

(b) The two villains must not be together.

Permutations with Repeating Objects

Warm-Up #2

Original Display

A store manager is making a fruit display at the entrance to the produce department. For the display she is arranging a pear, a banana, an apple, and an orange, as shown above.

1 ➡ **Determine the number of fruit displays that can be made in this manner.**

Later that day the manager realizes they've sold out of oranges. She decides to replace the orange with an apple, which is visually identical to the first apple.

New Display

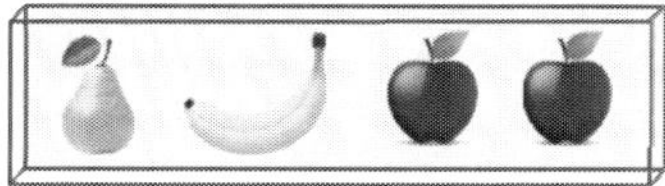

The manager realizes that with the original display, rearranging the last two items results in a different arrangement. However, with the new display, rearranging the last two items doesn't change anything:

Original Display **P B A O** → **P B O A**
Rearranging the last two items results in a different arrangement

New Display **P B A A** → **P B A A**
Rearranging the last two items **does not** result in a different arrangement

2 ➡ **Make a conjecture on whether more or less distinct arrangements can be made with the new display.**

With the Permutations formula we saw that from n objects, the number of arrangements that can be made using r of the objects, is given by:

$$\text{\# of PERMUTATIONS} = \frac{\text{\# of ways to arrange all } n \text{ items}}{\text{\# of ways to arrange items } \textit{not wanted}} \quad \Rightarrow \quad {}_nP_r = \frac{n!}{(n-r)!}$$

Determining the number of arrangements here, where some of the objects repeat, actually involves the same process:

$$\text{\# of ways to arrange } n \text{ objects where some objects repeat} = \frac{\text{\# of ways to arrange all } n \text{ items*}}{\text{\# of ways to arrange items } \textit{that are identical*}}$$

**We do these two calculations as if the objects were all different. The repeated arrangements in the numerator are accounted for by dividing the number of repetitions.*

3 ➡ **Use the method described above to determine the number of distinct arrangements that can be made in the new display, with two apples.**

If n objects are being arranged, where a of one kind are identical, as are b of a second kind, c of a third kind, and so on...

The number of distinct arrangements is given by: 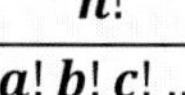 $\frac{n!}{a!\,b!\,c!\,...}$

Worked Example

(a) How many distinct arrangements can be made using the letters in the word: BLISSFULL

(b) In how many of the arrangements would the first letter be S?

(c) How many arrangements can be made if the ***L***s must be all together? *Disregard restriction from (b)*

Solution: (a) **Use:** $\frac{n!}{a!\,b!\,c!\,...}$ ➜ $= \frac{9!}{3!\,2!}$ ➜ $= \boxed{30\,240}$

There are **3** *L*s — And there are **2** *S*s

(b) **One of the S's is used up in the first spot. Arrange the remaining 8 letters: "BLISFULL"**

$= \frac{8!}{3!}$ ➜ $= \boxed{6\,720}$

There are **3** *L*s

(c) **Arrange:** B | I | S | S | F | U | LLL

Count the Ls as one letter, gives 7 to arrange

$= \frac{7!}{2!}$ ➜ $= \boxed{2\,520}$

There are **2** *S*s

Note that *there is only one way to arrange the **L**s among themselves, so no need to further multiply.*

note: In some problems one (or more) of the objects being arranged are assigned to a fixed position. The key is to disregard that object, and arrange the remaining, "free" objects!

Class Example 8.23 *Arranging letters in a Word – with Repetition*

For the questions below, refer to the word: SASKATOONS

(a) How many distinct arrangements can be made using all of the letters in the word?

(b) How many of the arrangements begin and end with an S?

(c) How many distinct arrangements have all three of the ***S***s together?

(d) How many arrangements do not have all three of the ***S***s together?

Class Example 8.24 *Arranging Items where some are Identical*

An outdoor supplies store has a display of patio tables that are identical except for colour. The display consists of a row of 3 green tables, 2 yellow tables, 1 brown table, and 2 white tables.

Determine the number of distinct displays that can be made where:

(a) A yellow table must be at both ends of the row.

(b) Both ends must have the same colour table.

(c) The two white tables must not be together.
Disregard earlier restrictions

8.3 Combinations

Exploration #1

Remember our group of students, from a previous section:

1. **A**bdel
2. **B**arack
3. **C**ecil
4. **D**haria
5. **E**ve
6. **F**rancis
7. **G**arreth

Let's suppose that each of these students are running for student council, where the positions are President, Secretary, and Treasurer.

1 ➡ **Determine the number of ways the student council positions can be filled from this group. Is the order in which the students are elected matter?**

Now let's suppose that after filling the positions above, the four leftover students were **A**bdel, **B**arack, **C**ecil, and **D**haria. From these four students, three student council advisors are needed.

2 ➡ **Does the order in which the three students are selected to fill the advisor groups matter?**

3 ➡ **List each of the possible three-member advisor groups that can be formed from the four leftover students.**
The first is listed for you, using the first letters of their names.

A, B, C

Exploration #2

Above you just listed all of the 3-letter combinations (where order doesn't matter) of A, B, C, and D.

Let's compare that to the list of all of the 3-letter permutations of the same letters:

A, B, C	A, B, D	A, C, D	B, C, D
A, C, B	A, D, B	A, D, C	B, D, C
B, A, C	B, A, D	C, A, D	C, B, D
B, C, A	B, D, A	C, D, A	C, D, B
C, A, B	D, A, B	D, A, C	D, B, C
C, B, A	D, B, A	D, C, A	D, C, B

1 ➡ **From this list and that you made above, we can see that:**

From 4 objects, the number of **combinations** of any 3 of them is ______.

From 4 objects, the number of **permutations** of any 3 of them is ______.

2 ➡ **How many times larger is the number of permutations compared to combinations?**
How does this result relate to the list of permutations above?

In a **permutation**, or arrangement of objects, the order is taken into account.

In a **combinations**, or selection of objects, order is not taken into account.

If from n objects, r are selected, The number of COMBINATIONS $= \dfrac{\text{\# of permutations}}{\text{\# of variants for each selection}} = \dfrac{_nP_r}{r!}$

Note that in warm-up 2 we found that the number of PERMS was $r!$ times larger than the number of COMBS.

You may also note that there's a similarity with the way we developed the permutations formula,

$$\text{\# of PERMS} = \frac{\text{\# of ways to arrange all } n \text{ items}}{\text{\# of ways to arrange items \textit{not wanted}}}$$

In both cases we divide by the number of redundancies counted in the numerator.

The Combinations Formula

On the previous page we saw how the number of COMBS is the number of PERMS, divided by *the redundancy* (the number of ways each selection can be arranged), $r!$ ➡ $_nC_r = \frac{_nP_r}{r!}$

Substituting the permutations formula gives: $_nC_r = \frac{\frac{n!}{(n-r)!}}{r!}$ Which rearranges to: $_nC_r = \frac{n!}{(n-r)!\,r!}$

The Formulas for Permutations and Combinations:

If from $\boldsymbol{n}$ objects, $\boldsymbol{r}$ are selected, $_n\boldsymbol{P}_r = \frac{\boldsymbol{n}!}{(\boldsymbol{n}-\boldsymbol{r})!}$ $_n\boldsymbol{C}_r = \frac{\boldsymbol{n}!}{(\boldsymbol{n}-\boldsymbol{r})!\,\boldsymbol{r}!}$ ← *The combs formula has an extra $\boldsymbol{r}!$* ***(# of combs is LESS)***

→Think PERMS where order matters and think COMBS where order need not be taken into account.

Worked Example A dance coach requires 5 volunteers for a demonstration. She wishes for three of the volunteers to be girls, and two to be boys, and there are 9 girls and 10 boys in the class.

(a) How many ways can the teacher select her volunteers, if one of the girls selected must be the top dancer, Sarah.

(b) How many ways can this be done if Sarah must still be selected, but the remaining volunteers can be boys or girls. *Disregard gender requirements from (a)*

Solution: (a) **If Sarah must be selected, only two more girls (plus the two boys) are needed.**

$_8C_2 \times {_{10}C_2}$

From the **8** girls (not incl. Sarah) "choose" **2** of them *AND* From the **10** boys choose **2**

$= \boxed{1260}$

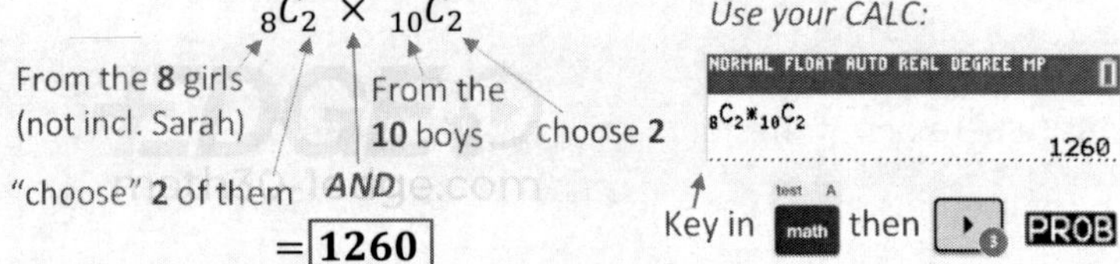

(b) **Here we'll need four more volunteers (since we have Sarah), but from the *whole class*, as we aren't concerned with "boys / girls")**

$_{18}C_4$

From the **18** students (subtract Sarah from the class of 19, she's already selected) choose **4**

$= \boxed{3060}$

Let's take a closer look at a combs term from the solution to (a) above, where we had $\boldsymbol{_8C_2}$ (from 8 girls, choose 2). Notice how we can reason out the calculation of number of combinations, and see how to relate to the formula :

$_nC_r = \frac{n!}{(n-r)!\,r!}$ ➡ $_8C_2 = \frac{8!}{6!\,2!}$

of **perms** of all 8 students
→ *First, this is the $\boldsymbol{n}!$ in the formula*

Make answer smaller by first dividing by # of students NOT used
→ *Second, this is the $(\boldsymbol{n}-\boldsymbol{r})!$*

And dividing by # of ways to arrange each group of 2 (since order doesn't matter)
→ *And this is the $\boldsymbol{r}!$*

We can evaluate this by expanding: $\frac{8!}{6!\,2!} = \frac{8 \times 7 \times \cancel{6!}}{\cancel{6!} \times 2 \times 1}$ ➡ $= \mathbf{28}$

Same as from our CALC!
NORMAL FLOAT AUTO REAL DEGREE MP
$_8C_2$
28

Class Example 8.31 *Selecting Objects - Records*

A record store has a promo shelf with various records on sale. In it are 5 different jazz records and 6 blues records. George is looking through the promo shelf items and decides he'll buy 4 records.

(a) Without concern for the music types, how many different selections of 4 records can George make?

(b) How many selections can George make if he'd like 2 jazz records and 2 blues records?

(c) How many selections can George make if he'd like 2 jazz records and 2 blues records, and a particular B.B. King album (blues) must be selected?

With combinations order doesn't matter! Suppose your locker combination is **5-10-15** ...as it would not open with **15-10-5**, it is in fact a **permutation lock**.

Same combination as 5-10-15 (!)

Class Example 8.32 *Problems involving Combinations – Cards*

A standard deck of cards is shown below.

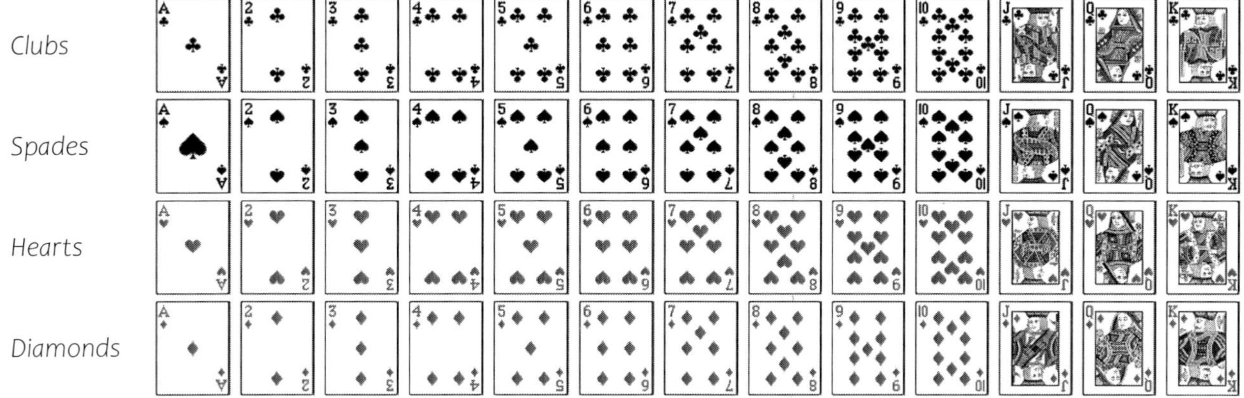

Determine the number of five card poker hands that can be formed containing:

(a) Any of the 52 standard playing cards.

(b) Only face cards. (Jack, Queen, King)

(c) Only cards of the same suit (all clubs, all spades, etc)

(d) Three aces and two kings (a "full house")

(e) Four kings

(d) Exactly two kings or exactly three kings

Class Example 8.33 *Problems involving Combinations – Committees*

The workforce of a small company includes 6 people who work in sales, 4 who work in finance, and 2 in human resources (HR). From these employees a strategy committee of 5 members must be chosen.

How - any ways can this committee be chosen if: Note: "Reset" for each part, treat as a separate questions

(a) The two HR members must both be included.

(b) Two members from sales, two from finance, and one from HR must be included. *Disregard restriction from (a)*

(c) The two HR members cannot both be included. *Disregard restriction from (b)*

(d) There must be exactly one or two people from sales. *Disregard restrictions from (a), (b), and (c).*

Combinations Problems Involving Cases

As we saw in the last example (as well as back in permutations), some problems involve breaking down into cases and adding the results. *(OR = "add")*

We'll next further explore these types of problems, including "at most" and "at least" scenarios.

Worked Example

A dance coach requires 5 volunteers for a demonstration. Her dance class consists of 9 girls and 10 boys.

(a) How many ways can the teacher select her volunteers, if there can be at most 2 boys?

(b) How many ways can the teacher select her volunteers, if there must be at least 1 boy?

Solution: (a) **Consider all cases... there can be:**

0 boys *(and 5 girls)* **or** 1 boy *(and 4 girls)* **or** 2 boys *(and 3 girls)*

$$_{10}C_0 \times {_9C_5} + {_{10}C_1} \times {_9C_4} + {_{10}C_2} \times {_9C_3} = \boxed{5166}$$

(b) **There are four cases here – *There can be*:** 1 boy **or** 2 boys **or** 3 boys **or** 4 boys **or** 5 boys

A better method than adding 5 calculations is to use an INDIRECT APPROACH

Total – Unwanted ➡ $_{19}C_5 - {_9C_5} \times {_{10}C_0} = \boxed{11\,502}$

Total: Number of groups of 5 volunteers, without regard to gender

Unwanted: Number of groups that are *all girls*

$_9C_5 \times {_{10}C_0}$: *Choose 5 girls and 0 boys*

*(Since $_{10}C_0$ is just **1**, you can simply do $_{19}C_5 - {_9C_5}$)*

Calculations involving combinations problems with many cases can instead be approached indirectly, using ***Total*** – ***Unwanted***

Class Example 8.34 *Problems Involving Combinations – Selecting Objects*

A child is taking some of her Disney figures a family trip. She has **four princesses** (Aurora, Tiana, Jasmine, and Moana), **three animals** (Simba, Mushu, and Louis the alligator), and **two villains** (Maleficent and Ursula).

For her trip she will select four of these figures to bring along. How many ways can she do this if:

(a) She wishes for at least one animal among the figures

(b) She wishes to bring at least one of each type (princesses, animals, and villains)

Patterns in Combinations

warm up $_nP_r = n!/(n-r)!$

Exploration #3

Horace is a basic cable subscriber, and he's considering enhancing his service. (He's been home a lot lately)

The cable company tells him he has the option of adding any of their available channel bundles, and he'll be billed accordingly.

Channel Bundles
- ☐ News Plus
- ☐ Total Sports
- ☐ More Movies
- ☐ Home and Garden
- ☐ History
- ☐ Sci-Fi

1 ➡ Given that Horace could choose to not enhance his service, or at the other end he may go full-out and add everything, fill out the blanks below for each possible case.

Horace can add:

___ *bundles or* ___ *bundles or* ___ *bundles or* ___ *bundles or* ___ *bundles or* ___ *bundles or* ___ *bundles*

2 ➡ For each of the cases above, indicate a combinations expression to determine the number of options.

3 ➡ Horace learns that the cable company is having a promotion where you can select any four bundles for a discounted price. To choose this promotion, he realizes that he could simply decide on the two bundles he doesn't want. Use combinations to show that Horace is correct.

Did you spot the pattern? You may have noticed the repetion in number of combinations $_nC_r$, as we progress from r down to zero.

When selecting r objects from a group of n, there will be "$n - r$" objects left over.
So determining the number ways to select either "r" or "$n - r$" objects from a group n ... *is the exact same problem.*

➡ $_nC_r = {_nC_{n-r}}$

4 ➡ **Determine each missing value, indicated by each ☐:** (a) $_9C_3 = {_9C_{☐}}$ (b) $_{50}C_{43} = {_{50}C_{☐}}$

Class Example 8.35 *The Combinations Formula*

Use the combinations formula to confirm that $_nC_{n-r}$ is equal to $_nC_r$.

Equation Solving with Combinations

The permutations and combinations formulas can be used to algebraically solve for an unknown value. Study the example below and consider how "guess and test" could also be used to determine n.

Worked Example Algebraically solve the equation $2({}_nP_2) = 144$

Solution: **Use the PERMS formula, substituting "2" for "r":**

$$\frac{n!}{(n-2)!} = 72$$

Apply formula, Divide both sides by 2

Goal: Rid this equation of factorials(!)

We do this by expanding the larger factorial term, until we can cancel out the smaller.

Remember, $n! = n \times (n-1) \times (n-2) \ldots \times 1$

$$\frac{n(n-1)\cancel{(n-2)!}}{\cancel{(n-2)!}} = 72$$

Expand top (it's the larger factorial), stopping at $(n-2)!$ To cancel with bottom

$$n(n-1) = 72$$

Solve resulting quadratic formula

$$n^2 - n - 72 = 0$$

$$(n+8)(n-9) = 0 \Rightarrow n = -8 \text{ or } n = 9$$

n must be positive, this solution is extraneous (pointing to $n = -8$)

$$\Rightarrow \boxed{n = 9}$$

Class Example 8.36 *The Combinations Formula – Equation Solving*

In a simpler time, people used to shake hands after a meeting. After one particular meeting all of the attendees shook hands with each other once, with 66 handshakes taking place.

Algebraically determine the number of attendees at this meeting.

An Alternate way to Express Combinations

Now that we're nearly done with our look at combinations, note that we can also express ${}_nC_r$ as: $\binom{n}{r}$

This is on your formula sheet

Say we wished to determine the number of combinations of 2 objects we can form from a set of 5 of them.
In Math 30-1, this would typically be expressed as ${}_5C_2$.

However, in some textbooks (or even say, a diploma exam!), we might see: $\binom{5}{2}$

Simply Remember This

${}_nC_r$ *is the same as* $\binom{n}{r}$

In the last two sections we've seen a lot of different counting problems! Part of the challenge of seeking mastery in this unit is to aptly identify problems that involve permutations and those that are combinations.

This can be overwhelming! As if the various types of problems form a vast confusing ocean. When in fact, it's more of a *gentle spring stream* of connected concepts....

Types of Problems

Fundamental counting principle *Multiply options by stage*

Ex - *License Plate questions*

Stage 1 options × Stage 2 options × Stage 3 options × ...

Permutation (arrangements) of n objects, or of r objects from n

Ex – *Arrange all, or any 3 of the letters in:* ***NUMBER*** $_nP_r = \frac{n!}{(n-r)!}$

$5!$ $5\times4\times3$

Permutations where *some of the objects must be together*

Ex – *Arrange all of the letters in* ***NUMBER****, if the consonants must be together.*

Count consonants as one: NMBR, U, A — 3 letters, so.... $3!\times4!$

Ways to arrange "together objects" among themselves

Permutations involving Repetition

Ex – *Arrange these objects: (in a single line)* ● ● ● ▲ ▲ ■

$= \frac{6!}{3!\,2!}$ ← Arrange all 6 objects

Identical circles (3!), Identical triangles (2!)

Combination Problems - selecting r objects from n

Use for problems where order doesn't matter...

Ex – *Select members for a committee*

$_nC_r = \frac{n!}{(n-r)!\,r!}$ And remember this can also be expressed: $\binom{n}{r} = \frac{n!}{(n-r)!\,r!}$

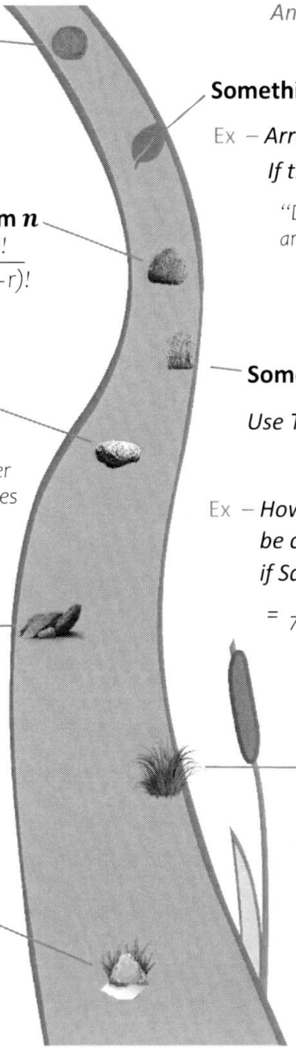

Common "twists"

And methods to un-twist!

Something must happen

Ex – *Arrange all the letters in* ***BOOKS****, If the first letter must be B*

"Disregard" the B as it's fixed, arrange remaining letters: OOKS

$= \frac{4!}{2!}$

Something must not happen

Use TOTAL – UNWANTED

Situation DOES happen

Ex – *How many committees of 3 can be chosen from a group of 7 people, if Sarah must NOT be included?*

$= {_7C_3} - {_6C_2}$

Sarah IS included, choose 2 more from 6 remaining

"At most", "At least", "OR" questions

ADD *each of the resulting cases*

Ex

How many committees of 3 or 4 people can be formed from a group of 7?

Select 3, Select 4

$= {_7C_3} + {_7C_4}$

Key Words: **Permutations** ➡ *Arrangements (setting up a display, arranging letters, lining up for a photograph)*

Combinations ➡ *Groups or Selections (or "committees" *)*

*Nuance alert: *Unless the members are being designated into specific roles, such as "president" and "treasurer" ... then it becomes an arrangement question*

Sometimes Questions Involve Perms *and* Combs

We've come a long way, and we're ready for our last topic. Naturally when you're working on problems in your exams, they won't be sorted into "Permutations" and "Combinations" as in this workbook. Although that's usually clear enough – simply identify the relevant keywords. (As on the bottom of the last page)

But some questions involve a hybrid method, which we'll end things off with!

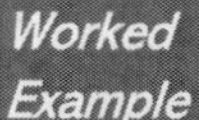

Determine the number of four-letter arrangements can be made from the letters in the word FORGIVE – using two vowels and two consonants.

Solution: Since we see the word "**arrangements**", we know this questions will involve **permutations.**

So while it seems tempting, this will NOT work:

# of 4 letter arrangements $= {}_4P_2 \times {}_3P_2$

From 4 consonants arrange 2 ($_4P_2$) — *From 2 vowels arrange 2* ($_3P_2$)

Why not? Note that the above calculation is the same as: $\mathbf{4 \times 3 \times 3 \times 2}$

- 4: *F, R, G, or V*
- 3: *Any of the consonant "leftovers"*
- 3: *O, I, or E*
- 2: *Any leftover vowel*

Do you see the issue? Every arrangement here would start with two consonants and end with two vowels, as in:

F R O E *or* **V F I O** *(and so on)*
C C V V   C C V V

CORRECT SOLUTION:

# of 4 letter arrangements = # of ways to **select** 2 const. and 2 vowels × # ways to **arrange** each resulting 4 letter group

$= {}_4C_2 \times {}_3C_2 \times 4!$

$= 18 \times 4!$

$= 18 \times 4!$

$\boxed{= \mathbf{432}}$

Class Example 8.37 *Applying Combinations with Permutations*

A large family, consisting of five adults (Mom, Dad, Grandpa, Grandma, and Uncle Chester) and six kids are taking a road trip with two vehicles. The first vehicle is a van which has two seats in the front row, two in the middle row, and three in the back row.

It is decided that the front row must consist of only adults, while the back two rows will seat four kids plus one adult. Determine the number of possible seating arrangements using all 7 seats in the van.

Warm-Up #1

Pascal's Triangle

For the first three rows of Pascal's Triangle, each value is given by the combs term below. The first value (for row 1) is indicated.

1 ➡ Indicate the values for rows 2, 3, and 4, by evaluating each combs term below.

2 ➡ Follow the **patterns** to fill out the values, and "n" and "r" for each comb term, for rows 5 and 6.

3 ➡ List the values (you can leave out the combs expressions) for the 7th and 8th row.

4 ➡ Use combs to state the terms for the "$n+1$"th row

5 ➡ Use the patterns you've observed state the first three terms for the 25th row of the triangle.

6 ➡ Predict the number of terms there will be at the 25th row.

1
$_0C_0$ ← 1st row

$_1C_0$ $_1C_1$ ← 2nd row

$_2C_0$ $_2C_1$ $_2C_2$ ← 3rd row

$_3C_0$ $_3C_1$ $_3C_2$ $_3C_3$ ← 4th row

C C C C C ← 5th row

C C C C C C ← 6th row

← 7th row

← 8th row

← "n+1"th row

Pascal's Triangle was described by French mathematician Blasé Pascal in 1653. There are many pattens that were used in the tasks above, including how combinations can be used to determine each value.

Warm-Up #2 **The Expansion of $(x+y)^n$**

We next consider the expansion of the binomial $(x+y)^n$, to various degrees:

You may notice the how the coefficients relate to the patters studied in warm-up #1. Use that and other patterns you notice to:

1 ➡ Fill in the blanks for the missing coefficient values of $(x+y)^4$

2 ➡ Fully expand $(x+y)^5$ ⟶

$(x+y)^0 = 1$

$(x+y)^1 = x+y$

$(x+y)^2 = x^2+2xy+y^2$

$(x+y)^3 = x^3+3x^2y+3xy^2+y^3$

$(x+y)^4 = x^4+\_x^3y+\_x^2y^2+\_xy^3+y^4$

$(x+y)^5 =$ ➡

3 ➡ Predict the number of terms there will be in the expansions of $(x+y)^6$ and $(x+y)^{12}$

4 ➡ Use combinations to predict the first three coefficients in the expansion of $(x+y)^{12}$

5 ➡ **Fill in the blanks:** In the expansion of $(x+y)^n$, there are _____ terms, and the sum of the exponents for each term is _____. The exponents of x go from _____ to _____, while the exponents of y go from _____ to _____.

The combs expressions for first three coefficients in the expansion of $(x+y)^n$ are: _____ (term 1), _____ (term 2), and _____ (term 3)

While for the last term the coefficient is: _____ (last term)

6 ➡ In the expansion of $(x+y)^n$, for the exponent of y is "0" for term # _____, it's "1" for term # _____, "2" for term # _____, and in general it is "k" for term # _______.

A binomial $(x+y)^n$, where $n \in N$, can be expanded, in *descending order of degree*, **to:**

$$_nC_0x^ny^0 + {}_nC_1x^{n-1}y^1 + {}_nC_2x^{n-2}y^2 + \ldots + {}_nC_{n-1}x^1y^{n-1} + {}_nC_nx^0y^n$$

- There are $n+1$ terms in the expansion
- The sum of the exponents for any term is n
- The exponents of x (or the first term) start at n, and for each term decrease by one down to 0.
- The exponents of y (or the second term) start at 0, and for each term increase by one up to n.

These patterns can be verified by studying the expansion of $(x+y)^5$:

$$(x+y)^5 = x^5 + 5x^4y + 10x^3y^2 + 10x^2y^3 + 5x^4y + y^5$$

Notice that....

- ✓ $n = 5$, and there are "$n+1$" or **6 terms**
- ✓ For the 3rd term, t_3, the exponents add up to 5. *(Value of n)*
- ✓ The exponents of n go from 5 to 0, while those of y go from 0 to n

Let's focus in on that third term (t_3) → $10x^3y^2$, which can be written $_5C_2\,x^3y^2$

Two things we can notice:

- The exponent of y is the same as the 2nd value in the combs term. In general, ***we'll call that value k.***
- The "term number" (here it's term **3**) is one higher than k. So, in general, the exponent of y is $\boldsymbol{k}$ for term number $\boldsymbol{k+1}$.

We're now ready to describe the general term in a binomial expansion, which we reference as the **binomial theorem.**

The **general term** in the expansion of $(x+y)^n$, $n \in N$, is: $t_{k+1} = {}_nC_k(x)^{n-k}(y)^k$

Note that if we start with $\boldsymbol{n = 0}$, and go up by one to $\boldsymbol{n}$, we'll get the *fully expanded form* at the top of this page!

Class Example 8.41 *Expanding a Binomial with Patterns*

Fully expand $(x+y)^7$ using the patterns observed on the previous page and noted above.

Class Example 8.42 *Applying the Fundamental Counting Principal*

For the expression $(x+y)^{10}$,

(a) State the number of terms there will be in the expansion.

(b) Determine the coefficients of the first three terms.

Worked Example Fully expand $(3x-2y)^4$

Method 1: Use Patterns

We know there will be $n+1$ or **5** terms in the expansion. Each term will have the form: **(coefficient)(x-term)$^{n \to 0}$(y-term)$^{0 \to n}$**

Step 1: Draw five sets of *triple brackets* ( )( )( ) + ( )( )( ) + ...

Step 2: In each 2nd bracket, place the first term ($3x$), in each 3rd bracket, place the 2nd term ($-2y$).

$$(\quad)(\mathbf{3x})(-\mathbf{2y}) + (\quad)(\mathbf{3x})(-\mathbf{2y}) + (\quad)(\mathbf{3x})(-\mathbf{2y}) + (\quad)(\mathbf{3x})(-\mathbf{2y}) + (\quad)(\mathbf{3x})(-\mathbf{2y})$$

Step 3: In each 1st bracket, place the coefficient. This can either be the combs term, starting with ${}_4C_0$ then for each term increasing by 1, or you can draw the first five rows of Pascal's Triangle, and use the 5th row.

$$\left({}_4\mathbf{C_0}\right)(3x)(-2y) + \left({}_4\mathbf{C_1}\right)(3x)(-2y) + \left({}_4\mathbf{C_2}\right)(3x)(-2y) + \left({}_4\mathbf{C_3}\right)(3x)(-2y) + \left({}_4\mathbf{C_4}\right)(3x)(-2y)$$

Step 4: The exponents of $\boldsymbol{n}$ will start at 4, then decrease by one down to 0, while those of $\boldsymbol{y}$ go from 0 to 4.

$$(1)(3x)^{\mathbf{4}}(-2y)^{\mathbf{0}} + (4)(3x)^{\mathbf{3}}(-2y)^{\mathbf{1}} + (6)(3x)^{\mathbf{2}}(-2y)^{\mathbf{2}} + (4)(3x)^{\mathbf{1}}(-2y)^{\mathbf{3}} + (1)(3x)^{\mathbf{0}}(-2y)^{\mathbf{4}}$$

Step 4: Simplify!

$$= (81x^4)(1) + (4)(27x^3)(-2y) + (6)(9x^2)(4y)^2 + (4)(3x)(-8y^3) + (1)(1)(16y^4)$$

$$= \boxed{81x^4 - 216x^3y + 216x^2y^2 - 96xy^3 + 16y^4}$$

Method 2: Use The Formula for each term

Use $t_{k+1} = {}_nC_k(x)^{n-k}(y)^k$, with $\boldsymbol{n} = \mathbf{4}$, $\boldsymbol{x} = \mathbf{3}\boldsymbol{x}$ and $\boldsymbol{y} = -\mathbf{2}\boldsymbol{y}$. For $\boldsymbol{k}$, start with 0 (for term 1), then increase by 1 until you get to 4 (for term 5.... Remember the term # is always one greater than k)

$t_1 \Rightarrow t_{0+1} = {}_4C_0(3x)^{4-0}(-2y)^0$

$t_2 \Rightarrow t_{1+1} = {}_4C_1(3x)^{4-1}(-2y)^1$

$t_3 \Rightarrow t_{2+1} = {}_4C_2(3x)^{4-2}(-2y)^2$

$t_4 \Rightarrow t_{3+1} = {}_4C_3(3x)^{4-3}(-2y)^3$

$t_5 \Rightarrow t_{4+1} = {}_4C_4(3x)^{4-4}(-2y)^4$

Each of these terms simplifies as above, to give us the same expansion:

$$= 81x^4 - 216x^3y + 216x^2y^2 - 96xy^3 + 16y^4$$

Class Example 8.43 *Fully Expanding a Binomial*

Fully expand $(2x-3y)^5$.

Worked Example Find the 3rd term in the expansion of: $(4m+1)^7$

Method 1: Use Patterns

Start listing the coefficients for each term, starting with $_7\boldsymbol{C_0}$. Stop when you get to the 3rd term.

$_7C_0$

$_7C_1$ ↙ *Next, add (first term) (second term)*

There it is! t_3.... ➧ $_7C_2\ (4m)\ (1)$

Now, the second # in your combs expression is the $\boldsymbol{k}$ value. (That is, it's the exponent of y) So we know that $\boldsymbol{k}=\mathbf{2}$. As for the exponent of $\boldsymbol{x}$, simply think .... *the exponents must add to 7.* (So x is 5) Indicate the exponents....

t_3.... ➧ $_7C_2\ (4m)^5(1)^2$

And simplify.... $= (21)(1024m^5)(1)$

$= \boxed{\mathbf{21\,504}\boldsymbol{m^5}}$

Method 2: Use The Formula

Use $\boldsymbol{t_{k+1}} = {_nC_k}(x)^{n-k}(y)^k$, with $\boldsymbol{n}=\mathbf{7}$, $\boldsymbol{x}=\mathbf{4}\boldsymbol{m}$ and $\boldsymbol{y}=\mathbf{1}$.

For $\boldsymbol{k}$, note that we want term 3, that is, t_3. Since the formula has "t_{k+1}", think *what must k be?* $t_{\square+1}$

↖ *k is what we put here, to make this equal to* $\boldsymbol{t_3}$ ➧ $\boldsymbol{k}$ is **2**

So, we have: $\boldsymbol{t_{2+1}} = {_7C_2}(4m)^{7-2}(1)^2$

$= \boxed{\mathbf{21\,504}\boldsymbol{m^5}}$

Class Example 8.44 *Finding a particular term in a binomial expansion*

(a) Find the 6th term in the expansion of $(x-4y)^8$

(b) Find the middle term in the expansion of $(2x+y^3)^8$

Worked Example Find the term involving x^5 in the expansion of $(3x - 2y)^8$

Use Patterns and the Formula

Think: If the exponent of x is 5, the exponent of $\mathbf{y}$ must be **3**. (Exponents of x and y terms add to "n", and $n = \mathbf{8}$)

$$t_{k+1} = {}_nC_k(x)^{n-k}(y)^k$$

Same number! (here, it's ***3****)*

This is your $\mathbf{k}$ *value in the formula*

$$t_{3+1} = {}_8C_3(3x)^{8-3}(-2y)^3$$

$$= (56)(243x^5)(-8y^3)$$

$$= \boxed{\mathbf{108\,864x^5y^3}}$$

Check – YES, the exponent of x is 5

Another way to determine the "k" value... Refer to the formula:

$$t_{k+1} = {}_nC_k(x)^{\boxed{n-k}}(y)^k$$

The exponent of $\mathbf{x}$ *is* $\mathbf{n - k}$
Here, $n = 8$, *so we have:*

$$8 - k = 5$$

$$\mathbf{k = 3}$$

Class Example 8.45 *Finding a particular term in a binomial expansion – given a characteristic*

(a) Find the term involving y^4 in the expansion of $(2x - 4y)^5$

(b) Find the term involving x^3 in the expansion of $(3x - y)^6$

Worked Example Find the term involving x^8 in the expansion of $(4y - x^2)^6$

Use Patterns and the Formula Think: If the exponent of x *is already 2....* $(x^2)^{\square}$

What power must this be to get an x^8 term?
➔ *Exponent must be* ***4***

Note that $\boldsymbol{x}$ is in the 2nd term, so for the formula $t_{k+1} = {}_nC_k(x)^{n-k}(y)^k$ ...

"$\mathbf{x}$" *(first term)* $= 4y$ "$\mathbf{y}$" *(second term)* $= -x^2$ "$\mathbf{n}$" $= 6$ "$\mathbf{k}$" $= 4$

$$t_{4+1} = {}_6C_4(4y)^{6-4}(-x^2)^4$$

$$= (15)(16y^2)(x^8) \quad ➔ \quad = \boxed{\mathbf{240x^8y^2}}$$

Remember ... k is the exponent the ***2nd term*** *(in this case "*$\mathbf{-x^2}$*") is taken to. And since we want an x^8 term, k must be 4.*

Class Example 8.46 *Finding a particular term in a binomial expansion*

(a) Find the term involving y^8 in the expansion of $(5x - 2y^2)^6$

(b) Find the term involving x^6 in the expansion of $(2x^2 + 2y)^7$

Challenge Questions – Finding a term in the Binomial

So far, we've used the binomial theorem to fully expand, or to find specific terms in an expansion. We'll next look at questions where you must work backwards – given a term in the expansion, find part of one of the terms in the binomial.

Worked Example One of the terms in the expansion of $(ax + 2y)^5$ is $8640x^3y^2$. Determine the value of a.

Set up an equation to find *a*

Think: Once again we start by wanting to know what "k" is – so once again refer to the formula! $t_{k+1} = {}_nC_k(x)^{n-k}(y)^{k}$

k is the exponent of y

"x" (first term) $= ax$ "y" (second term) $= 2y$ "n" $= 5$ "k" $= 2$

Look to the given term in the expansion: $8640x^3y^2$ — *k is 2*

Now, set up an equation: $t_{k+1} = {}_nC_k(x)^{n-k}(y)^k$

This is the given term, $8640x^3y^2$ — *Use the values identified above*

$$8640x^3y^2 = {}_5C_2(ax)^{5-2}(2y)^2$$

$$8640\cancel{x^3}\cancel{y^2} = 10(a^3\cancel{x^3})(4\cancel{y^2})$$ *Simplify / cancel terms*

$$8640 = 40a^3$$

$$a^3 = \frac{8640}{40}$$

$$a^3 = 216 \Rightarrow a = \sqrt[3]{216} \Rightarrow \boxed{a = 6}$$

Class Example 8.47 *Working backwards to find a term in the Binomial*

One of the terms in the expansion of $(ax - 3y)^6$ is $34\,560x^4y^2$. Determine the value of a.

Challenge Questions – Finding a CONSTANT TERM

Worked Example Find the constant term in the expansion of $(2x^2 + \frac{1}{x^3})^{10}$.

Use the formula to find the term where the exponent of x is 0

Step 1: *We are going to use the formula:* $t_{k+1} = {}_nC_k x^{n-k} y^k$

"x" (first term) $= 2x^2$ "y" (second term) $= 1/x^3$ "n" $= 10$ "k" $=??$ *Note:* Express $\frac{1}{x^3}$ using exponents → x^{-3}

Step 2: *Determine the value of* k*:* Think: The constant term has no x in it. *So we must find the k that makes the exponent of x is zero*

$$t_{k+1} = {}_{10}C_k(2x^2)^{10-k}(x^{-3})^k$$

Simplify x **terms only**... *(multiply exponents)*

$x^{20-2k} * x^{-3k}$ *add exponents*

x^{20-5k} ← *we want this exp. to be zero!*

$20 - 5k = 0$

$20 = 5k$ → $k = 4$

Step 3: *We just found the exponent of x is zero for* $k = 4$.

Now we're ready to substitute everything into the formula:

$$t_{4+1} = {}_{10}C_4(2x^2)^{10-4}(\frac{1}{x^3})^4$$

$$= 210(64\cancel{x^{12}})(\frac{1}{\cancel{x^{12}}})$$

$$= \boxed{13\ 440}$$

-or- Use patterns to start listing terms

Another method is to just start listing terms, starting with ${}_{10}C_0$

t_1 → ${}_{10}C_0(2x^2)^{10-0}(x^{-3})^0$

t_2 → ${}_{10}C_1(2x^2)^{10-1}(x^{-3})^1$

t_3 → ${}_{10}C_2(2x^2)^{10-2}(x^{-3})^2$

t_4 → ${}_{10}C_3(2x^2)^{10-3}(x^{-3})^3$

Notice when we get here, the exponents of x cancel out to zero!

t_5 → ${}_{10}C_4(2x^2)^{10-4}(x^{-3})^4 = {}_{10}C_4(2x^2)^6(x^{-12}) = 210(64)$ *(Same answer as above,* ***13 440****)*

Class Example 8.48 *Finding a Constant Term*

Find the constant term in the expansion of $(4x + \frac{1}{2x^2})^9$.

Relating n to the Number of Terms

We'll finish off this section, and the Perms and Combs unit, by exploring a few other Binomial Theorem questions we might encounter.

Worked Example In the expansion of $(5x + 3y)^{2a+1}$ there are 13 terms. Determine the value of a.

Solution Think: If there are 13 terms in the expansion, what must the exponent of the binomial ("n") be?

Recall that in the expansion of $(x + y)^n$, **there are $n + 1$ terms**

So here, $\boldsymbol{n = 12}$. (Exponent of the binomial is *one less* than the number of terms)

$2a + 1 = 13$ *Set up / solve an equation*

$2a = 12$ ➧ $\boxed{a = 6}$

Class Example 8.49 *Relating the Number of Terms*

The expanded form of $(3x - 4)^{m-4}$ has 7 terms. Determine:

(a) The value of m

(b) The coefficient of the first term.

(c) The constant term.

Manufactured by Amazon.ca
Acheson, AB